Lecture Notes in Artificial Intelligence   9310

Subseries of Lecture Notes in Computer Science

More information about this series at http://www.springer.com/series/1244

Christoph Beierle · Alex Dekhtyar (Eds.)

Scalable Uncertainty Management

9th International Conference, SUM 2015
Québec City, QC, Canada, September 16–18, 2015
Proceedings

 Springer

Editors
Christoph Beierle
Fakultät für Mathematik und Informatik
FernUniversität in Hagen
Hagen
Germany

Alex Dekhtyar
California Polytechnic State University
San Luis Obispo
USA

ISSN 0302-9743   ISSN 1611-3349   (electronic)
Lecture Notes in Artificial Intelligence
ISBN 978-3-319-23539-4   ISBN 978-3-319-23540-0   (eBook)
DOI 10.1007/978-3-319-23540-0

Library of Congress Control Number: 2015947418

LNCS Sublibrary: SL7 – Artificial Intelligence

Printed on acid-free paper

Springer International Publishing AG Switzerland is part of Springer Science+Business Media
(www.springer.com)

Preface

The International Conference on Scalable Uncertainty Management (SUM) is an annual conference that was launched in 2007 with the goal of exploiting and strengthening the connection between the artificial intelligence and database communities, aiming at bringing together all those researchers interested in the management of uncertain, incomplete, or inconsistent information. Such information originates commonly in applications where significant computational effort is needed to process data in a meaningful and semantically justifiable manner. Typical applications of that kind include databases, the Web, and the life sciences.

Previous SUM conferences have been held in Washington DC (USA) in 2007, in Naples (Italy) in 2008, in Washington DC (USA) in 2009, in Toulouse (France) in 2010, in Dayton (USA) in 2011, in Marburg (Germany) in 2012, again in Washington DC (USA) in 2013, and in Oxford (UK) in 2014.

This volume contains the papers presented at the Ninth International Conference on Scalable Uncertainty Management (SUM 2015) held in Québec City, Canada, September 16–18, 2015.

The call for papers for SUM 2015 solicited submissions in all areas of managing and reasoning with substantial and complex kinds of uncertain, incomplete, or inconsistent information. These include applications in decision support systems, risk analysis, machine learning, belief networks, logics of uncertainty, belief revision and update, argumentation, negotiation technologies, semantic web applications, search engines, ontology systems, information fusion, information retrieval, natural language processing, information extraction, image recognition, vision systems, data and text mining, and the consideration of issues such as provenance, trust, heterogeneity, and complexity of data and knowledge.

The call for papers resulted in 49 submissions, among which 41 were regular papers and 8 were short papers. In a rigorous reviewing process, each submitted article was reviewed by at least three Program Committee members. Based on the review reports and intense discussions, 25 regular papers (one was later withdrawn by the author) and 3 short papers were accepted for publication and presentation at the conference.

In addition, the conference greatly benefited from invited lectures by three world-leading researchers: Jean-Marie De Koninck, Lise Getoor, and Ronald R. Yager. This volume also contains the abstracts of the three invited talks as well as an article for one of them.

A conference such as this can only succeed as a team effort. We would like to thank several people and institutions. We thank all the authors of submitted papers, the invited speakers, and the conference participants. We are grateful to the members of the Program Committee and the external reviewers, to Alfred Hofmann and Springer for providing assistance and advice in the preparation of the proceedings, and to the creators and maintainers of the conference management system EasyChair. Special thanks go to Patrick Maupin and his team for being our hosts and for the wonderful days in Québec City.

July 2015   Christoph Beierle
                                                                    Alex Dekhtyar

Conference Organization

General Chair

Patrick Maupin   Defence Research and Development Canada

Program Committee Chairs

Christoph Beierle   University of Hagen, Germany
Alex Dekhtyar   Cal Poly, San Luis Obispo, USA

Program Committee

Chitta Baral	Arizona State University, USA
Nahla Ben Amor	Institut Supérieur de Gestion de Tunis, Tunisia
Leopoldo Bertossi	Carleton University, Canada
Cory Butz	University of Regina, Canada
Andrea Calì	University of London, Birkbeck College, UK
Reynold Cheng	The University of Hong Kong, Hong Kong, China
Laurence Cholvy	ONERA, Toulouse, France
Jan Chomicki	University at Buffalo, USA
Fabio Cozman	Universidade de Sao Paulo, Brazil
Madalina Croitoru	LIRMM, Univ. Montpellier II, France
Michael Dekhtyar	Tver State University, Russia
Thierry Denoeux	Université de Technologie de Compiègne, France
Jürgen Dix	Clausthal University of Technology, Germany
Didier Dubois	IRIT-CNRS, Toulouse, France
Thomas Eiter	Vienna University of Technology, Austria
Zied Elouedi	Institut Supérieur de Gestion de Tunis, Tunisia
Ronald Fagin	IBM Research - Almaden, USA
Mihai Cristian Florea	Thales Group, Canada
Lluis Godo	Artificial Intelligence Research Institute, IIIA - CSIC, Spain
John Grant	Towson University, USA
Sergio Greco	University of Calabria, Italy
Allel Hadjali	LIAS/ENSMA, Chasseneuil, France
Justin Hollands	DRDC, Canada
Anthony Hunter	University College London, UK
Gabriele Kern-Isberner	TU Dortmund University, Germany
Kristian Kersting	TU Dortmund University, Germany
Kathryn Laskey	George Mason University, Fairfax, USA
Sebastian Link	The University of Auckland, New Zealand

Peter Lucas	Radboud University Nijmegen, Netherlands
Thomas Lukasiewicz	University of Oxford, UK
Jianbing Ma	Bournemouth University, UK
Zongmin Ma	Northeastern University, China
Anders Madsen	HUGIN Expert, Denmark
Maria Vanina Martinez	University of Oxford, UK
Thomas Meyer	Centre for Artificial Intelligence Research, UKZN and CSIR Meraka, South Africa
Serafin Moral	University of Granada, Spain
Jeff Z. Pan	University of Aberdeen, UK
Simon Parsons	University of Liverpool, UK
Gabriella Pasi	Università degli Studi di Milano Bicocca, Italy
David Poole	University of British Columbia, Canada
Henri Prade	IRIT-CNRS, Toulouse, France
Andrea Pugliese	University of Calabria, Italy
Sebastian Rudolph	Technische Universität Dresden, Germany
Vítor Santos Costa	Universidade do Porto, Portugal
Elisa Shahbazian	OODA Technologies Inc., Canada
Guillermo R. Simari	Universidad Nacional del Sur, Argentina
Umberto Straccia	ISTI-CNR, Italy
V.S. Subrahmanian	University of Maryland, USA
Paul Weng	SYSU-CMU Joint Institute of Engineering, USA
Jef Wijsen	University of Mons, Belgium
Stefan Woltran	Vienna University of Technology, Austria
Ronald R. Yager	Iona College, USA
Vladimir Zadorozhny	University of Pittsburgh, USA

Additional Reviewers

Imen Boukhris	Luyi Mo	Gavin Rens
Zoé Faget	Cristian Molinaro	Cinzia Incoronata Spina
Maroua Haddad	Francesco Parisi	Slawek Staworko
Siqiang Luo	Sylwia Polberg	Yudian Zheng

Steering Committee

Didier Dubois	IRIT-CNRS, Toulouse, France
Henri Prade	IRIT-CNRS, Toulouse, France
V.S. Subrahmanian	University of Maryland, USA

Local Organization

Patrick Maupin   Defence Research and Development Canada
Anne-Laure Jousselme   CMRE, Italy

Invited Talks

The Hollow Universe of Mathematics

Jean-Marie De Koninck

Département de mathématiques et de statistique,
Université Laval, Québec G1V 0A6, Canada

In many fields of mathematics, the set of known results is very thin compared with the set of conjectures and hypothesis which have not yet been proved. Particularly remarkable is the hollow universe that sometimes seems to separate the world of the known from what we believe to be reality. Through various examples from number theory, we will attempt here to explore that hollow universe separating these two worlds.

Large-scale Collective Inference using Probabilistic Soft Logic

Lise Getoor

Computer Science Department, University of California, Santa Cruz, USA

One of the challenges in big data analytics is to efficiently learn and reason collectively about extremely large, heterogeneous, incomplete, noisy interlinked data. Collective reasoning requires the ability to exploit both the logical and relational structure in the data and the probabilistic dependencies. In this talk I will overview our recent work on probabilistic soft logic (PSL), a framework for collective, probabilistic reasoning in relational domains. PSL is able to reason holistically about both entity attributes and relationships among the entities. The underlying mathematical framework, which we refer to as a hinge-loss Markov random field, supports extremely efficient, exact inference. This family of graphical models captures logic-like dependencies with convex hinge-loss potentials. I will survey applications of PSL to diverse problems ranging from information extraction to computational social science. Our recent results show that by building on state-of-the-art optimization methods in a distributed implementation, we can solve large-scale problems with millions of random variables orders of magnitude faster than existing approaches.

Intelligent Technologies for Internet Social Applications

Ronald R. Yager

Machine Intelligence Institute, Iona College
New Rochelle, NY 10801, USA

The Internet has provided for a rapid growth of computer mediated social networks and other social interactions. One focus here is to discuss how to enrich the domain of social network modeling by introducing ideas from fuzzy sets and related intelligent technologies. We approach this extension in a number of ways. One is with the introduction of fuzzy graphs representing the networks. This allows a generalization of the types of connection between nodes in a network. A second and perhaps more interesting extension is the use of the fuzzy set based paradigm of computing with words to provide a bridge between a human network analyst's linguistic description of social network concepts and the formal model of the network. We also will describe some methods for sharing information obtained in these types of networks we used for computer mediated group decision making.

Contents

Applications

Invited Talk

The Mysterious World of Normal Numbers

Jean-Marie De Koninck$^{(\boxtimes)}$

Département de Mathématiques et de Statistique,
Université Laval, Québec G1V 0A6, Canada
jmdk@mat.ulaval.ca

Abstract. Given an integer $q \geq 2$, a *q-normal number* (or a *normal number*) is a real number whose q-ary expansion is such that any pre-assigned sequence of length $k \geq 1$, of base q digits from this expansion, occurs at the expected frequency, namely $1/q^k$. Even though there are no standard methods to establish if a given number is normal or not, it is known since 1909 that almost all real numbers are normal in every base q. This is one of the many reasons why the study of normal numbers has fascinated mathematicians for the past century. We present here a brief survey of some of the important results concerning normal numbers.

1 Introduction

Flip a coin. If you obtain heads, write 0; if you obtain tails, write 1. Keep flipping the coin, writing 0's and 1's depending on the outcome. After 100 times, count the number of 0's and 1's: you will most likely count approximately 50 of each. Then, count how many times you obtained two consecutive 0's: it will most likely be approximately 25 times, since the possible outcomes of two consecutive flips are 00, 01, 10 and 11, and the probability that any such particular outcome occurs is $1/4$. Similarly, if you keep flipping the coin many times, the probability that a given sequence of length k occurs will be around $1/2^k$; that's what you expect will happen: it would be perfectly normal! This is why we say that the sequence of 0's and 1's obtained by flipping a coin creates a *random sequence*, that is, a binary *normal sequence*. This is why if $a_1, a_2, a_3, \ldots$ is the infinite sequence of 0's and 1's obtained by flipping a coin (for ever!), we say that the expression $0.a_1 a_2 a_3 \ldots$ represents a *normal number*.

Humans have always been interested in creating random numbers. In fact, random number generators have applications in gambling, lotteries, computer simulation, cryptography, completely randomized design, and many other areas where producing an unpredictable result needs to be achieved. Normal numbers have their practical use in that they provide an infinite source of pseudorandom numbers. However, the real interest for the study of normal numbers lies in the fact tha they are extremely difficult to identify and that they are very mysterious in many other aspects.

2 Basic Definitions

Given an integer $q \geq 2$, a *q-normal number* (or a *normal number*) is a real number whose q-ary expansion is such that any preassigned sequence of length

C. Beierle and A. Dekhtyar (Eds.): SUM 2015, LNAI 9310, pp. 3–18, 2015.
DOI: 10.1007/978-3-319-23540-0_1

$k \geq 1$, of base q digits from this expansion, occurs at the expected frequency, namely $1/q^k$. Clearly, rational numbers cannot be normal since only a particular sequence of digits is repeated infinitely often.

Equivalently, given a positive irrational number η whose expansion is

$$\eta = \lfloor \eta \rfloor + 0.a_1a_2a_3\ldots = \lfloor \eta \rfloor + \sum_{j=1}^{\infty} \frac{a_j}{q^j}, \text{ with each } a_j \in \{0, 1, \ldots, q-1\},$$

where $\lfloor \eta \rfloor$ stands for the integer part of η, we say that η is a q-normal number if the sequence $\{q^m \eta\}$, $m = 1, 2, \ldots$ (here $\{y\}$ stands for the fractional part of y), is uniformly distributed in the interval $[0, 1)$.

Both definitions are equivalent, because the sequence $\{q^m \eta\}$, $m = 1, 2, \ldots$, is uniformly distributed in $[0, 1)$ if and only if for every integer $k \geq 1$ and $b_1 \ldots b_k \in \{0, 1, \ldots, q-1\}^k$, we have

$$\lim_{N \to \infty} \frac{1}{N} \#\{j \leq N : a_{j+1} \ldots a_{j+k} = b_1 \ldots b_k\} = \frac{1}{q^k}.$$

A real number is said to be *simply normal* in base q if each digit $d \in \{0, 1, \ldots, q-1\}$ occurs with frequency $1/q$. Of course, a number can be simply normal without being a normal number (such is the case of the binary number $0.1010101010101\ldots$).

A real number is said to be *absolutely normal* if it is normal in each base $q \geq 2$.

Normal numbers are mysterious for many reasons. For instance, the constant

$$\pi = 3.141592653589793238462643383279502884197169399375\underline{1}\ldots$$

has not yet been proved to be a normal number, although it is widely believed that it is. Similarly, the frequently used

$$\text{Euler constant } e = 2.718281828459045235360287471352662497757247\ldots$$
$$\sqrt{2} = 1.414213562373095048801688724209698078569671\ldots$$
$$\log 2 = 0.693147180559945309417232121458176568075500\ldots$$
$$\text{Apery number } \sum_{n=1}^{\infty} \frac{1}{n^3} = 1.202056903159594285399738161511449990764986\ldots$$
$$\text{Golden number } \frac{1 + \sqrt{5}}{2} = 1.618033988749894848204586834365638117720309\ldots$$

have not yet been proven to be normal numbers, although numerical evidence seems to indicate that they are. What is even more disturbing is the fact that none of the above numbers has been shown to be simply normal. For instance, it is possible that venturing along the decimals of π, from some point on, one could not any longer find the digit 0. Even though no one believes that could be the case, we can't disprove it.

On the other hand, it is widely believed that every irrational algebraic number is normal. Nevertheless, no algebraic irrational number has yet been proved to be normal (in any base).

Despite our inability to prove that any member of this large family of numbers is normal, Émile Borel [6] showed in 1909 that almost all real numbers (with respect to the Lebesgue measure) are absolutely normal.

3   A Story Line

Here is a story line of some of the key results obtained concerning normal numbers.

- 1909: Borel [6] introduces the concept of normal number and proves that almost all real numbers are absolutely normal.
- 1917: Sierpiński [23] provides an alternative proof that almost all real numbers are normal. It is an existence theorem, that is Sierpiński does not point out to any particular normal numbers. Here is the general idea of Sierpiński's proof. For each number $\varepsilon \in (0, 1]$, he first constructs a set $\Delta(\varepsilon)$ which is the union of countably many open intervals with rational endpoints, namely

$$\Delta(\varepsilon) := \bigcup_{q=2}^{\infty} \bigcup_{m=1}^{\infty} \bigcup_{n=n_{m,q}(\varepsilon)}^{\infty} \bigcup_{p=0}^{q-1} \Delta_{q,m,n,p} \, ,$$

where $\Delta_{q,m,n,p}$ is the set of all open intervals of the form

$$\left(\frac{b_1}{q} + \frac{b_2}{q^2} + \cdots + \frac{b_n}{q^n} - \frac{1}{q^n}, \frac{b_1}{q} + \frac{b_2}{q^2} + \cdots + \frac{b_n}{q^n} + \frac{2}{q^n} \right)$$

such that

$$\left| \frac{c_p(b_1, b_2, \ldots, b_n)}{n} - \frac{1}{q} \right| \geq \frac{1}{m},$$

where each $b_i \in \{0, 1, \ldots, q-1\}$ and where $c_p(b_1, b_2, \ldots, b_n)$ represents the number of times that the digit p appears amongst the digits $b_1, b_2, \ldots, b_n$. The idea is that $\Delta_{q,m,n,p}$ contains all the numbers that are not normal in base q. He then proves that every positive real number < 1 which is external to $\Delta(\varepsilon)$ is absolutely normal. Finally, he shows that $\mu(\Delta(\varepsilon)) < \varepsilon$ for every $\varepsilon \in (0, 1]$, that is that the Lebesgue measure of the set $\Delta(\varepsilon)$ tends to 0 with ε, thereby establishing that almost all numbers are normal.
- 1933: Champernowne [9], an undergraduate student, proves that the number

$$C_{10} = 0.1234567891011121314151617181920 21 \ldots,$$

made up from the concatenation of the positive integers, is normal in base 10. Observe that, by concatenating the sequence of integers written in any base $q \geq 2$, one can show that it provides a q-normal number.

- 1946: Copeland and Erdős [10] prove that the number $0.23571113171923293137\ldots$, obtained by the concatenation of the prime numbers, is normal in base 10. Observe that the same result holds by concatenating the sequence of prime numbers written in any base $q \geq 2$. More generally, they prove that if $a_1, a_2, a_3, \ldots$ is an increasing sequence of positive integers (expressed in base q) such that, for each positive $\theta < 1$, $\#\{a_i \leq x\} > x^\theta$ provided $x \geq x_0(\theta)$, then $0.a_1 a_2 a_3 \ldots$ is a q-normal number. Since $\pi(x) > \dfrac{x}{\log x}$ for all $x \geq 11$ (here $\pi(x)$ stands for the number of primes not exceeding x), then as a particular case we get that $0.235711131719\ldots$ is indeed normal in base 10.

  As another application of the general Copeland and Erdős result, we have that since each prime $p \equiv 1 \pmod 4$ can be written as $p = r^2 + s^2$ with $r, s \in \mathbb{N}$, and since $\#\{p \leq x : p \equiv 1 \pmod 4\} > cx/\log x$ for all $c < \frac{1}{2}$ provided x is large enough, it follows that $\#\{n_i \leq x : n_i = r^2 + s^2\} > cx/\log x$ for large x, thus implying that the number $0.n_1 n_2 n_3 \ldots = 0.5131729\ldots$ is normal.
- 1946: Copeland and Erdős [10] also conjecture that if $f(x)$ is any non constant polynomial whose values at $x = 1, 2, 3, \ldots$ are positive integers, then $0.f(1)f(2)f(3)\ldots$ is a normal number in base 10.
- 1952: Davenport and Erdős [11] prove this conjecture.
- 1956: Cassels [8] comes up with a large family of simply normal numbers by considering the function $f : [0,1] \to \mathbb{R}$ defined by

$$f(x) = \sum_{j=1}^{\infty} \frac{x_j}{3^j},$$

where $x_1, x_2, \ldots$ denote the binary digits of x. Then, one can easily establish that for almost all $x \in [0,1]$, $f(x)$ is simply normal with respect to every base $q \geq 2$ which is not a power of 3.
- 1992: Nakai and Shiokawa [21] prove that if $f \in \mathbb{R}[X]$ is such that $f(x) > 0$ for $x > 0$, then the real number $0.\lfloor f(1) \rfloor \lfloor f(2) \rfloor \lfloor f(3) \rfloor \ldots$, where $\lfloor f(n) \rfloor$ stands for the integer part of $f(n)$ expressed in base $q \geq 2$, is normal in base q. They also show that the same result holds if

$$f(x) = \alpha_0 x^{\beta_0} + \alpha_1 x^{\beta_1} + \cdots + \alpha_d x^{\beta_d},$$

where the α_i's and β_i's are real numbers with $\beta_0 > \beta_1 > \cdots > \beta_d \geq 0$ and $f(x) > 0$ for $x > 0$.
- 1997: Nakai and Shiokawa [22] prove that if $f \in \mathbb{Z}[X]$ is any noncon stant polynomial such that $f(x) > 0$ for $x > 0$, then the number $0.f(2)f(3)f(5)f(7)\ldots f(p)\ldots$ is normal in base 10.
- 2008: Madritsch, Thuswaldner and Tichy [19] extend the results of Nakai and Shiokawa by showing that, if f is an entire function of logarithmic order, then the numbers

$$0.\lfloor f(1) \rfloor_q \lfloor f(2) \rfloor_q \lfloor f(3) \rfloor_q \ldots \quad \text{and} \quad 0.\lfloor f(2) \rfloor_q \lfloor f(3) \rfloor_q \lfloor f(5) \rfloor_q \lfloor f(7) \rfloor_q \ldots,$$

where $\lfloor f(n) \rfloor_q$ stands for the base q expansion of the integer part of $f(n)$, are normal.

4 Series Representing Normal Numbers

In 1971, Stoneham considered constants represented by convergent series as possible candidates for normality.

As we mentioned in Sect. 1, no one has been able to show that $\log 2 = \sum_{n=1}^{\infty} \frac{1}{n2^n}$ is a normal number. Nevertheless, Stoneham [24] was able to show that the number

$$\alpha_{2,3} := \sum_{n=3^k>1}^{\infty} \frac{1}{n2^n} = \sum_{k=1}^{\infty} \frac{1}{3^k 2^{3^k}}$$

is normal in base 2. More generally, observe that $\log \frac{b}{b-1} = \sum_{n=1}^{\infty} \frac{1}{nb^n}$. In 2002, Bailey and Crandall [4] proved that, if $b, c \geq 2$ are coprime integers, then the number $\alpha_{b,c} := \sum_{n=c^k>1}^{\infty} \frac{1}{nb^n} = \sum_{k=1}^{\infty} \frac{1}{c^k b^{c^k}}$ is normal in base b. They even showed that if $r = 0.r_1 r_2 \ldots \in [0,1)$, then $\alpha_{2,3}(r) := \sum_{n=1}^{\infty} \frac{1}{3^n 2^{3^n + r_n}}$ is a normal number in base 2, thereby providing an uncountable class of normal numbers in base 2.

Is $\alpha_{2,3}$ normal in bases other than 2? Not always! In fact, in 2006, Bailey and Borwein [1] proved that $\alpha_{2,3}$ is not a 6-normal number. Their idea was based on the fact that since the expression

$$6^{3^m} \alpha_{2,3} \mod 1 \approx \frac{(3/4)^{3^m}}{3^{m+1}}$$

(Here $x = \theta \mod 1$ means that $\theta = x - \lfloor x \rfloor$) is very small for large m, this causes the number $\alpha_{2,3}$, in base 6, to have long stretches of 0's beginning at position $3^m + 1$, and as we know this is not acceptable for a normal number!

5 Equidistribution

A sequence of positive numbers $x_0, x_1, x_2, \ldots$, each smaller than 1, is said to be *equidistributed* if, for any $0 \leq c < d < 1$,

$$\lim_{N \to \infty} \frac{1}{N} \#\{0 \leq j < N : x_j \in [c,d)\} = d - c.$$

In 2001, Bailey and Crandall [3] considered the sequence $x_0, x_1, \ldots$ defined by $x_0 = 1$ and, for each $n \geq 1$, by

$$x_n = \left(2x_{n-1} + \frac{1}{n}\right) \mod 1.$$

They showed that if one could prove that this sequence is equidistributed in $[0,1]$, then it would imply that $\log 2$ is a binary normal number.

Similarly, consider the sequence $y_0, y_1, \ldots$ defined by $y_0 = 1$ and, for each $n \geq 1$, by

$$y_n = \left(16y_{n-1} + \frac{120n^2 - 89n + 16}{512n^4 - 1024n^3 + 712n^2 - 206n + 21}\right) \mod 1.$$

They showed that if one could prove that this sequence is equidistributed in $[0, 1]$, then it would imply that π is a 16-normal number (and hence a 2-normal number as well).

These results raise a natural question: Is it easier to prove the equidistribution of the sequence $(x_n)_{n \geq 1}$ or the normality of $\log 2$? What about the sequence $(y_n)_{n \geq 1}$ and its corresponding number π? Nobody knows!

6   Abnormal Numbers

Surely, if we have so much difficulty finding normal numbers, it should be easy to find many numbers which are not normal. It turns out that, except for the rational numbers, this task is not so easy!

A number is said to be *abnormal* in base q if it is not normal in base q. For instance, the binary number

$$\sum_{n=1}^{\infty} \frac{n}{2^{n^2}} = 0.101000011000010000000010100000000110000000000001110\ldots$$

is clearly abnormal since one can easily show that almost all of its digits in base 2 are zeros.

A less obvious example of a binary abnormal number is the amazing *Devil's staircase* number, namely the number $f(x) := \sum_{n=1}^{\infty} \frac{\lfloor nx \rfloor}{2^n}$ with $x \in [0, 1]$. Here is the graph of $f(x)$:

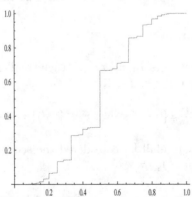

This function has amazing properties. Bailey and Crandall [4] studied this function and proved that, for $x \in (0, 1)$,

– f is monotone increasing,

- f is continuous at every irrational x, but discontinuous at every rational x,
- $f(x) \in \mathbb{R} \setminus \mathbb{Q}$ if and only if $x \in \mathbb{R} \setminus Q$,
- if x is irrational, then $f(x)$ is transcendental,
- the range of $f([0,1])$ is a set of measure zero,
- if $x = a/b$ with $a, b \in \mathbb{N}$ and $a/b < 1$, then $f(x) = \dfrac{1}{2^b - 1} + \displaystyle\sum_{m=1}^{\infty} \dfrac{1}{2^{\lfloor m/x \rfloor}}$, while

if x is irrational, then $f(x) = \displaystyle\sum_{m=1}^{\infty} \dfrac{1}{2^{\lfloor m/x \rfloor}}$.

But then the most interesting property of $f(x)$ shown by Bailey and Crandall is that it is never 2-normal.

7   Absolutely Abnormal Numbers

A number is said to be *absolutely abnormal* if it is not normal in every base $q \geq 2$. In May 2000, during a survey talk by Glynn Harman, Andrew Granville asked about a specific absolutely abnormal number. In response, Carl Pomerance suggested considering the Liouville number $\ell := \displaystyle\sum_{n=1}^{\infty} (n!)^{-n!}$. Recall that a number β is said to be a *Liouville number* if, given any large integer m, there exists a rational p/q such that

$$0 < \left| \beta - \frac{p}{q} \right| < \frac{1}{q^m}.$$

Observe that it is known that every Liouville number is transcendental. As of now, no one has proved that ℓ is absolutely abnormal. Intrigued by Granville's question, Martin [20] considered the very fast growing sequence

$$d_2 = 2^2, \ d_3 = 3^2, \ d_4 = 4^3, \ d_5 = 5^{16}, \ d_6 = 6^{30\,517\,578\,125}, \dots$$

with the recursive rule

$$d_j = j^{d_{j-1}/(j-1)} \qquad (j \geq 3).$$

Then he proved that the number

$$\prod_{j=2}^{\infty} \left(1 - \frac{1}{d_j} \right) = 0.6562499999956991 \ \underbrace{999 \dots 999}_{23,747,291,559 \text{ 9's}} \ 85284042016 \dots$$

is a Liouville number and in fact an absolutely abnormal normal.

More generally, given any sequence of positive integers $n_2, n_3, \dots$, set $d_2 = 2^{n_2}$ and

$$d_j = j^{n_j d_{j-1}/(j-1)} \qquad (j \geq 3)$$

and consider the number

$$\alpha := \prod_{j=2}^{\infty} \left(1 - \frac{1}{d_j} \right).$$

Martin proved that α is an absolutely abnormal number, thus providing an uncountable family of absolutely abnormal numbers.

8 Using the Prime Factorization to Construct Normal Numbers

As of 2011, all known normal numbers were essentially of one of the types described in Sects. 3 and 4. In 2011, a totally different approach was initiated. It is based on the idea that the prime factorization of integers is locally chaotic but globally very regular. Here is how it goes.

Let $q \geq 2$ be a fixed integer and let $\wp$ stand for the set of all primes. Let $\wp_0, \wp_1, \ldots, \wp_{q-1}$ be disjoint sets of primes such that

$$\wp = \mathcal{R} \cup \wp_0 \cup \wp_1 \cup \cdots \cup \wp_{q-1},$$

where $\mathcal{R}$ is a given finite (perhaps empty) set of primes. We call $\mathcal{R}, \wp_0, \wp_1, \ldots, \wp_{q-1}$ a *disjoint classification of primes*.

A simple example of a disjoint classification of primes is obtained by letting $q = 2$ and setting $\mathcal{R} = \{2\}$, $\wp_0 = \{p \in \wp : p \equiv 1 \pmod{4}\}$ and $\wp_1 = \{p \in \wp : p \equiv 3 \pmod{4}\}$.

Now, for each integer $q \geq 2$, let $A_q := \{0, 1, \ldots, q-1\}$. Given an integer $t \geq 1$, we say that an expression of the form $i_1 i_2 \ldots i_t$, where each $i_j \in A_q$, is a *word* of length t. The symbol Λ will denote the *empty word*. Now, given a disjoint classification of primes $\mathcal{R}, \wp_0, \wp_1, \ldots, \wp_{q-1}$, let the function $H : \wp \to A_q$ be defined by

$$H(p) = \begin{cases} j & \text{if } p \in \wp_j \text{ for some } j \in A_q, \\ \Lambda & \text{if } p \in \mathcal{R}. \end{cases}$$

Let A_q^* be the set of finite words over A_q and consider the function $T : \mathbb{N} \to A_q^*$ defined by

$$T(n) = T(p_1^{a_1} \cdots p_r^{a_r}) = H(p_1) \ldots H(p_r),$$

where we omit $H(p_i) = \Lambda$ if $p_i \in \mathcal{R}$. For convenience, we set $T(1) = \Lambda$. Finally, given a set of integers S, let $\pi(S) := \#\{p \in \wp \cap S\}$. In 2011, De Koninck and Kátai [13] proved the following result.

Theorem 1. *Let $q \geq 2$ be an integer and let $\mathcal{R}, \wp_0, \wp_1, \ldots, \wp_{q-1}$ be a disjoint classification of primes. Assume that, for a certain constant $c \geq 5$,*

$$\pi([u, u+v] \cap \wp_j) = \frac{1}{q} \pi([u, u+v]) + O\left(\frac{u}{\log^c u}\right)$$

uniformly for $2 \leq v \leq u$, $j = 0, 1, \ldots, q-1$, as $u \to \infty$. Moreover, let T be defined on $\mathbb{N}$ by

$$T(n) = T(p_1^{a_1} \cdots p_r^{a_r}) = H(p_1) \ldots H(p_r),$$

where

$$H(p) = \begin{cases} j & \text{if } p \in \wp_j \text{ for some } j \in A_q, \\ \Lambda & \text{if } p \in \mathcal{R}. \end{cases}$$

Then, $\xi = 0.T(1)T(2)T(3)T(4)\ldots$ is a q-normal number.

EXAMPLE: Let $q = 2$, $\mathcal{R} = \{2\}$, $\wp_0 = \{p : p \equiv 1 \pmod{4}\}$ and $\wp_1 = \{p : p \equiv 3 \pmod{4}\}$. In particular, $\{T(1), T(2), \ldots, T(15)\} = \{\Lambda, \Lambda, 1, \Lambda, 0, 1, 1, \Lambda, 1, 0, 1, 1, 0, 1, 10\}$. Then, it follows from Theorem 1 that $\xi = 0.T(1)T(2)T(3)T(4)\ldots = 0.101110110110\ldots$ is a binary normal number.

Although we will not give here a proof of Theorem 1, let us at least mention that a key element of its proof is a 1995 result of De Koninck and Kátai [12] which we state here as Theorem A.

Theorem A. *Let $\mathcal{R}, \wp_0, \wp_1, \ldots, \wp_{q-1}$ be a disjoint classification of primes such that*

$$\pi([u, u+v] \cap \wp_i) = \delta_i \pi([u, u+v]) + O\left(\frac{u}{(\log u)^{c_1}}\right)$$

holds uniformly for $2 \le v \le u$, $i = 0, 1, \ldots, q-1$, where $c_1 \ge 5$ is a constant, $\delta_0, \delta_1, \ldots, \delta_{q-1}$ are positive constants such that $\sum_{i=0}^{q-1} \delta_i = 1$. Assume that $\lim_{x \to \infty} w_x = +\infty$, $w_x = O(\log \log \log x)$, $\sqrt{x} \le Y \le x$ and $1 \le k \le c_2 \log \log x$, where c_2 is an arbitrary constant. Let $A \le \log \log x$ with $P(A) \le w_x$. Then, as $x \to \infty$, letting $\omega(n)$ stand for the number of distinct prime factors of n,

$$\#\{n = An_1 \le Y : p(n_1) > w_x, \ \omega(n_1) = k, \ H(n_1) = i_1 \ldots i_k\}$$

$$= (1+o(1))\delta_{i_1} \cdots \delta_{i_k} \frac{Y}{A \log Y} \frac{(\log \log x)^{k-1}}{(k-1)!} \varphi_{w_x}\left(\frac{k-1}{\log \log x}\right) F\left(\frac{k-1}{\log \log x}\right),$$

where

$$\varphi_w(z) := \prod_{p \le w} \left(1 + \frac{z}{p}\right)^{-1} \quad \text{and} \quad F(z) := \frac{1}{\Gamma(z+1)} \prod_p \left(1 + \frac{z}{p}\right)\left(1 - \frac{1}{p}\right)^z.$$

De Koninck and Kátai [15] also proved the following.

Theorem 2. *Let $q \ge 2$ be a fixed integer. Given a positive integer*

$$n = p_1^{e_1} \cdots p_{k+1}^{e_{k+1}}$$

(here, k can be zero), let

$$c_j(n) := \left\lfloor \frac{q \log p_j}{\log p_{j+1}} \right\rfloor \in A_q \quad (j = 1, \ldots, k).$$

Define the arithmetic function H by

$$H(n) = H(p_1^{e_1} \cdots p_{k+1}^{e_{k+1}}) = \begin{cases} c_1(n) \ldots c_k(n) & \text{if } k \ge 1, \\ \Lambda & \text{if } k \le 0. \end{cases}$$

Then the number $\xi = 0.H(1)H(2)H(3)\ldots$ is a q-normal number.

9 A Question Raised by Shparlinski

Let $P(n)$ stand for the largest prime factor of the integer $n \geq 2$. In 2010, Igor Shparlinski asked if the number

$$0.P(2)P(3)P(4)P(5)P(6)\ldots$$

is normal in base 10.

In 2011, De Koninck and Kátai [14] answered Shparlinski's question in the affirmative and actually proved more, as stated in Theorem 3 below.

But first some notation. Given a positive integer n, write its q-ary expansion as

$$n = \varepsilon_0(n) + \varepsilon_1(n)q + \cdots + \varepsilon_t(n)q^t,$$

where each $\varepsilon_i(n) \in A_q$ and $\varepsilon_t(n) \neq 0$. Then write

$$\overline{n} = \varepsilon_0(n)\varepsilon_1(n)\ldots\varepsilon_t(n).$$

Theorem 3. *Let $F \in \mathbb{Z}[x]$ be a polynomial with positive leading coefficient and positive degree, and such that $F(x) > 0$ if $x > 0$. Then the number*

$$\xi = 0.\overline{F(P(2))}\,\overline{F(P(3))}\,\overline{F(P(4))}\ldots\ldots$$

is normal.

We only give here a sketch of the proof of Theorem 3.

Let $L(n) := L_q(n) = \left\lfloor \dfrac{\log n}{\log q} \right\rfloor + 1$, that is, the number of digits of n in base q. Given a word $\theta = i_1 i_2 \ldots i_t \in A_q^t$, we write $\lambda(\theta) = t$. Also, let $\nu_\beta(\theta)$ stand for the number of times that the subword β occurs in the word θ. A key element of the proof of Theorem 3 is the following 1996 result of Bassily and Kátai [5].

Let $F \in \mathbb{Z}[x]$ be a polynomial with positive leading coefficient and of positive degree r. Let $\beta \in A_q^k$. Assume that κ_u is a function of u such that $\kappa_u > 1$ for all u. Setting

$$V_\beta(u) := \#\left\{ p \in \wp \cap [u, 2u] : \left| \nu_\beta(\overline{F(p)}) - \frac{L(u^r)}{q^k} \right| > \kappa_u \sqrt{L(u^r)} \right\},$$

then, there exists a positive constant c such that

$$V_\beta(u) \leq \frac{cu}{(\log u)\kappa_u^2}.$$

One can easily see that from this result it follows that given $\beta_1, \beta_2 \in A_q^k$ with $\beta_1 \neq \beta_2$ and setting

$$\Delta_{\beta_1,\beta_2}(u) := \#\left\{ p \in \wp \cap [u, 2u] : \left| \nu_{\beta_1}(\overline{F(p)}) - \nu_{\beta_2}(\overline{F(p)}) \right| > \kappa_u \sqrt{L(u^r)} \right\},$$

then, for some positive constant c,

$$\Delta_{\beta_1,\beta_2}(u) \leq \frac{cu}{(\log u)\kappa_u^2}. \tag{9.1}$$

Now, given a large number x, let $I_x = [x, 2x]$ and set $\theta = \overline{F(P(n_0))}\,\overline{F(P(n_1))} \ldots \overline{F(P(n_T))}$, where n_0 is the smallest integer in I_x, and n_T the largest.

It is clear that the proof of Theorem 3 will be complete if we can show that, given an arbitrary word $\beta \in A_q^k$, we have

$$\frac{\nu_\beta(\theta)}{\lambda(\theta)} \sim \frac{1}{q^k} \qquad (x \to \infty).$$

Since the number of digits of each prime $p \in I_x$ is of order $\log x$, it follows by the definition of θ that

$$\lambda(\theta) \approx r\,x\,\log x,$$

which reveals the true size of $\lambda(\theta)$.

Letting δ be a small positive number, one can easily show that the number of integers $n \in I_x$ for which either $P(n) < x^\delta$ or $P(n) > x^{1-\delta}$ is $\leq c\delta x$, implying that we may write

$$\nu_\beta(\theta) = \sum_{\substack{n \in I_x \\ x^\delta \leq P(n) \leq x^{1-\delta}}} \nu_\beta(\overline{F(P(n))}) + O(T) + O(\delta x \log x). \tag{9.2}$$

Let us now introduce the finite sequence $u_0, u_1, \ldots, u_H$ defined by $u_0 = x^\delta$ and thereafter by $u_j = 2u_{j-1}$ for each $1 \leq j \leq H$, where H is the smallest positive integer for which $2^H u_0 > x^{1-\delta}$, so that $H = \left\lfloor \dfrac{(1-2\delta)\log x}{\log 2} \right\rfloor + 1$.

Now, for each prime p, let $R(p) := \#\{n \in I_x : P(n) = p\}$. We have, in light of (9.2) and the fact that $T = O(x)$,

$$\nu_\beta(\theta) = \sum_{x^\delta \leq p \leq x^{1-\delta}} \nu_\beta(\overline{F(p)})R(p) + O(\delta x \log x). \tag{9.3}$$

Let $\beta_1, \beta_2 \in A_q^k$ with $\beta_1 \neq \beta_2$. Then, using (9.3), we have

$$\begin{aligned}
|\nu_{\beta_1}(\theta) - \nu_{\beta_2}(\theta)| &\leq \sum_{x^\delta \leq p \leq x^{1-\delta}} \left| \nu_{\beta_1}(\overline{F(p)}) - \nu_{\beta_2}(\overline{F(p)}) \right| R(p) + O(\delta x \log x) \\
&= \sum_{j=0}^{H-1} \sum_{u_j \leq p < u_{j+1}} \left| \nu_{\beta_1}(\overline{F(p)}) - \nu_{\beta_2}(\overline{F(p)}) \right| R(p) + O(\delta x \log x) \\
&= \sum_{j=0}^{H-1} S_j(x) + O(\delta x \log x), \tag{9.4}
\end{aligned}$$

say.

Set $\Psi(x, y) := \#\{n \leq x : P(n) \leq y\}$. Then, letting $z = \log x / \log y$, it is well known that

$$\Psi(x, y) = \rho(z) x + O\left(\frac{x}{\log y}\right) \qquad \text{uniformly for } 2 \leq y \leq x,$$

where ρ stands for the Dickman function (see for instance Theorem 9.14 in the book of De Koninck and Luca [18]).

We then have, as $x \to \infty$,

$$R(p) = \Psi\left(\frac{2x}{p}, p\right) - \Psi\left(\frac{x}{p}, p\right)$$

$$= \rho\left(\frac{\log(2x/p)}{\log p}\right)\frac{2x}{p} - \rho\left(\frac{\log(x/p)}{\log p}\right)\frac{x}{p} + O\left(\frac{x}{p \log p}\right)$$

$$= (1 + o(1))\rho\left(\frac{\log x}{\log p} - 1\right)\frac{x}{p},$$

from which it follows that

$$S_j(x) \leq \frac{2x}{u_j} \sum_{u_j \leq p < u_{j+1}} \left|\nu_{\beta_1}\left(\overline{F(p)}\right) - \nu_{\beta_2}\left(\overline{F(p)}\right)\right|. \tag{9.5}$$

Set $\kappa_u := \log \log u$. We will say that $p \in [u_j, u_{j+1})$ is a *good prime* if

$$\left|\nu_{\beta_1}\left(\overline{F(p)}\right) - \nu_{\beta_2}\left(\overline{F(p)}\right)\right| \leq \kappa_u \sqrt{L(u^r)},$$

and a *bad prime* otherwise.

Splitting the sum $S_j(x)$ into two sums, one running on the good primes and one running on the bad primes, it follows from (9.5) and the Bassily-Kátai result (9.1) that

$$S_j(x) \leq \frac{2x}{u_j}\kappa_{u_j}\sqrt{L(u_j^r)}\frac{u_j}{\log u_j} + \frac{2x}{u_j}\frac{u_j \log u_{j+1}}{(\log u_j)\kappa_{u_j}^2}$$

$$= 2x \cdot \left\{\frac{\kappa_{u_j}\sqrt{L(u_j^r)}}{\log u_j} + \frac{\log u_{j+1}}{(\log u_j)\kappa_{u_j}^2}\right\}$$

$$\leq 4x\left\{\frac{r \log \log u_j}{\sqrt{\log u_j}} + \frac{1}{(\log \log u_j)^2}\right\}.$$

Summing the above inequalities for $j = 0, 1, \ldots, H - 1$, we obtain that $\sum_{j=0}^{H-1} S_j(x) = o(x \log x)$ as $x \to \infty$ and thus that, in light of (9.4), for some constant $c > 0$,

$$|\nu_{\beta_1}(\theta) - \nu_{\beta_2}(\theta)| \leq c\delta x \log x + o(x \log x). \tag{9.6}$$

Now let ξ_N be the first N digits of the infinite word

$$\overline{F(P(2))} \ \overline{F(P(3))} \ \overline{F(P(4))} \ldots$$

and let m be the unique integer such that

$$\widetilde{\xi_N} := \overline{F(P(2))}\ \overline{F(P(3))}\ldots\overline{F(P(m))},$$

where $\lambda(\widetilde{\xi_N}) \le N < \lambda(\widetilde{\xi_N}\overline{F(P(m+1))})$, so that $\lambda(\overline{F(P(m+1))}) \ll \log m \ll \log N$, implying in particular that ξ_N and $\widetilde{\xi_N}$ have the same digits except for at most the last $\lfloor \log N \rfloor$ ones.

Let $2x = m$ and consider the intervals $I_x, I_{x/2}, I_{x/(2^2)}, \ldots, I_{x/(2^L)}$, where $L = 2[\log\log x]$, that is,

$$
\begin{array}{ccccc}
I_{x/2^L} & & I_{x/2^2} & I_{x/2} & I_x \\
|\!-\!| & \cdots & |\!-\!\!-\!\!-\!|\ |\!-\!\!-\!\!-\!\!-\!\!-\!\!-\!| & |\!-\!\!-\!\!-\!\!-\!\!-\!\!-\!\!-\!\!-\!\!-\!\!-\!\!-\!| & |\ 2x = m
\end{array}
$$

and write

$$\tau_j = \overline{F(P(a))}\ldots\overline{F(P(b))} \qquad (j = 0, 1, \ldots, L),$$

where a is the smallest and b the largest integer in $I_{x/(2^j)}$.

Moreover, let

$$\mu = \overline{F(P(2))}\ldots\overline{F(P(s))},$$

where s is the largest integer which is less than the smallest integer in $I_{x/(2^L)}$.

It is clear that

$$\left| \nu_{\beta_1}(\widetilde{\xi_N}) - \nu_{\beta_2}(\widetilde{\xi_N}) \right| \le \left| \nu_{\beta_1}(\mu) - \nu_{\beta_2}(\mu) \right| + \sum_{j=0}^{L} \left| \nu_{\beta_1}(\tau_j) - \nu_{\beta_2}(\tau_j) \right| \qquad (9.7)$$

and that

$$\nu_\beta(\mu) \le \lambda(\mu) \le \frac{x}{2^L} \cdot r \log x = o(x). \qquad (9.8)$$

Applying estimate (9.6) $L + 1$ times (with $\theta = \widetilde{\xi_N}$) by replacing successively $2x$ by $x, x/2, x/2^2, \ldots, x/2^L$, we obtain from (9.7) and in light of (9.8), that

$$\left| \nu_{\beta_1}(\widetilde{\xi_N}) - \nu_{\beta_2}(\widetilde{\xi_N}) \right| \le c\delta N + o(N) \qquad (N \to \infty). \qquad (9.9)$$

Now, one can easily see that

$$\sum_{\gamma \in A_q^k} \nu_\gamma(\theta) = \lambda(\theta) - k + 1,$$

from which it follows that

$$q^k \nu_\beta(\theta) - \lambda(\theta) = \sum_{\gamma \in A_q^k} (\nu_\beta(\theta) - \nu_\gamma(\theta)) + O(1),$$

implying that, setting $\theta = \xi_N$ and using (9.9),

$$\left| q^k \nu_\beta(\xi_N) - \lambda(\xi_N) \right| \le \sum_{\gamma \in A_q^k} \left| \nu_\beta(\xi_N) - \nu_\gamma(\xi_N) \right| + O(1)$$

$$\leq (c\delta N + o(N))q^k,$$

from which it follows that, observing that $\lambda(\xi_N) = N$,

$$\limsup_{N\to\infty} \left| \frac{\nu_\beta(\xi_N)}{N} - \frac{1}{q^k} \right| \leq c\delta.$$

Since $\delta > 0$ can be chosen arbitrarily small, it follows that

$$\limsup_{N\to\infty} \frac{\nu_\beta(\xi_N)}{N} = \frac{1}{q^k},$$

thus establishing that ξ is normal.

Later, in De Koninck and Kátai [16], we showed how the concatenation of the successive values of the smallest prime factor $p(n)$, as n runs through the positive integers, can also yield a normal number.

10   Using the Number of Prime Factors of an Integer to Create Normal Numbers

In the previous section, we showed that the number $0.P(2)P(3)P(4)\ldots$ is a normal number. What if we replace the function $P(n)$ by some other arithmetic function $f(n)$? Will we still get a normal number? Not always. Take for instance the function $\omega(n)$ which counts the number of distinct prime factors of n. One can easily show that the concatenation of the successive values of $\omega(n)$, say by considering the real number $\xi := 0.\overline{\omega(2)}\,\overline{\omega(3)}\,\overline{\omega(4)}\,\overline{\omega(5)}\ldots$, where each $\overline{m}$ stands for the q-ary expansion of the integer m, will not yield a normal number. Indeed, since the interval $I := [e^{e^{r-1}}, e^{e^r}]$, where $r := \lfloor \log\log x \rfloor$, covers most of the interval $[1, x]$ and since $\left| \frac{\omega(n)}{r} - 1 \right| < \frac{1}{r^{1/4}}$, say, with the exception of a small number of integers $n \in I$, it follows that ξ cannot be normal in basis q.

Recently, Vandehey [25] used another approach to yet create normal numbers using certain small additive functions. He considered irrational numbers formed by concatenating some of the base q digits from additive functions $f(n)$ that closely resemble the prime counting function $\Omega(n) := \sum_{p^\rho \| n} \rho$. More precisely, he used the concatenation of the last $\lceil y \frac{\log\log\log n}{\log q} \rceil$ digits of each $f(n)$ in succession and proved that the number thus created turns out to be normal in basis q if and only if $0 < y \leq 1/2$.

In De Koninck and Kátai [17], we showed that the concatenation of the successive values of $|\omega(n) - \lfloor \log\log n \rfloor|$, as n runs through the integers $n \geq 3$, yields a normal number in any given basis $q \geq 2$. Moreover, we showed that the same result holds if we consider the concatenation of the successive values of $|\omega(p+1) - \lfloor \log\log(p+1) \rfloor|$, as p runs through the prime numbers.

11 Final Remarks

In 2004, Bailey, Borwein, Crandall and Pomerance [2] proved that if x is an algebraic number of degree $d > 1$, then there exists a positive constant C such that the binary expansion of x through position n has at least $Cn^{1/d}$ ones, provided n is sufficiently large. For instance, choose $x = \sqrt{2}$. It is algebraic of degree 2. Hence according to this result, the first n digits of $\sqrt{2}$ must include at least $c\sqrt{n}$ ones (for some positive constant c). Of course, if we could prove that $\sqrt{2}$ is normal, then the first n digits should include approximately $n/2$ ones. This means that we are far from the truth.

Many authors have shown a great interest for the study of normal numbers. The recent book of Bugeaud [7] contains many other results concerning this fascinating topic along with many open problems on normal numbers.

References

1. Bailey, D.H., Borwein, J.M.: Nonnormality of the Stoneham constants (2011). http://crd.lbl.gov/dhbailey/dhbpapers/nonnormality.pdf
2. Bailey, D.H., Borwein, J.M., Crandall, R.E., Pomerance, C.: On the binary expansions of algebraic numbers. J. Number Theory Bordeaux **16**, 487–518 (2004)
3. Bailey, D.H., Crandall, R.E.: On the random character of fundamental constant expansions. Expe. Math. **10**(2), 175–190 (2001)
4. Bailey, D.H., Crandall, R.E.: Random generators and normal numbers. Exp. Math. **11**(4), 527–546 (2002)
5. Bassily, N.L., Kátai, I.: Distribution of consecutive digits in the q-ary expansions of some sequences of integers. J. Math. Sci. **78**(1), 11–17 (1996)
6. Borel, E.: Les probabilités dénombrables et leurs applications arithmétiques. Rend. Circ. Mat. Palermo **27**, 247–271 (1909)
7. Bugeaud, Y.: Distribution Modulo One and Diophantine Approximation, Cambridge Tracts in Mathematics, vol. 193. Cambridge University Press, New York (2012)
8. Cassels, J.W.S.: On a problem of Steinhaus about normal numbers. Colloquium Mathematicum **7**, 96–101 (1959)
9. Champernowne, D.G.: The construction of decimals normal in the scale of ten. J. London Math. Soc. **8**, 254–260 (1933)
10. Copeland, A.H., Erdős, P.: Note on normal numbers. Bull. Am. Math. Soc. **52**, 857–860 (1946)
11. Davenport, H., Erdős, P.: Note on normal decimals. Can. J. Math. **4**, 58–63 (1952)
12. De Koninck, J.M., Kátai, I.: On the distribution of subsets of primes in the prime factorization of integers. Acta Arith. **72**(2), 169–200 (1995)
13. De Koninck, J.M., Kátai, I.: Construction of normal numbers by classified prime divisors of integers. Functiones et Approximatio **45**(2), 231–253 (2011)
14. De Koninck, J.M., Kátai, I.: On a problem on normal numbers raised by Igor Shparlinski. Bull. Aust. Math. Soc. **84**, 337–349 (2011)
15. De Koninck, J.M., Kátai, I.: Some new methods for constructing normal numbers. Annales des Sciences Mathématiques du Québec **36**(2), 349–359 (2012)
16. De Koninck, J.M., Kátai, I.: Normal numbers generated using the smallest prime factor function. Annales mathématiques du Québec **38**(2), 133–144 (2014)

17. De Koninck, J.-M., Kátai, I.: The number of prime factors function on shifted primes and normal numbers. In: Rassias, T.M., Tóth, L. (eds.) Topics in Mathematical Analysis and Applications. Springer Optimization and Its Applications, vol. 94, pp. 315–326. Springer, Heidelberg (2014)

18. De Koninck, J.M., Luca, F.: Analytic Number Theory: Exploring the Anatomy of Integers, Graduate Studies in Mathematics, vol. 134. American Mathematical Society, Providence (2012)

19. Madritsch, M.G., Thuswaldner, J.M., Tichy, R.F.: Normality of numbers generated by the values of entire functions. J. Number Theory 128, 1127–1145 (2008)

20. Martin, G.: Absolutely abnormal numbers. Am. Math. Monthly 108(8), 746–754 (2001)

21. Nakai, Y., Shiokawa, I.: Discrepancy estimates for a class of normal numbers. Acta Arith. 62(3), 271–284 (1992)

22. Nakai, Y., Shiokawa, I.: Normality of numbers generated by the values of polynomials at primes. Acta Arith. 81(4), 345–356 (1997)

23. Sierpiński, W.: Démonstration élémentaire du théorème de M. Borel sur les nombres absolument normaux et détermination effective d'un tel nombre. Bull. Soc. Math. Fr. 45, 127–132 (1917)

24. Stoneham, R.: On absolute (j, ϵ)-normality in the rational fractions with applications to normal numbers. Acta Arith. 22, 277–286 (1973)

25. Vandehey, J.: The normality of digits in almost constant additive functions. Monatsh. Math. 171(3-4), 481–497 (2013)

Bayesian Networks

Probabilistic Query Answering in the Bayesian Description Logic $\mathcal{BEL}$

İsmail İlkan Ceylan[1]([✉]) and Rafael Peñaloza[2]

[1] Theoretical Computer Science, TU Dresden, Dresden, Germany
ceylan@tcs.inf.tu-dresden.de
[2] KRDB Research Centre, Free University of Bozen-Bolzano, Bolzano, Italy
rafael.penaloza@unibz.it

Abstract. $\mathcal{BEL}$ is a probabilistic description logic (DL) that extends the light-weight DL $\mathcal{EL}$ with a joint probability distribution over the axioms, expressed with the help of a Bayesian network (BN). In recent work it has been shown that the complexity of standard logical reasoning in $\mathcal{BEL}$ is the same as performing probabilistic inferences over the BN.

In this paper we consider conjunctive query answering in $\mathcal{BEL}$. We study the complexity of the three main problems associated to this setting: computing the probability of a query entailment, computing the most probable answers to a query, and computing the most probable context in which a query is entailed. In particular, we show that all these problems are tractable w.r.t. data and ontology complexity.

1 Introduction

Description Logics (DLs) [3] are a family of knowledge representation formalisms that have been successfully employed for modeling the knowledge of many application domains. Its success has been specially clear in the bio-medical sciences, with the development and use of very large ontologies [29]. Very briefly, an ontology is simply a collection of *axioms* that provide some explicit knowledge of the application domain; different reasoning tasks are then used to extract additional knowledge that is implicit within this ontology.

As with most logic-based formalisms, one of the issues that limit the applicability of DLs to real-world ontologies is their incapability to model and handle uncertainty in their statements. To address this limitation, many extensions of DLs for reasoning with uncertainty have been proposed over the last two decades; see e.g. [24] for a thorough, although slightly outdated, survey. A very relevant modeling choice that needs to be made is how to represent and handle the joint probability of axioms. Most probabilistic extensions of DLs avoid this problem by implicitly assuming that all axioms are (probabilistically) independent from

This work was partially supported by DFG within the Research Training Group "RoSI" (GRK 1907) and the Cluster of Excellence 'cfAED.' Most of the work was developed while R. Peñaloza was still affiliated with TU Dresden and the Center for Advancing Electronics Dresden, Germany.

© Springer International Publishing Switzerland 2015
C. Beierle and A. Dekhtyar (Eds.): SUM 2015, LNAI 9310, pp. 21–35, 2015.
DOI: 10.1007/978-3-319-23540-0_2

each other. Unfortunately, this is a very strong assumption that cannot be guaranteed to hold in general. Very recently, it was proposed to represent the logical and probabilistic dependencies of the axioms in an ontology through a Bayesian network (BN) ranging over a class of sub-ontologies, called contexts. This idea gave rise to the family of Bayesian DLs [9].

To understand the properties of Bayesian DLs, the complexity of standard reasoning on $\mathcal{BEL}$, the Bayesian extension of the light-weight DL $\mathcal{EL}$ [2,6], was studied in detail. In particular, it was shown that standard reasoning in this logic remains tractable w.r.t. the size of the logical component of the input, although intractable w.r.t. the BN [11,12]. These analysis have also shown their impact in practice, we refer the reader to the recent prototypical reasoner BORN [8] for such details (available at http://lat.inf.tu-dresden.de/systems/born).

In this paper we build on top of previous work [7], and study the complexity of answering conjunctive queries over a probabilistic knowledge base expressed in $\mathcal{BEL}$. Given the probabilistic knowledge, we focus on computing the probability of entailing a given query. Moreover, we study the problem of finding the most probable answers to a query, and the most probable contexts that entail a query. As is standard in query answering, we parameterize the complexity measures according to different input parameters. Among our results, we show that all the reasoning problems that we study remain tractable w.r.t. the size of the ontology. This means that it is possible to handle large ontologies efficiently, assuming that the probabilistic component and the query remain relatively small.

2   Preliminaries

We first briefly introduce the basic notions for query answering in the lightweight DL $\mathcal{EL}$ and its Bayesian extension $\mathcal{BEL}$, and the complexity measures that we will study throughout this paper.

As with all DLs, the main components of $\mathcal{EL}$ are concepts, that are built from concept- and role-names using a set of constructors. Let N_I, N_C and N_R be mutually disjoint sets of *individual-*, *concept-* and *role-names*, respectively. $\mathcal{EL}$ *concepts* are built by the grammar rule $C ::= A \mid \top \mid C \sqcap C \mid \exists r.C$, where $A \in N_C$ and $r \in N_R$. The *semantics* of $\mathcal{EL}$ is given by interpretations. An *interpretation* is a tuple $\mathcal{I} = (\Delta^{\mathcal{I}}, \cdot^{\mathcal{I}})$ where $\Delta^{\mathcal{I}}$ is a non-empty *domain* and $\cdot^{\mathcal{I}}$ is an *interpretation function* that maps every individual name a to an element $a^{\mathcal{I}} \in \Delta^{\mathcal{I}}$, every concept name A to a set $A^{\mathcal{I}} \subseteq \Delta^{\mathcal{I}}$, and every role name r to a binary relation $r^{\mathcal{I}} \subseteq \Delta^{\mathcal{I}} \times \Delta^{\mathcal{I}}$. The interpretation function $\cdot^{\mathcal{I}}$ is extended to $\mathcal{EL}$ concepts as shown in the upper part of Table 1.

The domain knowledge is encoded through a set of axioms that restrict the class of interpretations considered. A *TBox* $\mathcal{T}$ is a finite set of *general concept inclusions (GCIs)* of the form $C \sqsubseteq D$, where C, D are concepts. An *ABox* is a finite set of *concept assertions* $C(a)$ and *role assertions* $r(a,b)$, where $a,b \in N_I$, C is a concept, and $r \in N_R$. An *ontology* is a pair $\mathcal{O} = (\mathcal{T}, \mathcal{A})$ where $\mathcal{T}$ is a TBox and $\mathcal{A}$ an ABox. We use the term *axiom* as a general expression for GCIs and assertions. The interpretation $\mathcal{I}$ *satisfies* an axiom λ iff it satisfies the

Table 1. Syntax and semantics of $\mathcal{EL}$

Name	Syntax	Semantics
Top	$\top$	$\Delta^{\mathcal{I}}$
Conjunction	$C \sqcap D$	$C^{\mathcal{I}} \cap D^{\mathcal{I}}$
Exist. Rest.	$\exists r.C$	$\{d \mid \exists e \in \Delta^{\mathcal{I}} : (d,e) \in r^{\mathcal{I}}, e \in C^{\mathcal{I}}\}$
GCI	$C \sqsubseteq D$	$C^{\mathcal{I}} \subseteq D^{\mathcal{I}}$
Concept assertion	$C(a)$	$a^{\mathcal{I}} \in C^{\mathcal{I}}$
Role assertion	$r(a,b)$	$(a^{\mathcal{I}}, b^{\mathcal{I}}) \in r^{\mathcal{I}}$

conditions on the lower part of Table 1. It is a *model* of the ontology $\mathcal{O} = (\mathcal{T}, \mathcal{A})$ iff it satisfies all the axioms in $\mathcal{T}$ and $\mathcal{A}$. For the rest of this paper we will denote as $\mathsf{N_I}(\mathcal{A})$ the set of all individual names that appear in the ABox $\mathcal{A}$.

In the presence of an ontology, one is often interested in deciding entailment and finding answers to a (conjunctive) query. Let $\mathsf{N_V}$ be a set of *variables*, which is disjoint from $\mathsf{N_C}$, $\mathsf{N_R}$, and $\mathsf{N_I}$. An *atom* is an expression of the form $A(\chi)$ or $r(\chi, \psi)$, where $A \in \mathsf{N_C}$, $r \in \mathsf{N_R}$, and $\chi, \psi \in \mathsf{N_I} \cup \mathsf{N_V}$. A *conjunctive query* (CQ) q is a non-empty set of atoms associated to a set $\mathsf{DV}(\mathsf{q}) \subseteq \mathsf{N_V}$ of *distinguished variables*. If $\mathsf{DV}(\mathsf{q}) = \emptyset$, then q is called a *Boolean CQ*. A special case of a CQ is an *instance query*, which consists of only one atom $A(\chi)$ with $A \in \mathsf{N_C}$.

Let q be a Boolean CQ and $\mathsf{IV}(\mathsf{q})$ be the set of all individual names and variables appearing in q. The interpretation $\mathcal{I}$ *satisfies* q if there exists a function $\pi : \mathsf{IV}(\mathsf{q}) \to \Delta^{\mathcal{I}}$ such that (i) $\pi(a) = a^{\mathcal{I}}$ for all $a \in \mathsf{N_I} \cap \mathsf{IV}(\mathsf{q})$, (ii) $\pi(\chi) \in A^{\mathcal{I}}$ for all $A(\chi) \in \mathsf{q}$, and (iii) $(\pi(\chi), \pi(\psi)) \in r^{\mathcal{I}}$ for all $r(\chi, \psi) \in \mathsf{q}$. In this case, we call π a *match* for $\mathcal{I}$ and q. The ontology $\mathcal{O}$ *entails* q ($\mathcal{O} \models \mathsf{q}$) iff every model of $\mathcal{O}$ satisfies q. For an arbitrary CQ q, a function $\mathfrak{a} : \mathsf{DV}(\mathsf{q}) \to \mathsf{N_I}(\mathcal{A})$ is an *answer* to q w.r.t. $\mathcal{O}$ iff $\mathcal{O}$ entails the Boolean CQ $\mathfrak{a}(\mathsf{q})$ obtained by replacing every distinguished variable $\chi \in \mathsf{DV}(\mathsf{q})$ with $\mathfrak{a}(\chi)$. *Conjunctive query answering* (CQA) is the task of finding all answers of a CQ, and query entailment is the problem of deciding whether an ontology entails a given Boolean CQ.

It is known that in $\mathcal{EL}$ query entailment is tractable if the query is fixed, but NP-complete if the query is considered as part of the input [27]. $\mathcal{EL}$ does not enjoy the so-called full *first order rewritability* which has been considered as a key feature for CQA, since it allows one to reduce the problem to standard tasks in relational database management systems. However, other methods like the combined approach [26] have been successfully used in this setting.

The Bayesian DL $\mathcal{BEL}$ [11] has been introduced as a probabilistic extension of $\mathcal{EL}$. In $\mathcal{BEL}$ probabilities are encoded through a *Bayesian network* (BN) [17]; that is, a pair $\mathcal{B} = (G, \Phi)$, where $G = (V, E)$ is a finite directed acyclic graph (DAG) whose nodes represent Boolean random variables, and Φ contains, for every node $x \in V$, a conditional probability distribution $P_{\mathcal{B}}(x \mid \pi(x))$ of x given its parents $\mathsf{pa}(x)$. If V is the set of nodes in G, we say that $\mathcal{B}$ is a BN *over* V. In a BN, every variable $x \in V$ is considered to be conditionally independent of its non-descendants given its parents. Thus, every BN $\mathcal{B}$ defines a unique joint probability distribution over V given by the so-called *chain rule*, defined as

$$P_{\mathcal{B}}(V) = \prod_{x \in V} P_{\mathcal{B}}(x \mid \pi(x)).$$

In $\mathcal{BEL}$ concepts are constructed as for $\mathcal{EL}$. The difference appears in encoding the domain knowledge through axioms. $\mathcal{BEL}$ generalizes classical ontologies by annotating the axioms with a context, defined by a set of literals from a BN.

Let V be a finite set of Boolean variables. A V-*context* is a conjunction of literals from V. A $(V - GCI)(resp.\ V - assertion)$ is an expression of the form $\langle \lambda : \kappa \rangle$ where λ is a GCI (resp. an assertion) and κ is a V-context. A V-*TBox (resp. V-ABox)* is a finite set of V-GCIs (resp. V-assertions). A $\mathcal{BEL}$ *knowledge base* (KB) is a tuple $\mathcal{K} = (\mathcal{B}, \mathcal{T}, \mathcal{A})$ where $\mathcal{B}$ is a BN over V, $\mathcal{T}$ is a V-TBox and $\mathcal{A}$ is a V-ABox.

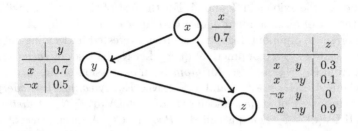

Fig. 1. The BN $\mathcal{B}_{\mathrm{ABC}}$ over the variables $\{x, y, z\}$

Example 1. The tuple $\mathcal{K} = (\mathcal{T}_{\mathrm{ABC}}, \mathcal{A}_{\mathrm{ABC}}, \mathcal{B}_{\mathrm{ABC}})$ where

$$\mathcal{T}_{\mathrm{ABC}} := \{ \langle \mathsf{A} \sqsubseteq \exists \mathsf{r}.\mathsf{B} : \{y\} \rangle, \langle \mathsf{B} \sqsubseteq \mathsf{C} : \{x\} \rangle \}$$
$$\mathcal{A}_{\mathrm{ABC}} := \{ \langle \mathsf{A}(a) : \{x\} \rangle, \langle \mathsf{r}(a, b) : \{z\} \rangle, \langle \mathsf{C}(b) : \{x, z\} \rangle, \langle \mathsf{A}(c) : \{y\} \rangle \}$$

and $\mathcal{B}_{\mathrm{ABC}}$ is the BN given in Fig. 1 represents a $\mathcal{BEL}$ KB.

Intuitively, a $\mathcal{BEL}$ KB provides a propositional abstraction over an $\mathcal{EL}$ KB. More formally, given a $\mathcal{BEL}$ KB $\mathcal{K} = (\mathcal{B}, \mathcal{T}, \mathcal{A})$ and a context κ, we define the *restriction of $\mathcal{K}$ w.r.t. κ* as an $\mathcal{EL}$ ontology $\mathcal{K}_\kappa = (\mathcal{T}_\kappa, \mathcal{A}_\kappa)$ by setting

$$\mathcal{T}_\kappa := \{C \sqsubseteq D \mid \langle C \sqsubseteq D : \mu \rangle \in \mathcal{T}, \ \kappa \models \mu\},$$
$$\mathcal{A}_\kappa := \{C(a) \mid \langle C(a) : \mu \rangle \in \mathcal{A}, \kappa \models \mu\} \cup \{r(a, b) \mid \langle r(a, b) : \mu \rangle \in \mathcal{A}, \kappa \models \mu\}.$$

We will usually speak of *contextual axioms*, or V-*axioms* to address both V-GCIs and V-assertions; if it is clear from the context, we will also drop the prefix V. The intuition behind the contextual axioms is to enforce an axiom to hold within a given context, but not necessarily in others. To formalize this intuition, we extend the notion of an interpretation, to also consider the context variables. A V-*interpretation* is a tuple $\mathcal{I} = (\Delta^{\mathcal{I}}, \cdot^{\mathcal{I}}, \mathcal{V}^{\mathcal{I}})$ where $(\Delta^{\mathcal{I}}, \cdot^{\mathcal{I}})$ is a classical $\mathcal{EL}$ interpretation, and $\mathcal{V}^{\mathcal{I}}$ is a valuation of the variables in V. The V-interpretation $\mathcal{I}$ *satisfies* the axiom $\langle \lambda : \kappa \rangle$ $(\mathcal{I} \models \langle \lambda : \kappa \rangle)$ iff either (i) $\mathcal{V}^{\mathcal{I}} \not\models \kappa$, or (ii) $(\Delta^{\mathcal{I}}, \cdot^{\mathcal{I}}) \models \lambda$. It is a *model* of the TBox $\mathcal{T}$ (resp. ABox $\mathcal{A}$) iff it satisfies all the axioms in $\mathcal{T}$ (resp. $\mathcal{A}$).

There is a strong link between the restrictions and the contextual interpretations. For any valuation $\mathcal{W}$ of the variables in V, $\mathcal{K}_{\mathcal{W}}$ represents all the $\mathcal{EL}$ axioms that must be satisfied by any contextual interpretation of the form $(\Delta^{\mathcal{I}}, \cdot^{\mathcal{I}}, \mathcal{W})$.

In $\mathcal{BEL}$, uncertainty is represented through a BN that describes a joint probability distribution over the context variables. $\mathcal{BEL}$ is linked to this distribution using *multiple world semantics*: a probabilistic interpretation defines a probability distribution over a set of (contextual) interpretations; this distribution is required to be consistent with the joint probability distribution provided by the BN. Formally, a *probabilistic interpretation* is a pair $\mathcal{P} = (\mathfrak{I}, P_{\mathfrak{I}})$, where $\mathfrak{I}$ is a set of V-interpretations and $P_{\mathfrak{I}}$ is a probability distribution over $\mathfrak{I}$ such that $P_{\mathfrak{I}}(\mathcal{I}) > 0$ only for finitely many interpretations $\mathcal{I} \in \mathfrak{I}$. $\mathcal{P}$ is a *model* of the TBox $\mathcal{T}$ (resp. ABox $\mathcal{A}$) if every $\mathcal{I} \in \mathfrak{I}$ is a model of $\mathcal{T}$ (resp. $\mathcal{A}$). $\mathcal{P}$ is *consistent* with the BN $\mathcal{B}$ if for every valuation $\mathcal{W}$ of the variables in V it holds that

$$\sum_{\mathcal{I} \in \mathfrak{I}, \mathcal{V}^{\mathcal{I}} = \mathcal{W}} P_{\mathfrak{I}}(\mathcal{I}) = P_{\mathcal{B}}(\mathcal{W}).$$

The probabilistic interpretation $\mathcal{P}$ is a *model* of the KB $(\mathcal{B}, \mathcal{T}, \mathcal{A})$ iff it is a probabilistic model of $\mathcal{T}$, $\mathcal{A}$ and consistent with $\mathcal{B}$.

In previous work, the standard reasoning problems for $\mathcal{EL}$ have been extended to their probabilistic variant in $\mathcal{BEL}$, leading to tight complexity bounds for several problems [11,12]. Particularly, it has been shown that the complexity of these tasks is bounded by the complexity of reasoning in $\mathcal{EL}$ and in the BN.

In the next sections we will study the complexity of different query-related reasoning tasks in $\mathcal{BEL}$. As is customary in the context of conjunctive queries, we will consider the complexity w.r.t. different parameters. The measures we consider here are: (i) *data complexity*, where only the ABox is considered as part of the input; (ii) *ontology complexity*, which considers both, the ABox and the TBox; (iii) *network complexity*, w.r.t. the size of the BN; (iv) *KB complexity*, which uses the whole KB as input; and (v) *combined complexity* in which the input is measured in terms of the KB and the query.

3   Probabilistic Query Entailment

The problem of deciding whether a Boolean CQ is entailed by a $\mathcal{BEL}$ KB is not interesting, since it ignores the probabilistic information encoded in the BN. Recall that a $\mathcal{BEL}$ KB describes a probability distribution over different worlds, in which some conditions must hold. In this setting, we are interested in finding the probability of observing a world in which the query is entailed.

Definition 2 (Probabilistic Entailment). Let $\mathcal{K}$ be a $\mathcal{BEL}$ KB, $\mathcal{P} = (\mathfrak{I}, P)$ a probabilistic interpretation and q a Boolean CQ. The *probability of* q w.r.t. $\mathcal{P}$ is

$$P_{\mathcal{P}}(\mathsf{q}) := \sum_{\mathcal{I} \in \mathfrak{I},\ (\Delta^{\mathcal{I}}, \cdot^{\mathcal{I}}) \models \mathsf{q}} P(\mathcal{I}).$$

The probability of q w.r.t. $\mathcal{K}$ is $P_{\mathcal{K}}(\mathsf{q}) := \inf_{\mathcal{P} \models \mathcal{K}} P_{\mathcal{P}}(\mathsf{q})$. The query q is entailed with probability $p \in (0,1]$ iff $P_{\mathcal{K}}(\mathsf{q}) \geq p$.

Recall that for a given $\mathcal{EL}$ KB $\mathcal{K}$ and a valuation $\mathcal{W}$, $\mathcal{K}_{\mathcal{W}}$ defines an $\mathcal{EL}$ ontology that contains all the axioms that must be satisfied by any contextual interpretation using the valuation $\mathcal{W}$. We show that considering the restrictions $\mathcal{K}_{\mathcal{W}}$ over valuations $\mathcal{W}$ is enough to decide probabilistic query entailment.

Theorem 3. *For every $\mathcal{BEL}$ KB $\mathcal{K} = (\mathcal{B}, \mathcal{T}, \mathcal{A})$ and Boolean CQ q it holds that $P_{\mathcal{K}}(\mathsf{q}) = \sum_{\mathcal{K}_{\mathcal{W}} \models \mathsf{q}} P_{\mathcal{B}}(\mathcal{W})$.*

Proof. We define the probabilistic interpretation $\mathcal{R} = (\mathfrak{J}_{\mathcal{R}}, P_{\mathfrak{J}_{\mathcal{R}}})$ where

(i) $\mathfrak{J}_{\mathcal{R}} = \bigcup_{i=0}^{2^n - 1} \mathcal{I}_i = (\Delta^{\mathcal{I}_i}, \cdot^{\mathcal{I}_i}, \mathcal{V}^{\mathcal{I}_i})$
(ii) $P_{\mathfrak{J}_{\mathcal{R}}}(\mathcal{I}_i) = P_{\mathcal{B}}(\mathcal{W}_i)$ with $\mathcal{W}_i = \mathcal{V}^{\mathcal{I}_i}$ for all $0 \leq i \leq 2^n - 1$
(iii) $(\Delta^{\mathcal{I}_i}, \cdot^{\mathcal{I}_i}) \models \mathcal{K}_{\mathcal{W}_i}$ for all $0 \leq i \leq 2^n - 1$
(iv) $(\Delta^{\mathcal{I}_i}, \cdot^{\mathcal{I}_i}) \models \mathsf{q}$ iff $\mathcal{K}_{\mathcal{W}_i} \models \mathsf{q}$ for all $0 \leq i \leq 2^n - 1$

Notice that, we can ensure (iii) by the fact that every $\mathcal{EL}$ ontology has a model.

It follows from the construction that $\mathcal{R} \models (\mathcal{T}, \mathcal{A})$ and $\mathcal{R}$ is consistent with $\mathcal{B}$. Hence, $\mathcal{R}$ is a model of $\mathcal{K}$. We show the probability of q w.r.t. $\mathcal{R}$ to be

$$P_{\mathcal{R}}(\mathsf{q}) := \sum_{\substack{\mathcal{I}_i \in \mathfrak{J}_{\mathcal{R}} \\ (\Delta^{\mathcal{I}_i}, \cdot^{\mathcal{I}_i}) \models \mathsf{q}}} P_{\mathfrak{J}_{\mathcal{R}}}(\mathcal{I}_i) = \sum_{\mathcal{K}_{\mathcal{W}_i} \models \mathsf{q}} P_{\mathcal{B}}(\mathcal{W}_i),$$

which concludes $P_{\mathcal{K}}(\mathsf{q}) \leq \sum_{\mathcal{K}_{\mathcal{W}} \models \mathsf{q}} P_{\mathcal{B}}(\mathcal{W})$.
Assume now that the inequality is strict. This implies the existence of a model $\mathcal{S} = (\mathfrak{J}_{\mathcal{S}}, P_{\mathfrak{J}_{\mathcal{S}}})$ such that

$$P_{\mathcal{S}}(\mathsf{q}) = \sum_{\substack{\mathcal{I} \in \mathfrak{J}_{\mathcal{S}} \\ (\Delta^{\mathcal{I}}, \cdot^{\mathcal{I}}) \models \mathsf{q}}} P_{\mathfrak{J}_{\mathcal{S}}}(\mathcal{I}) < \sum_{\mathcal{K}_{\mathcal{W}} \models \mathsf{q}} P_{\mathcal{B}}(\mathcal{W}).$$

This holds iff for some $\mathcal{W}$ where $\mathcal{K}_{\mathcal{W}} \models \mathsf{q}$ and $P_{\mathcal{B}}(\mathcal{W}) > 0$ it holds that

$$\sum_{\substack{(\Delta^{\mathcal{I}}, \cdot^{\mathcal{I}}, \mathcal{W}) \in \mathfrak{J}_{\mathcal{S}} \\ (\Delta^{\mathcal{I}}, \cdot^{\mathcal{I}}) \models \mathsf{q}}} P_{\mathfrak{J}_{\mathcal{S}}}(\mathcal{I}) < P_{\mathcal{B}}(\mathcal{W}).$$

Since $\sum_{\mathcal{I} \in \mathfrak{J}_{\mathcal{S}}, \mathcal{V}^{\mathcal{I}} = \mathcal{W}} P_{\mathfrak{J}_{\mathcal{S}}}(\mathcal{I}) = P_{\mathcal{B}}(\mathcal{W})$ by the definition of a model, there exists a contextual interpretation $(\Delta^{\mathcal{I}'}, \cdot^{\mathcal{I}'}, \mathcal{V}^{\mathcal{I}'}) \in \mathfrak{J}_{\mathcal{S}}$ where $\mathcal{V}^{\mathcal{I}'} = \mathcal{W}$ and $(\Delta^{\mathcal{I}'}, \cdot^{\mathcal{I}'}) \not\models \mathsf{q}$ while $\mathcal{K}_{\mathcal{W}'} \models \mathsf{q}$. It follows that $(\Delta^{\mathcal{I}'}, \cdot^{\mathcal{I}'}) \not\models \mathcal{K}_{\mathcal{W}}$ and $(\Delta^{\mathcal{I}'}, \cdot^{\mathcal{I}'}, \mathcal{V}^{\mathcal{I}'}) \not\models (\mathcal{T}, \mathcal{A})$, which contradicts with the assumption that $\mathcal{S}$ is a model. $\qquad\square$

Theorem 3 provides a simple method for computing the probability of a query q w.r.t. a $\mathcal{BEL}$ KB: one needs only to compute, for each valuation $\mathcal{W}$, the $\mathcal{EL}$ ontology $\mathcal{K}_\mathcal{W}$ and decide whether this ontology entails q, adding the probabilities (w.r.t. $\mathcal{B}$) of all the worlds for which this test is positive. We illustrate probabilistic query entailment on our running example.

Example 4. Consider the $\mathcal{BEL}$ KB provided in Example 1 and the Boolean CQ q = $\{A(\chi), r(\chi, \psi), C(\psi)\}$. Clearly, $\mathcal{K}_\mathcal{W} \models$ q only for worlds $\mathcal{W}$ such that $\mathcal{W} \models (x \wedge y) \vee (x \wedge z)$. Hence, we get $P_\mathcal{K}(\mathsf{q}) = P_{\mathcal{B}_{ABC}}((x \wedge y) \vee (x \wedge z)) = 0.511$.

Since there are $2^{|V|}$ valuations, $\mathcal{EL}$ query entailment is decidable in polynomial time in ontology complexity, and computing the probability of a valuation is polynomial in $|V|$, we obtain the following result.

Theorem 5. *Probabilistic query entailment is polynomial w.r.t. data and ontology complexity and in* ExpTime *w.r.t. network, KB, and combined complexity.*

Notice that the algorithm sketched above iterates over all the possible scenarios described by the BN and performs an entailment test in each of them. The positive complexity results w.r.t. data and ontology complexity arise from the fact that in these settings the size of the BN is assumed to be constant. In order to obtain a better upper bound w.r.t. network complexity, we can dualize this idea; i.e., iterate over all the sub-ontologies performing standard probabilistic inferences at each iteration.

Theorem 6. *Probabilistic query entailment is* PP-*complete w.r.t. network complexity.*

Proof. The lower complexity bound follows from the complexity of standard reasoning in $\mathcal{BEL}$ [12]. To show membership, we define a *sub-ontology* of a given $\mathcal{BEL}$ KB $\mathcal{K} = (\mathcal{B}, \mathcal{T}, \mathcal{A})$ as a pair $\mathcal{O} = (\mathcal{T}', \mathcal{A}')$ such that $\mathcal{T}' \subseteq \mathcal{T}$ and $\mathcal{A}' \subseteq \mathcal{A}$. Each sub-ontology $\mathcal{O} = (\mathcal{T}', \mathcal{A}')$ defines a context

$$\mathsf{con}(\mathcal{O}) = \bigwedge_{\langle \lambda : \kappa \rangle \in \mathcal{T}'} \kappa \wedge \bigwedge_{\langle \lambda : \kappa \rangle \in \mathcal{A}'} \kappa,$$

and an $\mathcal{EL}$ ontology $\mathcal{O}_{\mathcal{EL}} = (\mathcal{T}'_{\mathcal{EL}}, \mathcal{A}'_{\mathcal{EL}})$

$$\mathcal{T}'_{\mathcal{EL}} := \{C \sqsubseteq D \mid \langle C \sqsubseteq D : \kappa \rangle \in \mathcal{T}' \text{ for some context } \kappa\},$$
$$\mathcal{A}'_{\mathcal{EL}} := \{C(a) \mid \langle C(a) : \kappa \rangle \in \mathcal{A}' \text{ for some context } \kappa\} \cup$$
$$\{r(a, b) \mid \langle r(a, b) : \kappa \rangle \in \mathcal{A}' \text{ for some context } \kappa\}.$$

For every contextual interpretation $\mathcal{I} = (\Delta^\mathcal{I}, \cdot^\mathcal{I}, \mathcal{V}^\mathcal{I})$ with $\mathcal{I} \models (\mathcal{T}, \mathcal{A})$, we observe that if $\mathcal{V}^\mathcal{I} \models \mathsf{con}(\mathcal{O})$, then $(\Delta^\mathcal{I}, \cdot^\mathcal{I}) \models \mathcal{O}_{\mathcal{EL}}$. For a given Boolean CQ q, we define

$$\mathsf{con}_\mathcal{K}(\mathsf{q}) := \bigvee_{\mathcal{O}_{\mathcal{EL}} \models \mathsf{q}} \mathsf{con}(\mathcal{O}).$$

From Theorem 3, we know that $P_{\mathcal{K}}(\mathsf{q}) = P_{\mathcal{B}}(\mathsf{con}(\mathsf{q}))$. Thus, it suffices to compute the probability of the DNF formula $\mathsf{con}(\mathsf{q})$ to obtain the probability of the query q. Since Bayesian network inferences are PP-complete [28], and the class PP is closed under intersection and complementation [4], it follows that probabilistic query entailment is also in PP w.r.t. network complexity. $\qquad\square$

We consider now the case of combined complexity, in which the ontology, the BN, and the query are all considered as part of the input. We show that in this case, the complexity of probabilistic query entailment is at most PSPACE.

Theorem 7. *Probabilistic query entailment is in* PSPACE *w.r.t. combined complexity.*

Proof. Theorem 3 ensures that to compute $P_{\mathcal{K}}(\mathsf{q})$ it suffices to check for every valuation $\mathcal{W}$, whether $\mathcal{K}_{\mathcal{W}} \models \mathsf{q}$, and in case it does, compute $P_{\mathcal{B}}(\mathcal{W})$. $\mathcal{K}_{\mathcal{W}}$ can be constructed by adding all axioms λ to $\mathcal{K}_{\mathcal{W}}$ where $\langle \lambda : \kappa \rangle \in \mathcal{K}$ and $\mathcal{W} \models \kappa$. This requires only linear time on both $|\mathcal{K}|$ and $|V|$. Deciding whether $\mathcal{K}_{\mathcal{W}} \models \mathsf{q}$ is an NP-complete problem w.r.t. the sizes of $\mathcal{K}$ and q. Finally, $P_{\mathcal{B}}(\mathcal{W})$ can be computed in time polynomial on the size of $\mathcal{B}$, using the chain rule for BNs. A PSPACE algorithm avoids storing exponentially many valuations of the variables in V simultaneously; instead iterates for each valuation independently. $\qquad\square$

Obviously, this result also yields a PSPACE upper bound for this problem w.r.t. KB complexity. In terms of lower bounds, Theorem 6 shows that probabilistic query entailment is also PP-hard w.r.t. KB and combined complexity. Unfortunately, we were unable to obtain tight complexity bounds for these measures.

4   Probabilistic Query Answering

In query answering we do not restrict to Boolean CQs anymore, but consider queries that may contain distinguished variables. As described before, in this case we are interested in finding the possible substitutions of these distinguished variables into individuals appearing in the ontology such that the resulting Boolean CQ is entailed; these substitutions are called *answers*. To find all these answers, one could simply perform a query entailment test for each of the possible substitutions. There are exponentially many such substitutions, measured on the number of individuals in the ontology, and potentially all of them can be answer to a given query, and receiving so many results might be uninformative to a user. Rather than providing all possible answers to a query, we are interested in finding a limited number of them having the highest probability of being entailed.

Let q be a query with the distinguished variables $\mathsf{DV}(\mathsf{q})$, and $\mathcal{K} = (\mathcal{B}, \mathcal{T}, \mathcal{A})$ a $\mathcal{BEL}$ KB. Recall that every function $\mathfrak{a} : \mathsf{DV}(\mathsf{q}) \to \mathsf{N}_{\mathsf{I}}(\mathcal{A})$ defines a Boolean CQ obtained by replacing every $\chi \in \mathsf{DV}(\mathsf{q})$ in q with $\mathfrak{a}(\chi)$. Abusing of the notation, we call this query $\mathfrak{a}(\mathsf{q})$. We call any function $\mathfrak{a} : \mathsf{DV}(\mathsf{q}) \to \mathsf{N}_{\mathsf{I}}(\mathcal{A})$ an *answer* to q w.r.t. $\mathcal{K}$, and define its probability as $P_{\mathcal{K}}(\mathfrak{a}) := P_{\mathcal{K}}(\mathfrak{a}(\mathsf{q}))$. Clearly, since an answer defines a Boolean CQ, computing the probability of such an answer is exactly as hard as probabilistic query entailment in all measures considered. We use this probability as a means to identify the most relevant answers, returning only those that are most likely to be observed.

Definition 8 (top-k answer). Let q be a query, $\mathcal{K}$ be a $\mathcal{BEL}$ KB, and $k \in \mathbb{N}$. A *top-k answer* to q w.r.t. $\mathcal{K}$ is a tuple $(\mathfrak{a}_1, \ldots, \mathfrak{a}_k)$ of different answers to q w.r.t. $\mathcal{K}$ such that (i) for all $i, 1 \leq i < k$, $P_{\mathcal{K}}(\mathfrak{a}_i) \geq P_{\mathcal{K}}(\mathfrak{a}_{i+1})$, and (ii) for every other answer $\mathfrak{a}$, $P_{\mathcal{K}}(\mathfrak{a}_k) \geq P_{\mathcal{K}}(\mathfrak{a})$.

In other words, a top-k answer is an ordered tuple of the k answers that have the highest probability. We assume that k is a constant that is fixed *a priori*. Thus, it is not considered part of the input of the problem. Obviously, since different answers may have the same degree, top-k answers are not unique. Here we are only interested in finding one of these tuples. Stating it as a decision problem, we want to verify whether a given tuple is a top-k answer.

Example 9. Consider the $\mathcal{BEL}$ KB $\mathcal{K} = (\mathcal{T}_{\text{ABC}}, \mathcal{A}_{\text{ABC}}, \mathcal{B}_{\text{ABC}})$ provided in Example 1 and the query $q = \{A(\chi)\}$ with $\chi \in \mathsf{DV}$. We are interested in identifying the top-1 answer to q w.r.t. $\mathcal{K}$. Notice that both $\mathfrak{a}_0 : \chi \mapsto a$ and $\mathfrak{a}_1 : \chi \mapsto c$ are answers to q with positive probability. Clearly, $\mathfrak{a}_0$ is the top-1 answer since $P_{\mathcal{K}}(\mathfrak{a}_0) > P_{\mathcal{K}}(\mathfrak{a}_1)$.

Assuming that the size of q and the BN $\mathcal{B}$ are fixed, then there are polynomially many answers to q w.r.t. $\mathcal{K}$, and for each answer $\mathfrak{a}$, we can compute $P_{\mathcal{B}}(\mathfrak{a})$ performing constantly many $\mathcal{EL}$ query entailment tests. Thus, it is possible to verify whether $(\mathfrak{a}_1, \ldots, \mathfrak{a}_k)$ is a top-k answer in polynomial time w.r.t. ontology complexity. Likewise, if the ontology and the query are constant, then we can compute $P_{\mathcal{B}}(\mathfrak{a})$ through constantly many probabilistic inferences in the BN, as described in the previous section. Overall, we obtain the following result.

Corollary 10. *Deciding top-k answers is in* PTime *w.r.t. data and ontology complexity and* PP-*complete w.r.t. network complexity.*

We consider now the case of the combined complexity, in which all the elements are considered as part of the input and show that our problem is at least hard to the level coNP^{PP}; that is a class known to be between PH (the limit of the polynomial hierarchy) and PSPACE [30]

Theorem 11. *Deciding whether a tuple $\mathfrak{A}$ is a top-k answer is* coNP^{PP}-*hard w.r.t. KB complexity.*

Proof. We provide a reduction from the decision version of the *maximum a-posteriori* (D-MAP) problem for BNs [17]. Formally, given a BN $\mathcal{B}$ over V, a set $Q \subseteq V$, a context κ, and $p > 0$, the D-MAP problem consists of deciding whether there exists a valuation μ of the variables in Q such that $P_{\mathcal{B}}(\kappa \wedge \mu) > p$. Consider an arbitrary but fixed instance of D-MAP described by the BN $\mathcal{B} = ((V, E), \Phi)$, the context κ, $Q \subseteq V$, and $p > 0$. We introduce a new Boolean random variable z not appearing in V. Using this variable, we construct a new DAG (V', E) with $V' = V \cup \{z\}$ and a new BN $\mathcal{B}' = ((V', E), \Phi')$, where

$P_{\mathcal{B}'}(v \mid \mathsf{pa}(v)) = P_{\mathcal{B}}(v \mid \mathsf{pa}(x))$ for all $v \in V$, and $P_{\mathcal{B}'}(z) = p$. Consider the $\mathcal{BEL}$ KB $\mathcal{K} = (\mathcal{B}', \emptyset, \mathcal{A})$ where

$$\mathcal{A} := \{\langle A_x(a_x) : x \rangle, \langle A_x(b_x) : \neg x \rangle, \langle A_x(c) : z \rangle \mid x \in Q\} \cup \{\langle B(a) : \kappa \rangle, \langle B(c) : z \rangle\},$$

and query $\mathsf{q} := \{A_x(\chi_x) \mid x \in Q\} \cup \{B(\chi)\}$, where all the variables are distinguished; i.e., $\mathsf{DV}(\mathsf{q}) = \{\chi_x \mid x \in Q\} \cup \{\chi\}$. It is easy to see that the mapping $\mathfrak{a}_0 : \mathsf{DV}(\mathsf{q}) \to \{c\}$ that maps all the distinguished variables to the individual name $c \in \mathsf{N_I}(\mathcal{A})$ is an answer to this query and $P_{\mathcal{K}}(\mathfrak{a}_0) = p$. Moreover, any other answer that maps any variable to c will have probability at most p, since it can only be entailed in contexts satisfying z. Suppose that there is an answer $\mathfrak{a}$ such that $P_{\mathcal{K}}(\mathfrak{a}) > p$. This answer must map every variable χ_x to either a_x or b_x and χ to a. Let $\mu_{\mathfrak{a}} := \bigwedge_{\mathfrak{a}(\chi_x)=a_x} x \wedge \bigwedge_{\mathfrak{a}(\chi_x)=b_x} \neg x$. By construction, $\mu_{\mathfrak{a}}$ is a valuation of the variables in Q, $P_{\mathcal{B}}(\kappa \wedge \mu_{\mathfrak{a}}) > p$, and $\mathfrak{a}(\mathsf{q})$ is only entailed by valuations satisfying the context $\kappa \wedge \mu_{\mathfrak{a}}$. Overall this means that $\mathfrak{a}_0$ is *not* a top-1 answer iff there is a valuation μ of the variables in Q such that $P_{\mathcal{B}}(\kappa \wedge \mu) > p$. $\square$

In the previous section we have shown that probabilistic query entailment is decidable in PSPACE w.r.t. combined complexity. Since PSPACE is a deterministic complexity class, we can in fact compute the precise probability of an entailment using only polynomial space. To show that a tuple is *not* a top-k answer, we can guess a new answer and show that its probability is strictly larger than some answer in the tuple. Overall, this means that top-k query answering remains in PSPACE w.r.t. combined complexity.

Obtaining most probable answers for a query is a crucial task for the domains, where imprecise characterizations of knowledge is necessary. The next section is dedicated to another reasoning task that can be seen dual to top-k answers, namely top-k contexts.

5   Most Likely Contexts for a Query

Dually to finding the most likely answers to a query, we are also interested in finding the k most likely contexts that entail a given Boolean query q. More precisely, suppose that we have already observed that the query q holds; then, we are interested in finding out which is the current context. As in the previous section, we do not consider one, but search for a fixed number of contexts that are the most likely to hold.

As explained before, $\mathcal{K}_{\kappa}$ specifies the minimal conditions that must be satisfied in any contextual interpretation that satisfies the context κ. If $\mathcal{K}_{\kappa}$ entails the Boolean query q, then we say that q holds in context κ. We are interested in finding out the most likely contexts in which a given query holds.

Definition 12 (top-k mlc). Let q be a CQ, $\mathcal{K}$ a $\mathcal{BEL}$ KB, and $k \in \mathbb{N}$. $\kappa_1, \ldots, \kappa_k$ are *top-k most likely contexts* (top-k mlc) for q w.r.t. $\mathcal{K}$ if $\mathcal{K}_{\kappa_i}$ entails q for all $i, 1 \leq i \leq k$; $P_{\mathcal{B}}(\kappa_i) \geq P_{\mathcal{B}}(\kappa_{i+1})$ for all $i, 1 \leq i \leq k$; and there is no other context κ such that $\mathcal{K}_{\kappa} \models \mathsf{q}$ and $P_{\mathcal{B}}(\kappa) > P_{\mathcal{B}}(\kappa_k)$.

We illustrate top-k mlc with our continuing example. In this case, we are interested in finding out the 2 most likely context that entail the query.

Example 13. Consider the $\mathcal{BEL}$ KB $\mathcal{K}$ and query q provided in Example 1. Clearly all contexts κ that entail q are such that $\kappa \models \{x, y\} \vee \{x, z\}$. The top-2 contexts are then $\langle \{x, y\}, \{x, z\} \rangle$ since $P_{\mathcal{B}_{\mathrm{ABC}}}(\{x, y\}) > P_{\mathcal{B}_{\mathrm{ABC}}}(\{x, z\}) > P_{\mathcal{B}_{\mathrm{ABC}}}(\kappa)$ for any other context κ.

We show that deciding top-k mlc is tractable w.r.t. ontology complexity. Furthermore, we obtain a coNP^{PP} lower bound for the combined complexity as an analogous result to top-k answer. Differently from top-k answer; for this reasoning problem, we are able to show that this complexity bound is tight.

Theorem 14. *Top-k mlc is polynomial w.r.t. data, and ontology complexity, and coNP^{PP}-complete w.r.t. KB and combined complexity.*

Proof. If the BN is fixed, then the number of contexts is constant, and they can be ordered w.r.t. their complexity in constant time. The top-k mlc problem is then solved by applying a constant number of $\mathcal{EL}$ CQ entailment tests, yielding a polynomial upper bound w.r.t. ontology complexity.

For the combined complexity, coNP^{PP}-hardness is immediate since deciding one most likely context for simple queries is already coNP^{PP}-hard w.r.t. KB complexity [12]. We prove that top-k mlc is in coNP^{PP} : If a tuple is not a top-k mlc, then guess a new context κ and show using a PP oracle that $\mathcal{K}_\kappa \models$ q and $P_{\mathcal{B}}(\kappa) > P_{\mathcal{B}}(\kappa_k)$. □

In terms of network complexity, a PP-hardness follows easily from the complexity of probabilistic entailment in BNs. The upper bound w.r.t. network complexity requires polynomially many calls to a PP oracle.

Theorem 15. *Top-k mlc is PP-hard and in P^{PP} w.r.t. network complexity.*

Proof. We show that top-k mlc is in P^{PP} w.r.t. networks complexity. Recall that if $\mathcal{T}$, $\mathcal{A}$ and q are fixed, then there is a constant number of contexts that entail the consequence, using only the Boolean variables that appear in $\mathcal{T}$ and $\mathcal{A}$; call this number ℓ. However, the BN $\mathcal{B}$ may also contain other variables. If $\ell < k$, then we need to expand the previously found contexts with new literals from $\mathcal{B}$ until enough contexts have been found. In the worst case, this would require a polynomial number (in the size of $\mathcal{B}$) of probabilistic entailments. Thus, this algorithm only yields a P^{PP} upper bound w.r.t. network complexity. □

To reduce the complexity of finding the most likely contexts, we consider a special case of the problem in which we are interested in full valuations of all the variables in the BN $\mathcal{B}$. We call this problem *top-k worlds*. In this case, deciding $P_{\mathcal{B}}(\mathcal{W}) > P_{\mathcal{B}}(\mathcal{W}_k)$ requires only polynomial time w.r.t. network complexity, since the chain rule of BNs yields the probability of a valuation in polynomial time. The problem is also easier than the top-k mlc w.r.t. the combined complexity: simply check whether $P_{\mathcal{K}}(\mathcal{W}) > P_{\mathcal{K}}(\mathcal{W}_k)$ and decide $\mathcal{K}_{\mathcal{W}} \models$ q, where the former can be done in time polynomial and the latter is complete for the class NP.

Table 2. $\mathcal{BEL}$ reasoning problems and their complexity

Problem	Data	ont.	Network	KB	Combined
Probabilistic entailment	P	P	PP-c	PP-h	in PSPACE
Probability of an answer	P	P	PP-c	PP-h	in PSPACE
top-k answer	P	P	PP-c	PP-h	coNP^{PP}/PSPACE
top-k mlc	P	P	PP/P^{PP}	coNP^{PP}-c	coNP^{PP}-c
top-k worlds	P	P	P	coNP-c	coNP/Π_2^p

Notice that, top-k contexts and top-k answers are dual to each other, but they do not necessarily overlap. Consider for instance the case, where all top-k answers to a query q are retrieved from the same context κ. In this case, top-k contexts for q will contain other contexts than κ with the assumption that $k > 1$. Top-k contexts is particularly informative where the diversity of knowledge is important.

6   Related Work

Probabilistic query answering is an important reasoning task that has been widely studied in different domains such as relational databases [15,18,20], RDF graphs [21] and XML databases [1,22]. As mentioned before, there are many DL-based probabilistic ontology languages [24]. Surprisingly, only few of them concentrate on query answering.

In the probabilistic extension of Datalog+/- [19] authors are interested in retrieving the answers that are above a threshold value that is set *a priori*. In contrast to $\mathcal{BEL}$, in probabilistic Datalog+/- the underlying semantics is based on Markov logic networks. The Prob-DL family [25] extends classical DLs with subjective probabilities, also known as Type II probabilities [23]. The main difference with our logic is that Prob-$\mathcal{EL}$ introduces probabilities as a concept constructor, whereas we allow only probabilities over axioms.

More closely related to $\mathcal{BEL}$ is BDL-Lite [16]. As is in $\mathcal{BEL}$, BDL-Lite only allows probabilities over axioms and conditional dependencies are represented faithfully. However, as it has been pointed before [12], the authors use a closed world assumption, which easily leads to inconsistencies.

7   Conclusions

In this paper we continued the analysis of the complexity of reasoning in the Bayesian DL $\mathcal{BEL}$, and considered tasks associated to conjunctive queries. Specifically, we have studied the complexity of deciding probabilistic entailment of a Boolean CQ, and of verifying that a tuple of answers to a CQ are those with the highest probability of being entailed. Dually, we consider also the problem of finding the most likely contexts that entail a Boolean query. All these complexity results are summarized in Table 2.

As it can be seen from the table, if one considers only the purely logical components of the problem (ontology complexity), then reasoning is tractable, which is consistent with the complexity of reasoning in classical $\mathcal{EL}$. The network complexity is also typically the same as performing standard probabilistic inferences over a BN. However, the complexity tends to increase if we combine these factors and consider also the query. Unfortunately, to the best of our efforts, we were unable to close all the gaps in the complexity results. Our conjecture is that the KB and the combined complexity coincide in all the problems considered here; in particular, we expect all the problems described in Table 2, with the exception of top-k worlds, to be (co)NP^{PP}-complete w.r.t. these two complexity measures.

The algorithm for deciding query entailment through the computation of con(q) provides a tight upper bound for this problem w.r.t. network complexity. However, it would be impractical to implement as it iterates over all possible subontologies. Arguably, techniques such as weighted model counting [14] would lead towards more practical algorithms for this problem. We will explore the possibility of extending the Bayesian ontology reasoner BORN with an efficient query entailment service using such techniques.

The proof of hardness for top-k query answering w.r.t. combined complexity uses a very simple query which is in fact acyclic. Thus, contrary to classical $\mathcal{EL}$ [5], restricting to acyclic queries does not suffice for reducing the complexity of reasoning. On the other hand, for simple instance queries the combined complexity should not be higher than the network complexity. This claim can be shown by adapting the proof structures from [10] to the completion-based algorithm for $\mathcal{ELO}$ as pointed in [7]. It would be interesting to find other meaningful restrictions that reduce the complexity of these reasoning tasks.

One important open issue is the use of partial information in our reasoning problems, through conditioning. For example, one could be interested in finding the context κ with the highest probability of occurring, given that a query q holds. Notice that this problem is different from finding the most likely context since in this case, we do not require that $\mathcal{K}_\kappa$ entails the query q.

Another future direction id to extend the framework to consider also temporal queries over dynamic ontologies in which the probabilistic knowledge evolves over time as described in [13].

Most of the notions and ideas presented here are independent of the logical formalism used. Indeed, although the specific complexity bounds found are specific to the properties of the DL $\mathcal{EL}$, the reasoning algorithms presented usually require only classical query entailment tests, and hence can be adapted to other ontological languages where these tests are decidable, without major trouble.

References

1. Abiteboul, S., Senellart, P.: Querying and updating probabilistic information in XML. In: Ioannidis, Y., Scholl, M.H., Schmidt, J.W., Matthes, F., Hatzopoulos,

M., Böhm, K., Kemper, A., Grust, T., Böhm, C. (eds.) EDBT 2006. LNCS, vol. 3896, pp. 1059–1068. Springer, Heidelberg (2006)

2. Baader, F., Brandt, S., Lutz, C.: Pushing the EL. In: Proceedings of the IJCAI 2005. Morgan Kaufmann Publishers (2005)

3. Baader, F., Calvanese, D., McGuinness, D.L., Nardi, D., Patel-Schneider, P.F. (eds.): The Description Logic Handbook: Theory, Implementation, and Applications, 2nd edn. Cambridge University Press, Cambridge (2007)

4. Beigel, R., Nick, R., Spielman, D.A.: PP Is closed under intersection. J. Comput. Syst. Sci. **50**(2), 191–202 (1995)

5. Bienvenu, M., Ortiz, M., Šimkus, M., Xiao, G.: Tractable queries for lightweight description logics. In: Proceedings of the IJCAI 2013. AAAI (2013)

6. Brandt, S.: Polynomial time reasoning in a description logic with existential restrictions, GCI axioms, and–what else? In: Proceedings of the ECAI 2004, vol. 110. IOS Press (2004)

7. Ceylan, İ.İ.: Query answering in Bayesian description logics. In: Proceedings of the DL 2015. CEUR Workshop Proceedings, vol. 1350. CEUR-WS (2015)

8. Ceylan, İ.İ., Mendez, J., Peñaloza, R.: The Bayesian ontology reasoner is BORN! In: Proceedings of ORE 2015. CEUR Workshop Proceedings, vol. 1387. CEUR-WS (2015)

9. Ceylan, İ.İ., Peñaloza, R.: Bayesian description logics. In: Proceedings of DL 2014. CEUR Workshop Proceedings, vol. 1193. CEUR-WS (2014)

10. Ceylan, İ.İ., Peñaloza, R.: Reasoning in the description logic $\mathcal{BEL}$ using Bayesian networks. In: Proceedings of StarAI 2014. AAAI Workshops, vol. WS-14-13. AAAI (2014)

11. Ceylan, İİ., Peñaloza, R.: The Bayesian description logic $\mathcal{BEL}$. In: Demri, S., Kapur, D., Weidenbach, C. (eds.) VSL 2014. LNCS, vol. 8562, pp. 480–494. Springer, Heidelberg (2014)

12. Ceylan, İİ., Peñaloza, R.: Tight complexity bounds for reasoning in the description logic $\mathcal{BEL}$. In: Fermé, E., Leite, J. (eds.) Logics in Artificial Intelligence. LNCS, vol. 8761, pp. 77–91. Springer, Heidelberg (2014)

13. Ceylan, İ.İ., Peñaloza, R.: Dynamic Bayesian ontology languages. CoRR abs/1506.08030 (2015)

14. Chavira, M., Darwiche, A.: On probabilistic inference by weighted model counting. Artif. Intell. **172**(6–7), 772–799 (2008)

15. Dalvi, N., Suciu, D.: Efficient query evaluation on probabilistic databases. VLDB J. **16**(4), 523–544 (2007)

16. d'Amato, C., Fanizzi, N., Lukasiewicz, T.: Tractable reasoning with Bayesian description logics. In: Greco, S., Lukasiewicz, T. (eds.) SUM 2008. LNCS (LNAI), vol. 5291, pp. 146–159. Springer, Heidelberg (2008)

17. Darwiche, A.: Modeling and Reasoning with Bayesian Networks. Cambridge University Press, Cambridge (2009)

18. Fuhr, N., Rölleke, T.: A probabilistic relational algebra for the integration of information retrieval and database systems. ACM TOIS **15**(1), 32–66 (1997)

19. Gottlob, G., Lukasiewicz, T., Martinez, M.V., Simar, G.L.: Query answering under probabilistic uncertainty in datalog +/- ontologies. Ann. Math. AI **69**(1), 131–159 (2013)

20. Grädel, E., Gurevich, Y., Hirsch, C.: The complexity of query reliability. In: Proceedings of the ACM SIGACT-SIGMOD-SIGART 1998 (1998)

21. Huang, H., Liu, C.: Query evaluation on probabilistic RDF databases. In: Vossen, G., Long, D.D.E., Yu, J.X. (eds.) WISE 2009. LNCS, vol. 5802, pp. 307–320. Springer, Heidelberg (2009)

22. Hung, E., Getoor, L., Subrahmanian, V.S.: PXML: a probabilistic semistructured data model and algebra. In: Proceedings of the ICDE 2003 (2003)
23. Joseph, H.: An analysis of first - order logics of probability. In: Proceedings of the IJCAI 1989. Morgan Kaufmann Publishers (1989)
24. Lukasiewicz, T., Straccia, U.: Managing uncertainty and vagueness in description logics for the semantic web. Web Semant. Sci. Serv. Agents World Wide Web **6**(4), 291–308 (2008)
25. Lutz, C., Schröder, L.: Probabilistic description logics for subjective uncertainty. In: Proceedings of the KR 2010. AAAI (2010)
26. Lutz, C., Toman, D., Wolter, F.: Conjunctive query answering in the description logic calel using a relational database system. In: Proceedings of the IJCAI 2009. AAAI (2009)
27. Rosati, R.: On conjunctive query answering in EL. In: Proceedings of the DL 2007. CEUR Workshop Proceedings, vol. 250. CEUR-WS (2007)
28. Roth, D.: On the hardness of approximate reasoning. Artif. Intell. **82**(1–2), 273–302 (1996)
29. Schulz, S., Suntisrivaraporn, B., Baader, F., Boeker, M.: SNOMED reaching its adolescence: ontologists' and logicians' health check. Int. J. Med. Inf. **78**(Supplement 1), 86–94 (2009)
30. Toda, S.: On the Computational power of PP and +P. In: Proceedings of the SFCS 1989, pp. 514–519. IEEE (1989)

The Complexity of Plate Probabilistic Models

Fabio G. Cozman[(✉)] and Denis D. Mauá

Universidade de São Paulo, São Paulo, Brazil
{fgcozman,denis.maua}@usp.br

Abstract. Plate-based probabilistic models combine a few relational constructs with Bayesian networks, so as to allow one to specify large and repetitive probabilistic networks in a compact and intuitive manner. In this paper we investigate the combined, data and domain complexity of plate models, showing that they range from polynomial to #P-complete to #EXP-complete.

1 Introduction

The desire to tackle complex decision scenarios, where many variables interact and vast quantities of data are collected, has produced various modeling languages based on graphs. For instance, Bayesian and Markov networks offer visually pleasant tools by which one can represent interacting variables [5,10,15]. In practice, several scenarios display repetitive patterns that can be best encoded using relations, domains, and individuals. To address this reality, formalisms have been proposed that combine features of Bayesian and Markov networks with relational languages; for instance, plates [12], Markov logic networks [17], relational Bayesian networks [8].

Plate-based probabilistic models are possibly the simplest and most successful of these "probabilistic relational models". By capturing symmetries in the model, plates make communication and modelling much more efficient. Plates are simple to draw, easy to understand, and quite powerful in what they can represent. Plate models have been extensively used in statistical practice [11] since they were introduced with the BUGS project [7,12]. In machine learning, they have been used (often informally) to convey several models since their first appearance [3]. One example is the smoothed Latent Dirichlet Allocation (sLDA) model [1], usually represented with plates as in Fig. 1 (explained later).

In this paper we present results on the inferential complexity of plate models, a topic that seems not to have received due attention. There are (at least) three kinds of complexity results that are of interest in this context [4]: first, the *combined* complexity of inferences, where model, evidence and domain are given as input; second, the *data* complexity, where query, evidence and domain are given as input (and we fix a model); finally, the *domain* complexity, where only the domain is given as input (and model, query and evidence are fixed).

We start with the original plate models (Sect. 2), where nodes in a plate can only have children inside the same plate, and we investigate their inferential complexity (Sect. 3). We first look into the combined complexity, which we show

© Springer International Publishing Switzerland 2015
C. Beierle and A. Dekhtyar (Eds.): SUM 2015, LNAI 9310, pp. 36–49, 2015.
DOI: 10.1007/978-3-319-23540-0_3

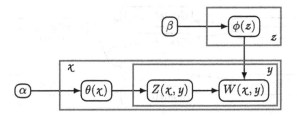

Fig. 1. Smoothed Latent Dirichlet Allocation, a plate model.

to be #P-complete; this is a surprising result given that plates can generate grounded models that are exponentially large on their description size (hence a naive approach to inference would take exponential time). We show data complexity also to be #P-complete, and domain complexity to be constant. We then move to more general plate models where a node can have children in any plate (Sect. 4). Here it is necessary to allow "aggregation functions" that specify probability values. We focus on the simplest combination functions, and show that combined complexity leads to #EXP-complete inference, while data complexity leads to #P-complete inference.

2 Plates

In this section we define plate models and some related concepts; because the literature does not have a standard formalization, we start from somewhat basic notions. We only deal with finite spaces, so every variable has finitely many values, leaving for the future a study of continuous variables [19].

A Bayesian network consists of a directed acyclic graph where each one of its n nodes is a random variable X_i, and where the following Markov condition holds: any X_i is independent of its nondescendants given its parents. Additionally, a Bayesian network contains a set of conditional probability distributions: for each X_i, we have $\mathbb{P}(X_i = x_i | \mathrm{pa}(X_i) = \pi_i)$ for all values x_i and π_i ($\mathrm{pa}(X_i)$ denotes the parents of X_i in the graph). The Markov condition implies the factorization $\mathbb{P}(X_1 = x_1, \ldots, X_n = x_n) = \prod_{i=1}^{n} \mathbb{P}(X_i = x_i | \mathrm{pa}(X_i) = \pi_i)$, where π_i is the projection of $x_1, \ldots, x_n$ on $\mathrm{pa}(X_i)$ [15]. As a simple example [10], assume we want to infer the performance of John (a student), as given by his grade; the performance is affected by John's intelligence and by the course's difficulty. This is graphically represented as $\boxed{\mathsf{Difficulty}} \rightarrow \boxed{\mathsf{Grade}} \leftarrow \boxed{\mathsf{Intelligence}}$. Now we may be interested in a set of students; we use $\mathsf{Grade}(c_i, s_j)$ to denote the grade of student s_j in course c_i; similarly, $\mathsf{Difficulty}(c_i)$ denotes the difficult of the ith course and $\mathsf{Intelligence}(s_j)$ the intelligence of the jth student. A Bayesian network for two students and two courses is shown in Fig. 2. The very same model can be described using plates as in Fig. 3 [7].

To define plate models more formally, we adopt the following concepts, mixing definitions by Koller and Friedman [10, Chap. 6] and terminology by Poole [16].

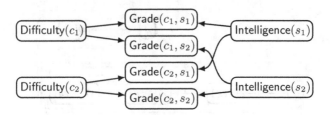

Fig. 2. Bayesian network with repetitive structure.

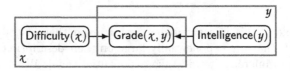

Fig. 3. Plate model for the network in Fig. 2.

We first take any *logical variable (logvar)* to be typed, ranging over a finite set, called its *domain*. Note: in the language of plates each logical variable is uniquely attached to a symbol, from which we can deduce its domain. A *parameterized variable (parvariable)* X is a function that yields a random variable $X(\chi_1, \ldots, \chi_k)$ for each instantiation of the logvars $\chi_1, \ldots, \chi_k$ (all typed). All the random variables in the image of a parvariable take values in the same space (hence we can unambiguously talk about the space of values of parvariables). Denote by $\mathsf{logvar}(X)$ the tuple of logvars associated with parvariable X. A *plate graph* consists of a directed acyclic graph where each node X_i is a parvariable, and such that for any X_i and $Y \in \mathsf{pa}(X_i)$, $\mathsf{logvar}(Y) \subseteq \mathsf{logvar}(X_i)$.[1] If two nodes share a logvar they are said to belong to the same plate. Each plate is usually indicated by a rectangle, containing the parvariables that belong to the plate, and information about the logvars in the plate. A *plate model* consists of a plate graph and, additionally, a *template conditional probability distribution* for each parvariable X_i, that yields $\mathbb{P}(X_i = x_i | \mathsf{pa}(X_i) = \pi_i)$ for all possible values x_i and π_i. Template distributions are often called *parfactors* [16]; the latter word is also used to refer to arbitrary functions over parvariables.

To specify the semantics of plate models, we need an additional piece of notation. Suppose we have an ordered set of logvars $\vec{\chi} = (\chi_1, \ldots, \chi_k)$ and one of its possible instantiations, $\vec{a} = (a_1, \ldots, a_k)$; then, given another ordered set of logvars, $\vec{\chi}'$, denote by $\vec{\chi}'[\vec{\chi}/\vec{a}]$ the ordered set where, for each possible i, χ_i is replaced by a_i.

Concerning semantics, a plate model represents a (possibly large) Bayesian network, constructed as follows. First, for each parvariable X, generate all instantiations of $\mathsf{logvar}(X)$ (as they range over their domains), and for each instantiation $\vec{a}$ create a node $X(\vec{a})$. Second, for each parvariable X, generate again

---

[1] Note that the definition of plate model in Ref. [10] does not require acyclicity, but this seems to be a necessary requirement in all the relevant literature.

all instantiations of its logvars; for each instantiation $\overrightarrow{a} = (a_1, \ldots, a_k)$ and for each Y in $pa(X)$, add an edge from $Y(logvar(Y)[logvar(X)/\overrightarrow{a}])$ to $X(\overrightarrow{a})$. Third, associate each grounded parvariable with the corresponding template conditional distribution.

The graph in Fig. 2 is the "grounding" of the plates model in Fig. 3: χ runs over a set containing courses c_1 and c_2, and y runs over a set containing students s_1 and s_2. To complete specification of the grounded Bayesian network, we must have template probabilities. Suppose, for instance, that all variables are binary, and $\mathbb{P}(\mathsf{Difficulty}(\chi) = 0) = 1/3$. Then we have both $\mathbb{P}(\mathsf{Difficulty}(c_1) = 0) = 1/3$ and $\mathbb{P}(\mathsf{Difficulty}(c_2) = 0) = 1/3$ in the grounded Bayesian network.

The plates we have defined so far are the "classic" plates that first appeared with the BUGS system [12]. One of their limitations is that a node only has children inside the same plate (we assume there is a "base plate" containing all nodes). In practice plate models go beyond this, by letting a node to have children in other plates. See for instance the sLDA model in Fig. 1: here $\phi(z)$ has a child $W(\chi, y)$. The semantics of these enhanced plates is discussed in Sect. 4.

3   The Complexity of "Classic" Plate Models

We now examine the complexity of inference with classic plate models. In this context, an inference typically refers to the calculation of a conditional probability $\mathbb{P}(\mathbf{Q}|\mathbf{E})$, where $\mathbf{Q}$ and $\mathbf{E}$ are sets of assignments $\{X(\overrightarrow{a}) = x\}$ (understood as conjunctions). We assume that $\mathbf{Q}$ and $\mathbf{E}$ are not contradictory (i.e., that they do not contain different assignments to the same variable) and that $\mathbb{P}(\mathbf{E}) > 0$ so that the inference is well-defined. Note that checking whether this last assumption holds is again an inference problem. As an example of inference, consider computing $\mathbb{P}(\mathsf{Difficulty}(c_1) = 1 | \mathsf{Intelligence}(s_2) = 1, \mathsf{Grade}(c_1, s_2) = 0, \mathsf{Grade}(c_2, s_2) = 1)$. The set $\mathbf{Q}$ is the *query* and the set $\mathbf{E}$ is the *evidence*.

To discuss the complexity of inferences, we need a few concepts. The complexity class #P is the class of integer-valued functions computed by counting Turing machines in polynomial time; a counting Turing machine is a standard nondeterministic Turing machine that prints in binary notation, on a separate tape, the number of accepting computations induced by the input [21]. Thus, #P contains the counting versions of NP-complete problems. If a problem is #P-hard and can be solved with one call to a #P oracle and with polynomial-time computations (i.e., if it belongs to $\mathsf{FP}^{\#P[1]}$), it is said to be #P[1]-equivalent [6]. Roth [18] showed that inference in Bayesian networks is #P-hard. Because inference can be reduced to counting solutions of NP-complete problems followed by a normalization step [5], the problem is #P[1]-equivalent [6]. Note that one cannot assert #P-completeness of such inferences, as #P produces integers and inference produces rationals. Later we will need the class #EXP; that is, the class of functions computed by counting Turing machines in exponential time (the number of accepting paths may have exponential size) [13]. A problem is #EXP[1]-equivalent if it is #EXP-hard and can be solved with one call to a #EXP oracle whose result x is processed through a function $h(x)$ that requires polynomial time with respect to the size of x.

A plate model can specify any Bayesian network: just define all parvariables without logvars. Hence the calculation of $\mathbb{P}(\mathbf{Q}|\mathbf{E})$ for plate models is #P-hard. Now consider a generic plate model, with as many logvars (as many plates) as needed. For each plate, a domain must be specified; suppose the domain is given as a list of elements. This assumption means, in essence, that we are not contemplating compact ways to specify the domain (for instance, one might just give a number N in binary notation, with the understanding that the domain is $\{1, 2, \ldots, N\}$; we do not deal with this case here).

Even an explicit specification for domains can lead to exponentially large grounded Bayesian networks. By taking M nested plates, each with a domain consisting of N elements, a single parvariable can specify N^M random variables. Here is a simple example. Suppose we are interested in groups of individuals; for instance, we wish to model the parvariable Family that indicates whether or not M individuals are related. We draw:

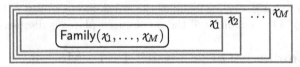

Assuming we have N individuals, the grounding of this plate model has N^M grounded nodes. In this example inference is trivial as all random variables are independent. If we instead had several parvariables connected in complex ways, we might face a very dense, exponentially large Bayesian network. This might suggest that calculation of probability values would take us to #EXP in the worst case. The nice fact about plate models is that inference has the same complexity of Bayesian networks, despite the possibly exponential size of grounded Bayesian networks:

Theorem 1. *Inference in plate models is #P[1]-equivalent.*

Proof. Clearly, the problem is #P-hard, so it suffices to show it can be computed with one call to #P plus polynomial-time post-processing. We achieve this by showing that, in any given inference, only a polynomially-large fragment of the (possibly exponentially-large) grounded Bayesian network is necessary, and inference in this fragment is known to be #P[1]-equivalent.

We are to compute $\mathbb{P}(\mathbf{Q}|\mathbf{E})$. Suppose we produce the complete grounded Bayesian network. All nodes that appear in $\mathbf{Q}$ and in $\mathbf{E}$ appear in this network. The fragment consisting of these grounded nodes and their (grounded) ancestors is the sub-network that contains all information needed for inference; other grounded nodes can be discarded [15]. This fragment may contain disconnected sub-networks; the only sub-network that matters for the computation of $\mathbb{P}(\mathbf{Q}|\mathbf{E})$ is the sub-network that contains the grounded parvariables in $\mathbf{Q}$. Refer to this fragment of the original grounded as the *requisite* network.

Suppose there are m grounded nodes in $\mathbf{Q} \cup \mathbf{E}$. For any grounded node W in this set, the number of ancestors of W is less than the number of (parvariable) nodes in the plate model. For instance, in our last example, the ancestors of a grounding of Grade contain exactly one grounding of Difficulty and one grounding of Intelligence. Now if there are n parvariables in the plate model, there are at most mn grounded nodes in the requisite network. By running inference in this requisite network, we obtain the desired result.   □

This result focuses on the *combined complexity* of plate models; that is, the complexity of inference when plate model, query, evidence, and domains are given as input. However, in practice we may be more interested in the complexity of inferences when the model is fixed. This is justified when we expect the model to be small but the data to be abundant. For instance, we may be interested in modeling relations in a social network; we may have a few relations (friendship, marriage, etc.), but an enormous amount of data. The complexity of inference when query, evidence and domains are inputs (and the model is fixed) is called the *data* complexity of inference; similarly, we may be interested in a fixed model and fixed query/evidence, with only the domains as the input; in this case we have *domain* complexity [4]. Data and domain complexities are directly related to the concepts of lqe-liftability and domain-liftability that are often employed in the literature on lifted inference of probabilistic relational models [9]. Lqe-liftability means that data complexity is polynomial, and domain-liftability means that domain complexity is polynomial.

Concerning the data complexity of plates, we have:

Theorem 2. *Inference in plate models when the model is fixed is* #P[1]-*equivalent.*

Proof. Given Theorem 1, we only need to show hardness. Consider a monotone 2-CNF formula on propositional variables $a_1, \ldots, a_n$ with m clauses. We call a_i and a_j left and right variables, respectively, of the clause $a_i \vee a_k$. We assume an ordering of the clauses, so that we can refer to the left (right) variable of ith clause. Counting the number of assignments to the variables that make the sentence true is a #P-complete problem [21].

Build the plate model in Fig. 4. Both the logvars x and y index the propositions in the CNF formula; their domains are $\{1, \ldots, n\}$. A grounded variable Left(i) represents the proposition a_i when it appears as the left variable in a clause. Similarly, Right(j) represents a_j when it appears as the right variable. Impose $\mathbb{P}(\mathsf{Left}(x)) = 1/2$, $\mathbb{P}(\mathsf{Right}(y)) = 1/2$. The Equivalence($x, y$) parvariable enforces that Left(i) and Right(j) must take on the same value whenever they represent the same proposition; this is achieved by imposing

$$\mathbb{P}\big(\mathsf{Equivalence}(x, y) = 1 | \mathsf{Left}(x), \mathsf{Right}(y)\big) = \begin{cases} 1, & \text{if } \mathsf{Left}(x) = \mathsf{Right}(y), \\ 0, & \text{if } \mathsf{Left}(x) \neq \mathsf{Right}(y). \end{cases}$$

Finally, Disjunction(x, y) encodes a clause with propositions Left(x) and Right(y): $\mathbb{P}\big(\mathsf{Disjunction}(x, y) = 1 | \mathsf{Left}(x), \mathsf{Right}(y)\big) = 0$ if $\mathsf{Left}(x) = \mathsf{Right}(y) = 0$ and $\mathbb{P}\big(\mathsf{Disjunction}(x, y) = 1 | \mathsf{Left}(x), \mathsf{Right}(y)\big) = 1$ otherwise.

Now create the evidence **E** that contains for each $i = 1, \ldots, N$ (that is, for each proposition), the assignment $\{\mathsf{Equivalence}(i, i) = 1\}$. Likewise, create the

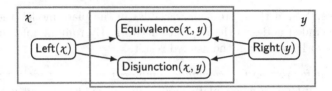

Fig. 4. A plate model that counts satisfying assignments of monotone 2-CNF formulas.

query $\mathbf{Q}$ containing for each clause the assignment $\{\mathsf{Disjunction}(i,j) = 1\}$, where a_i and a_j are, respectively, the left and right variables of the ith clause. Building this plate model takes polynomial time in the size of the CNF formula. One can check that $\mathbb{P}(\mathbf{Q}|\mathbf{E})$ equals the number of satisfying assignments of the formula up to a (polynomial-time computable) constant.   □

The proof of Theorem 1 shows that the size of domain is irrelevant once the model, query and evidence are fixed. Hence, domain complexity is constant:

Theorem 3. *Inference in plate models, when the model, query and evidence are fixed, takes constant time.*

4   Enhanced Plate Models

We now consider plate models where a node in a plate can have children in other plates; we refer to these as *enhanced plate models*. As before, to guarantee that such a definition works for all cases, we assume that there is a "base plate" encompassing all nodes, so that a node outside of all drawn plates is already in the base plate, and it can have children in other plates. We do not draw the base plate in our plate models.

A popular model that employs enhanced plate models is sLDA, depicted in Fig. 1. Here the logvar z runs over a set of topics, while x runs over a set of documents, and y runs over a designated set of strings. The node $W(x,y)$ is the child of $\phi(z)$; grounding produces:

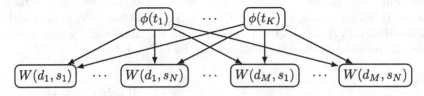

Now consider a particular grounded variable $W(d,s)$; the number of parents of this variable in the grounded graph depends on the number of topics (that is, on the size of domain of z). Hence the template probability $\mathbb{P}\big(W(x,y)|\phi(z)\big)$ must specify which procedure is to be used to produce probability values given the domains. The typical solution is to allow *aggregation functions* to be given [10], where an aggregation function takes a set of groundings, and assignments for them, and produces a probability value out of them. Another strategy is used in relational Bayesian networks, where *combination functions* are adopted to compute probability values [8]. Now we need to be careful when selecting a family of aggregation functions lest the complexity of inference may be dictated by the complexity of a single aggregation function. For example, consider allowing a function that counts the number of solutions of an EXP-complete problem. Then clearly the inference problem is #EXP-hard. On the other extreme, consider aggregation functions that return constant values; these functions act as if disconnecting the parents from the child, and do not add any expressivity over classic plate models.

We choose to investigate the simplest possible language, where all parvariables are binary (have two values), and aggregation functions are specified using existential quantifiers. Such quantifiers can describe many common phenomena; for instance, they can be used to specify Noisy-Or models [15]. Moreover, existential quantifiers are easily computed, and concisely specified. Say we have a parvariable Y with parent $X(\chi)$, and χ takes values in $\{a_1, \ldots, a_N\}$. The grounded Bayesian network contains the variable Y with parents $X(a_1), \ldots, X(a_N)$. Then the corresponding conditional distribution is $\mathbb{P}(Y = 1|X(a_1), \ldots, X(a_N)) = 0$ if $X(a_1) = \cdots = X(a_N) = 0$ and $\mathbb{P}(Y = 1|X(a_1), \ldots, X(a_N)) = 1$ otherwise. This can be stated more concisely as $Y = \exists \chi X(\chi)$.

Now suppose the model has another variable Z with $\mathbb{P}(Z|Y) = 1$ if $Z \neq Y$ and $\mathbb{P}(Z|Y) = 0$ if $Z = Y$. That is, $Z = \neg Y$. Then $\mathbb{P}(Z|X(\chi)) = \neg \exists \chi X(\chi) = \forall \chi \neg X(\chi)$. Thus we assume, as a syntactic sugar, that we can specify aggregation functions containing arbitrary logical formulas with existential and universal quantifiers, and that we specify the aggregation function as a first-order logical expression: for example, $Y = (\exists \chi \forall y Z(\chi, y) \wedge \neg X(\chi) \rightarrow W(y))$.

Given that we may have a polynomial number of nested plates, and this produces an exponential number of groundings, one might suspect that inference with enhanced plates requires exponential effort. However, it is not obvious how to prove this, because the language of plate models does not allow us to directly build standard complete problems for exponential classes. We have the restriction that each logvar is tied to a plate/domain; hence we cannot write a logical expression such as $X(\chi) \rightarrow \neg X(y)$, where the same parvariable X appears with distinct logvars. However, our main result in this section shows that exponential behavior is actually realized:

Theorem 4. *Inference in enhanced plate models is #EXP[1]-equivalent.*

Proof. To prove pertinence, note that an enhanced plate model can be grounded into an exponentially larger Bayesian network, and inference can be carried out in that network (which implies it can be solved with one call to a #EXP machine and some exponential-time post-processing).

To prove hardness, we resort to bounded domino problems; indeed, we will build a plate model (Fig. 5) that encodes a domino problem. A *domino system* consists of a finite set D of *tiles* and a pair of *compatibility relations* H and V, both on $D \times D$, respectively expressing horizontal and vertical constraints between tiles. The idea is that tiles are to be placed in points of a *torus* $\mathbb{N}_s \times \mathbb{N}_t$, where $\mathbb{N}_s$ denotes the integers modulo s, and that adjacent tiles need to satisfy the constraints H and V. Such a torus is denoted compactly by $U(s,t)$.

Some tiles, the *initial conditions*, are assigned to the first n points in the bottom row of torus $U(s,t)$. We denote by d_i^0 the ith initial condition; that is, the initial condition for point $(i,0)$. The torus has a *tiling* if it is possible to cover the whole torus while respecting the compatibility relations and the initial conditions. That is, there must be a mapping $\tau : U(s,t) \to D$ such that for all $(x,y) \in U(s,t)$, (i) $(\tau(x,y), \tau(x+1 \mod s, y)) \in H$; (ii) $(\tau(x,y), \tau(x, y+1 \mod t)) \in V$; (iii) $\tau(i,0) = d_i^0$ for $0 \le i < n$.

Börger et al. showed that given a (time/space) bounded Turing machine one can construct a bounded domino system that reproduces its behavior [2, Theorem 6.1.2]. Unfortunately in their construction the number of accepting paths in the Turing machine and the number of tilings in the domino system may differ, and this is inappropriate for a counting class such as #EXP. We need to produce a *parsimonious* reduction [14, Sect. 18.1]; that is, a reduction that preserves the number of accepting paths in the Turing machine. To do it, we must recapitulate the construction by Börger et al. They start by assuming that we have a *simple* nondeterministic Turing machine M over alphabet Σ containing a *blank* character. That is, the Turing machine works on a single semi-infinite tape where cells are numbered from 0 on; the machine never tries to move to the left of the first cell, and at every stage of the computation there is some integer n such that cells 0 to n contain non-blank characters and all other cells contain blanks; finally, the machine has a unique accepting state q_a, in which the tape contains only blanks and the head is in the first cell. Given any Turing machine, we can enlarge it polynomially so that it satisfies these restrictions, as described by the following result (the proof is omitted due to space constraints, but it can be produced by an explicit construction):

Lemma 1. *Let M be a simple nondeterministic Turing machine with alphabet Σ, input alphabet Σ', and set of states Q. An input x is a sequence $\sigma_0' \sigma_1' \ldots \sigma_{n-1}'$. Then there exists a domino system and a linear-time reduction that takes any input x to a sequence d^0 of n tiles such that:*
(i) if M accepts x in time t_0 and space s_0 then for any accepting computation there is a single tiling for torus $U(s,t)$ with initial condition d^0 where s and t are polynomials on s_0, t_0, and M;
(ii) if M does not accept x then the torus $U(s,t)$ is not tiled with initial condition d^0 for all $s,t \ge 2$.

Hence, counting the number of tilings is a #EXP-complete problem. From now on we assume that we have a domino system with m tiles ($|D| = m$) specifying a torus $U(2^n, 2^n)$ and initial conditions $d_i^0 \ldots d_{n-1}^0$. Our goal is to reduce the problem of counting tilings to an inference in an enhanced plate model. Our reduction is inspired by a similar result by Tobies [20].

First we need to represent the positions of the torus; we do so by creating $2n$ logical variables $x_{0,0}, \ldots, x_{0,n-1}$ and $x_{1,0}, \ldots, x_{1,n-1}$. All these variables have the same binary domain $\{0,1\}$. The idea is that these variables represent the coordinates of a position (x,y) in the torus in the following way: an assignment

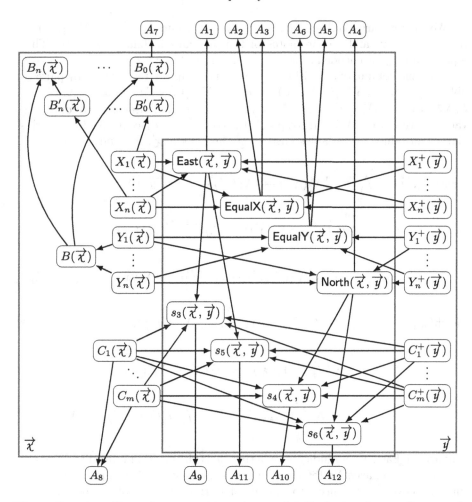

Fig. 5. Counting tilings of a torus with a plate model. Each plate is actually a set of n nested plates, one for each logvar in the indicated vector of logvars. Nodes s_3, s_4, s_5 and s_5 encode auxiliary logical expressions indicated in the text.

$\vec{a}$ to $\vec{x}_0 = (x_{0,0}, \ldots, x_{0,n-1})$ represents the value of x (the column) in binary notation, while an assignment $\vec{b}$ to $\vec{x}_1 = (x_{1,0}, \ldots, x_{1,n-1})$ represents the value of y (the row) in binary notation. To make the presentation more clear and succinct, we treat all these logical variables as a single variable $\vec{x}$ whose domain are the natural numbers between zero and 2^{2n}. One should bear in mind that this is simply syntactic sugar (so the reduction is polynomial in the input).

The proof builds two torus, which we force to be identical. The positions of the second torus are represented by the logical variables $\vec{y}$ whose combined domain are the natural numbers between zero and $2^{2n} - 1$. Here again, this is simply syntactic sugar to avoid writing $2n$ logical variables with binary domains.

We create parvariables $X_0(\overrightarrow{x}),\ldots,X_{n-1}(\overrightarrow{x})$ and $Y_0(\overrightarrow{x}),\ldots,Y_{n-1}(\overrightarrow{x})$ to represent the x- and y-coordinates of the positions in binary notation. Thus a position (x,y) is encoded as $x = \sum_{i=0}^{n-1} X_i(x,y)\cdot 2^i$ and $y = \sum_{i=0}^{n-1} Y_i(x,y)\cdot 2^i$. The row and column of a position in the second torus are represented by parvariables $X_0^+(\overrightarrow{y}),\ldots,X_{n-1}^+(\overrightarrow{y})$ and $Y_0^+(\overrightarrow{y}),\ldots,Y_{n-1}^+(\overrightarrow{y})$. We impose $\mathbb{P}(X_i = 1) = \mathbb{P}(X_i^+ = 1) = \mathbb{P}(Y_i = 1) = \mathbb{P}(Y_i^+ = 1) = 1/2$ (here and elsewhere we omit logvars to save space). We need to specify the concept of adjacent positions; to this end we introduce parvariables $\mathsf{East}(\overrightarrow{x},\overrightarrow{y})$ and $\mathsf{North}(\overrightarrow{x},\overrightarrow{y})$, and specify:

$$\mathsf{East}(\overrightarrow{x},\overrightarrow{y}) = \bigwedge_{k=0}^{n-1} (\wedge_{j=0}^{k-1} X_j(\overrightarrow{x})) \to (X_k(\overrightarrow{x}) \leftrightarrow \neg X_k^+(\overrightarrow{y})) \wedge$$

$$\bigwedge_{k=0}^{n-1} (\vee_{j=0}^{k-1} \neg X_j(\overrightarrow{x})) \to (X_k(\overrightarrow{x}) \leftrightarrow X_k^+(\overrightarrow{y})),$$

$$\wedge \bigwedge_{k=0}^{n-1} ((Y_k(\overrightarrow{x}) \to Y_k^+(\overrightarrow{y})) \wedge (\neg Y_k(\overrightarrow{x}) \to \neg Y_k^+(\overrightarrow{y}))),$$

$$\mathsf{North}(\overrightarrow{x},\overrightarrow{y}) = \bigwedge_{k=0}^{n-1} (\wedge_{j=0}^{k-1} Y_j(\overrightarrow{x})) \to (Y_k(\overrightarrow{x}) \leftrightarrow \neg Y_k^+(\overrightarrow{y})) \wedge$$

$$\bigwedge_{k=0}^{n-1} (\vee_{j=0}^{k-1} \neg Y_j(\overrightarrow{x})) \to (Y_k(\overrightarrow{x}) \leftrightarrow Y_k^+(\overrightarrow{y}))$$

$$\wedge \bigwedge_{k=0}^{n-1} ((X_k(\overrightarrow{x}) \to X_k^+(\overrightarrow{y})) \wedge (\neg X_k(\overrightarrow{x}) \to \neg X_k^+(\overrightarrow{y}))).$$

Parvariable $\mathsf{East}(\overrightarrow{x},\overrightarrow{y})$ indicates whether $\overrightarrow{y}$ is the position immediately to the right of $\overrightarrow{x}$; similarly, $\mathsf{North}(\overrightarrow{x},\overrightarrow{y})$ indicates whether $\overrightarrow{y}$ is the position immediately above $\overrightarrow{x}$. We need to enforce that the positions of $\overrightarrow{x}$ and $\overrightarrow{y}$ with $\overrightarrow{x} = \overrightarrow{y}$ have the same encoding: $\mathsf{EqualX}(\overrightarrow{x},\overrightarrow{y}) = \bigwedge_{k=0}^{n-1} X_k(\overrightarrow{x}) \leftrightarrow X_k^+(\overrightarrow{y})$, $\mathsf{EqualY}(\overrightarrow{x},\overrightarrow{y}) = \bigwedge_{k=0}^{n-1} Y_k(\overrightarrow{x}) \leftrightarrow Y_k^+(\overrightarrow{y})$. We can now create variables to define the adjacency of every position: $A_1 = \forall \overrightarrow{x} : \exists \overrightarrow{y} : \mathsf{East}(\overrightarrow{x},\overrightarrow{y})$, $A_2 = \forall \overrightarrow{x} : \exists \overrightarrow{y} : \mathsf{EqualX}(\overrightarrow{x},\overrightarrow{y})$, $A_3 = \forall \overrightarrow{y} : \exists \overrightarrow{x} : \mathsf{EqualX}(\overrightarrow{x},\overrightarrow{y})$, $A_4 = \forall \overrightarrow{x} : \exists \overrightarrow{y} : \mathsf{North}(\overrightarrow{x},\overrightarrow{y})$, $A_5 = \forall \overrightarrow{x} : \exists \overrightarrow{y} : \mathsf{EqualY}(\overrightarrow{x},\overrightarrow{y})$, $A_6 = \forall \overrightarrow{y} : \exists \overrightarrow{x} : \mathsf{EqualY}(\overrightarrow{x},\overrightarrow{y})$.

We then need to represent the base row (so that we can establish an origin and initial conditions for the torus). We create a parvariable $B_i'(\overrightarrow{x})$, for each $i = 0,\ldots,n-1$, such that $B_i'(\overrightarrow{x})$ reflects the binary encoding of i, as follows: $B_0'(\overrightarrow{x}) = \bigwedge_{k=0}^{n-1} \neg X_k(\overrightarrow{x})$, $B_1'(\overrightarrow{x}) = X_0(\overrightarrow{x}) \wedge \bigwedge_{k=1}^{n-1} \neg X_k(\overrightarrow{x})$, $B_2'(\overrightarrow{x}) = \neg X_0(\overrightarrow{x}) \wedge X_1(\overrightarrow{x}) \wedge \bigwedge_{k=2}^{n-1} \neg X_k(\overrightarrow{x})$, and so on. Now specify $B(\overrightarrow{x}) = \bigwedge_{k=0}^{n-1} \neg Y_k(\overrightarrow{x})$ and, for each $i \in \{0,\ldots,n-1\}$, $B_i(\overrightarrow{x}) = B_i'(\overrightarrow{x}) \wedge B(\overrightarrow{x})$. The parvariable $B(\overrightarrow{x})$ indicates that the position is in the base row, and the parvariable $B_i(\overrightarrow{x})$ indicates that the position is in the ith column. Together, they specify the relevant part of the base row that we need to specify the initial tiles. We must enforce an origin for the torus, so we introduce: $A_7 = \exists \overrightarrow{x} : B_0(\overrightarrow{x})$.

At this point, by fixing $\bigwedge_{i=1}^{7} A_i$ we build a torus of size $2^n \times 2^n$. It remains to represent the horizontal, vertical and initial constraints.

We introduce a pair $C_j(\overrightarrow{x})$ and $C_j^+(\overrightarrow{y})$ for each possible tile. For each tile $j = 1, \ldots, m$, we impose $\mathbb{P}(C_j(\overrightarrow{x})) = \mathbb{P}(C_j^+(\overrightarrow{y})) = 1/2$. Each position of the torus must have one and only one tile:

$$A_8 = \forall \overrightarrow{x} : \left(\bigvee_{j \in \mathcal{C}} C_j(\overrightarrow{x}) \right) \wedge \left(\bigwedge_{j \in \mathcal{C}, k \in \mathcal{C}, j \neq k} \neg(C_j(\overrightarrow{x}) \wedge C_k(\overrightarrow{x})) \right).$$

Moreover, the tiles must satisfy the horizontal and vertical restrictions:

$$A_9 = \forall \overrightarrow{x} : \bigwedge_{j \in \mathcal{C}} C_j(\overrightarrow{x}) \rightarrow (\forall \overrightarrow{y} : \mathsf{East}(\overrightarrow{x}, \overrightarrow{y}) \rightarrow \vee_{k:(j,k) \in H} C_k^+(\overrightarrow{y})),$$

$$A_{10} = \forall \overrightarrow{x} : \bigwedge_{j \in \mathcal{C}} C_j(\overrightarrow{x}) \rightarrow (\forall \overrightarrow{y} : \mathsf{North}(\overrightarrow{x}, \overrightarrow{y}) \rightarrow \vee_{k:(j,k) \in V} C_k^+(\overrightarrow{y})).$$

Now make both sets of parvariables related to tiles behave the same:

$$A_{11} = \bigwedge_{j \in \mathcal{C}} \forall \overrightarrow{x} : \forall \overrightarrow{y} : C_j^+(\overrightarrow{y}) \wedge \mathsf{East}(\overrightarrow{x}, \overrightarrow{y}) \rightarrow C_j(\overrightarrow{x}),$$

$$A_{12} = \bigwedge_{j \in \mathcal{C}} \forall \overrightarrow{x} : \forall \overrightarrow{y} : C_j^+(\overrightarrow{y}) \wedge \mathsf{North}(\overrightarrow{x}, \overrightarrow{y}) \rightarrow C_j(\overrightarrow{x}).$$

Finally, we impose the initial conditions: $A_{13} = \forall x : \bigwedge_{i=0}^{n-1} B_i(\overrightarrow{x}) \rightarrow C_i^0(\overrightarrow{x})$, where C_i^0 represents the ith tile as given by initial condition.

Computing the probability of $A_{14} = \bigwedge_{i=1}^{13} A_i$ produces the probability that a tiling is built satisfying all horizontal and vertical restrictions and the initial condition. If we can recover the number of tilings of the torus from this probability, we obtain the number of accepting computations of the exponentially-bounded Turing machine we started with. Assume we have $\mathbb{P}(A_{14} = 1)$. Then $\mathbb{P}(A_{14} = 1) \times 2^\delta$ is the number of truth assignments that build the torus satisfying horizontal and vertical relations and initial conditions, where $\delta = 2^{2n}(2n + 2^{2n+1} + m)$. However, this number is *not* equal to the number of tilings of the torus. To see this, consider the grounded Bayesian network where each a in the domain is associated with a "slice" containing groundings $X_i(a)$, $Y_i(a)$, $C_j(a)$ and so on. If a particular configuration of these indicator variables corresponds to a tiling, then we can produce the same tiling by permuting all elements of the domain with respect to the slices of the network. Intuitively, we can fix a tiling and imagine that we are labelling each point of the torus with an element of the domain; clearly every permutation of these labels produces the same tiling (this intuition is appropriate because each a corresponds to a different point in the torus). So, in order to produce the number of tilings of the torus, we must compute $\mathbb{P}(A_{14} = 1) \times 2^\delta / (2^{2n}!)$, where we divide the number of satisfying truth assignments by the number of repeated tilings. Consequently we obtain the number of accepting computations of the original Turing machine just by processing the inference $\mathbb{P}(A_{14} = 1)$, proving the desired result.   □

Concerning the data complexity of enhanced plates, we have:

Theorem 5. *Inference in enhanced plate models when the model is fixed is* #P[1]-*equivalent.*

Proof. Hardness follows from Theorem 2. To show pertinence, consider that, once the plate model is fixed, the arity of any relation is fixed. And given the domains as input, the combined domain has size that is a polynomial on the domains (where the maximum arity appears in the exponent). So one can then produce a grounded Bayesian network of size polynomial in the input. The result follows as inference in the grounded Bayesian network belongs to $\mathsf{FP}^{\#\mathsf{P}[1]}$.   □

Concerning domain complexity, the fact that one can build complex logical expressions using plates (see the proof of Theorem 4) suggests that polynomial behavior cannot be expected [9]. However, we have not been able to provide precise lower and upper bounds on domain complexity, so we leave this as a challenge for future work.

5   Conclusion

Plates allow large Bayesian networks to be concisely described, and are particularly useful when one faces scenarios with many variables and intricate relations. Despite the popularity of plate models, few results on their complexity are available. We have presented here a number of results concerning the complexity of "classic" and enhanced plates; the former display #P[1]-equivalent combined/data complexity (despite the fact that they may induce exponentially large groundings), while the latter display #EXP[1]-equivalent combined complexity and #P[1]-equivalent data complexity. The results on enhanced plates are obtained when all relations are binary and aggregation functions are based on existential quantification. It is not difficult to see that exponential complexity there stems from the nesting of plates; in fact, if the level of nesting is limited, the combined complexity goes down to #P[1]-equivalent.

There are several avenues open for future work. The domain complexity of enhanced plate models is an open problem. Also, plate models are often augmented with additional resources to allow recursive descriptions and structural uncertainty [10]; the complexity of these more sophisticated languages deserves analysis. Finally, it would be interesting to examine more restricted languages; for instance, languages where evidence can only be "positive", or where aggregation functions can only have some bounded complexity.

Acknowledgements. The first author was partially supported by CNPq and the second author was partially supported by FAPESP.

References

1. Blei, D.M., Ng, A.Y., Jordan, M.I.: Latent Dirichlet allocation. J. Mach. Learn. Res. **3**(3), 993–1022 (2003)
2. Börger, E., Grädel, E., Gurevich, Y.: The Classical Decision Problem. Cambridge University Press, Cambridge (1997)
3. Buntine, W.L.: Operations for learning with graphical models. J. Artif. Intell. Res. **2**, 159–225 (1994)
4. Cozman, F.G., Mauá, D.D.: Bayesian networks specified using propositional and relational constructs: combined, data, and domain complexity. In: AAAI Conference on Artificial Intelligence, pp. 3519–3525 (2015)
5. Darwiche, A.: Modeling and Reasoning with Bayesian Networks. Cambridge University Press, Cambridge (2009)
6. de Campos, C.P., Stamoulis, G., Weyland, D.: A structured view on weighted counting with relations to quantum computation and applications. Technical report 133, Electronic Colloquium on Computational Complexity (2013)
7. Gilks, W., Thomas, A., Spiegelhalter, D.: A language and program for complex Bayesian modelling. The Stat. **43**, 169–178 (1993)
8. Jaeger, M.: Relational Bayesian networks. In: Conference on Uncertainty in Artificial Intelligence, pp. 266–273 (1997)
9. Jaeger, M., van den Broeck, G.: Upper and lower bounds. In: Statistical Relational AI Workshop, Liftability of Probabilistic Inference (2012)
10. Koller, D., Friedman, N.: Probabilistic Graphical Models: Principles and Techniques. MIT Press, Cambridge (2009)
11. Lunn, D., Jackson, C., Best, N., Thomas, A., Spiegelhalter, D.: The BUGS Book: A Practical Introduction to Bayesian Analysis. CRC Press/Chapman and Hall, Boca Raton (2012)
12. Lunn, D., Spiegelhalter, D., Thomas, A., Best, N.: The BUGS project: evolution, critique and future directions. Stat. Med. **28**, 3049–3067 (2009)
13. Papadimitriou, C.H.: A note on succinct representations of graphs. Inf. Control **71**, 181–185 (1986)
14. Papadimitriou, C.H.: Computational Complexity. Addison-Wesley Publishing, Reading (1994)
15. Pearl, J.: Probabilistic Reasoning in Intelligent Systems: Networks of Plausible Inference. Morgan Kaufmann, San Mateo (1988)
16. Poole, D.: First-order probabilistic inference. In: International Joint Conference on Artificial Intelligence (IJCAI), pp. 985–991 (2003)
17. Richardson, M., Domingos, P.: Markov logic networks. Mach. Learn. **62**(1–2), 107–136 (2006)
18. Roth, D.: On the hardness of approximate reasoning. Artif. Intell. **82**(1–2), 273–302 (1996)
19. Sontag, D., Roy, D.: Complexity of inference in Latent Dirichlet Allocation. Adv. Neural Inf. Process. Syst. **24**, 1008–1016 (2011)
20. Tobies, S.: The complexity of reasoning with cardinality restrictions and nominals in expressive description logics. J. Artif. Intell. Res. **12**, 199–217 (2000)
21. Valiant, L.G.: The complexity of enumeration and reliability problems. SIAM J. Comput. **8**(3), 410–421 (1979)

DL-Lite Bayesian Networks: A Tractable Probabilistic Graphical Model

Denis D. Mauá[✉] and Fabio G. Cozman

Universidade de São Paulo, São Paulo, Brazil
{denis.maua,fgcozman}@usp.br

Abstract. The construction of probabilistic models that can represent large systems requires the ability to describe repetitive and hierarchical structures. To do so, one can resort to constructs from description logics. In this paper we present a class of relational Bayesian networks based on the popular description logic DL-Lite. Our main result is that, for this modeling language, marginal inference and most probable explanation require polynomial effort. We show this by reductions to edge covering problems, and derive a result of independent interest; namely, that counting edge covers in a particular class of graphs requires polynomial effort.

1 Introduction

The search for an expressive *and* tractable formalism that can represent uncertainty *and* repetitive structures or hierarchical terminologies, is not an easy one. Most probabilistic models are propositional [14,24], while combinations of logic and probabilities are typically quite flexible but intractable [3,18]. However, there are proposals that try to balance expressivity and complexity by mixing logical constructs with graphs and independence relations [17,32,33]; for instance, *probabilistic relational models* [16] and *relational Bayesian networks* [21]. A few variants of these latter models even allow for polynomial time inferences by significantly restricting the syntax [15,29].

In this paper we investigate the computational complexity of a modeling language that combines features of relational Bayesian networks with constructs of the popular description logic DL-Lite [1,7]. In essence, we consider Bayesian networks which can be concisely specified using terminological assertions expressed in DL-Lite and marginal probability assertions on basic concepts and roles. For instance, we obtain a Bayesian network through the assertions

$$\textsf{Employee} \equiv \textsf{Person} \sqcap \exists\textsf{salary}, \qquad \mathbb{P}(\textsf{Person}) = 1/3, \qquad \mathbb{P}(\textsf{salary}) = 1/4,$$

which encodes knowledge that an employee is defined as a person who receives a salary, an object is a person with probability $1/3$ and two objects are connected through the relation salary with probability $1/4$.

Our main contribution here is to show that marginal inferences and most probable explanations can be generated in polynomial time in our modeling

© Springer International Publishing Switzerland 2015
C. Beierle and A. Dekhtyar (Eds.): SUM 2015, LNAI 9310, pp. 50–64, 2015.
DOI: 10.1007/978-3-319-23540-0_4

language. So, we identify an island of tractability with non-trivial expressivity, offering a language that can be easily meshed with ontologies and relational schema.

The paper is organized as follows. We with some necessary background in Sect. 2. We then present DL-Lite Bayesian networks (Sect. 3) and their complexity with respect to marginal inferences and most probable explanations (Sects. 4 and 5). Some of our results depend on a polynomial algorithm for counting edge covers, a result of independent interest that is briefly presented in Sect. 6. The connections with related work is discussed in Sect. 7. Section 8 comments on possible extensions and concludes the paper.

2   Bayesian Networks, and DL-Lite

A Bayesian network consists of an acyclic directed graph whose nodes are random variables $X_1, \ldots, X_n$, and a collection of conditional probability distributions, one distribution for each random variable given its parents. In this work, we consider only Boolean variables: we assume that each variable X_i takes on values 1 ("true") and 0 ("false"). The product of all conditional probability distributions determines a joint probability distribution over all variables, such that $\mathbb{P}(X_1 = x_1, \ldots, X_n = x_n) = \prod_{i=1}^{n} \mathbb{P}(X_i = x_i | \mathrm{pa}(X_i) = \pi_i)$, where $\mathrm{pa}(X_i)$ denotes the parents of X_i and π_i is the projection of $\{x_1, \ldots, x_n\}$ onto $\mathrm{pa}(X_i)$. A Bayesian network is *extensively specified* when its probability distributions are specified through tables of rational numbers.

A *marginal inference* is the computation of the probability of a number of assignments $\{X_i = x_i\}$ (*query*) given other assignments (*evidence*). This is a #P-complete problem [36], and NP-hard even to approximate [11].[1] Other common inference is *most probable explanation* (MPE), where one seeks an assignment to all variables that maximizes their joint probability given some evidence. Polynomial-time inference in extensively specified networks seems to require, under widely accepted assumptions about complexity classes, a bound on graph treewidth [25,26], hence the interest in networks with restricted expressivity [9,13,15,19,31,35].

To study the inferential complexity of various classes of Bayesian networks beyond the treewidth barrier, we have proposed a convenient framework in which to specify networks with binary variables [10]. In this framework, a directed acyclic graph is given where each node is a random variable; each root variable X is associated with a marginal probability $\mathbb{P}(X = 1) = \alpha$, and each non-root variable Y is associated with a formula $Y \Leftrightarrow \phi$, where ϕ is a well-formed formula on the parents of Y; the latter is equivalent to specifying that $\mathbb{P}(Y = 1 | \phi) = 1$ if ϕ is true and zero otherwise. By restricting the language from which ϕ can be selected, one obtains a class of Bayesian networks (a language is simply a

---

[1] Recall that #P is the class of integer-valued functions computed by counting Turing machines in polynomial time; a counting Turing machine is a standard nondeterministic Turing machine that prints in binary notation, on a separate tape, the number of accepting computations induced by the input.

set of well-formed formulas). For instance, if the language consists of all propositional sentences, we can represent any joint probability distribution (perhaps introducing fresh variables). Or we may employ a sub-Boolean language with only conjunction and disjunction, as in the next network:

$$\mathbb{P}(X_1) = 1/4, \qquad \mathbb{P}(X_2) = 1/2,$$
$$X_3 \Leftrightarrow X_1 \wedge X_2, \quad X_4 \Leftrightarrow X_2 \vee X_3.$$

A *relational Bayesian network* consists of a directed acyclic graph whose nodes are relations $r_1, \ldots, r_n$ [21,22], plus a set of real-valued functions soon to be explained. To interpret a relational Bayesian network, first take a set of *individuals* $\mathcal{D}$, called a *domain*. A *grounding* of k-ary relation r is denoted by $r(a_1, \ldots, a_k)$, where $a_1, \ldots, a_k \in \mathcal{D}$. Given a relational Bayesian network and a domain, one can build a directed acyclic graph where each possible grounding is a node, and where an arc is added between two groundings if there is an arc between their corresponding relations in the network. The real-valued functions we have mentioned specify the probability of each grounding given its parents' grounding. In Jaeger's original proposal [21], these real-valued functions are restricted to a few basic forms. Here we focus on the restricted syntax proposed in [10]: for each root relation r we have an assessment $\mathbb{P}(r) = \alpha$, where α is a rational in $[0, 1]$. And for each non-root k-ary relation s we have an equivalence $s(x_1, \ldots, x_k) \Leftrightarrow \phi(x_1, \ldots, x_k)$, where each x_i denotes a logical variable and ϕ is a well-formed formula in a first-order language. Our strategy in this paper is to restrict ϕ to constructs from the DL-Lite description logic.

DL-Lite is particularly interesting because it captures a great deal of features found in conceptual modeling by ER or UML diagrams, and yet common inference services have polynomial complexity [7]. A whole family of variants of DL-Lite has been developed [1], and in fact this family is the basis of one of the OWL QL profile (http://www.w3.org/TR/owl2-profiles/). To recap, DL-Lite is a description logic that deals with concepts, roles, and individuals; we treat those as unary relations, binary relations, and constants. Some of the concepts are marked as *primitive* ones. Given a primitive concept s, both s and ¬s are formulas, to be interpreted respectively as $s(x)$ and $\neg s(x)$. Given a role r, both $\exists r$ and $\exists r^-$ are formulas, to be interpreted respectively as $\exists y : r(x, y)$ and $\exists y : r(y, x)$ (r^- is an *inverse role*). Also, if ϕ and φ are formulas, then $\phi \sqcap \varphi$ is a formula (interpreted as $\phi \wedge \varphi$). Finally, a *concept definition* $s \equiv \phi$ is interpreted as $\forall x : s(x) \Leftrightarrow \phi(x)$. Note that any formula ϕ can have only one free logical variable. A primitive concept cannot appear in the left-hand side of a concept definition (indeed this characterizes primitive concepts). Inverse roles are defined as $\forall x, y : r^-(y, x) \Leftrightarrow r(x, y)$. The semantics of DL-Lite uses a domain $\mathcal{D}$ and an interpretation $\mathbb{I}$ that maps each individual to an element of $\mathcal{D}$, each concept to a subset of $\mathcal{D}$, and each role to a set of pairs in $\mathcal{D} \times \mathcal{D}$. The semantics of a formula in essence reads the formula as a first-order formula and uses $\mathcal{D}$ and $\mathbb{I}$ in the usual semantics of first-order logic [7].

Example 1. The following concept definitions express simple facts about families: first, female ≡ ¬male; also, father ≡ male ⊓ ∃parentOf, mother ≡ female ⊓ ∃parentOf, son ≡ male ⊓ ∃parentOf⁻, daughter ≡ female ⊓ ∃parentOf⁻.

3   DL-Lite Bayesian Networks

We now consider the class of relational Bayesian networks over binary variables where each conditional probability is specified through a DL-Lite formula. A *DL-Lite Bayesian network* is a relational Bayesian network that consists of a directed acyclic graph where each node is a unary or binary relation, and such that

- each root relation r is associated with an assessment $\mathbb{P}(r) = \alpha$, for a rational $\alpha \in [0,1]$, and
- each non-root relation r is either a unary relation associated with a concept definition r ≡ ϕ, where ϕ is a formula in DL-Lite only mentioning parent relations, or an inverse role s⁻ with s as its single parent.

Example 2. The graph and assessments in Fig. 1, plus the concept definitions in Example 1, specify a DL-Lite Bayesian network.

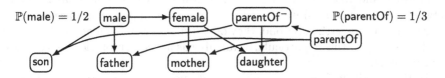

Fig. 1. A DL-Lite Bayesian network.

The semantics of DL-Lite Bayesian networks is given by a simple combination of semantics for relational Bayesian networks and for DL-Lite. That is, consider a domain $\mathcal{D}$ containing individuals; in this paper we assume every domain to be finite and given as a list of elements. We also assume, as most first-order probabilistic logics do, that interpretations are *rigid* [3] in that an element corresponds to the same individual in every possible interpretation of relations. For each concept s and individual a, produce the grounding s(a); likewise, for each role r and each pair of individuals (a, b), produce the grounding r(a, b). A set with all possible interpretations is obtained by considering all possible truth assignments for these groundings. We can associate each grounding with a random variable that takes each possible interpretation either to 1 (the grounding is true in that interpretation) or to 0 (otherwise). To simplify the notation, we use the same symbol for a grounding and its associated random variable. Now construct a grounded graph. First, each grounding is a node. Second, take each concept definition s ≡ ϕ; for each grounding s(a), specify as its parents the

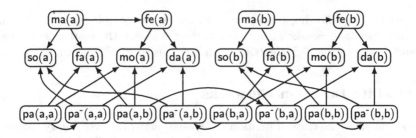

Fig. 2. Grounded network, domain $\mathcal{D} = \{a, b\}$.

groundings that appear in $\phi(a)$. Third, associate with each root grounding the corresponding grounded probabilistic assessment. For instance, suppose we have assessments $\mathbb{P}(s) = \alpha'$ and $\mathbb{P}(r) = \alpha''$ for concept s and role r; then, attach to grounding s(a) the assessment $\mathbb{P}(s(a)) = \alpha'$, and similarly, attach to grounding r(a, b) the assessment $\mathbb{P}(r(a, b)) = \alpha''$.

Example 3. Take $\mathcal{D} = \{a, b\}$. Then the relational Bayesian network in Example 2 induces the Bayesian network in Fig. 2 (where names of relations have been shortened, e.g. parentOf has become pa).

DL-Lite Bayesian networks can be argued for in two ways. First, they offer an intuitive and disciplined language in which to express relational Bayesian networks; in essence, DL-Lite is used to reduce the complexity of Jaeger's combination functions [21,22]. Second, they offer a simple way to create probabilistic acyclic ontologies; this is particularly valuable as acyclic ontologies are common in practice [2].

To these inviting features we add a third, most important one: useful inferences in DL-Lite Bayesian networks require polynomial effort. The inferences of interest are as follows. Suppose we have a DL-Lite Bayesian network $\mathbb{B}$ and a domain $\mathcal{D}$ (as a list of individuals). To compute a marginal inference $\mathbb{P}(\mathbf{Q}|\mathbf{E})$ for sets of assignments $\mathbf{Q}$ and $\mathbf{E}$, we calculate $\mathbb{P}(\mathbf{Q} \wedge \mathbf{E})/\mathbb{P}(\mathbf{E})$. So, our central *inference* problem is to compute $\mathbb{P}(\mathbf{E})$ for a given set of assignments (the evidence). That is, inference in DL-Lite Bayesian networks boils down to computing marginals, as usual in Bayesian networks.

For reasons to be clear, we say that evidence is *positive* when all assignments attach value 1 (true). For instance, $\{\mathsf{male}(\mathsf{John}) = 1, \mathsf{female}(\mathsf{Mary}) = 1\}$ is positive evidence. Similarly, $\{\mathsf{father}(\mathsf{John}) = 0, \mathsf{mother}(\mathsf{Mary}) = 0\}$ is *negative* evidence.

Another problem is to find a *most probable explanation (MPE)*; that is, to find an interpretation for *all* groundings of the DL-Lite Bayesian network $\mathbb{B}$ with respect to domain $\mathcal{D}$, that maximizes the probability and is consistent with a given set of assignments (evidence) $\mathbf{E}$.

As a digression, note that we argue in Sect. 8 that results in the next section can be adapted to produce fully-polynomial time approximations in a larger set of languages that can be directly useful in conceptual modeling.

4 Marginal Inferences

Our main result is that (marginal and MPE) inference in a DL-Lite Bayesian network requires polynomial effort as long as evidence is positive. One consequence of this result is that DL-Lite Bayesian networks are dqe-/domain-liftable for positive evidence [4,23]. Another point is that our results offer explicit algorithms for a class of two-variable logics [5].

The focus on positive assignments is justified, as conjunction (a subset of DL-Lite) leads to #P-hardness with arbitrary evidence [10]. Our first result is:

Theorem 1. *With a DL-Lite Bayesian network, a domain, and positive evidence as input, inference is polynomial-time computable in the size of the input.*

We prove this theorem by a quadratic-time reduction to multiple independent problems of counting weighted edge covers with uniform weights in a very particular class of graphs. Then we show (in Sect. 6) that the latter problem can be solved in quadratic time (hence the total time is quadratic).

We first transform the relational network into an equal-probability model. Collapse each role r and its inverse r^- into a single node r. For each (collapsed) role r, insert variables $e_r \equiv \exists r$ and $e_r^- \equiv \exists r^-$; replace each appearance of the formula $\exists r$ by the variable e_r, and each appearance of $\exists r^-$ by e_r^-. This transformation does not change the probability of $\mathbf{E}$, and it allows us to easily refer to groundings of formulas $\exists r$ and $\exists r^-$ as groundings of e_r and e_r^-, respectively.

Observe that only the nodes with assignments in $\mathbf{E}$ and their ancestors are relevant for the computation of $\mathbb{P}(\mathbf{E})$, as every other node in the Bayesian network is barren [14]. Hence, we can assume without loss of generality that $\mathbf{E}$ contains only leaves of the network. If $\mathbf{E}$ contains only root nodes, then $\mathbb{P}(\mathbf{E})$ can be computed trivially as the product of marginal probabilities which are readily available from the specification. Thus assume that $\mathbf{E}$ assigns a positive value to at least one leaf grounding $s(a)$, where a is some individual in the domain. Then by construction $s(a)$ is associated with a logical sentence $X_1 \wedge \cdots \wedge X_k$, where each X_i is either a grounding of non-primitive unary relation in individual a, a grounding of a primitive unary relation in a, or the negation of a grounding of a primitive unary relation in a. It follows that $\mathbb{P}(\mathbf{E}) = \mathbb{P}(s(a) = 1 | X_1 = 1, \ldots, X_k = 1)\mathbb{P}(\mathbf{E}') = \mathbb{P}(\mathbf{E}')$, where $\mathbf{E}'$ is $\mathbf{E}$ after removing the assignment $s(a) = 1$ and adding the assignments $\{X_1 = 1, \ldots, X_k = 1\}$. Now it might be that $\mathbf{E}'$ contains both the assignments $\{X_i = 1\}$ and $\{X_i = 0\}$. Then $\mathbb{P}(\mathbf{E}) = 0$ (this can be verified efficiently). So assume there are no such inconsistencies. The problem of computing $\mathbb{P}(\mathbf{E})$ boils down to computing $\mathbb{P}(\mathbf{E}')$; in the latter problem the node $s(a)$ is discarded for being barren. Moreover, we can replace any assignment $\{\neg r(a) = 1\}$ in $\mathbf{E}'$ for some primitive concept r with the equivalent assignment $\{r(a) = 0\}$. By repeating this procedure for all internal nodes which are not groundings of e_r or e_r^-, we end up with a set $\mathbf{A}$ containing positive assignments of groundings of roles and of concepts e_r and e_r^-, and (not necessarily positive) assignments of groundings of primitive concepts. Each grounding of a primitive concept or role is (a root node hence) marginally independent from all other groundings in $\mathbf{A}$; hence

$\mathbb{P}(\mathbf{A}) = \mathbb{P}(\mathbf{B}|\mathbf{C})\prod_i \mathbb{P}(A_i)$, where each A_i is an assignment to a root node, $\mathbf{B}$ are (positive) assignments to groundings of concepts e_r and e_r^- for relations r, and $\mathbf{C} \subseteq \{A_1, A_2, \ldots\}$ are groundings of roles (if $\mathbf{C}$ is empty then assume it expresses a tautology). Since the marginal probabilities $\mathbb{P}(A_i)$ are available from the specification the joint $\prod_i \mathbb{P}(A_i)$ can be computed in linear time in the input. We thus focus on computing $\mathbb{P}(\mathbf{B}|\mathbf{C})$ as defined (if $\mathbf{B}$ is empty, we are done). To recap, $\mathbf{B}$ is a set of assignments $e_r(a) = 1$ and $e_r^-(b) = 1$ and $\mathbf{C}$ is a set of assignments $r(c, d) = 1$ for arbitrary roles r and individuals a, b, c and d.

For a role r, let $\mathcal{D}_r$ be the set of individuals $a \in \mathcal{D}$ such that $e_r(a) = 1$ is in $\mathbf{B}$, and let $\mathcal{D}_r^-$ be the set of individuals $a \in \mathcal{D}$ such that $\mathbf{B}$ contains $e_r^-(a) = 1$. Let gr(r) be the set of all groundings of relation r, and let $r_1, \ldots, r_k$ be the roles in the (relational) network. By the factorization property of Bayesian networks it follows that

$$\mathbb{P}(\mathbf{B}|\mathbf{C}) = \sum_{\text{gr}(r_1)} \cdots \sum_{\text{gr}(r_k)} \prod_{i=1}^{k} \prod_{a \in \mathcal{D}_{r_i}} \mathbb{P}(e_{r_i}(a) = 1 | \text{pa}(e_{r_i}(a)), \mathbf{C}) \times$$

$$\prod_{a \in \mathcal{D}_{r_i}^-} \mathbb{P}(e_{r_i}^-(a) = 1 | \text{pa}(e_{r_i}^-(a)), \mathbf{C})\mathbb{P}(\text{gr}(r_k)|\mathbf{C}),$$

which by distributing the products over sums is equal to

$$\prod_{i=1}^{k} \sum_{\text{gr}(r_i)} \prod_{a \in \mathcal{D}_r} \mathbb{P}(e_r(a) = 1 | \text{pa}(e_r(a)), \mathbf{C}) \prod_{a \in \mathcal{D}_r^-} \mathbb{P}(e_r^-(a) = 1 | \text{pa}(e_r^-(a)), \mathbf{C})\mathbb{P}(\text{gr}(r_k)|\mathbf{C}).$$

Consider an assignment $r(a, b) = 1$ in $\mathbf{C}$. By construction, the children of the grounding $r(a, b)$ are $e_r(a)$ and $e_r^-(b)$. Moreover, the assignment $r(a, b) = 1$ implies that $\mathbb{P}(e_r(a) = 1 | \text{pa}(e_r(a)), \mathbf{C}) = 1$ (for any assignment to the other parents) and $\mathbb{P}(e_r^-(b) = 1 | \text{pa}(e_r(a)), \mathbf{C}) = 1$ (for any assignment to the other parents). This is equivalent in the factorization above to removing $r(a, b)$ from $\mathbf{C}$ (as it is independent of all other groundings), and removing individuals a from $\mathcal{D}_r$ and b from $\mathcal{D}_r^-$. So repeat this procedure for every grounding in $\mathbf{C}$ until this set is empty (this can be done in polynomial time). The inference problem becomes one of computing $\gamma(r) = \sum_{\text{gr}(r_i)} \prod_{a \in \mathcal{D}_r} \mathbb{P}(e_r(a) = 1 | \text{pa}(e_r(a))) \prod_{a \in \mathcal{D}_r^-} \mathbb{P}(e_r^-(a) = 1 | \text{pa}(e_r^-(a)))\mathbb{P}(\text{gr}(r_k))$ for every relation r_i, $i = 1, \ldots, k$. We will show that this problem can be reduced to a tractable instance of counting weighted edge covers.

To this end, consider the graph G whose node set V can be partitioned into sets $V_1 = \{e_r^-(a) : a \in \mathcal{D} \setminus \mathcal{D}_r^-\}$, $V_2 = \{e_r(a) : a \in \mathcal{D}_r\}$, $V_3 = \{e_r^-(a) : a \in \mathcal{D}_r^-\}$, $V_4 = \{e_r(a) : a \in \mathcal{D} \setminus \mathcal{D}_r\}$, and for $i = 1, 2, 3$ the graph obtained by considering nodes $V_i \cup V_{i+1}$ is bipartite complete. An edge with endpoints $e_r(a)$ and $e_r^-(b)$ represents the grounding $r(a, b)$; we identify every edge with its corresponding grounding. We call this graph the *intersection graph* of $\mathbf{B}$ with respect to r and $\mathcal{D}$. The parents of a node in the graph correspond exactly to the parents of the node in the Bayesian network. For example, the graph in Fig. 3 represents the assignments $\mathbf{B} = \{e_r(a) = 1, e_r(b) = 1, e_r(d) = 1, e_r^-(b) = 1, e_r^-(c) = 1\}$,

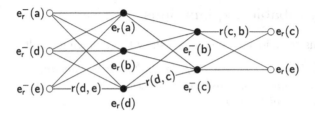

Fig. 3. Representing assignments by graphs.

with respect to domain $\mathcal{D} = \{a, b, c, d, e\}$. The black nodes (resp., white nodes) represent groundings in (resp., not in) **B**. For clarity's sake, we label only a few edges.

Before showing the equivalence between the inference problem and counting edges covers, we need to introduce some graph-theoretic notions and notation. Consider a (simple, undirected) graph $G = (V, E)$. Denote by $E_G(u)$ the set of edges incident on a node $u \in V$, and by $N_G(u)$ the open neighborhood of u. For $U \subseteq V$, we say that $C \subseteq E$ is a U-cover if for each node $u \in U$ there is an edge $e \in C$ incident in u (i.e., $e \in E_G(u)$). For any fixed real λ, we say that $\lambda^{|C|}$ is the weight of cover C. The *partition function* of G is $Z(G, U, \lambda) = \sum_{C \in EC(G,U)} \lambda^{|C|}$, where $U \subseteq V$, $EC(G, U)$ is the set of U-covers of G and λ is a positive real. If $\lambda = 1$ and $U = V$, the partition function is the number of edge covers. The following result connects counting edge covers to marginal inference in DL-Lite Bayesian networks.

Proposition 1. *Let $G = (V_1, V_2, V_3, V_4, E)$ be the intersection graph of* **B** *with respect to a relation* r *and domain* $\mathcal{D}$. *Then* $\gamma(\mathsf{r}) = Z(G, V_2 \cup V_3, \alpha/(1-\alpha))/(1-\alpha)^{|E|}$, *where* $\alpha = \mathbb{P}(\mathsf{r}(\mathbf{x}, \mathbf{y}))$.

Proof. Let $B = V_2 \cup V_3$, and consider a B-cover C. The assignment that sets to true all groundings $\mathsf{r}(\mathsf{a}, \mathsf{b})$ corresponding to edges in C, and sets to false the remaining groundings of r makes $\mathbb{P}(e_\mathsf{r}(\mathsf{a}) = 1|\mathrm{pa}(e_\mathsf{r}(\mathsf{a}))) = \mathbb{P}(e_\mathsf{r}^-(\mathsf{b}) = 1|\mathrm{pa}(e_\mathsf{r}^-(\mathsf{b}))) = 1$ for every $\mathsf{a} \in \mathcal{D}_\mathsf{r}$ and $\mathsf{b} \in \mathcal{D}_\mathsf{r}^-$; it makes $\mathbb{P}(\mathsf{gr}(\mathsf{r})) = \mathbb{P}(\mathsf{r})^{|C|}(1 - \mathbb{P}(\mathsf{r}))^{|E|-|C|} = (1-\alpha)^{|E|}\alpha^{|C|}/(1-\alpha)^{|C|}$, which is the weight of the cover C scaled by $(1 - \alpha)^{|E|}$. Now consider a set of edges C which is not a B-cover and obtains an assignment to groundings $\mathsf{gr}(\mathsf{r})$ as before. There is at least one node in B that does not contain any incident edges in C. Assume that node is $e(\mathsf{a})$; then all parents of $e(\mathsf{a})$ are assigned false, which implies that $\mathbb{P}(e_\mathsf{r}(\mathsf{a}) = 1|\mathrm{pa}(e_\mathsf{r}(\mathsf{a}))) = 0$. The same is true if the node not covered is a grounding $e^-(\mathsf{a})$. Hence, for each B-cover C the probability of the corresponding assignment equals its weight up to the factor $(1 - \alpha)^{|E|}$. And for each edge set C which is not a B-cover its corresponding assignment has probability zero. □

We have thus established that, if a particular class of edge cover counting problems is polynomial, then marginal inference in DL-Lite Bayesian networks is also polynomial. Since the former is shown to be true in Sect. 6, this concludes the proof of Theorem 1.

5   Most Probable Explanations

Using previous techniques, we can also show the following result:

Theorem 2. *With a DL-Lite Bayesian network, a domain, and positive evidence as input, finding a most probable explanation is polynomial-time computable in the size of the input.*

In this theorem we are interested in finding an assignment $\mathbf{X}$ to all groundings that maximizes $\mathbb{P}(\mathbf{X} \wedge \mathbf{E})$, where $\mathbf{E}$ is a set of positive assignments. Perform the substitution of formulas $\exists r$ and $\exists r^-$ by logically equivalent concepts e_r and e_r^- as before. Consider a non-root grounding $s(a)$ in $\mathbf{E}$ which is not the grounding of e_r or e_r^-; by construction, $s(a)$ is logically equivalent to a conjunction $X_1 \wedge \cdots \wedge X_k$, where $X_1, \ldots, X_k$ are unary groundings. Because $s(a)$ is assigned to true, any assignment $\mathbf{X}$ with nonzero probability assigns $X_1, \ldots, X_k$ to true. Moreover, since $s(a)$ is an internal node, its corresponding probability is one. Hence, if we include all the assignments $X_i = 1$ to its parents in $\mathbf{E}$, the MPE value does not change. As in the computation of inference, we might generate an inconsistency when setting the values of parents; in this case halt and return zero (and an arbitrary assignment). So assume we repeated this procedure until $\mathbf{E}$ contains all ancestors of the original groundings which are groundings of unary relations, and that no inconsistency was found. Note that at this point we only need to assign values to nodes which are either not ancestors of any node in the original set $\mathbf{E}$, and to groundings of (collapsed) roles r.

Consider the groundings of primitive concepts r which are not ancestors of any grounding in $\mathbf{E}$. Setting its value to maximize its marginal probability does not introduce any inconsistency with respect to $\mathbf{E}$. Moreover, for any assignment to these groundings, we can find a consistent assignment to the remaining groundings (which are internal nodes and not ancestors of $\mathbf{E}$), that is, an assignment which assigns positive probability. Since this is the maximum probability we can obtain for these groundings, this is a partial optimum assignment.

We are thus only left with the problem of assigning values to the groundings of relations r which are ancestors of $\mathbf{E}$. Consider a relation r such that $\mathbb{P}(r) \geq 1/2$. Then assigning all groundings of r to true maximizes their marginal probability and satisfies the logical equivalences of all groundings in $\mathbf{E}$. Hence, this is a maximum assignment (and its value can be computed efficiently). So assume there is a relation r with $\mathbb{P}(r) < 1/2$ such that a grounding of e_r or e_r^- appear in $\mathbf{E}$. In this case, the greedy assignment sets every grounding of r; however, such an assignment is inconsistent with the logical equivalence of e_r and e_r^-, hence obtains probability zero. Now consider an assignment that assigns exactly one grounding $r(a,b)$ to true and all the other to false. This assignment is consistent with $e_r(a)$ and $e_r(b)$, and maximizes the probability; any assignment that sets more groundings to true has a lower probability since it replaces a term $1 - \mathbb{P}(r) \geq 1/2$ with a term $\mathbb{P}(r) < 1/2$ in the joint probability. More generally, to maximize the joint probability we need to assign to true as few groundings $r(a,b)$ which are ancestors of $\mathbf{E}$ as possible. This is equivalent to a minimum cardinality edge covering problem as follows.

For every relation r in the relational network, construct the bipartite complete graph $G_r = (V_1, V_2, E)$ such that V_1 is the set of groundings $e_r(a)$ that appears and have no parent $r(a, b)$ in $\mathbf{E}$, and V_2 is the set of groundings $e_r^-(a)$ that appears and have no parents in $\mathbf{E}$. We identify an edge connecting $e_r(a)$ and $e_r^-(b)$ with the grounding $r(a, b)$. For any set $C \subseteq E$, construct an assignment by attaching true to the groundings $r(a, b)$ in C and false to every other grounding $r(a, b)$. This assignment is consistent with $\mathbf{E}$ if and only if C is an edge cover; hence the minimum cardinality edge cover maximizes the joint probability (it is consistent with $\mathbf{E}$ and attaches true to the least number of groundings of $\mathbf{r}$). This concludes the proof of Theorem 2.

6   Counting Edge Covers

In this section we discuss the fact that, for graphs such as those representing formulas in DL-Lite, the partition function can be computed in polynomial time. Specifically, we consider graphs $G = (V, E)$ whose nodes can be partitioned into four disjoint sets V_1, V_2, V_3, V_4 such that the subgraph obtained by considering only edges V_i and V_{i+1} is complete bipartite $(i = 1, 2, 3)$. We call such graphs stepwise bipartite complete. For lack of space, we only present the main ideas; details and proofs can be found elsewhere [30].

We partition the nodes into *white* nodes $W = V_1 \cup V_4$ and *black* nodes $B = V_2 \cup V_3$. As we will be interested only in B-covers, we will refer to them simply as covers. An edge $e = (u, v)$ is classified into one of three categories with respect to the partition W, B: it is a *free edge* if $u, v \in W$; a *dangling edge* if $u \in W, v \in B$, or a *regular edge* if $u, v \in B$. For convenience, we fix λ and write $Z(G)$ to denote $Z(G, \lambda)$. Computing $Z(G)$ for general graphs is #P-complete even for $\lambda = 1$ [6], and admits a FPTAS for bounded λ [27, 28]. We will show that for stepwise bipartite complete graphs, the problem can be solved in polynomial time.

Let e be an edge and u be a vertex in $G = (W, B, E)$. We define the following operations and notation: $G - e = (W, B, E \setminus \{e\})$ and $G - u = (W \cup \{u\}, B \setminus \{u\}, E)$. These operations do not change the vertex set (only the partition), and are associative (e.g., $G - e - f = G - f - e$, $G - u - v = G - v - u$, and $G - e - u = G - u - e$). Hence, if $E = \{e_1, \cdots, e_d\}$ is a set of edges, we can write $G - E$ to denote $G - e_1 - \cdots - e_d$ applied in any arbitrary order. The same is true for any combination of these operations.

The following results, easily derived from the work of Lin, Liu and Lu [27], show that the partition function can be computed recursively on smaller graphs and solved efficiently when no black nodes exist:

Proposition 2. *Let $e = (u, v)$ be an edge. (1) If e is dangling edge with u colored black then $Z(G) = (1 + \lambda)Z(G - e - u) - Z(G - E_G(u) - u)$. (2) If e is a free edge of G then $Z(G) = (1 + \lambda)Z(G - e)$. (3) If u is an isolated white node (i.e., $N_G(u) = \emptyset$) then $Z(G) = Z(G - u)$.*

The result above allows us to decompose the problem of computing $Z(G)$ into two smaller problems until the the problems are simple enough to be solved

---

procedure PF(G, w, k_1, k_2):
if $(w \in V_1$ **and** $k_1 + k_2 < n)$ **or** $(w \in V_4$ **and** $k_1 + k_2 < m)$:
    if $w \in V_1$ **do** $v \leftarrow u_{k_1+k_2}$ **else do** $v \leftarrow v_{k_1+k_2}$;
    return $(1+\lambda) * \mathsf{PF}(G - (v, w) - v, w, k_1, k_2+1) - \mathsf{PF}(G - E_G(v) - v, w, k_1+1, k_2)$;
else:
    remove free edges and isolate nodes; **let** k be the number of edges removed;
    do $w \leftarrow w \in V_1$; $V_1 \leftarrow V_1 \cup V_3$; $V_3 \leftarrow \emptyset$;
    return $(1 + \lambda)^k * \mathsf{PF}(G, w, 0, 0)$;

Fig. 4. Algorithm PF. Takes a graph $G = (V_1, V_2, V_3, V_4, E)$ with $V_2 = \{u_1, \ldots, u_n\}$ and $V_3 = \{v_1, \ldots, v_m\}$, a node $w \in V_1 \cup V_4$, and nonnegative integers k_1 and k_2.

by a simple count (of free edges). That approach however generates an exponential number of recursions. A polynomial-time algorithm can be obtained by exploiting the symmetries in the graphs, obtained through graph isomorphism.

Two graphs are isomorphic if there is an edge-preserving bijection between the nodes of the two graphs that also preserves their color. Two isomorphic graphs have the same value of the partition function. The next result shows that the order in which operations of edge removal and and node whitening are performed among isomorphic nodes does not affect the value of $Z(G)$:

Proposition 3. *Let* $u_1, \ldots, u_n$ *be the nodes in* V_2 *(V_3) and* w *be a node in* V_1 *(V_4). Given any permutation* σ *on* V_2 *(V_3) and nonnegative integers* $k_1 + k_2 \leq n$ *the graphs* $G' = G - E_G(u_1) - \cdots - E_G(u_{k_1}) - (w, u_{k_1+1}) - \cdots - (w, u_{k_1+k_2}) - u_1 - \cdots - u_{k_1+k_2}$ *and* $G'' = G - E_G(\sigma(u_1)) - \cdots - E_G(\sigma(u_{k_1})) - (w, \sigma(u_{k_1+1})) - \cdots - (w, \sigma(u_{k_1+k_2})) - \sigma(u_1) - \cdots - \sigma(u_{k_1+k_2})$ *are isomorphic.*

Using these facts, Algorithm PF (Fig. 4) produces $Z(G)$.

Theorem 3. *Let* G *be a stepwise bipartite complete graph. Then Algorithm* PF *with an arbitrary node* w *in* V_4 *and* $k_1 = k_2 = 0$ *outputs* $Z(G)$ *in time and memory polynomial in the number of nodes and edges of* G *if the calls are cached (so that no two calls with same arguments are performed).*

The algorithm PF requires the existence of dangling edges. Now it might be that the graph contains no white nodes (hence no dangling edges), that is, that G is complete bipartite graph. The next result shows how to decompose the problem of computing the partition function into problems of computing the partition function in smaller graphs.

Proposition 4. *Let* G *be a bipartite complete graph and* $e = (u, v)$ *be some edge. Then* $Z(G) = (1+\lambda)Z(G - e - u - v) - Z(G - E_G(v) - v) - Z(G - E_G(u) - u) - Z(G - E_G(u) - E_G(v) - u - v)$. *The graphs in the right-hand side are either bipartite complete or stepwise bipartite complete with a dangling edge.*

7   Related Work

In previous work [10] we have shown that marginal inference can be computed in polynomial-time in models described by a restricted version of the language considered here, one that does not admit inverse roles. The absence of inverse roles leads to Bayesian networks composed of disconnected components, where each component contains all concepts related to an individual; the complexity of inferences with "positive" evidence is then easily seem to be polynomial by applying d-separation. The use of inverse roles connects components related to different individuals, so the same argument cannot be used.

There have been many attempts at combining description logics with probabilities. Heinsohn [20] was one of the first to propose modeling languages that allow uncertainty into terminological descriptions. Much of the work in probabilistic description logics is however hindered by intractability of inferences. The DL-Lite language was conceived as a lightweight knowledge representation scheme to represent large bases of relational data with very efficient reasoning services. The simplicity and computational efficiency of the DL-Lite language have led many researchers to use it as a building block of modeling languages that combine description logics and Bayesian networks. For instance, D'Amato et. al [12] proposed a variant of DL-Lite where the interpretation of each sentence is conditional on a context. The context is specified by a Bayesian network, and is hence probabilistic. The probability of concepts can then be extended to determine the probability of logical expressions. A similar approach was taken by Ceylan and Peñalosa in their Bayesian Description Logic [8], with minor semantic differences. A different approach is to extend the syntax of DL-Lite sentences with probabilistic subsumption connectives, as in the Probabilistic DL-Lite [34]. Differently from our proposal here, none of those works use DL-Lite to specify (large) Bayesian networks.

8   Extensions, and Conclusion

The previous results can be directly extended in some important ways. For example, if we allow negative groundings of roles in the evidence, then most of the proof of Theorem 1 follows; the difference is that the intersection graphs obtained do not satisfy the same symmetries. We can then resort to approximations for weighted edge cover counting [28], so as to develop a fully polynomial-time approximation scheme (FPTAS) for inference. For most probable explanations, the problem remains polynomial. Similarly, we could allow for different groundings of the same relation to be associated with different probabilities; the proofs given here can be modified to develop a FPTAS for inference. This implies that both *probabilistic relational models (PRMs)* [17] and *recursive relational Bayesian networks (RRBNs)* [22], when appropriately restricted to DL-Lite constructs, allow for inference through FPTAS. We intend to pursue details of such conceptual modeling tools in the future.

Other possible extensions of our results merit attention. First, one might investigate whether there are similar polynomial/FPTAS results not only for the many existing variants of DL-Lite [1], but also for networks specified through other popular description logics such as $\mathcal{EL}$ and $\mathcal{ALC}$ [2], or even other languages such as temporal logics.

To conclude, DL-Lite Bayesian networks offer a flexible and effective language, that can be used to specify probabilistic acyclic ontologies or entity-relationship diagrams. Usual services, such as inference and explanations, have tractable algorithms that can be used directly or called during learning.

Acknowledgements. The second author was partially supported by CNPq.

References

1. Artale, A., Calvanese, D., Kontchakov, R., Zakharyashev, M.: The DL-Lite family and relations. J. Artif. Intell. Res. **36**, 1–69 (2009)
2. Baader, F., Nutt, W.: Basic description logics. In: Baader, F., McGuinness, D.L., Nardi, D., Patel-Schneider, P.F. (eds.) Description Logic Handbook, pp. 47–100. Cambridge University Press, Cambridge (2002)
3. Bacchus, F.: Representing and Reasoning with Probabilistic Knowledge: A Logical Approach. MIT Press, Cambridge (1990)
4. van den Broeck, G.: On the completeness of first-order knowledge compilation for lifted probabilistic inference. In: Advances in Neural Processing Information Systems (2011)
5. van den Broeck, G., Wannes, M., Darwiche, A.: Skolemization for weighted first-order model counting. In: Proceedings of the International Conference on Principles of Knowledge Representation and Reasoning (2014)
6. Cai, J.Y., Lu, P., Xia, M.: Holographic reduction, interpolation and hardness. Comput. Complex. **21**(4), 573–604 (2012)
7. Calvanese, D., De Giacomo, G., Lembo, D. abd Lenzerini, M., Rosati, R.: DL-Lite: tractable description logics for ontologies. In: Proceedings of the AAAI Conference, pp. 602–607 (2005)
8. Ceylan, I., Peñaloza, R.: The Bayesian description logic ⌊⌉⌢. In: Proceedings of the 7th International Joint Conference on Automated Reasoning, pp. 480–494 (2014)
9. Chavira, M., Darwiche, A.: Compiling Bayesian networks with local structure. In: Proceedings of the Nineteenth International Joint Conference on Artificial Intelligence, pp. 1306–1312 (2005)
10. Cozman, F.G., Mauá, D.D.: Bayesian networks specified using propositional and relational constructs: combined, data, and domain complexity. In: Proceedings of the 29th AAAI Conference on Artificial Intelligence, pp. 3519–3525 (2015)
11. Dagum, P., Luby, M.: Approximating probabilistic inference in Bayesian belief networks is NP-hard. Artif. Intell. **60**(1), 141–53 (1993)
12. d'Amato, C., Fanizzi, N., Lukasiewicz, T.: Tractable reasoning with Bayesian description logics. In: Greco, S., Lukasiewicz, T. (eds.) SUM 2008. LNCS (LNAI), vol. 5291, pp. 146–159. Springer, Heidelberg (2008)
13. D'Ambrosio, B.: Local expression languages for probabilistic dependence. Int. J. Approximate Reasoning **13**(1), 61–1 (1995)
14. Darwiche, A.: Modeling and Reasoning with Bayesian Networks. Cambridge University Press, Cambridge (2009)

15. Domingos, P., Webb, W.: A tractable first-order probabilistic logic. In: Proceedings of the AAAI Conference on Artificial Intelligence (2012)
16. Friedman, N., Getoor, L., Koller, D., Pfeffer, A.: Learning probabilistic relational models. In: Proceedings of the International Joint Conference on Artificial Intelligence, pp. 1300–1309 (1999)
17. Getoor, L., Taskar, B.: Introduction to Statistical Relational Learning. MIT Press, Cambridge (2007)
18. Halpern, J.Y.: Reasoning About Uncertainty. MIT Press, Cambridge (2003)
19. Heckerman, D.: A tractable inference algorithm for diagnosing multiple diseases. In: Proceedings of the 5th Conference on Uncertainty in Artificial Intelligence, pp. 174–181 (1989)
20. Heinsohn, J.: Probabilistic description logics. In: Proceedings of the 10th International Conference on Uncertainty in Artificial Intelligence, pp. 311–318 (1994)
21. Jaeger, M.: Relational Bayesian networks. In: Procedings of the Conference on Uncertainty in Artificial Intelligence, pp. 266–273 (1997)
22. Jaeger, M.: Complex probabilistic modeling with recursive relational Bayesian networks. Ann. Math. Artif. Intell. **32**, 179–220 (2001)
23. Jaeger, M., van Den Broeck, G.: Liftability of probabilistic inference: upper and lower bounds. In: Proceedings of the 2nd Statistical Relational AI Workshop (2012)
24. Koller, D., Friedman, N.: Probabilistic Graphical Models. MIT press, Cambridge (2009)
25. Kwisthout, J.: Treewidth and the computational complexity of MAP approximations. In: van der Gaag, L.C., Feelders, A.J. (eds.) PGM 2014. LNCS, vol. 8754, pp. 271–285. Springer, Heidelberg (2014)
26. Kwisthout, J.H.P., Bodlaender, H.L., van der Gaag, L.C.: The necessity of bounded treewidth for efficient inference in Bayesian networks. In: Proceedings of the 19th European Conference on Artificial Intelligence, pp. 237–242 (2010)
27. Lin, C., Liu, J., Lu, P.: A simple FPTAS for counting edge covers. In: Proceedings of the 8th Annual ACM-SIAM Symposium on Discrete Algorightms, pp. 341–348 (2014)
28. Liu, J., Lu, P., Zhang, C.: FPTAS for counting weighted edge covers. In: Proceedings of the 22nd Annual European Symposium on Algorithms, pp. 654–665 (2014)
29. Lowd, D., Rooshenas, A.: Learning Markov networks with arithmetic circuits. In: Proceedings of the 16th International Conference on Artificial Intelligence and Statistics, pp. 406–414 (2013)
30. Mauá, D.D., Cozman, F.G.: A tractable class of model counting problems. Technical report, Decision Making Laboratory, University of São Paulo (2015)
31. Poon, H., Domingos, P.: Sum-product networks: a new deep architecture. In: Proceedings of the Twenty-Seventh Conference on Uncertainty in Artificial Intelligence, pp. 337–346 (2011)
32. De Raedt, L.: Logical and Relational Learning. Springer, New York (2008)
33. De Raedt, L., Kersting, K.: Probabilistic inductive logic programming. In: De Raedt, L., Frasconi, P., Kersting, K., Muggleton, S.H. (eds.) Probabilistic Inductive Logic Programming. LNCS (LNAI), vol. 4911, pp. 1–27. Springer, Heidelberg (2008)
34. Ramachandran, R., Qi, G., Wang, K., Wang, J., Thornton, J.: Probabilistic reasoning in DL-Lite. In: Proceedings of the 12th Pacific Rim International Conference on Trends in Artificial Intelligence, pp. 480–491 (2012)

35. Rosenkrantz, D.J., Marathe, M.V., s. Ravi, S., Vullikanti, A.K.: Bayesian inference in treewidth-bounded graphical models without indegree constraints. In: Proceedings of the 30th Conference on Uncertainty in Artificial Intelligence, pp. 702–711 (2014)
36. Roth, D.: On the hardness of approximate reasoning. Artif. Intell. **82**(1–2), 273–302 (1996)

Probabilistic Models

State Space Search with Stochastic Costs and Risk Aversion

Anisse Ismaili[(✉)]

Sorbonne Universités, UPMC Univ Paris 06, LIP6,
UMR 7606, 75005 Paris, France
anisse.ismaili@lip6.fr

Abstract. In this paper we study state space search problems where the costs of transitions are uncertain. Cost uncertainty can be due to the existence of several scenarios impacting the entire set of transitions; it can also result from local random factors impacting each transition independently, or from more complex combinations of these two cases. This leads us to consider three different settings for handling cost uncertainty in state space graphs. For each of them, we recall some key properties of first-order and second-order stochastic dominance. Then we propose dominance-based heuristic search algorithms to determine the set of possibly optimal solutions with respect to the expected utility model and Yaari's model, with and without assuming risk aversion. Finally, to preserve scalability on large-size instances, we adapt these algorithms for the fast determination of an ε-covering of the potentially optimal solutions.

Keywords: State space search · Uncertainty · Stochastic dominance

1 Introduction

State-space search is a general formal framework that can be used to solve various practical optimization problems with a combinatorial structure, including, for instance, shortest path determination, actions planning, scheduling, game search and puzzle solving. The usual representation of a problem is a state space graph, the nodes of which represent different possible states of a system, and the arcs represent the possible transitions. Hence the problem consists in finding a shortest path in this graph (explicitly or implicitly known), from a given initial node to a goal node [11,13].

Baseline algorithms such as A* [7] proposed for finding the shortest path in a state-space graph standardly assume that costs are deterministic. However, in practical applications, many external factors may impact on transition costs (e.g., travel times may depend on the local weather) which makes costs uncertain. In this paper we consider state space search problems with uncertain transition costs[1], with or without probabilistic dependencies among

---

[1] Note that transitions are still assumed to be deterministic which makes a significant difference with Markov Decision Processes [2,16].

© Springer International Publishing Switzerland 2015
C. Beierle and A. Dekhtyar (Eds.): SUM 2015, LNAI 9310, pp. 67–82, 2015.
DOI: 10.1007/978-3-319-23540-0_5

the costs. To every transition in the graph, and therefore to every path in the graph, is associated a random variable defining its uncertain prospect in terms of cost. In this context of search with stochastic costs, we are interested in determining the preferred paths for a decision maker (DM) whose preferences are represented by standard decision models such as the expected utility [18,19] or the dual expected utility [22]. Since we are in a cost minimization context, these models switch to expected disutility functions to be minimized.

Some prior works in this area concern the determination of the preferred paths, assuming the decision model is given, see e.g. [3–5,8,10,12]. Compared to this line, here we do not assume that the decision model is completely known. In particular, the DM's attitude towards risk may not be known precisely which precludes any precise preference-based optimization. However, most decision models considered in decision making under risk are monotonic with respect to the *first order stochastic dominance*, or even with the *second order stochastic dominance* whenever the decision maker is risk-averse. Consequently, we are interested in determining *the set* of non-dominated solution paths with respect to one or both of these dominance relations. For instance, first order non-stochastically dominated solutions correspond exactly to *the set* of all possibly optimal solutions for an expected disutility minimizer. It even works when no information about the disutility function is available. This computation provides the DM with the set of uncertain prospects that are non-dominated. It is a pre-selection step towards the final decision. Similar remarks hold for the second-order stochastic dominance and risk-averse DMs, motivating the development of algorithms for determining the set of stochastically non-dominated solution paths in state space graphs.

This topic has been addressed in different settings by the past, see e.g., [14,20,21]. However, the impact of multiple uncertainties on the costs make new challenges arise. It complicates the computation of non-dominated solution paths, while this computation appears to be a critical issue. Indeed, the number of non-dominated solutions may grow with the size of the problem. Our aim here is to present a general setting encompassing different contexts of uncertainty and to propose exact and approximate algorithms for the determination of stochastically non-dominated solution paths. More precisely, the paper is organized as follows[2]: in Sect. 2 we recall some preliminary definitions and properties that will be used throughout the paper. In Sect. 3 we introduce a search algorithm for the determination of the set of non-dominated solution paths, and we establish its admissibility in different settings. Then, a near admissible algorithm for approximating the set of non-dominated solution paths is provided in Sect. 4. Finally we provide in Sect. 5 the results of numerical experiments[3] performed on random instances of state space graphs.

---

[2] For the deep readers, the detailed proofs of our lemmas and theorems can be downloaded at:
https://infotomb.com/yh2mt.pdf.

[3] One can reproduce this paper's results by using the following C++ programs:
https://infotomb.com/2b07w.zip.

2   Preliminaries

2.1   State Space Graph Under Cost Uncertainty

Let $G = (N, A)$ denote a *state space graph* where N is a finite set of *nodes*[4], and A is the set of arcs representing feasible *transitions*. This set of transitions A denotes $A = \{(n, n') : n \in N, n' \in S(n)\}$ where $S(n) \subseteq N$ is the set of all *successors* n' of node n. A *path* $P = \langle a_1, \ldots, a_x \rangle$ is a sequence of feasible transitions $\langle (n_0, n_1), (n_1, n_2), \ldots, (n_{x-1}, n_x) \rangle$ from a node to the next one. Let $\mathcal{P}(n, n')$ denote the *set of all paths* from the node n to the node n'. Given the source node s and the goal node t, we call a *solution-path*, a path from s to t (i.e. an element of $\mathcal{P}(s, t)$). Let us now introduce the uncertainty settings.

Let $R = \{r_1, \ldots, r_s\}$ denote the *set of regions*. Transition $a \in A$ belongs to the region $r(a) \in R$. Indeed, the set of transitions is partitioned into the regions by $A = \cup_{r \in R} A_r$, where $A_r = \{a \in A \mid r(a) = r\}$. For each region $r \in R$, let $\Omega^r = \{\omega_1^r, \ldots, \omega_{m_r}^r\}$ denote the set of *regional scenarios* and let $\mathbb{P}_{\Omega^r}$ denote the independent probability measure over the scenarios of this region r. In each region r, each scenario ω^r occurs with a probability $\mathbb{P}_{\Omega^r}(\omega^r) \in [0, 1]$ and $\sum_{\omega^r \in \Omega^r} \mathbb{P}_{\Omega^r}(\omega^r) = 1$. The set of *overall scenarios* is $\Omega = \times_{r \in R} \Omega^r$, and an overall scenario $(\omega^r \mid r \in R) \in \Omega$ is denoted by ω and gives for each region r, the scenario occurring there. The overall probability measure is $\mathbb{P}_\Omega = \otimes_{r \in R} \mathbb{P}_{\Omega^r}$. That is: each $\omega = (\omega^r \mid r \in R)$ occurs with probability $\mathbb{P}_\Omega(\omega) = \prod_{r \in R} \mathbb{P}_{\Omega^r}(\omega^r)$.

Example 1 (Rabban Bar Sauma's Travel from Beijing to Paris). The path-planning of a travel on a graph is a problem as old as man himself [1,6,15,17]. Year 1287 AD; Rabban [17] wants to travel from Beijing to Paris. For this purpose, he needs to cross several regions, where different scenarios can occur. In the Pamir region, the weather can be snow (blocking mountain passes), rain (flooding some valleys), or sun (drying the desert). In Syria, the war can put at risk the Chinese-Christians (like Rabban) of being hated on some roads. In Italy, a rare volcanic eruption would cut some roads. What path to choose in this setting where for several roads the cost is uncertain?

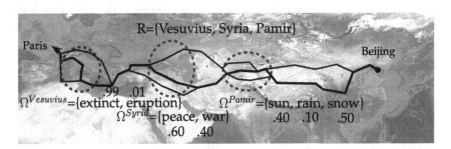

Recall that one can define a *random variable* (r.v.) X in three manners. Firstly, X can be defined as an integer-valued function of the scenarios $X : \Omega \to \mathbb{N}$, giving birth to a density $(\mathbb{P}_\Omega(X(\omega) = k) \mid k \in \mathbb{N})$ that gives for each *cost*

---

[4] In dynamic programming, nodes are referred as *states*.

$k \in \mathbb{N}$, its probability $\mathbb{P}(X = k)$. Secondly, X is completely characterized by this density $(\mathbb{P}(X = k) \mid k \in \mathbb{N})$. Thirdly, recall that two r.v.s X and Y can be added into $X + Y$, which is functionally defined by $(X + Y)(\omega) = X(\omega) + Y(\omega)$ and characterized by the density $\mathbb{P}(X + Y = k) = \mathbb{P}(\cup_{t=0}^{k}\{X = t\} \cap \{Y = k - t\})$. The *scope* of X is $\mathcal{S}(X) = \{k \in \mathbb{N} \mid \mathbb{P}(X = k) > 0\}$.

In this paper, the cost of each transition $a \in A$ depends on the regional scenario of the region $r(a)$ it belongs to. For each transition $a \in A$, it is a random variable $X^a : \Omega^{r(a)} \to \mathbb{N}$ giving for each regional scenario $\omega^{r(a)} \in \Omega^{r(a)}$, the random cost $X^a(\omega^{r(a)}) \in \mathbb{N}$ of using transition a when the regional scenario $\omega^{r(a)}$ occurs. Similarly, for a path P and an overall scenario ω, let $X^P(\omega) = \sum_{a \in P} X^a(\omega^{r(a)}) = \sum_{r \in R} \sum_{a \in P \cap A_r} X^a(\omega^r)$ denote the cost of using path P when the overall scenario $\omega = (\omega^r \mid r \in R) \in \Omega$ occurs.

More particularly, our framework generalizes three sub-settings. The two firsts are well known in the literature and are extreme cases.

1. In Setting (S1), we have $|R| = 1$, hence $\Omega^r \equiv \Omega = \{\omega_1, \ldots, \omega_m\}$ with m bounded by a constant, and the random variables are totally dependent. In the literature under risk this setting is referred to as Savage's framework [18]. The paths of the graph have random costs that can be seen as Savage's acts, i.e. functions of the form $X^P : \Omega \to \mathbb{N}$ that give for any state $\omega \in \Omega$ the cost $X^P(\omega)$ of path P. Here, the comparison of paths amounts to comparing acts and the preferences over paths are inherited from preferences over acts.

2. In Setting (SA), we have $R = A$, hence the random variables of each transition are independent. The costs of arcs (transitions) are seen as probabilistic lotteries (random variables with a finite support). Hence, the cost of each path is a lottery as well, obtained by summation over the arcs composing the path. Hence a lottery is associated to any path. Here the comparison of paths amounts to comparing lotteries like in the framework of von Neumann and Morgenstern [19] and the preferences over paths are inherited from preferences over lotteries.

3. Between these two settings, the Setting (SR) is legitimate in many concrete problems, when uncertainty occurs in independent regions for distinct political/meteorological/geological reasons (see e.g. Example 1). In Setting (SR), we just assume that the regions are in a number $|R|$ bounded by a constant and are rather contiguous (Contiguousness will be specified later).

While Setting (S1) models a total dependence of the transitions' costs towards a set of scenarios, Setting (SA) models the opposite, that is for each transition, a total independence between the scenarios. Setting (SR) is in-between.

In each setting, deciding a path P yields a *random variable* (r.v.) $X^P = \sum_{a \in P} X^a$, that is also a probability density on the finite set of costs $[\![0, M]\!]^5$. Therefore, in this paper, the preferences and dominances will be stated on r.v.s

---

[5] Assume that the costs of each solution-path is bounded below by 0 and above by $M \in \mathbb{N}$, and that each integer-valued random variable has the same bounds throughout this paper.

For a set of paths $\mathcal{P}$, we will denote the set of r.v.s associated to each path by $\mathcal{X}(\mathcal{P}) = \{X^P \mid P \in \mathcal{P}\}$.

Given a probability-space $(\Omega, 2^\Omega, \mathbb{P})$ and an integer-valued random variable $X : \Omega \rightarrow [\![0, M]\!]$, let $G_X : [\![0, M]\!] \rightarrow [0, 1]$ denote the *decumulative function* defined for $t \in [\![0, M]\!]$, by $G_X(t) = \mathbb{P}(X > t)$. The decumulation can be operated a second time, to define the *second order decumulative function* G_X^2 for $t \in [0, M]$, by $G_X^2(t) = \int_t^M G_X(x)dx$. Note that the present variables X are *discrete*, hence the function G_X, extending its scope from $[\![0, M]\!]$ to $\mathbb{R}$ is a step function such that $G_X([t - 1, t[) = \{G_X(t - 1)\}$, for each $t \in [\![-1, M]\!]$. More precisely, going forward on the integers from -1 to M, we have $G_X(-1) = 1$ and for each $t \in [\![0, M]\!], G_X(t) = G_X(t - 1) - \mathbb{P}(X = t)$. Consequently, G_X^2 is piecewise linear. The (characterizing) breakpoints of G_X^2 have their abscissa in $[\![0, M]\!]$ and their ordinate defined by: $G_X^2(M) = 0$ and for each $t \in [\![1, M]\!], G_X^2(t - 1) = \int_{t-1}^t G_X(x)dx + G_X^2(t) = G_X^2(t) + G_X(t - 1)$, hence for each $t \in [\![0, M - 1]\!], G_X^2(t) = \sum_{k=t}^{M-1} G_X(k)$.

2.2 Decision Criteria Under Risk

A risk-averse decision maker can be modelled as follows [18,19]. Between a random variable and a mean preserving spread of this random variable, the risk-averse DM will prefer the former. Alternatively, when the (non-decreasing) disutility function $w : [0, M] \rightarrow \mathbb{R}$ represents the subjective or psychological evaluation of the costs, a risk averse DM puts more weight on the marginal improvements of higher costs. That is: assuming a convex function w models *risk aversion*.

To model decision criteria under risk and for risk-aversion, let us now define *dominance relations* between $[\![0, M]\!]$-valued random variables. The optimization of the expected utility is a standard criterion under risk and for risk aversion. In this cost-minimization setting, it writes as an expected disutility that we aim to minimize.

Definition 1 (Expected disutility and Risk-aversion). *Given a random variable X and a non-decreasing disutility $w : [0, M] \rightarrow \mathbb{R}$, the expected disutility of X is:*

$$\mathbb{E}[w(X)] = \sum_{x=0}^{M} P(X = x)w(x)$$

For two random variables X and Y, it induces the following dominance relations:

$$X \precsim_{EW} Y \Leftrightarrow \mathbb{E}[w(X)] \leq \mathbb{E}[w(Y)] \quad and \quad X \prec_{EW} Y \Leftrightarrow \mathbb{E}[w(X)] < \mathbb{E}[w(Y)]$$

Assuming a convex function w models risk aversion.

The optimization of Yaari's model is a dual criterion under risk and for risk aversion, using a non-decreasing transformation $\varphi : [0, 1] \rightarrow [0, 1]$ of decumulatives.

Definition 2 (Yaari's Criterion and Risk-aversion). *Given a random variable X and a non-decreasing transformation $\varphi : [0,1] \to [0,1]$ such that $\varphi(0) = 0$ and $\varphi(1) = 1$, Yaari's evaluation of X is defined by:*

$$\mathbb{E}_\varphi[X] = \sum_{x=0}^{M} \left(\varphi(G_X(x-1)) - \varphi(G_X(x)) \right) x$$

For two random variables X and Y, it induces the following dominance relations:

$$X \precsim_{YAA} Y \Leftrightarrow \mathbb{E}_\varphi[X] \le \mathbb{E}_\varphi[Y] \qquad \text{and} \qquad X \prec_{YAA} Y \Leftrightarrow \mathbb{E}_\varphi[X] < \mathbb{E}_\varphi[Y]$$

Assuming $\varphi(p) \ge p$ models weak risk aversion.

Though standard, the dominances $\prec_{EW}$ and $\prec_{YAA}$ leave the disutility and transformation functions subjected to *undetermined arbitrary choices*. To avoid this arbitrariness, the partial dominance relations $\prec_{FD}$, $\prec_{FSD}$ and $\prec_{SSD}$ below produce many optima, do not assume any disutility function, are compatible[6] with $\prec_{EW}$ and $\prec_{YAA}$, and hence encompass all the optima of $\prec_{EW}$ and $\prec_{YAA}$, whatever the disutility or transformation functions might be.

Definition 3 (Functional Dominance). *Given a set of scenarios Ω and two random variables X and Y from Ω to $[0,M]$, X weakly functionally dominates Y, denoted by $X \precsim_{FD} Y$, if and only if:*

$$\forall \omega \in \Omega, \quad X(\omega) \le Y(\omega)$$

The functional dominance $\prec_{FD}$ is the assymetric part of $\precsim_{FD}$ and then also requires that there exists at least one scenario $w \in \Omega$ such that $X(\omega) < Y(\omega)$.

This simplest dominance is a scenario-wise comparison of the cost outcomes, and does not take account for probabilities. In setting (S1), deciding if $X \precsim_{FD} Y$ takes time $O(m)$. However, in setting (SA), deciding $X \precsim_{FD} Y$ on $m = O(\prod_{a \in A} m_a)$ overall scenarios would be intractable. Also, the functional dominance is a non-discriminative partial order, producing many incomparable random variables and many optima.

Definition 4 (First Order Stochastic Dominance). *Given two random variables X and Y on $[0,M]$, X weakly stochastically dominates Y (to the first order), denoted by $X \precsim_{FSD} Y$, if and only if:*

$$\forall t \in [0,M], \quad G_X(t) \le G_Y(t)$$

The first order stochastic dominance $\prec_{FSD}$ is the assymetric part of $\precsim_{FSD}$ and then also requires that there exists at least one t in $[0,M]$ such that $G_X(t) < G_Y(t)$.

---

[6] That is, $\prec_{YAA}$ and $\prec_{EW}$ are monotonic with respect to $\prec_{FSD}$ and $\prec_{SSD}$.

The intuition behind this standard definition is that for each cost threshold $t \in [0, M]$, Y puts more probability on the more costly outcomes than X, making X preferred to Y. Under setting (S1), since $X \precsim_{\text{FSD}} Y$ can be tested by comparing only the steps, it can be tested in $O(|\mathcal{S}(X) \cup \mathcal{S}(Y)|) = O(2m)$ time. Under setting (SA), since the support of X and Y is finite, it is straightforward to test $X \precsim_{\text{FSD}} Y$ in $O(M)$ time.

Definition 5 (Second Order Stochastic Dominance). *Given two random variables X and Y on $[0, M]$, X weakly stochastically dominates Y to the second order, denoted by $X \precsim_{SSD} Y$, if and only if:*

$$\forall t \in [0, M], \quad G_X^2(t) \leq G_Y^2(t)$$

The second order stochastic dominance $\prec_{SSD}$ is the assymetric part of $\precsim_{SSD}$ and then also requires that there exists at least one t in $[0, M]$ such that $G_X^2(t) < G_Y^2(t)$.

The second order stochastic dominance models risk aversion. For instance, given a variable X, a mean preserving spread Y of X is more risky and hence less preferred. To compute if $X \precsim_{SSD} Y$, one only has to test the Pareto dominance on the $O(|\mathcal{S}(X) \cup \mathcal{S}(Y)|) = O(M)$ breakpoints. Let us now define formally the subset of the best random variables, with respect to a weak dominance relation $\precsim$.

Definition 6 (The $\precsim$-non-dominated subset). *Given a multi-set L of random variables and a weak dominance relation $\precsim$, the complete $\precsim$-non-dominated sub(multi)set is:*

$$\overline{\mathcal{M}}(L, \precsim) = \{X \in L : \forall Y \in L, \ Y \precsim X \Rightarrow X \precsim Y\}$$

Denoting the asymmetric-part of $\precsim$ by $\prec$, it is equally: $\overline{\mathcal{M}}(L, \precsim) = \{X \in L : \forall Y \in L, \ not(Y \prec X)\}$. The $\precsim$-non-dominated subset $\mathcal{M}(L, \precsim)$ exactly contains one element per equivalence class in the quotient set $\overline{\mathcal{M}}(L, \precsim)/ \sim$. (It is not a multi-set.)

Theorem 1 below justifies the computation of $\mathcal{M}(L, \precsim_{\text{FSD}})$ and $\mathcal{M}(L, \precsim_{\text{SSD}})$ as a pre-selection tool for the optimization of $\prec_{\text{EW}}$ and $\prec_{\text{YAA}}$, whatever the unknown disutility function or whatever the unknown probability transformation.

Theorem 1 [9,19,22]. *For two r.v.s X and Y, one has: $X \precsim_{FD} Y \Rightarrow X \precsim_{FSD} Y \Rightarrow X \precsim_{SSD} Y$. Consequently, given a set L of random variables, one has:*

$$\mathcal{M}(L, \precsim_{FD}) \quad \supseteq \quad \mathcal{M}(L, \precsim_{FSD}) \quad \supseteq \quad \mathcal{M}(L, \precsim_{SSD})$$

- *Moreover, denoting by $\mathcal{F}^{\uparrow}$ the set of non-decreasing disutility functions, we have:*
$X \prec_{FSD} Y \Leftrightarrow \forall w \in \mathcal{F}^{\uparrow}, X \prec_{EW} Y$
hence: $\mathcal{M}(L, \precsim_{FSD}) = \bigcup_{w \in \mathcal{F}^{\uparrow}} \mathcal{M}(L, \precsim_{EW})$

- *Denoting the set of convex non-decreasing disutility functions by $\mathcal{C}^{\uparrow} \subseteq \mathcal{F}^{\uparrow}$, we have:*

$X \prec_{SSD} Y \Leftrightarrow \forall w \in \mathcal{C}^\uparrow : X \prec_{EW} Y$

hence: $\mathcal{M}(L, \precsim_{SSD}) = \bigcup_{w \in \mathcal{C}^\uparrow} \mathcal{M}(L, \precsim_{EW})$

- *Concerning Yaari's model, denoting by Φ the set of increasing probability transformations φ such that $\varphi(0) = 0$ and $\varphi(1) = 1$, then we have:*

$X \prec_{FSD} Y \Leftrightarrow \forall \varphi \in \Phi, X \prec_{YAA} Y$

hence: $\mathcal{M}(L, \precsim_{FSD}) = \bigcup_{\varphi \in \Phi} \mathcal{M}(L, \precsim_{YAA})$

- *More particularly, denoting by $\Phi^C \subseteq \Phi$ the subset of transformations φ that satisfy $\varphi(p) \geq p$ (weak risk aversion), then we have that:*

$X \prec_{FSD} Y \Rightarrow \forall \varphi \in \Phi^C, X \prec_{YAA} Y$

hence: $\mathcal{M}(L, \precsim_{FSD}) \supseteq \bigcup_{\varphi \in \Phi^C} \mathcal{M}(L, \precsim_{YAA})$

Applying this Theorem, a risk averse DM, in order to make his choice, can just focus on the subset of the $\precsim_{FSD}$ or $\precsim_{SSD}$ optima.

Definition 7 ($\precsim$-Dominance Based Search Problem). *Given $\precsim \in \{\precsim_{FSD}, \precsim_{SSD}\}$, a state space graph $G = (N, A)$ under uncertainty, and two nodes $s \in N$ and $t \in N$, one wants to find one path per equivalence class in the quotient set $\overline{\mathcal{M}}(\mathcal{X}(\mathcal{P}(s,t)), \precsim)/\sim$.*

3   DBA* Algorithms

In this section, we introduce a dominance-based algorithm framework for the $\precsim$-DBS problems, and we establish its admissibility for settings (S1), (SA) and (SR).

Given a state space graph $G = (N, A)$ under uncertainty, a $\precsim$-*dominance based search algorithm* for the DBS Problem proceeds in all generality as follows. It starts from the source node s and develops partial solution paths (labels). At each iteration, it selects the most promising partial solution-path in memory - a so called *label* - and develops it, by using the transitions starting from its state-node (the last node of the partial path), into new paths extended by a new transition. When such a label reaches the goal node t, it is memorized as a complete solution-path. The iterations (select a label and develop it) stop when the set of solution-paths that were already discovered, dominate each promising label remaining in memory.

Theorem 2 (Correctness of Algorithm 1). *Given a state space graph under the uncertainty setting (S1), (SA) or (SR), and $\precsim \in \{\precsim_{FSD}, \precsim_{SSD}\}$, Algorithm 1 solves $\precsim$-DBS:*

- *by using $\precsim_{LOC} := \precsim_{FD}$ for (S1),*
- *by using $\precsim_{LOC} := \precsim_{FSD}$ for (SA),*
- *or for (SR) by using $\precsim_{LOC} := \precsim_{FD}$ or $\precsim_{LOC} := \precsim_{FSD}$ (depending on the node..).*

More precisely, each *label* $\ell = [P, n, X]$ represents a partial solution, hence a path P from the source node $s \in N$ to the state node $n \in N$, which yields the cost r.v. X. Along with X, more informations can be memorized in a label ℓ,

---

Algorithm 1: Dominance Based Search

Input: A State Space Graph $G = (N, A)$ under uncertainty, with a source node
s and a goal node t, overall dominance $\precsim$, local dominance $\precsim_{LOC}$
Output: The subset of $\precsim$-non-dominated solution paths.

1. INITIALIZATION
insert label $[\langle\rangle, s, 0]$ into OPEN and $\mathcal{L}(s)$

2. CHECK TERMINATION
if OPEN $= \emptyset$ **then return** $\mathcal{L}(t)$

3. LABEL EXPANSION
$\ell = [P, n, X^P] \longleftarrow$ pop(OPEN)
for *each node* $n' \in S(n)$, *denoting* $a = (n, n')$ **do**
    create $\ell' := [\langle P, a \rangle, n', X^P + X^a]$
    if $LocalPruning(\mathcal{L}(n'), \precsim_{LOC}, \ell')$ **then discard** ℓ'
    else if $OverallPruning(\mathcal{L}(t), \precsim, \ell')$ **then discard** ℓ'
    else if $n' = t$ **then** UpdateInsert$(\ell', \precsim, \mathcal{L}(t))$ **else**
        UpdateInsert$(\ell', \precsim_{LOC}, \mathcal{L}(n'))$
        OPEN $\leftarrow$ OPEN $\cup \ell'$
Go to *2.*

---

like the first and second order decumulative functions G_X and G_X^2, and further
heuristic evaluations. At each step, the *set of open labels*, denoted by OPEN,
contains references to labels (or partial paths) which will still possibly generate
an $\precsim$-non-dominated solution path. Since we are dealing with partial orders,
note that there are many non-dominated solutions to each node. Consequently,
on each state-node $n \in N$, we maintain a set $\mathcal{L}(n)$, the currently existing set of
labels (partial paths) ending on n. The procedure UpdateInsert$(\ell', \precsim_{LOC}, \mathcal{L}(n'))$
removes from $\mathcal{L}(n')$ all the labels $\ell'' \in \mathcal{L}(n')$ which r.v. is $\precsim_{LOC}$-dominated by
the r.v. of ℓ', and finally adds ℓ' to $\mathcal{L}(n)$.

In the following subsections, for each setting, we specify Algorithm 1. We
precise the representation and computation (using $+$) of the random variables
X. Then, we precise how Bellman's principle is used by $\precsim_{LOC}$ in LocalPruning.
Then, we use lower bounding heuristics in OverallPruning. Finally, we specify
the search priority used for the selection, development and removal of a label ℓ
in pop(OPEN).

3.1 Specification of Algorithm 1 under Setting (S1)

For a label $\ell = [P, n, X]$, the representation of the random variable X depends
on the setting, and so is the case for a transition $a \in A$ and the computation of
$X + X^a$.

In setting (S1), the probability $\mathbb{P}_\Omega$ is overall. Hence, one simply has to represent the function $X : \Omega \to \mathbb{N}$ as an m-dimensional vector $(X(\omega) \mid \omega \in \Omega)$. And
$X + Y$ is defined for $\omega \in \Omega$ by $(X + Y)(\omega) = X(\omega) + Y(\omega)$, hence it is the
addition of two m-dimensional vectors, which takes time $\Theta(m)$.

For the local pruning rules, Bellman's principle would state that "the partial sub-paths of optimal solution paths are optimal too". For the purpose of computing the set of $\precsim_{\text{FSD}}$ or $\precsim_{\text{SSD}}$-optimal solutions, it adapts as follows, in the procedures LocalPruning and UpdateInsert, when a label ℓ reaches a node n and can be compared with other paths of $\mathcal{P}(s,n)$ in memory (in $\mathcal{L}(n)$). The dominance $\precsim_{\text{SSD}}$, in addition of being partial and producing many optima, is well known to *not* satisfy Bellman's principle, as it is not additive. Indeed, one can have three random variables[7] X, Y and Z such that $X \precsim_{\text{SSD}} Y$ and not $X + Z \precsim_{\text{SSD}} Y + Z$. With $\precsim_{\text{FSD}}$, the same issue occurs under Setting (S1). Such a local pruning requires a compatible and additive dominance $\precsim_{\text{LOC}}$. The local preference $\precsim_{\text{LOC}}$ must satisfy additivity: $X \precsim_{\text{LOC}} Y \Rightarrow X + Z \precsim_{\text{LOC}} Y + Z$, and must be compatible with the global preference $\precsim$, that is: $X \precsim_{\text{LOC}} Y \Rightarrow X \precsim Y$, in order to conclude that $X + Z$ dominates $Y + Z$. Here, it is straightforward that $\precsim_{\text{FD}}$ can always be used for $\precsim_{\text{LOC}}$:

Lemma 1 (Additivity of $\precsim_{FD}$). *Let X, Y and Z be three random variables.*
$$\text{If} \quad X \precsim_{FD} Y, \quad \text{then} \quad X + Z \precsim_{FD} Y + Z.$$

In order to save computational efforts, let us now discard the partial-solutions that will obviously be dominated by a known complete solution. In a $\precsim$-dominance based search algorithm for $\precsim \in \{\precsim_{\text{FSD}}, \precsim_{\text{SSD}}\}$, we define a heuristic as a function $H : N \to \mathcal{X}$ from nodes N to random variables $\mathcal{X}$, which satisfies:

$$\forall P \in \mathcal{P}(n,t), \quad H(n) \precsim_{\text{FD}} X^P \tag{1}$$

It is used in OverallPruning($\mathcal{L}(t), \precsim, \ell$), as follows. Suppose that you generated a new label ℓ ending on the node n and yielding the r.v. X^ℓ and suppose that you already know a complete solution $\ell^* \in \mathcal{L}(t)$ that yields the r.v. Z. If $Z \precsim X^\ell + H(n)$, then one is sure that all the solutions that could be generated from ℓ would be $\precsim$-dominated by ℓ^*. Indeed, since $\precsim_{\text{FD}}$ is (always) additive, then $H(n) \precsim_{\text{FD}} X^P$ implies $X^\ell + H(n) \precsim_{\text{FD}} X^\ell + X^P$ for each path P in $\mathcal{P}(n,t)$. Consequently, by the compatibility of $\precsim \in \{\precsim_{\text{FSD}}, \precsim_{\text{SSD}}\}$ with $\precsim_{\text{FD}}$, and by the transitivity of $\precsim \in \{\precsim_{\text{FSD}}, \precsim_{\text{SSD}}\}$ we have $Z \precsim X^\ell + X^P$, hence ℓ can be discarded.

In setting (S1), $H(n)$ can be defined by satisfying Equation (1), for each $\omega \in \Omega$:

$$H(n)(\omega) = \min\{X^P(\omega) \quad : \quad P \in \mathcal{P}(n,t)\}$$

This amounts to focus on each scenario $\omega \in \Omega$ separately, and then to compute the shortest path for this scenario.

3.2   Specification of Algorithm 1 under Setting (SA)

In setting (SA), we just know that X is a random variable on $[0, M]$ (or on its *scope* $\mathcal{S}(X) = \{x \in [0, M] \mid \mathbb{P}(X = x) \neq 0\} \subseteq [0, M]$). We represent it as a

---

[7] Where X and Y are thought as the cost r.v.s of paths of $\mathcal{P}(s,n)$, and Z of a path of $\mathcal{P}(n,t)$.

map, which for each $x \in \mathcal{S}(X)$ maps to $\mathbb{P}(X = x)$. With such maps, for two r.v.s X, Y, the r.v. $Z = X + Y$ is defined for $z \in [0, M]$ by $\mathbb{P}(Z = z) = \mathbb{P}(X + Y = z) = \sum_{t=0}^{z} \mathbb{P}(X = t)\mathbb{P}(Y = z - t)$ because of the independence $X \perp Y$ assumed in setting (SA). This is easily computed in a forward manner, by operating for each $(x, y) \in \mathcal{S}(X) \times \mathcal{S}(Y)$ the addition $\mathbb{P}(Z = x + y) + = \mathbb{P}(X = x)\mathbb{P}(Y = y)$.

Concerning the local pruning with Setting (SA), the functional dominance seems intractable, as it is formulated on $m = \prod_{a \in A} m_a$ scenarios. Fortunately, the stochastic dominance $\precsim_{FSD}$ becomes possible because the r.v.s are independent in (SA). Even more pruning than $\precsim_{FD}$, $\precsim_{FSD}$ is additive and can be tested in time $O(|\mathcal{S}(X) \cup \mathcal{S}(Y)|)$.

Lemma 2 (Additivity of $\precsim_{FSD}$ under Independence). *Let X, Y and Z be three random variables such that $X \perp Z$ and $Y \perp Z$.*

$$\text{If } X \precsim_{FSD} Y, \text{ then } X + Z \precsim_{FSD} Y + Z.$$

Corollary: *In Setting (SA), one can use $\precsim_{FSD}$ as a local pruning rule.*

To bound below the r.v.s of going from node n to the goal t, with Setting (SA), the most optimistic overall scenario is when each transition realizes its lowest cost $c^a = \min\{X^a(\omega^a) \mid \omega^a \in \Omega^a\}$. We defined $H(n)$ as the deterministic variable which always costs the shortest path from n to t in the graph where each transition a costs $c^a \in \mathbb{N}$.

3.3   Specification of Algorithm 1 under Setting (SR)

With Setting (SR), assuming a bounded number of regions, for each path P, the r.v. X^P is decomposed into the regions by $X^P = \sum_{r \in R} X^{P,r}$ where $X^{P,r} = \sum_{a \in P \cap A_r} X^a$. In order to represent X^P, knowing for each region $r \in R$ the probability $\mathbb{P}_{\Omega^r}$, in the same manner as for Setting (S1), one only has to maintain the $|R|$ $|\Omega^r|$-dimensional vectors $(X^{P,r}(\omega^r) \mid \omega^r \in \Omega^r)$ associated to each region $r \in R$. Then, computing $\sum_{r \in R} X^{P,r}$ is easily added in the same manner as in setting (SA), since the regions are independent and $|R|$ is bounded.

For the local-dominances, on a state-node n, Lemmas 1 and 2 suggest to use the local dominances $\precsim_{FD}$ or $\precsim_{FSD}$, depending on the independence between the cost r.v.s of $\mathcal{P}(s, n)$ and $\mathcal{P}(n, t)$. Indeed, if the cost r.v.s of $\mathcal{P}(s, n)$ and $\mathcal{P}(n, t)$ are independent, one can apply the dominance $\precsim_{FSD}$ for LocalPruning and UpdateInsert. Otherwise, when there are dependencies, one has to use $\precsim_{FD}$ which amounts to a $| \times_{r \in R} \Omega^r|$-dimensional Pareto-dominance on vectors $(\sum_{r \in R} X^{P,r}(\omega^r) \mid \omega \in \times_{r \in R}\Omega^r)$. The next definition gives a sufficient condition on node n, to apply $\precsim_{FSD}$ as a local pruning rule.

Definition 8 (Separating Node). *Given a state space graph $G = (N, A)$ under the uncertainty setting (SR) with the regions R and the partition of transitions $A = \cup_{r \in R} A^r$, a separating node, is a node n such that:*

$$\forall P \in \mathcal{P}(s, n), \quad \forall P' \in \mathcal{P}(n, t), \quad r(P) \cap r(P') = \emptyset$$

For instance, under the Setting (SA), all nodes are separating, while under (S1), none are. On an acyclic layer graph where each region is contiguous, the nodes of a layer separating two regions are separating nodes, while the nodes inside of a region are not.

To define a heuristic $H(n)$, recall that we need to satisfy Equation (1). To fill this purpose, we define H for each $n \in N$ and each $\omega = (\omega^r \mid r \in R) \in \Omega$, as follows:

$$H(n)(\omega) = \sum_{r \in R} \min\{X^{P,r}(\omega^r) \mid P \in \mathcal{P}(n,t)\} \quad \leq \min\{X^P(\omega) \mid P \in \mathcal{P}(n,t)\}$$

Hence, computing H requires to solve $\sum_{r \in R} |\Omega^r|$ shortest path problems.

3.4 Search Priority

For the purpose of guiding the search quickly to the goal, one must define search priorities telling what labels to develop first. For a given label, the priority information must be easy to compute and cannot be perfectly informed. Here, we decide to develop the label $\ell = [P, n, X^\ell]$ that minimizes $\mathbb{E}[X^\ell + H(n)]$. However, such a search priority may fail to *early* provide complete solutions. So we slightly modified the priority to shoot greedy developments towards the goal (depth-first), once in a while. This provides a few complete solution-paths, in order to trigger `OverallPruning`.

4   Near Admissible DBA* Algorithms

In this section, we provide a near admissible generalization of Algorithm 1 for the DBS Problem. The aim is to improve the computation times of Algorithm 1 by computing a representative sample of non-dominated solutions. To this end, for $\varepsilon > 0$, we generalize the dominance relations to the following approximate $(1 + \varepsilon)$-dominance relations:

Definition 9 (Approximate Dominances). *For two random variables X, Y and an $\varepsilon > 0$, the expected-disutility (for a non-decreasing disutility function $w : [0, M] \to \mathbb{R}$), Yaari's (for a transformation φ), first order and second order ε-approximate dominances are respectively defined by:*

$$X \precsim_{EW}^{(\varepsilon)} Y \Leftrightarrow \mathbb{E}[w(X)] \leq (1 + \varepsilon)\mathbb{E}[w(Y)]$$
$$X \precsim_{YAA}^{(\varepsilon)} Y \Leftrightarrow \mathbb{E}_\varphi[X] \leq (1 + \varepsilon)\mathbb{E}_\varphi[Y]$$
$$X \precsim_{FSD}^{(\varepsilon)} Y \Leftrightarrow \forall t \in [\![0, M]\!], \;\; G_X(t) \leq (1 + \varepsilon)G_Y(t)$$
$$X \precsim_{SSD}^{(\varepsilon)} Y \Leftrightarrow \forall t \in [\![0, M]\!], \;\; G_X^2(t) \leq (1 + \varepsilon)G_Y^2(t)$$

Definition 10 ($\precsim^{(\varepsilon)}$-Covering). *For a set of random variables L given $\varepsilon > 0$, a $\precsim^{(\varepsilon)}$-covering $L^{(\varepsilon)}$ is a subset $L^{(\varepsilon)} \subseteq L$, such that:*

$$\forall X \in L, \quad \exists Y \in L^{(\varepsilon)}, \;\; : \;\; Y \precsim^{(\varepsilon)} X$$

For instance, $L \precsim^{(\varepsilon)}$-covers itself, but in order to simplify computation, we are more interested in *small* coarse coverings. That is, the approximate $\precsim$-DBS problem we address aims to compute a $\precsim^{(\varepsilon)}$-covering of what the exact problem's solution-set would be. For this purpose, we make Algorithm 1 use $\precsim^{(\varepsilon)}$ in the OverallPruning.

Interestingly, Theorem 1 generalizes to approximations: Theorems 3 and 4 show how a coarse optimization with $\precsim^{(\varepsilon)}_{SSD}$ or $\precsim^{(\varepsilon)}_{FSD}$, enables to approximately encompass the set of all the optima for $\precsim_{EW}$ or $\precsim_{YAA}$, providing a succinct set of solutions for risk-averse decision-makers.

Theorem 3. *For $\varepsilon > 0$ and random variables X and Y:*

$$X \precsim^{(\varepsilon)}_{FSD} Y \Rightarrow X \precsim^{(\varepsilon)}_{SSD} Y \tag{2}$$

$$X \precsim^{(\varepsilon)}_{FSD} Y \Rightarrow \forall w \in \mathcal{F}^{\uparrow}, \quad X \precsim^{(\varepsilon)}_{EW} Y \tag{3}$$

$$X \precsim^{(\varepsilon)}_{SSD} Y \Rightarrow \forall w \in \mathcal{C}^{\uparrow}, \quad X \precsim^{(\varepsilon)}_{EW} Y \tag{4}$$

Similarly, it follows that if $L^{(\varepsilon)}$ is an $\precsim^{(\varepsilon)}_{FSD}$-covering of L, then for each non-decreasing function $w \in \mathcal{F}^{\uparrow}$ and a corresponding optimum r.v. $X \in \mathcal{M}(L, \precsim_{EW})$, we have:

$$\exists Y \in L^{(\varepsilon)} \text{ such that: } Y \precsim^{(\varepsilon)}_{EW} X$$

and if $L^{(\varepsilon)}$ is an $\precsim^{(\varepsilon)}_{SSD}$-covering of L, then for each non-decreasing convex function $w \in \mathcal{C}^{\uparrow}$ and a corresponding optimum r.v. $X \in \mathcal{M}(L, \precsim_{EW})$, we have:

$$\exists Y \in L^{(\varepsilon)} \text{ such that: } Y \precsim^{(\varepsilon)}_{EW} X$$

Concerning Yaari's dominance, as the approximate dominance $\precsim^{(\varepsilon)}_{FSD}$ is stated on the probabilities, one must require some regularity on φ. Let $\Phi^C_\kappa \subseteq \Phi^C$ denote the set of κ-Lipschitz (weakly risk averse) probability transformations.

Theorem 4. *For $\varepsilon > 0$ and random variables X and Y:*

$$X \precsim^{(\varepsilon)}_{FSD} Y \Rightarrow \forall \varphi \in \Phi^C_\kappa, \quad X \precsim^{(\kappa\varepsilon)}_{YAA} Y \tag{5}$$

Similarly, it follows that if $L^{(\varepsilon)}$ is an $\precsim^{(\varepsilon)}_{FSD}$-covering of L, then for each regular and weakly risk-averse transformation $\varphi \in \Phi^C_\kappa$ and a corresponding optimum r.v. $X \in \mathcal{M}(L, \precsim_{YAA})$, we have:

$$\exists Y \in L^{(\varepsilon)} \text{ such that: } Y \precsim^{(\kappa\varepsilon)}_{YAA} X$$

For instance, if L^* is a 0.01 %-covering of the $\precsim_{FSD}$-optimal solutions L, then for $\varphi(p) = \sin((\pi/2)p)$, we are sure that there exists $Y \in L^*$ such that $\mathbb{E}_\varphi[.]$ is at most $1 + (0.01 \times 1.58)$ times the optimum of $\mathbb{E}_\varphi[.]$ on L, since φ is κ-Lipschitz with $\kappa = 1.58$.

5   Numerical Experiments

In order to simulate a significant branching width, we did experiments [8] on layer graphs composed of $L + 1 \in \mathbb{N}$ layers with 10 nodes-per-layer, and the source

---

[8] One can reproduce this paper's results by using the following C++ programs: https://infotomb.com/2b07w.zip.

Table 1. Computation times under Setting (S1) and Setting (SA)

	Setting (S1)						Setting (SA)					
	$\succsim_{\mathrm{FSD}}$			$\succsim_{\mathrm{SSD}}$			$\succsim_{\mathrm{FSD}}$			$\succsim_{\mathrm{SSD}}$		
L	exact	5 %	10 %	exact	5 %	10 %	exact	5 %	10 %	exact	5 %	10 %
20	7.9	7.3	6.6	3.0	2.7	2.6	9.8	7.2	6.6	6.8	6.6	6.4
25	30.6	28.0	25.2	9.4	8.6	7.9	17.5	14.5	14.0	14.2	13.6	13.3
30	102.0	92.4	82.5	27.2	24.0	22.3	27.0	24.4	23.6	23.6	23.4	22.7
35	-	-	-	73.4	64.5	59.5	39.2	36.0	35.2	35.4	34.7	33.9
40	-	-	-	179.2	156.9	146.2	55.6	51.5	49.9	49.3	48.0	47.8

Table 2. Computation times under Setting (SR)

	2 regions				3 regions				4 regions			
	$\succsim_{\mathrm{FSD}}$			$\succsim_{\mathrm{SSD}}$	$\succsim_{\mathrm{FSD}}$			$\succsim_{\mathrm{SSD}}$	$\succsim_{\mathrm{FSD}}$			$\succsim_{\mathrm{SSD}}$
L	exact	5 %	10 %	exact	exact	5 %	10 %	exact	exact	5 %	10 %	exact
10	0.7	0.6	0.5	0.3	2.2	1.7	1.4	0.9	6.8	4.9	4.1	2.9
15	6.6	5.1	4.2	2.0	44.2	30.7	24.0	11.7	-	-	-	-
20	18.4	13.9	11.2	4.2	-	-	-	45.7	-	-	-	-

and goal at the ends. (Hence, there are $10(L+1)+2$ nodes.) From each layer to the next one, there are all the 10×10 transitions. (Hence there are $100L + 20$ transitions.) For Setting (S1, $|R| = 1$) there are 3 scenarios and the costs are drawn in $\{1, \ldots, 100\}$. For Setting (S2, $R = A$), the random variable X^a of each transition is obtained by drawing 3 Dirac masses on the costs $\{1, \ldots, 100\}$. For Setting (SR), there are $|R| = 2, 3, 4$ contiguous regions, and there are 3 scenarios-per-region, with costs drawn in $\{1, \ldots, 100\}$ for each transition and each scenario. We tested Algorithm 1 for $\succsim_{\mathrm{FSD}}$ and $\succsim_{\mathrm{SSD}}$, exactly, for $\varepsilon = 5\%$, and for $\varepsilon = 10\%$. Some results are summarized in Tables 1 and 2, depicting the average cpu-times (seconds) on 100 random instances.

Observations. Under Setting (S1), the computation times grow exponentially with respect to the layer graph's length, due to the fact that the branching width is significant. Focusing on $\succsim_{\mathrm{SSD}}$ enables to improve the length of the instances solved. Under Setting (SA), the computation times seem to increase polynomially with respect to the graph's length. Under Setting (SR), Algorithm 1 enables to solve instances up to 200 nodes, for 2 to 4 regions. Computing $\succsim_{\mathrm{SSD}}$ is significantly faster. Approximations seem to just slightly improve the computation times.

6   Conclusion

In order to model the preferences of a risk averse decision-maker *with no arbitrary choice*, we applied the first and second order stochastic dominances.

For these preferences, we provided an algorithm on state space graphs with stochastic costs. Our general framework encompasses three cost uncertainty settings ranging from total dependence to total independence. Our programs scale-up to graphs with 400 nodes. While our theorems show that approximating the stochastic dominances would indeed transfer the precision to the expected utility or Yaari's criterion, the corresponding schemes mildly improve the computation times.

Prospects. We have reasons to think that a FPTAS is not possible in these settings, due to the fact that approximations ought to be both on costs and probabilities. We strongly suspect that using bidirectional search instead of unidirectional search would significantly improve the computation times.

Acknowledgements. I wish to thank Patrice Perny and Olivier Spanjaard for their interesting discussions on the problem, and the reviewers for their comments.

References

1. Battuta, I.: The Rihla of Ibn Battuta. Ibn Juzayy (1355)
2. Bonet, B., Geffner, H.: Faster heuristic search algorithms for planning with uncertainty and full feedback. In: Proceedings of IJCAI, pp. 1233–1238 (2003)
3. Fu, L., Rilett, L.R.: Expected shortest paths in dynamic and stochastic traffic networks. Transp. Res. **32**, 317–322 (1998)
4. Galand, L., Perny, P.: Search for Choquet-optimal paths under uncertainty. In: Proceedings of UAI, pp. 125–132 (2007)
5. Gao, Y.: Shortest path problem with uncertain arc lengths. Comput. Math. Appl. **62**(6), 2591–2600 (2011)
6. Hanno: The Voyage of Hanno, commander of the Carthaginians, round the parts of Libya beyond the Pillars of Heracles, which he deposited in the Temple of Kronos. A wall in the temple of Ba'al Hammon in Carathage (500 BC)
7. Hart, P.E., Nilsson, N.J., Raphael, B.: A formal basis for the heuristic determination of minimum cost paths. IEEE Trans. Syst. Cyb. **SSC–4**(2), 100–107 (1968)
8. Loui, R.P.: Optimal paths in graphs with stochastic or multidimensional weights. Commun. ACM **26**(9), 670–676 (1983)
9. Muliere, P., Scarsini, M.: A note on stochastic dominance and inequality measures. J. Econ. Theory **49**(2), 314–323 (1989)
10. Nikolova, E., Brand, M., Karger, D.R.: Optimal route planning under uncertainty. In: ICAPS, vol. 6, pp. 131–141 (2006)
11. Nilsson, N.J.: Problem-Solving Methods in Artificial Intelligence. McGraw-Hill Pub. Co., New York (1971)
12. Papadimitriou, C.H., Yannakakis, M.: Shortest paths without a map. In: Ausiello, G., Dezani-Ciancaglini, M., Rocca, S.R.D. (eds.) Automata, Languages and Programming. LNCS, vol. 372, pp. 610–620. Springer, Heidelberg (1989)
13. Pearl, J.: Heuristics: Intelligent Search Strategies for Computer Problem Solving. Addison-Wesley, Reading (1984)
14. Perny, P., Spanjaard, O., Storme, L.X.: State space search for risk-averse agents. In: IJCAI, pp. 2353–2358 (2007)
15. Polo, M.: Devisement du Monde, The Travels of Marco Polo. Rustichello da Pisa (1298)

16. Puterman, M.L.: Markov Decision Processes: Discrete Stochastic Dynamic Programming, 1st edn. Wiley, New York (1994)
17. Sauma, R.: History of Mar Yahballaha III and of Rabban Sauma. A clergyman (1281–1317)
18. Savage, L.J.: The Foundations of Statistics. Wiley, New York (1954)
19. Von Neumann, J., Morgenstern, O.: Theory of Games and Economic Behavior (60th Anniversary Commemorative Edition). Princeton University Press, Princeton (2007)
20. Wellman, M., Larson, K., Ford, M., Wurman, P.: Path planning under time-dependent uncertainty. In: Proceedings of the Eleventh Conference on Uncertainty in Artificial Intelligence, pp. 532–539 (1995)
21. Wurman, P., Wellman, M.: Optimal factory scheduling using stochastic dominance A*. In: Proceedings of the Twelfth Conference on Uncertainty in Artificial Intelligence, pp. 554–563 (1996)
22. Yaari, M.E.: The dual theory of choice under risk. Econometrica J. Econometric Soc. 55, 95–115 (1987)

On the Impact of Junction-Tree Topology on Weighted Model Counting

Batya Kenig[(⊠)] and Avigdor Gal

Technion, Israel Institute of Technology, Haifa, Israel
batya.kenig@gmail.com, avigal@ie.technion.ac.il

Abstract. We present and evaluate the power of a new framework for weighted model counting and inference in graphical models, based on exploiting the topology of the junction tree representing the formula. The proposed approach uses the junction tree topology in order to craft a reduced set of partial assignments that are guaranteed to decompose the formula. We show that taking advantage of the junction tree structure, along with existing optimization methods borrowed from the CNF-SAT domain, can translate into significant time savings for weighted model counting algorithms.

1 Introduction

Weighted Model Counting (WMC) on a propositional knowledge base in Conjunctive Normal Form (CNF) is an effective and popular approach to solve problems of exact probabilistic inference [1, 4, 21], conformant planning [10], and the study of hard combinatorial problems [11] by taking advantage of local structures. WMC is based on the model counting or #SAT problem [11], where the objective is to count the number of assignments that satisfy the propositional formula. WMC generalizes model counting by assigning a weight to each literal, and computing the weighted sum of satisfying assignments.

Model counting (and WMC) is #P-hard in general [23]. However, much work is devoted to create methods that capitalize on local structure in the form of determinism and context specific independence to enable significant speedups compared to classic inference approaches [9, 17].

In this work we continue this line of research and propose a novel approach for performing WMC that is based on message passing in junction trees. We observe that the topology of a formula's junction tree reveals structure that can be utilized for enhancing the performance of WMC. The algorithm we propose in this work generates compact factors that contain a small set of mutual exclusive and exhaustive partial assignments that are guaranteed to decompose the formula.

We evaluate the proposed approach on three benchmarks, comparing it to c2d [6], a leading compiler for WMC. The empirical analysis leads to interesting observations about the pros and cons of each of the methods.

The rest of the paper is organized as follows. Junction trees are introduced in Sect. 2 followed by the introduction of CNF-trees and their role in modeling

© Springer International Publishing Switzerland 2015
C. Beierle and A. Dekhtyar (Eds.): SUM 2015, LNAI 9310, pp. 83–98, 2015.
DOI: 10.1007/978-3-319-23540-0_6

the underlying conditional independences between formula variables (Sect. 3). Next, we outline the main idea of the paper, where formula decomposition is performed by partial assignments (Sect. 4). We show how to generate the reduced set of partial assignments (modeled as tree-CPTs) in Sect. 5. Section 6 presents the empirical evaluation. We conclude with a discussion of related work (Sect. 7) and concluding remarks (Sect. 8).

2 Background: The Junction Tree Algorithm

In what follows, we denote variables in upper case letters (*e.g.*, X) and their instantiations in lower case (*e.g.*, x). Sets of variables are denoted using bold upper case letters (*e.g.*, $\mathbf{X}$) and their instantiations in bold lower case letters (*e.g.*, $\mathbf{x}$).

A Probabilistic Graphical Model (PGM) is a graph $G(V, E)$ in which nodes represent random variables $\mathbf{X} = \{X_i : i \in V\}$, and edges represent direct dependencies between them. The graphical model contains a set of discrete functions $\mathbf{F}$, termed *factors*, that are defined over a subset of its variables. Factors are typically represented as tables, indexed by variable instantiations. Formally, a factor is a function $F(\mathbf{Y}) : \mathbf{y} \to [0, 1]$ where $\mathbf{y}$ is an instantiation of $\mathbf{Y}$. The probability distribution defined by the graphical model is $\Pr(\mathbf{X}) = \frac{1}{\mathcal{Z}} \prod_{F_i \in \mathbf{F}} F_i(\mathbf{X}_i)$ where $\mathbf{X}_i \subseteq \mathbf{X}$, and $\mathcal{Z}$, termed *partition function*, normalizes the probability to sum to one.

One of the prominent methods for performing exact probabilistic inference in graphical models is the Junction Tree algorithm [12,17]. Let $G(\mathbf{X}, E)$ be a PGM. A *Junction Tree* for G is a tree $T(\mathbf{C})$, defined over a set of nodes $\mathbf{C}$ that satisfy the following properties:

1. Each node $C_i \in \mathbf{C}$ is associated with a set of variables $\mathbf{Y}_i \subseteq \mathbf{X}$ from the PGM and a factor $G_i(\mathbf{Y}_i)$ (not to be confused with the PGM factors denoted F_i).
2. For each factor $F_k(\mathbf{X}_k)$ in the PGM, there exists a tree node $C_i \in \mathbf{C}$ such that $\mathbf{X}_k \subseteq \mathbf{Y}_i$.
3. If nodes $C_i, C_j \in \mathbf{C}$ are both associated with a variable $X \in \mathbf{X}$, then every node on the path connecting them in T is also associated with X.

The edges of the junction tree are labeled with the intersection of their endpoints. A separator, $\mathbf{S}_{i,j}$, connects nodes C_i and C_j and is referred to as a *separator node*.

Inference in junction trees is performed by passing messages between adjacent clique nodes. Evidence, $\mathbf{E} = \mathbf{e}$ is materialized by eliminating inconsistent factor entries. The message passing is carried out in two phases, inward - from the leaves towards the root, and outward - from the root towards the leaves. A node C_i sends a message to its neighbor, C_j, only after it has received messages from the rest of its neighbors $Nbr_i \setminus C_j$. The message $\mu_{i \to j}(\mathbf{S}_{ij})$ from node C_i to C_j is a tabular factor defined over their intersection, $\mathbf{S}_{ij} = \mathbf{Y}_i \cap \mathbf{Y}_j$, as follows: $\mu_{i \to j}(\mathbf{S}_{ij}) = \sum_{\mathbf{Y}_i \setminus \mathbf{S}_{ij}} G_i \prod_{k \in Nbr_i \setminus C_j} \mu_{k \to i}$. Once message propagation completes, each tree-node factor holds the marginal distribution *i.e.*, $G_i(\mathbf{Y}_i) = \Pr(\mathbf{Y}_i, \mathbf{e})$.

The *width* of a junction tree is the size of its largest node minus one. The treewidth $tw(G)$ of a graph G is the minimum width among all possible junction trees for G. In general, minimizing the graph width is known to be NP-complete. Since the junction-tree algorithm relies on tabular factors for performing the marginalization operation required for message-passing, the runtime of the algorithm depends, exponentially, on its width. Therefore, bounded width implies tractability in graphical models.

3   CNF-Trees: Junction Trees for CNFs

In this section we introduce common notation and define CNF-trees, which are specialized junction-trees for Boolean formulas in CNF. The proposed algorithm, described in Sect. 4.1, operates over this structure.

A *literal l* of a binary variable X is either a variable or its negation, which are denoted by $x, \bar{x}$, respectively. The variable corresponding to a literal l is denoted by var(l). Each literal, l, is associated with a weight $p_l \in [0, 1]$. An assignment is a function $\gamma : \mathbf{V} \rightarrow \{0, 1\}$ and will be denoted by its set literals $\gamma = \{l_1, l_2, \ldots, l_k\}$. An assignment's weight is defined as the product of its literal weights. The *projection* of an assignment γ over a subset of its vars $\mathbf{Y} \subseteq$ var(γ) is denoted $\gamma|\mathbf{Y}$. For example, given the assignment $\gamma = \{x_1, \bar{x}_2, x_3\}$, then $\gamma|\{X_2, X_3\} = \{\bar{x}_2, x_3\}$.

A Boolean formula f over variables $\mathbf{X}$ maps each instantiation $\mathbf{x}$ to either *true* or *false*. $f(\mathbf{X})$ is in Conjunctive Normal Form (CNF), constructed from a conjunction of *clauses*, each a disjunction of literals. We denote by $\phi_1, \phi_2, \ldots, \phi_n$ the set of unique clauses in f, where every ϕ_i represents a set of literals. The variables in a clause ϕ_i are denoted var(ϕ_i), and the clauses of f that contain a literal l are denoted clauses(l). We assume that the formula f is simplified, meaning, for every pair of clauses $\phi_i, \phi_j \in f$, $\phi_i \not\subseteq \phi_j$. Conditioning a CNF formula f on literal l, denoted $f|l$, consists of removing the literal $\bar{l}$ from all clauses, and dropping the clauses that contain l. Conditioning a formula on an assignment, or a set of literals $\gamma = \{l_1, l_2, \ldots, l_k\}$, denoted $f|\gamma$, amounts to conditioning it on every literal $l \in \gamma$. We say that an assignment γ is *consistent* if $f|\gamma \neq 0$. We say that a variable X *affects* the formula's outcome if $f|x \neq f|\bar{x}$. We denote by var(f) the set of variables that affect the formula. A pair of formulae f_1, f_2 are disjoint if var(f_1) $\cap$ var(f_2) $= \emptyset$. The *weighted model count* or probability that f is satisfied is denoted by $\Pr(f)$ and the two terms may be used interchangeably.

Let $G_f(\mathbf{X}, E)$ denote the primal graph of $f(\mathbf{X})$, where nodes represent variables and there is an edge between pairs of variables that belong to a common clause.

Definition 1 (CNF-tree). *Let $f(\mathbf{X})$ be a Boolean formula in CNF with primal graph $G_f(\mathbf{X}, E)$. A CNF-tree for f is a rooted junction tree, $T_r(\mathbf{C})$, for $G_f(\mathbf{X}, E)$*

where each clause $\phi_i \in f$ is represented as a leaf node with factor $F_{\phi_i}(\boldsymbol{y})$, $\boldsymbol{Y} \subseteq$ var(ϕ_i):

$$F_{\phi_i}(\boldsymbol{y}) = \begin{cases} 0 & \text{if } \phi_i|\boldsymbol{y} = 0 \\ 1 & \text{if } \phi_i|\boldsymbol{y} = 1 \\ 1 - \prod_{l \in \phi_i|\boldsymbol{y}} \Pr(\bar{l}) & \text{otherwise} \end{cases} \qquad (1)$$

According to the junction tree properties, each clause, $\phi_i \in f$, is associated with a node $C_i \in \mathbf{C}$ such that var(ϕ_i) $\subseteq \mathbf{X}_i$. This node-clause relationship is reflected in the tree by attaching a leaf, representing the clause, to its associated tree-node. For example, consider the CNF formula, its junction and CNF-trees in Fig. 1. The shaded leaf nodes represent clauses.

Configuring the leaf-node factors to return the probability that their respective clause is satisfied is equivalent to introducing evidence which prohibits assignments that falsify the formula. Thereby, the weighted model count of f can be performed by message-passing on the CNF-tree.

In the general setting, a separator set $\mathbf{S}_{i,j} = \mathbf{X}_i \cap \mathbf{X}_j$, between junction tree nodes C_i, C_j, enables inducing independence between variables on different sides of the edge only when *all* of the variables in $\mathbf{S}_{i,j}$ were assigned a value [8]. We observe, however, that in CNF-trees this requirement may be too strict as illustrated in Example 1.

Example 1. Consider the CNF-tree in Fig. 1. The *partial* assignment $\gamma_1 = \{x_1\}$ renders the disjoint variable-sets on the two sides of the edge, (C_0, C_1) (marked) independent, even though variables X_2 and X_4 remain unassigned. The reason for this is that given x_1, the original formula is reduced to $f|x_1 = \underbrace{\phi_6}_{f_1} \underbrace{\phi_5 \phi_7}_{f_2}$.

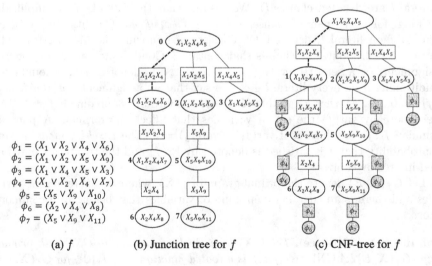

$$\phi_1 = (X_1 \lor X_2 \lor X_4 \lor X_6)$$
$$\phi_2 = (X_1 \lor X_2 \lor X_5 \lor X_9)$$
$$\phi_3 = (X_1 \lor X_4 \lor X_5 \lor X_3)$$
$$\phi_4 = (X_1 \lor X_2 \lor X_4 \lor X_7)$$
$$\phi_5 = (X_5 \lor X_9 \lor X_{10})$$
$$\phi_6 = (X_2 \lor X_4 \lor X_8)$$
$$\phi_7 = (X_5 \lor X_9 \lor X_{11})$$

(a) f (b) Junction tree for f (c) CNF-tree for f

Fig. 1. A formula, $f = \wedge_{i=1}^{7} \phi_i$, and its corresponding junction and CNF trees

Variables X_3, X_6, and X_7 become irrelevant to f's outcome following the partial assignment x_1, and can be disregarded. The variables that belong to the disjoint components in the reduced formula $f|x_1$, namely $\text{var}(f_1)$ and $\text{var}(f_2)$, are conditionally independent given x_1.

Example 1 motivates the search of a set of partial assignments to the factors of a CNF-tree that will render their subtrees independent. Partial, rather than complete assignments, may reduce factor sizes, enabling more efficient inference.

4   Decomposition by Partial Assignments

The next two sections lay out the main contribution of the paper. We first explain how the CNF-tree structure can be utilized for generating tree-Conditional-Probability-Tables (tree-CPTs) [2] consisting of a small set of mutual exclusive and exhaustive partial assignments, which are guaranteed to decompose a formula. We then define tree-CPT cardinality and suggest optimizations for size reduction. Tree-CPTs, introduced in [2], is a representation which captures Context-Specific-Independence which can be exploited for probabilistic inference. We adopt this structure in order to represent partial assignments of CNF-tree-node members, but give it a different semantics.

Let $T_r(\mathsf{C})$ be a CNF-tree rooted at node r. For each tree node $C_i \in \mathsf{C}$, $F_i(\mathbf{X}_i)$ is the factor associated with this node, T_i is the subtree rooted at C_i, and f_i is the subformula induced by the clauses in this subtree. For example, the formulae represented by subtrees T_1, T_2 rooted at nodes C_1, C_2, respectively, in Fig. 1c are $f_1 = \phi_1 \phi_4 \phi_6$, and $f_2 = \phi_2 \phi_5 \phi_7$. The children of tree-node C_i are denoted ch_i.

Definition 2. *Let $T_r(\mathsf{C})$ be a CNF-tree rooted at node r and $C_i \in \mathsf{C}$ a node in T_r with $\text{ch}_i = \{C_1, C_2, \ldots, C_m\}$. Let $\mathbf{Y} = \mathbf{y}$ be a partial assignment to $\mathbf{X}_i$. $\mathbf{y}$ is called a* valid *(partial) assignment to $\mathbf{X}_i$ if the following two conditions are satisfied:*

1. *$\mathbf{Y} = \mathbf{y}$ is consistent ($f_i|\mathbf{y} \neq 0$)*
2. *$f_i|\mathbf{y}$ is decomposed to sub-formulas $f_1|\mathbf{y}, f_2|\mathbf{y}, \ldots, f_m|\mathbf{y}$, which are pairwise disjoint.*

Our goal is to generate the smallest factor for each CNF-tree node. Namely, per each node, we would like to identify the smallest set of mutual exclusive and exhaustive *valid* partial assignments (Definition 2). The factors will be represented by tree-CPTs [2] where non-terminal vertices represent variables and terminal vertices correspond to the assignment defined by the path from the root. The variable corresponding to a vertex v in the tree-CPT is denoted $\text{var}(v)$, its parent $p(v)$, and its right and left children corresponding to assignment $\text{var}(v) = 1/0$ as v^r/v^l, respectively. The set of assignments represented by terminal nodes in F_i will be denoted γ_i, their cardinality $k_i = |\gamma_i|$, and for each $\gamma \in \gamma_i$, the assignment's marginal probability will be denoted $\Pr(\gamma)$. For each vertex v in the tree-CPT, we denote the path from the root to v, and the assignment it dictates, by P_v. The assignment P_v will be referred to as v's *context*. We say that literal $l \in P_v$ if $l|P_v = 1$.

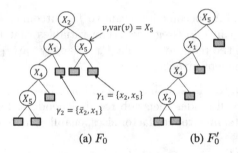

Fig. 2. Two possible tree-CPTs for root-node C_0 ($\mathbf{X}_0 = \{X_1, X_2, X_4, X_5\}$) of the junction tree in Fig. 1c.

Example 2. An example of two possible tree-CPTs for root-node C_0 in the CNF-tree of Fig. 1c appear in Fig. 2. Note the terminal vertex in Fig. 2a that represents assignment $\gamma_2 = \{\bar{x}_2, x_1\}$. The partial assignment γ_2 induces subformulae $f_1|\gamma_2 = \phi_6$, $f_2|\gamma_2 = \phi_5\phi_7$, and $f_3|\gamma_2 = \emptyset$, which correspond to subtrees T_1, T_2, T_3, respectively. Given assignment γ_2, these subformulae are consistent and pairwise disjoint, *e.g.*, $\text{var}(f_1|\gamma_2) \cap \text{var}(f_2|\gamma_2) = \emptyset$, thus the partial assignment γ_2 is valid.

4.1 Message-Passing in CNF-Trees

The procedure for performing WMC over CNF-trees is presented in Algorithm 1, taking a CNF-tree T_r, which represents f_r, and a partial (possibly empty) assignment e, and returning the probability that $f_r|e$ is satisfied. The algorithm avoids repeated computation of equivalent CNFs using a cache whose key represents the CNF. The function $nextValid$ (Line 6) retrieves the next valid assignment to process. We detail the generation of valid assignments in Sect. 5.

Algorithm 1. $\text{MP}(T_r, e)$, returns $\Pr(f_r|e)$

---

1 **if** *r is a leaf-node* **then**
2 $\quad$ **return** $F_r(e)$ // By Eq. 1
3 **if** $cache(f_r|e) \neq nil$ **then**
4 $\quad$ **return** $cache(f_r|e)$
5 $\Pr(f_r|e) \leftarrow 0.0$ // init the return value
6 **while** $\gamma \leftarrow nextValid(\gamma, f_r|e, \mathbf{X}_i \setminus \text{var}(e)) \neq nil$ **do**
7 $\quad$ $\Pr(f_r|e\gamma) \leftarrow \prod_{l \in \gamma} p_l$ // assignment weight
8 $\quad$ **foreach** *node* $n \in \text{ch}_r$ **do**
9 $\quad\quad$ $\Pr(f_n|e\gamma) \leftarrow \text{MP}(T_n, e\gamma)$ // recurse
10 $\quad\quad$ $\Pr(f_r|e\gamma) \leftarrow \Pr(f_r|e\gamma) \cdot \Pr(f_n|e\gamma)$// Thm. 1
11 $\quad$ $\Pr(f_r|e) \leftarrow \Pr(f_r|e) + \Pr(f_r|e\gamma)$ // Thm. 1
12 $cache(f_r|e) \leftarrow \Pr(f_r|e)$
13 **return** $\Pr(f_r|e)$

Theorem 1 establishes the soundness of the algorithm. Its proof is inductive and follows from the validity (Definition 2) of the tree-CPT assignments. Due to space constraints proofs are omitted.

Theorem 1. *Let T_r be a CNF-tree representing CNF f_r. The call MP(T_r, e) returns $\Pr(f_r|e)$.*

5   Generating Small Tree-CPTs

Algorithm 1 motivates the search for small tree-CPTs. Each tree-CPT internal vertex induces an instantiation of the variable it represents. Therefore, we begin by observing the conditions that forgo the requirement to instantiate a variable.

Definition 3 (safe variable). *Let C_i be a CNF-tree node with arguments $\mathbf{X}_i$, and let γ be an assignment. A variable $X \in \mathbf{X}_i$ is called* safe *if there is at most a single node, $C_j \in \mathrm{ch}_i$, such that $X \in \mathrm{var}(f_j|\gamma)$. The set of variables in $\mathbf{X}_i$ that are safe under assignment γ are denoted $\mathrm{safe}_i(\gamma)$.*

We first note that by Definition 3, instantiated variables are safe because they cannot appear in any induced sub-formula associated with a node's subtrees.

To relate safe variables to compact CPT-trees let $F_i(\mathbf{X}_i)$ be C_i's CPT-tree, and let vertex $v \in F_i$ have context P_v. If $X \in \mathrm{safe}_i(P_v)$, and P_v is consistent (recall, $f_i|P_v \neq 0$) then, by Definition 3, there is a valid assignment that contains P_v, but not X. Furthermore, a context P_v for which all the node arguments are safe, is a valid assignment by definition.

The variation between tree-CPTs, and hence the efficiency of the WMC algorithm, stems from the different ordering of variable instantiation (see Fig. 2). Definition 4 gives the ordering constraints between the arguments of a node C_i, which will be used to derive its tree-CPT.

Definition 4 (Conditioning graph). *The* conditioning graph *of a CNF-tree-node C_i is a directed graph $D_i(L_i, E_i)$, where $L_i = \{x, \bar{x} : X \in \mathbf{X}_i\}$ is the set of literals of $\mathbf{X}_i$. There is an edge $(l_1, l_2) \in E_i$ if $\exists \phi_1, \phi_2 \in f_i$ such that*

1. $\phi_1, \phi_2 \in \mathrm{clauses}_i(\mathrm{var}(l_1)) \setminus \mathrm{clauses}_i(l_2)$
2. Node C_i is their Least Common Ancestor (LCA) in the CNF-tree.

The compliment of D_i is denoted $\bar{D}_i$.

The intuition behind the conditioning graph becomes apparent when looking at absent edges, or at the conditioning-graph's compliment, $\bar{D}_i$. If, for example, $x_1 \to x_2 \in \bar{D}_i$, that is, $x_1 \to x_2 \notin D_i$, then, by Definition 4, any two clauses that contain X_1 (*i.e.*, x_1 or $\bar{x}_1$), but not the literal x_2, are confined to the same subtree of node C_i. Practically, this means that given an assignment in which $x_2 = 1$, the set of unsatisfied clauses containing variable X_1 are confined to (at most) a single subformula represented by one of C_i's subtrees. In other words, variable X_1 is *safe* (Definition 3) for any assignment where x_2 is set. Essentially, given two literals, l_1 and l_2, the conditioning graph answers the following question:

"Given $l_2 = 1$ is var(l_1) safe ?". If $l_1 \to l_2 \in \bar{D}_i$ then the answer is affirmative, otherwise negative.

We also note the following about the conditioning graph and its compliment. First, for each variable $X \in \mathbf{X}_i$, out$_i(x)$ = out$_i(\bar{x})$ because the out-edges are determined by the existence of a variable (*i.e.*, literal x or $\bar{x}$) in the clauses of Definition 4. Also, since assigning a variable makes it safe, then neither the conditioning graph nor its compliment contain edges from a literal to its compliment or self-loops.

Example 3. Figure 3 presents the conditioning graph of the root C_0 of the CNF-tree in Fig. 1. The edge $x_5 \to x_2$ ($\bar{x}_5 \to x_2$) is due to clauses ϕ_7 and ϕ_3. Both clauses contain x_5 but not x_2, and their least common ancestor in the CNF-tree is C_0.

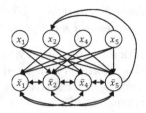

Fig. 3. The conditioning graph, D_0 of node C_0 of the CNF-tree in Fig. 1c

We now characterize a subclass of safe variables in context P_v (Definition 3), denoted $\mathbf{Y}_v$, using the conditioning graph.

Definition 5. *Let C_i denote a node in a rooted CNF-tree with conditioning graph D_i, and tree-CPT $F_i(\mathbf{X}_i)$. Let v be a vertex in the tree-CPT F_i, with context P_v, then:*

$$\mathbf{Y}_v = \{X \in \mathbf{X}_i : \exists l \in P_v : x \to l \in \bar{D}_i\}$$

Theorem 2. *Let v be a vertex in F_i with context P_v. Then $\mathbf{Y}_v \subseteq \text{safe}_i(v)$.*

Theorem 2 characterizes the set of arguments whose clauses are confined to a single subformula induced by the subtrees of node C_i, given the assignment dictated by P_v. By Theorem 2, X will not require instantiation in order to extend P_v to a valid assignment. The importance of the variable-set $\mathbf{Y}_v$ stems from the fact that it can be *statically* identified by considering only the structure of the CNF-tree.

The complement of the variable set $\mathbf{Y}_v$, described in Definition 5, has incoming edges to all of the literals set by the assignment P_v. We define this set of variables, $\mathbf{Z}_v$, recursively as follows.

Definition 6. *Let v be a vertex in CPT-tree F_i, with parent $p(v)$, and variable* $\mathrm{var}(p(v))$. *The variable-set $\mathbf{Z}_v$ is:*

$$\mathbf{Z}_v = \begin{cases} \mathbf{X}_i & \text{if } P_v = \emptyset \text{ or } v \text{ is the root} \\ \mathbf{Z}_{p(v)} \cap \mathrm{in}_i(\mathrm{var}(p(v))) & \text{if } \mathrm{var}(p(v)) = 1 \\ \mathbf{Z}_{p(v)} \cap \mathrm{in}_i(\overline{\mathrm{var}(p(v))}) & \text{if } \mathrm{var}(p(v)) = 0 \end{cases}$$

According to Definitions 5 and 6, we have that $\mathbf{Y}_v \cap \mathbf{Z}_v = \emptyset$, and for every vertex $v \in F_i$, $\mathbf{X}_i = \mathbf{Y}_v \cup \mathbf{Z}_v \cup \mathrm{var}(P_v)$.

Example 4. Let v refer to the right child of X_2 in the tree-CPT of Fig. 2a ($\mathrm{var}(v) = X_5$). Then $\mathbf{Z}_v = \mathbf{Z}_{p(v)} \cap \mathrm{in}_0(x_2) = \{x_1, x_2, x_4, x_5\} \cap \{x_5\} = \{x_5\}$. In this case variable X_5 requires instantiation in context x_2 in order to extend the partial assignment $\{x_2\}$ to one that is valid.

5.1 Tree-CPT Cardinality

To express the size of a node's tree-CPT $F_i(\mathbf{X}_i)$ with children ch_i and conditioning graph D_i we denote (with a slight abuse of notation) the variable associated with each tree-CPT vertex v, $\mathrm{var}(v) = V$, and its literals v and $\bar{v}$ respectively. $T : \mathbf{Z}_v \to \mathbb{N}$ maps $\mathbf{Z}_v$ to the number of valid assignments in the subtree rooted at vertex v:

$$T(\mathbf{Z}_v) = \begin{cases} 1 & \text{if } \mathbf{Z}_v = \emptyset \\ T(\mathbf{Z}_v \cap \mathrm{in}_i(v)) + T(\mathbf{Z}_v \cap \mathrm{in}_i(\bar{v})) & \text{o.w} \end{cases} \tag{2}$$

This expression considers only the variable set $\mathbf{Z}_v$ because by Theorem 2, the members of the compliment set, $\mathbf{Y}_v$, are safe (Definition 3), and thus do not require instantiation.

When $\mathbf{Z}_v = \emptyset$ then no variable requires instantiation, and a single terminal node can represent the valid assignment. Otherwise, the total size of the tree-CPT rooted at vertex v is determined by the size of the tree-CPTs rooted at its left and right children v^l, v^r respectively. By Definition 6, $\mathbf{Z}_{v^l} = \mathbf{Z}_v \cap \mathrm{in}_i(\bar{v})$ and $\mathbf{Z}_{v^r} = \mathbf{Z}_v \cap \mathrm{in}_i(v)$. It is easy to see that repeated expansion of Eq. 2 can lead to the known exponential bound for the number of valid assignments in the tree-CPT, whenever for each tree-CPT vertex v:

$$\mathbf{Z}_v \cap \mathrm{in}_i(v) = \mathbf{Z}_v \cap \mathrm{in}_i(\bar{v}) = \mathbf{X}_i \setminus \{\mathrm{var}(P_v) \cup \{V\}\}$$

That is, in the worst case the complexity of Algorithm 1 is exponential in the size of the largest node in the CNF-tree, or the width of the formula's primal graph. In Sect. 6 we show that despite this worse-case behavior, and with the assistance of the optimization discussed next, Algorithm 1 performs well on known benchmarks.

5.2   Optimizations for Generating Small Tree-CPTs

We broadly address two types of optimizations that we apply to Algorithm 1, aimed at minimizing the size of the tree-CPTs. The first is a heuristic that selects the next node-member to assign, using the analysis in Sect. 5.1. The second is Unit Propagation and conflict directed clause learning, adapted from the CNF-SAT domain.

A tree-CPT can be viewed as a binary decision tree where terminal nodes identify valid assignments. Constructing an optimal decision tree, one with fewest nodes, is generally an NP-hard problem [15]. We apply a heuristic strategy that, at each stage, selects the variable that minimizes the cardinality of the variable-set that is common to its left and right tree-CPTs. Formally:

$$\arg\min_{v} \left(|\mathbf{Z}_v \cap \text{in}_i(v)) \cap (\mathbf{Z}_v \cap \text{in}_i(\bar{v})| \right)$$

Ties may be broken by selecting the variable that further minimizes the set of unsafe variables at either of its sub trees. That is:

$$\arg\min_{v} \left[\max \left(|\text{in}_i(v) \cap \mathbf{Z}_v|, |\text{in}_i(\bar{v}) \cap \mathbf{Z}_v| \right) \right]$$

Unit Propagation (UP) refers to the process of iteratively assigning literals of unit clauses until none are left. It is part of both DPLL-based model counters [21,22] and compilers that generate d-DNNF circuits [6,19]. Specifically, if $\phi = \{l\}$ is a unit clause of a CNF formula f, then the UP process deletes all occurrences of $\bar{l}$, and all clauses containing l, which are now satisfied. Each valid assignment generated by Algorithm 1 is extended by applying unit propagation. That is, the valid assignments are guaranteed to decompose the formula and ensure that no unit clauses are present. Unit propagation is applied after every variable assignment during the tree-CPT construction.

If UP results in a conflict, then a new clause is learned by applying the first Unique Implication Point schema [18]. The newly learned clause is added to the subformula being processed, f_i. We note that the learned clauses are used only during unit propagation, in order to detect conflicts early. They are not represented as leaves in the CNF-tree, and are not considered during caching. Also, since different nodes represent different subformulas, then each CNF-tree node holds its own local set of conflict clauses.

6   Empirical Evaluation

We evaluate the proposed approach on a set of benchmark networks from the UAI probabilistic inference challenge.[1] We compare our results with the C2D compiler [6], part of the Ace system.[2] Besides evaluating the efficiency of the proposed approach, we discuss the properties of networks that benefit from it. The experimental setup is given in Sect. 6.1, followed by results and analysis in Sect. 6.2. We implemented our algorithm in C++[3] and carried out the experiments

---

[1] Available online at http://www.cs.huji.ac.il/project/PASCAL/showNet.php.

[2] Available online at http://reasoning.cs.ucla.edu/ace/.

[3] Code is available at: https://github.com/batyak/PROSaiCO/.

on a 2.33 GHz quad-core `AMD64` with 8 GB of RAM running `CentOS` Linux 6.6. Individual runs were limited to a 2000-s time-out.

6.1   Overview and Methodology

The compilation process of `C2D` is guided by a binary tree, termed *dtree* whose leaves are associated with the clauses of f. The dtree determines the instantiation order materialized in the d-DNNF [5,6]. Figure 4 depicts a dtree of the CNF of Fig. 1a. Each internal node, T, is associated with a variable-set, called *separator* [11], which is the variable-set common to the left and right subtrees of the node. Once these variables have been assigned, the formulae represented by the two subtrees become disjoint. Darwiche [6] observed that there is no need to set all variables in the dtree-node T in order to decompose the formula. That is, after setting a subset of the dtree-node variables, enough clauses may become satisfied such that the rest of the T's variables are no longer shared between the formulas represented by its left and right children. For this reason the `C2D` compiler *recomputes* the separator for T each time a variable of T is decided [6]. Within each separator, the `C2D` compiler chooses the variable that appears in the largest number of unsatisfied clauses.

  Darwiche shows that the clusters of a dtree satisfy the junction-tree property ([8], Theorem 9.10). That is, the maximal clusters of a dtree can be connected such that they constitute a junction-tree. Once the junction-tree is created, we can attach the clauses as leaf nodes to obtain the CNF-tree. Applying our algorithm to a junction tree corresponding to the dtree generated by the `C2D` compiler, enables comparing the two approaches on an even ground, although our proposed approach is not limited to binary junction trees. Furthermore, we can gain insight into the types of networks that benefit from our proposed approach, which requires more analysis at each junction tree node.

  There is a wide range of settings for `C2D`, and in particular for generating the dtree. We experimented with the default provided by `Ace`, termed `dtBnMinfill`. This option instructs the program to generate a dtree for the original Bayesian

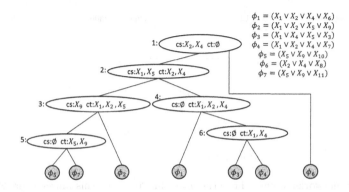

Fig. 4. CNF and corresponding dtree.

network using the *minfill* heuristic [16], which is widely known for generating small induced width elimination orders. Each leaf in the resulting dtree corresponds to one of the network CPTs. Then, each leaf is replaced with the dtree that represents the corresponding CPT.

6.2 Experimental Results

We report the results obtained for Grid, Promedas, and Segmentation networks. The evaluation is presented using scatter plots. Each instance is represented as a point in the chart whose x, y coordinates represent runtime, in seconds, of WMC_JT and C2D, respectively. Points above the $y = x$ line represent problem instances where WMC_JT performs better. Axes are log-scale. We also mark a linear trendline (which translates to exponential trendline due to the log-scale) with it R^2 value.

Grid Networks. The nodes in random grid networks represent binary variables that are arranged in an $N \times N$ square. Each CPT is generated uniformly at random. The fraction of the nodes that are assigned deterministic CPTs, having only 0 and 1 probability entries, is captured by the *deterministic ratio*.

Figure 5 shows the results over the grid networks for three deterministic ratios. All networks solved by at least one of the solvers are present. On the grids with a 50 % deterministic ratio, WMC_JT outperformed C2D on 39 out of the 60 instances. On one instance, C2D did not complete within the designated timeout. On the grids with a 75 % deterministic ratio, WMC_JT outperformed C2D on 43 out of the 110 instances, while C2D outperformed WMC_JT on 63 instances. On

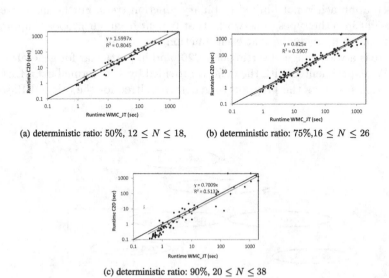

(a) deterministic ratio: 50%, $12 \leq N \leq 18$,   (b) deterministic ratio: 75%, $16 \leq N \leq 26$

(c) deterministic ratio: 90%, $20 \leq N \leq 38$

Fig. 5. Grid networks: points above the $y = x$ line represent instances where WMC_JT is better.

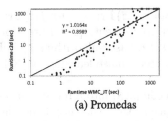
(a) Promedas

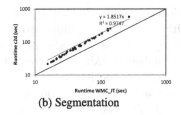
(b) Segmentation

Fig. 6. Points above the $y = x$ line represent instances where WMC_JT is better.

the 90 % benchmark, WMC_JT outperformed C2D on only 13 of the 100 instances. Therefore, we can conclude that in general, with less determinism WMC_JT tends to outperform C2D. When the underlying network contains a large percentage of deterministic factors (90 %), a small fraction of the node members determine the rest through UP and the time invested by WMC_JT in the generation of conditioning graphs and small tree-CPTs may be too costly.

Promedas and Segmentation. Promedas stands for "PRObabilistic MEdical Diagnostic Advisory System". The Promedas benchmark contains 238 Markov networks, consisting of binary variables, which were converted from layered noisy-or Bayesian networks that represent real-world medical diagnosis cases. The networks' treewidth is up to 60, and many of them are considered too difficult for exact algorithms[4]. Results are plotted in Fig. 6a. Out of the 238 networks, WMC_JT processed 102 networks within the designated timeout, while C2D completed 89. Out of the 89 networks processed by both algorithms, C2D outperformed WMC_JT on 65, while WMC_JT outperformed C2D on 24. On the Segmentation benchmark, Fig. 6b, WMC_JT outperformed C2D on all 50 instances.

Overall, we observe that WMC_JT tends to outperform C2D over instances with a low to medium percentage of deterministic factors. Furthermore, the results and trendlines of the Promedas and Segmentation benchmarks (Fig. 6) suggest that the relative performance of WMC_JT improves on "harder" instances, those which require more CPU cycles for both algorithms. That said, we note that the algorithm execution time is determined by many variables. These include the d-tree used, its orientation and the variable order generated by the heuristic described in Sect. 5.2.

7   Related Work

We position our work along two dimensions: *exhaustive search* vs. *knowledge compilation*, and *dynamic* vs. *static decomposition*.

DPLL-based algorithms exhaustively explore the search-tree for a formula, while pruning unsatisfiable branches. At the heart of the search-based techniques for weighted model counting are two operations, a Shannon expansion

---

[4] http://graphmod.ics.uci.edu/uai08/Evaluation/Report/Benchmarks.

on a decision variable Z, that is, $\Pr(f) = \Pr(f|\bar{z})\Pr(\bar{z}) + \Pr(f|z)\Pr(z)$ and the partitioning of the formula into disjoint components [11]. Extensions that tremendously improve the performance of DPLL-based algorithms include non-chronological backtracking, [1], conflict directed clause-learning (CDCL), and variable branching heuristics [21].

In *knowledge compilation*, the formula is compiled into a representation that enables computing the probability of evidence in time that is polynomial in its size [4,19]. These representations are based on Negation Normal Form (NNF) circuits [7] where internal nodes represent either conjunctions or disjunctions and leaf nodes represent constants or literals. Circuits that enable tractable model counting, termed deterministic-DNNF (d-DNNF), must be decomposable and deterministic. The former requires children of conjunction nodes to share no variables, and the latter requires children of disjunction nodes to be mutual exclusive. State-of-the-art model counting compilers, C2D [6] and DSharp [19], generate Decision-DNNF circuits that ensure determinism as follows. Each *or* node, n, is associated with a variable X such that n's right and left children represent subformulas $f_n|x$, and $f_n|\bar{x}$ respectively. This method of ensuring determinism is closely related to the instantiation step of DPLL-based algorithms [13,14]. Our proposed approach fits knowledge compilation, where the CNF-tree may be reused to answer different queries.

Static variable instantiation order is used to compile formulas to Ordered Binary Decision Diagrams (OBDDs) [3]. In contrast, a fully dynamic order, applied in DPLL-based algorithms, becomes effective in formulae that can be decomposed by a small number of well selected variables. DPLL-based algorithms attempt to decompose the formula into disjoint components after each instantiation. Nevertheless, despite the use of clever heuristics [20], there is no guarantee to the effectiveness of the instantiation in terms of partitioning the residual formula into disjoint components [11], making fully-dynamic variable instantiation inefficient when applied to heavily connected formulae.

The approach presented in this paper, as well as the dtree-guided C2D approach, may be considered semi-dynamic because the variable instantiation order is largely determined by the structure and orientation of the CNF-tree (or dtree). Our approach, however, takes a more holistic view and identifies the set of valid *assignments* that are guaranteed to decompose the formula. It also makes a deliberate effort to minimize the cardinality of this set by careful ordering of the node members.

8 Conclusions

We present CNF-trees of Boolean formulae to reveal structure that can be used to enhance the performance of WMC algorithms. We present a method for utilizing this structure in order to generate small tree-CPTs, and evaluate it over a set of known benchmarks. As part of future research we intend to characterize CNF-trees that enable efficient WMC.

Acknowledgments. The work was carried out in and partially supported by the Technion–Microsoft Electronic Commerce research center.

References

1. Bacchus, F., Dalmao, S., Pitassi, T.: Algorithms and complexity results for #sat and Bayesian inference. In: FOCS, pp. 340–351 (2003)
2. Boutilier, C., Friedman, N., Goldszmidt, M., Koller, D.: Context-specific independence in Bayesian networks. In: UAI, pp. 115–123 (1996)
3. Bryant, R.E.: Symbolic Boolean manipulation with ordered binary-decision diagrams. ACM Comput. Surv. **24**, 293–318 (1992)
4. Chavira, M., Darwiche, A.: On probabilistic inference by weighted model counting. Artif. Intell. **172**(6–7), 772–799 (2008)
5. Darwiche, A.: Decomposable negation normal form. J. ACM **48**, 608–647 (2001)
6. Darwiche, A.: New advances in compiling CNF into decomposable negation normal form. In: ECAI, pp. 328–332 (2004)
7. Darwiche, A., Marquis, P.: A knowledge compilation map. J. Artif. Intell. Res. **17**, 229–264 (2002)
8. Darwiche, P.A.: Modeling and Reasoning with Bayesian Networks, 1st edn. Cambridge University Press, New York (2009)
9. Dechter, R.: Bucket elimination: a unifying framework for reasoning. Artif. Intell. **113**(1–2), 41–85 (1999)
10. Domshlak, C., Hoffmann, J.: Probabilistic planning via heuristic forward search and weighted model counting (2011). CoRR, abs/1111.0044
11. Gomes, C.P., Sabharwal, A., Selman, B.: Model counting. In: Biere, A., Heule, M., Maaren, H.V., Walsh, T. (eds.) Handbook of Satisfiability, pp. 633–654. IOS Press, Amsterdam (2009)
12. Huang, C., Darwiche, A.: Inference in belief networks: a procedural guide. Int. J. Approximate Reasoning **15**(3), 225–263 (1996)
13. Huang, J., Darwiche, A.: DPLL with a trace: from sat to knowledge compilation. In: Kaelbling, L.P., Saffiotti, A. (eds.) IJCAI, pp. 156–162. Professional Book Center (2005)
14. Huang, J., Darwiche, A.: The language of search. J. Artif. Intell. Res. **29**, 191–219 (2007)
15. Hyafil, L., Rivest, R.L.: Constructing optimal binary decision trees is NP-complete. Inf. Process. Lett. **5**(1), 15–17 (1976)
16. Kjaerulff, U.: Triangulation of graphs: algorithms giving small total state space. Technical report, Department of Mathematics and Computer Science, March 1990
17. Lauritzen, S.L., Spiegelhalter, D.J.: Readings in Uncertain Reasoning. Morgan Kaufmann Publishers Inc., San Francisco (1990)
18. Marques Silva, J.P., Sakallah, K.A.: Conflict analysis in search algorithms for satisfiability. In: Proceedings Eighth IEEE International Conference on Tools with Artificial Intelligence, pp. 467–469. IEEE (1996)
19. Muise, C.J., McIlraith, S.A., Beck, J.C., Hsu, E.I.: Dsharp: fast d-DNNF compilation with sharpsat. In: Proceedings of the Canadian Conference on Artificial Intelligence, pp. 356–361 (2012)
20. Sang, T., Beame, P., Kautz, H.: Heuristics for fast exact model counting. In: Bacchus, F., Walsh, T. (eds.) SAT 2005. LNCS, vol. 3569, pp. 226–240. Springer, Heidelberg (2005)

21. Sang, T., Beame, P., Kautz, H.A.: Performing Bayesian inference by weighted model counting. In: AAAI, pp. 475–482 (2005)
22. Thurley, M.: sharpSAT – counting models with advanced component caching and implicit BCP. In: Biere, A., Gomes, C.P. (eds.) SAT 2006. LNCS, vol. 4121, pp. 424–429. Springer, Heidelberg (2006)
23. Valiant, L.G.: The complexity of enumeration and reliability problems. SIAM J. Comput. 8(3), 410–421 (1979)

A System for Probabilistic Inductive Answer Set Programming

Matthias Nickles[✉] and Alessandra Mileo

Insight Centre for Data Analytics,
National University of Ireland, Galway, Galway, Ireland
{matthias.nickles,alessandra.mileo}@deri.org

Abstract. We describe a prototypical software framework for probabilistic inductive logic programming which supports the seamless combination of non-monotonic reasoning, probabilistic inference and parameter learning. While building upon existing as well as new approaches to probabilistic Answer Set Programming, our framework distinguishes itself from related works by placing virtually no restrictions on the annotation of knowledge with probabilities. User-configurable algorithms provide for general as well as specialized, scalable approaches to inference and parameter learning, allowing for adaptability with regard to complex reasoning and weight learning tasks.

1 Introduction and Related Work

This short paper presents PrASP (Probabilistic Answer Set Programming) which is both a probabilistic logic programming language and a software for probabilistic inference and machine learning (parameter learning) based on *Answer Set Programming* (ASP). Reasoning in the presence of uncertainty and relational structures (such as social networks and Linked Data) is an important aspect of knowledge discovery and representation for the Web, the Internet Of Things, and other heterogeneous and complex domains. Probabilistic logic programing, and the ability to learn probabilistic logic programs from data, can provide an attractive approach to uncertainty reasoning and statistical relational learning, since it combines the deduction power and declarative nature of logic programming with probabilistic inference abilities traditionally known from graphical models, such as Bayesian and Markov networks. The main enhancement provided by PrASP over (non-probabilistic) ASP as well as existing probabilistic approaches to ASP is the possibility to annotate any formulas with probabilities (including formulas in full FOL syntax, albeit over finite domains of discourse only), while providing, in addition to general inference algorithms, specialized, scalable inference algorithms for special cases where certain assumptions hold (e.g., independence of probabilistic events).

Related approaches include, e.g., PRISM, P-Log, ProbLog, Markov Logic Networks (MLN) and others - [1–5] support probabilistic inference based on monotonic reasoning, whereas [6–9] are based on non-monotonic logics. Probabilistic logic programming belongs to the wider area of probabilistic programming with further

© Springer International Publishing Switzerland 2015
C. Beierle and A. Dekhtyar (Eds.): SUM 2015, LNAI 9310, pp. 99–105, 2015.
DOI: 10.1007/978-3-319-23540-0_7

approaches such as [10,11]. Our approach is influenced by P-log [7], which, like our framework, computes probability distributions over answer sets (that is, possible worlds are identified with stable models). However, P-log as well as [8] do not allow for annotating arbitrary formulas (including FOL formulas) with weights. [9] allows to associate probabilities with abducibles (only) and to learn both rules and probabilistic weights from given data (in form of literals). ProbLog [3] allows for probabilistic facts and definite clauses, and approaches to probabilistic rule and parameter learning (from interpretations) also exist for ProbLog. Inference is based on weighted model counting, which is similarly to our approach, but uses Boolean satisfiability instead of stable model semantics. Another important approach outside the area of ASP are *Markov Logic Networks* (MLN) [5]. A Markov Logic Network consists of first-order formulas annotated with weights (which are, in contrast to our approach, not probabilities).

2   Probabilistic Inference

In this section, we briefly describe the formal language and inference approaches of PrASP. Compared to [12], the syntax of PrASP programs has been extended with interval and non-ground weights, and new inference algorithms have been added.

Let Φ be a set of function, predicate and object symbols and $\mathcal{L}(\Phi)$ a first-order language over Φ with the usual connectives (including both strong negation "-" and default negation "not") and first-order quantifiers. It can be assumed that this language covers both ASP and FOL syntax (ASP "specialties" such as choice constructs can be seen as syntactic sugar which we omit here in order to keep things simple). A PrASP program (background knowledge) is a non-empty finite set $\Lambda = \{[l_i; u_i]f_i\} \cup \{[l_i; u_i|c_i]f_i\} \cup \{indep(\{f_1^i, ..., f_n^i\})\}$ of annotated formulas, $f_i, c_i, f_j^i \in \mathcal{L}(\Phi)$, and optional independence constraints. $[l; u]f$ asserts that the probability of f is within interval $[l, u]$ (i.e., $l \leq Pr(f) \leq u$) whereas $[l; u|c]f$ states that the probability of f conditioned on formula c is within interval $[l, u]$ ($l \leq Pr(f|c) \leq u$). Formulas can be non-ground (including existentially or universally quantified variables in FOL formulas). For the purpose of this paper, weights need to be ground (real numbers), however, the prototype implementation also allows for certain non-ground weights. An independence constraint $indep(\{f_1^i, ..., f_n^i\})$ specifies that the set of formulas $\{f_1^i, ..., f_n^i\}$ is mutually independent in the probabilistic sense (independence can also be discovered by analyzing the background knowledge, but this is computationally more costly of course).

If the weight is omitted, weight $[1; 1]$ is assumed. Point probability weights $[p]$ are translated into weights of the form $[p; p]$ (analogously for conditional probabilities). Weighted formulas can intuitively be seen as constraints which specify which possible worlds (in the form of answer sets) are indeed possible, and with which probability. $w(f)$ denotes the weight of formula f. The f_i and c_i are formulas either in FOL syntax (by means of a transformation into ASP syntax [13]) or ASP/AnsProlog syntax.

The semantics of a program Λ is defined in terms of a probability distribution over the answer sets (possible worlds) of the so-called *spanning program* $\rho(\Lambda)$ of Λ, which is defined as the disjunctive program generated by removing all weights and transforming each formerly weighted formula f or $\neg f$ into a disjunction $f|\neg f$, where $\neg$ stands for default negation. FOL syntax (optional) is converted into ASP syntax [13]. We write $\theta \models_\Lambda f$ iff θ is an answer set of $\rho(\Lambda) \cup f$. For further formal details please refer to [12].

We define the parameterized probability distribution $\mu^l(\Lambda, \Theta, q)$ over set $\Theta = \{\theta_i \in \Theta\}$ of answer sets of $\rho(\Lambda)$, a PrASP program $\Lambda = \{([p_i]f_i, i = 1..n)\} \cup \{([p_i|c_i]f_i^c)\} \cup \{indep(\{f_1^i, ..., f_k^i\})\}$ and query formula q as maximum entropy solution $\{Pr(\theta_i) : \theta_i \in \Theta\}$ of the following inequalities (*constraints*) such that $Pr^l(q) = \sum_{\theta_i \in \Theta : \theta_i \models_\Lambda q} Pr(\theta_i)$ is minimized (analogously, μ^u denotes the maximizing distribution).

$$l(f_1) \le \sum_{\theta_i \in \Theta : \theta_i \models_\Lambda f_1} Pr(\theta_i) \le u(f_1) \ ... \ l(f_n) \le \sum_{\theta_i \in \Theta : \theta_i \models_\Lambda f_n} Pr(\theta_i) \le u(f_n) \quad (1)$$

$$\sum_{\theta_i \in \Theta} \theta_i = 1 \quad (2)$$

$$\forall \theta_i \in \Theta : 0 \le Pr(\theta_i) \le 1 \quad (3)$$

At this, $l(f_i)$ and $u(f_i)$ denote the endpoints of the probability interval (weight) of unconditional formula f_i (analogous for endpoints $l(f_i^c|c_i)$ and $u(f_i^c|c_i)$ of conditional probabilities). In addition to the constraints above, *indep*-declaration of the form $indep(\{f_1^i, ..., f_r^i\})$ in the program induce constraints of the following form:

$$\prod_{f_k^i = \{1..r\}} l(f_k^i) \le \sum_{\theta_j \in \Theta : \theta_j \models_\Lambda \bigwedge f_k^i = \{1..r\}} Pr(\theta_j) \le \prod_{f_k^i = \{1..r\}} u(f_k^i) \quad (4)$$

and any conditional probability formula $[p_i|c_i]f_i^c$ in the program induces constraints

$$\sum_{\theta_j \in \Theta} Pr(\theta_j)\nu(\theta_j, f_i^c \wedge c_1) + \sum_{\theta_j \in \Theta} -l(f_i^c|c_i)Pr(\theta_j)\nu(\theta_j, c_i) > 0$$

$$\sum_{\theta_j \in \Theta} Pr(\theta_j)\nu(\theta_j, f_i^c \wedge c_i) + \sum_{\theta_j \in \Theta} -u(f_i^c|c_i)Pr(\theta_j)\nu(\theta_j, c_i) < 0$$

At this, we define $\nu(\theta, f) = \begin{cases} 1, & \text{if } \theta \models_\Lambda f \\ 0, & \text{otherwise} \end{cases}$

For small systems, PrASP can compute minimizing and maximizing probability distributions using linear programming and a maximum entropy solution amongst a number of candidate distributions (solutions of an underdetermined system) can be discovered using gradient descent. However, to make distribution search tractable, we need to use different algorithms, as described in the next section. That is, the linear system above could normally not be used directly for inference (except for very small systems), it serves mainly as a means to define the semantics of PrASP formulas.

The result of query [?] q is defined as the interval $[Pr^l(q), Pr^u(q)]$ (analogously for conditional queries [?|c] f, where we compute $Pr(f|c)$ using $Pr(f \wedge c)/Pr(c)$).

An example PrASP program (background knowledge):

```
coin(1..10).
[0.4;0.6] coin_out(1,heads).
[[0.5]] coin_out(N,heads) :- coin(N), N != 1.
1{coin_out(N,heads), coin_out(N,tails)}1 :- coin(N).
n_win :- coin_out(N,tails), coin(N).
win :- not n_win. [0.8|win] happy. :- happy, not win.
```

The line starting with [[0.5]]... is syntactic sugar for a set of weighted rules where variable N is instantiated with all its possible values (i.e.,
[0.5] coin_out(2,heads) :- coin(2), 2 != 1 and
[0.5] coin_out(3,heads) :- coin(3), 3 != 1). It would also be possible to use [0.5] as annotation of this rule, in which case the weight 0.5 would specify the probability of the entire non-ground formula instead.
1{coin_out(N,heads), coin_out(N,tails)}1 (Gringo ASP syntax) denotes that a coin comes up with either heads or tails but not both. Our system accepts query formulas in format [?] a, which asks PrASP for the marginal probability of a and [?|b] a which computes $Pr(a|b)$.

2.1   Sampling and Inference Algorithms

PrASP (as a software system) contains a variety of exact and approximate sampling and inference algorithms. Using command line options, the user selects a *pipeline* of alternative simplification, sampling and inference or learning steps (depending on the nature of the respective problem). E.g., the user might chose to sample possible worlds using uniform sampling and to pass on the resulting models to a simulated annealing algorithm which computes a probability distribution over the sampled possible worlds. Finally, this distribution is used to compute the probabilities of the query formulas. Inference algorithms available in PrASP version 0.7:

Linear Programming: Direct solution for the linear inequalities system described before. Very fast for very small systems, intractable otherwise.

Various Sampling Algorithms ("initial sampling"): Can sometimes directly compute a distribution which complies with the constraints expressed in the PrASP program.

Parallel Simulated Annealing: Can be used in combination with an initial sampling stage (e.g., Algorithm 1). This approach performs simulated annealing for inference problems where no assumptions can be made about independence or other properties of the program (except consistency).

Iterative Refinement: An adaptation of the algorithm described in [14] which reaches minimal Kullback–Leibler divergence to the uniform distribution (i.e., max. entropy).

Direct Counting: Weights are transformed into unweighted formulas and queries are then solved by mere counting of models (see [12] for details).

For lack of space, we describe only one initial sampling algorithm. The interesting property of this algorithm is its ability to provide a suitable distribution over possible worlds directly if all weighted formulas in the PrASP program are mutually independent (analogously to the independence assumption typically made by distribution semantics-based approaches). More concretely, Algorithm 1 samples answer sets and computes a probability distribution over these models which reflects the weights provided in the PrASP program, provided that all uncertain formulas in the program describe a stochastically independent set of events. Other user-provided constraints (such as conditional probabilities in the PrASP program) are ignored here. Also, Algorithm 1 does in general not compute a maximum entropy solution.

Algorithm 1. Sampling from models of spanning program (point probabilities only)

Require: max number of samples n, set of uncertain formulas $uf = \{[w(uf_i)]uf_i \text{ with } 0 < w(uf_i) < 1\}$, set of certain formulas $cf = \{cf_i : w(uf_i) = 1\}$ (i.e., with probability 1)

1: $i \leftarrow 1$
2: **for** $i \leq |uf|$ **do**
3:     $r^i \leftarrow$ random element of $Sym(\{1, ..., n\})$ (permutations of $\{1, ..., n\}$)
4:     $i \leftarrow i + 1$
5: **end for**
6: $m \leftarrow \{\}, j \leftarrow 1$
7: **parfor** $j \in \{1, ..., n\}$ **do**
8:     $p \leftarrow \{\}, k \leftarrow 1$
9:     **for** $k \leq |uf|$ **do**
10:         **if** $r_j^k \leq n \cdot w(uf_k)$ **then** $p \leftarrow p \cup uf_k$ **else** $p \leftarrow p \cup \neg uf_k$ **endif**
11:         $k \leftarrow k + 1$
12:     **end for**
13:     $s \leftarrow$ model sampled uniformly from models of program $cf \cup p$ ($\emptyset$ if UNSAT)
14:     $m \leftarrow m \uplus \{s\}$
15: **end parfor**
Ensure: Multiset m contains samples from all answer sets of spanning program such that
16: $\forall uf_i : w(uf_i) \approx \frac{|\{s \in m : s \models uf_i\}|}{|m|}$ *iff* set uf mutually independent.

Algorithm 1 can optionally be combined in a pipeline-like fashion with simulated annealing or iterative refinement: if Algorithm 1 doesn't already compute a suitable probability distribution, e.g., simulated annealing discovers this (as part of its energy calculation) and iteratively adjusts the distribution.

While for space-related reasons this paper focuses on inference, PrASP also allows for inductive reasoning (parameter learning). Please refer to [12] for details.

2.2 Initial Experiment

The main goal of PrASP is not to outperform existing approaches in terms of speed but to provide a flexible, scalable and highly configurable framework which puts as few restrictions as possible on what users can express while being competitive with more specialized approaches if the respective conditions (e.g., event independence) are met.

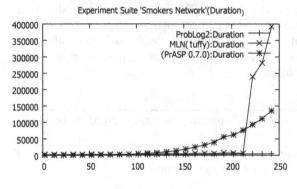

The following shows how PrASP copes with a typical benchmark task (the well-known friends-and-smokers problem [5]) which can be tractably solved using the algorithms described above. In this scenario, a randomly chosen number of persons are friends, a randomly chosen subset of all people smoke, there is a certain probability of being stressed ([[0.3]] stress(X)), stress causes smoking (smokes(X) :- stress(X)), some friends influence each other with a certain probability ([[0.2]] influences(X,Y) and smokes(X) :- friend(X,Y), influences(Y,X), smokes(Y)?). Smoking might lead to asthma ([[0.4]] h(X). asthma(X) :- smokes(X), h(X)). The query comprises of asthma(X) for all persons.

We compared the performance (duration in dependency of the number of people in the social network) of the current prototype of PrASP with that of Tuffy 0.3 (http://i.stanford.edu/hazy/hazy/tuffy/), a recent implementation of Markov Logic Networks which uses databases to increase scalability) and ProbLog2 2.1. The results (see figure) have been averaged over five trials. ProbLog2 scores best in this scenario. PrASP, using Algorithm 1, does quite well for most of episodes but looses on ProbLog2. Tuffy does very well below 212 persons, then performance breaks in (possibly due to some cache overflow). Times are in ms using a i7 4-cores/3.4Ghz processor.

3   Conclusion

We have presented a new software framework for uncertainty reasoning and parameter estimation based on Answer Set Programming. In contrast to most other approaches to probabilistic logic programming, the philosophy of PrASP is to provide a very expressive formal language on the one hand and a variety of inference algorithms which are able to take advantage of certain problem domains which facilitate "fast track" reasoning and learning (in particular inference in the presence of event independence) on the other. Ongoing work focuses mainly on the development and integration of further inference algorithms. This work was sponsored by SFI grant n. SFI/12/RC/2289.

References

1. Sato, T., Kameya, Y.: Prism: a language for symbolic-statistical modeling. In: Proceedings of the 15th International Joint Conference on Artificial Intelligence (IJCAI 97) (1997)
2. Kersting, K., Raedt, L.D.: Bayesian logic programs. In: Proceedings of the 10th International Conference on Inductive Logic Programming (2000)
3. Raedt, L.D., Kimmig, A., Toivonen, H.: Problog: A probabilistic prolog and its application in link discovery. In: IJCAI, pp. 2462–2467 (2007)
4. Poole, D.: The independent choice logic for modelling multiple agents under uncertainty. Artif. Intell. **94**, 7–56 (1997)
5. Richardson, M., Domingos, P.: Markov logic networks. Mach. Learn. **62**, 107–136 (2006)
6. Ng, R.T., Subrahmanian, V.S.: Stable semantics for probabilistic deductive databases. Inf. Comput. **110**, 42–83 (1994)
7. Baral, C., Gelfond, M., Rushton, N.: Probabilistic reasoning with answer sets. Theory Pract. Log. Program. **9**, 57–144 (2009)
8. Saad, E., Pontelli, E.: Hybrid probabilistic logic programs with non-monotonic negation. In: Gabbrielli, M., Gupta, G. (eds.) ICLP 2005. LNCS, vol. 3668, pp. 204–220. Springer, Heidelberg (2005)
9. Corapi, D., Sykes, D., Inoue, K., Russo, A.: Probabilistic rule learning in non-monotonic domains. In: Leite, J., Torroni, P., Ågotnes, T., Boella, G., van der Torre, L. (eds.) CLIMA XII 2011. LNCS, vol. 6814, pp. 243–258. Springer, Heidelberg (2011)
10. Goodman, N.D., Mansinghka, V.K., Roy, D.M., Bonawitz, K., Tenenbaum, J.B.: Church: a language for generative models. In: Proceedings of Uncertainty in Artificial Intelligence (2008)
11. Pfeffer, A.: Figaro: An object-oriented probabilistic programming language. In: Charles River Analytics Technical report (2009)
12. Nickles, M., Mileo, A.: Probabilistic inductive logic programming based on answer set programming. In: 15th International Workshop on Non-Monotonic Reasoning (NMR 2014) (2014)
13. Lee, J., Palla, R.: System f2lp – Computing Answer Sets of First-Order Formulas. In: Erdem, E., Lin, F., Schaub, T. (eds.) LPNMR 2009. LNCS, vol. 5753, pp. 515–521. Springer, Heidelberg (2009)
14. Rodder, W., Meyer, C.: Coherent knowledge processing at maximum entropy by spirit. In: Proceedings of 12th Conference on Uncertainty in Artificial Intelligence (UAI 1996) (1996)

Towards Large-Scale Probabilistic OBDA

Joerg Schoenfisch[✉] and Heiner Stuckenschmidt

Data- and Web Science Group, University of Mannheim,
B6 26, 68159 Mannheim, Germany
{joerg,heiner}@informatik.uni-mannheim.de

Abstract. Ontology-based Data Access has intensively been studied as
a very relevant problem in connection with semantic web data. Often it
is assumed, that the accessed data behaves like a classical database, i.e.
it is known which facts hold for certain. Many Web applications, especially those involving information extraction from text, have to deal with
uncertainty about the truth of information. In this paper, we introduce
an implementation and a benchmark of such a system on top of relational
databases. Furthermore, we propose a novel benchmark for systems handling large probabilistic ontologies. We describe the benchmark design and
show its characteristics based on the evaluation of our implementation.

1 Motivation

Ontology-based Data Access (ODBA) has received a lot of attention in the
Semantic Web Community. In particular, results on light weight description
logics that allow efficient reasoning and query answering provide new possibilities for using ontologies in data access. One approach for ontology-based data
access is to rewrite a given query based on the background ontology in such a
way that the resulting – more complex – query can directly be executed on a
relational database. This is possible for different light-weight ontology languages,
in particular the *DL-Lite* family [1].

At the same time, it becomes more and more clear that many applications
in particular on the (Semantic) Web have to deal with uncertainty in the data.
Examples are large-scale information extraction from text or the integration of
heterogeneous information sources. To cope with uncertainty, the database community has investigated probabilistic databases where each tuple in the database
is associated with a probability indicating the belief in the truth of the respective
statement. Querying a probabilistic database requires not only to retrieve tuples
that match the query, but also to compute a correct probability for each answer.

The goal of our work is to develop data access methods that can use background knowledge in terms of a light weight ontology and also deal with uncertainty in the data. A promising idea for efficiently computing probabilistic query
answers is use existing approaches for OBDA based on query rewriting and
pose the resulting query against a probabilistic database that computes answers
with associated probabilities.

A number of approaches have been proposed for combining description logics with probabilistic reasoning. An overview of early approaches is [2], more

© Springer International Publishing Switzerland 2015
C. Beierle and A. Dekhtyar (Eds.): SUM 2015, LNAI 9310, pp. 106–120, 2015.
DOI: 10.1007/978-3-319-23540-0_8

recent approaches include DISPONTE/BUNDLE [3,4] or Pronto [5], and Log-linear Description Logics [6]. On the other hand, the logic programming and statistical relational learning community has developed probabilistic versions of datalog-style languages (e.g. Problog [7]) that can be used to partially model ontological background knowledge. While for many of these languages efficient subsets have been identified (e.g. [4,5]) and optimized reasoning algorithms have been proposed, none of the existing approaches is really designed to handle large amounts of data as we find on the Web. For example, the full dataset extracted from Web pages by the NELL (Never Ending Language Learning) Project [8] currently contains about 50 million statements with associated probabilities.

Jung et al. have shown that query rewriting for OBDA can directly be lifted to the probabilistic case [9]. Furthermore, they prove that the complexity results and the dichotomy of safe (data complexity in PTIME) and unsafe (in #P-hard) queries also carries over. To the best of our knowledge, no evaluation on the performance and scalability of the approach was conducted, and there exists no implemented system. We believe, that combining the power of probabilistic database systems with the *DL-Lite* approach to ODBA – namely rewriting the query using the background ontology in such a way that the resulting query posed against a database returns the correct results – is a way to scale up to datasets of the size of NELL and beyond.

The main contributions presented in the paper are the following:

- a preliminary implementation of a system that can answer safe probabilistic queries over large probabilistic knowledge bases up to several hundred millions of facts.
- a synthetic benchmark dataset for probabilistic OBDA on the basis of the LUBM benchmark that can be scaled to an arbitrary number of probabilistic statements
- a comparison of the prototype to a state of the art system using that dataset and a real world knowledge base

The paper is structured as follows: We show how the distribution semantics is applied to *DL-Lite* in Sect. 2. Section 3 is concerned with implementing reasoning on top of probabilistic databases. In Sect. 4 we describe the datasets that can be used for benchmarking probabilistic OBDA. An experimental evaluation of a prototype for large-scale probabilistic OBDA is given in Sect. 5. Related work is discussed in Sect. 6 and we conclude and give an outlook in Sect. 7.

Motivating Example. Consider the following assertions about Arnold Schwarzenegger and his wife Maria Shriver from the NELL dataset:

1.00 $Politicanus(Arnold\_Schwarzenegger)$

1.00 $Actor(Arnold\_Schwarzenegger)$

0.50 $hasoffice(Arnold\_Schwarzenegger, President)$

1.00 $husbandof(Arnold\_Schwarzenegger, M\_Shriver)$

0.75 $agentcontrols(NBC, M\_Shriver)$

1.00 $acquired(NBC, Telemundo)$

While some of the statements are considered to be definitely true – either because they have been part of the seeds used for training the extractor or because enough evidence has been found – some of them, however, are only believed to be true with a certain probability. Knowledge bases like NELL [8] or ReVerb [10] contain huge amounts of information. The NELL project, for example, has extracted more than 50 million possible facts about the world. Accessing the stored information in a meaningful way requires to query the knowledge base. In this work, we consider positive conjunctive queries that allow us for example to ask for all politicians that have been president and actor or for all politicians married to someone who is under control of a company:

$$\exists X. \; (politician(X), actor(X), hasoffice(X, president))$$
$$\exists X.\exists Y.\exists Z. \; (politician(X), spouse(X, Y), agentcontrolledby(Y, Z), company(Z))$$

It has been shown that there is a fundamental difference between these two queries as the data complexity of the first query is in PTIME whereas the second is #P-hard. Answering these queries requires background knowledge about the terminology used in the query and the data. In particular, we can derive the following $DL\text{-}Lite_R$ axioms from the metadata in the NELL knowledge base, stating that US politicians are politicians, agentcontrolledby is the inverse of agentcontrols and that the domain of the acquired relation is company.

$$politicianus \sqsubseteq politician$$
$$agentcontrolledby \sqsubseteq agentcontrols^-$$
$$\exists acquired \sqsubseteq company$$

Using these axioms, we can compute the probability that Arnold Schwarzenegger is a correct answer to these queries. By benchmarking our prototypical implementation we want to address the question the question whether answering such queries scales serviceably.

2 $DL\text{-}Lite_R$ and the Distribution Semantics

In this section we briefly introduce $DL\text{-}Lite$, the description logic underlying the OWL 2 QL profile. Then we detail how the distribution semantics for probabilistic description logics [4] is applied to $DL\text{-}Lite_R$.

In $DL\text{-}Lite_R$ concepts and roles are formed in the following syntax [11]:

$$B \to A \mid \exists R \qquad\qquad C \to B \mid \neg B$$
$$R \to P \mid P^- \qquad\qquad E \to R \mid \neg R$$

where A denotes an atomic concept, P an atomic role, and P^- the inverse of the atomic role P. B denotes a basic concept, i.e. either an atomic concept or a concept of the form $\exists R$, where R denotes a basic role, that is, either an atomic role or the inverse of an atomic role. C denotes a general concept, which can be a basic concept or its negation, and E denotes a general role, which can be a basic role or its negation.

A $DL\text{-}Lite_R$ knowledge base (KB) $\mathcal{K} = \langle \mathcal{T}, \mathcal{A} \rangle$ models a domain in terms of a TBox $\mathcal{T}$ and an ABox $\mathcal{A}$. A TBox is formed by a finite set of inclusion assertions of the form $B \sqsubseteq C$ or $R \sqsubseteq E$. An ABox is formed by a finite set of membership assertions on atomic concepts and on atomic roles, of the form $A(a)$ or $P(a, b)$ stating respectively that the object denoted by the constant a is an instance of A and that the pair of objects denoted by the pair of constants (a, b) is an instance of the role P.

The semantics of a DL is as an interpretation $\mathcal{I} = (\Delta^{\mathcal{I}}, \cdot^{\mathcal{I}})$, consisting of a nonempty interpretation domain $\Delta^{\mathcal{I}}$ and an interpretation function $\cdot^{\mathcal{I}}$ that assigns to each concept C a subset $C^{\mathcal{I}}$ of $\Delta^{\mathcal{I}}$, and to each role R a binary relation $R^{\mathcal{I}}$ over $\Delta^{\mathcal{I}}$:

$$A^{\mathcal{I}} \subseteq \Delta^{\mathcal{I}}$$
$$P^{\mathcal{I}} \subseteq \Delta^{\mathcal{I}} \times \Delta^{\mathcal{I}}$$
$$(P)^{\mathcal{I}} = \{(o_2, o_1) | (o_1, o_2) \in P^{\mathcal{I}}\}$$
$$(\exists R)^{\mathcal{I}} = \{o \mid \exists o'.(o, o') \in R^{\mathcal{I}}\}$$
$$(\neg B)^{\mathcal{I}} = \Delta^{\mathcal{I}} \setminus B^{\mathcal{I}}$$
$$(\neg R)^{\mathcal{I}} = \Delta^{\mathcal{I}} \times \Delta^{\mathcal{I}} \setminus R^{\mathcal{I}}$$

An interpretation $\mathcal{I}$ is a model of $C_1 \sqsubseteq C_2$, where C_1, C_2 are general concepts if $C_1^{\mathcal{I}} \subseteq C_2^{\mathcal{I}}$. Similarly, I is a model of $E_1 \sqsubseteq E_2$, where E_1, E_2 are general roles if $E_1^{\mathcal{I}} \subseteq E_2^{\mathcal{I}}$.

To specify the semantics of membership assertions, the interpretation function is extended to constants, by assigning to each constant a a distinct object $a^{\mathcal{I}} \in \Delta^{\mathcal{I}}$. This enforces the unique name assumption on constants. An interpretation $\mathcal{I}$ is a model of a membership assertion $A(a)$ (resp., $P(a, b)$), if $a^{\mathcal{I}} \in A^{\mathcal{I}}$ (resp., $(a^{\mathcal{I}}, b^{\mathcal{I}}) \in P^{\mathcal{I}}$).

Given an assertion α and an interpretation $\mathcal{I}$, $\mathcal{I} \vDash \alpha$ denotes the fact that $\mathcal{I}$ is a model of α. Given a (finite) set of assertions λ, $\mathcal{I} \vDash \lambda$ denotes the fact that $\mathcal{I}$ is a model of every assertion in λ. A model of a KB $K = \langle \mathcal{T}, \mathcal{A} \rangle$ is an interpretation $\mathcal{I}$ such that $\mathcal{I} \vDash \mathcal{T}$ and $\mathcal{I} \vDash \mathcal{A}$. A KB is satisfiable if it has at least one model. A KB K logically implies an assertion α, written $K \vDash \alpha$, if all models of K are also models of α. Similarly, a TBox T logically implies an assertion α, written $T \vDash \alpha$, if all models of T are also models of α.

A TIP-OWL knowledge base $\mathcal{TKB} = \langle \mathcal{T}, \mathcal{A}, P \rangle$ consists of a $DL\text{-}Lite_R$ T-Box $\mathcal{T}$, an ABox $\mathcal{A}$, and a probability distribution $P : \mathcal{A} \to [0, 1]$. Abusing terminology, we say that $\mathcal{KB} \subseteq \mathcal{TKB}$ if they have the same T-Box and the A-Box of $\mathcal{KB}$ is a subset of the A-Box of $\mathcal{TKB}$. We adopt the independent tuple semantics in the spirit of the probabilistic semantics for logic programs proposed in [7] and define the probabilistic semantics of a TIP-OWL knowledge base in terms of a distribution over possible knowledge bases as follows:

Definition 1. *Let $\mathcal{TKB} = \langle \mathcal{T}, \mathcal{A}, P \rangle$ be a TIP-OWL knowledge base. Then the probability of a DL-Lite$_R$ Knowledge Base $\mathcal{KB} = \langle \mathcal{T}, \mathcal{A}' \rangle \subseteq \mathcal{TKB}$ is given by:*

$$P(\mathcal{KB}|\mathcal{TKB}) = \prod_{a' \in \mathcal{A}'} P(a') \cdot \left(1 - \prod_{a \in \mathcal{A} \setminus \mathcal{A}'} P(a) \right)$$

Based on this semantics, we can now define the probability of existential queries over TIP-OWL knowledge bases as follows. First, the probability of a query over a possible knowledge base is one if the query follows from the knowledge base and zero otherwise:

$$P(Q|\mathcal{KB}) = \begin{cases} 1 & \mathcal{KB} \models Q \\ 0 & \text{otherwise} \end{cases}$$

Taking the probability of possible knowledge bases into account, the probability of an existential query over a possible knowledge base becomes the product of the probability of that possible knowledge base and the probability of the query given that knowledge base. By summing up these probabilities over all possible knowledge bases, we get the following probability for existential queries over TIP-OWL knowledge bases:

$$P(Q|\mathcal{TBK}) = \sum_{\mathcal{KB} \subseteq \mathcal{TKB}} P(Q|\mathcal{KB}) \cdot P(\mathcal{KB}|\mathcal{TKB})$$

This defines a complete probabilistic semantics for TIP-OWL knowledge bases and queries.

3   Implementing Reasoning on Top of Probabilistic Databases

In this section, we first briefly recall the idea of first-order rewritability of queries in *DL-Lite* and then show that the query rewriting approach proposed by [11] can be used on top of tuple independent probabilistic databases for answering queries in TIP-OWL without changing the semantics of answers.

3.1   Query Rewriting

Query processing in $DL\text{-}Lite_R$ is based on the idea of first-order reducibility. This means that for every conjunctive query q we can find a query q' that produces the same answers as q by just looking at the A-Box. Calvanese et al. also define a rewriting algorithm that computes a q' for every q by applying transformations that depend on the T-Box axioms.

Given a consistent T-Box $\mathcal{T}$ the algorithm takes a conjunctive query q_0 and expands it into a union of conjunctive queries U starting with $U = \{q_0\}$. The algorithm successively extends U by applying the following rule:

$$U = U \cup \{q[l/r(l,I)]\}$$

where q is a query in U, l is a literal in q, I is an inclusion axiom from $\mathcal{T}$ and r is a replacement function that is defined as follows:

l	I	$r(l,I)$	l	I	$r(l,I)$
$A(x)$	$A' \sqsubseteq A$	$A'(x)$	$P(\_,x)$	$A \sqsubseteq \exists P^-$	$A(x)$
$A(x)$	$\exists P \sqsubseteq A$	$P(x,\_)$	$P(\_,x)$	$\exists P' \sqsubseteq P^-$	$P'(x,\_)$
$A(x)$	$\exists P^- \sqsubseteq A$	$P(\_,x)$	$P(\_,x)$	$\exists P'^- \sqsubseteq \exists P^-$	$P'(\_,x)$
$P(x,\_)$	$A \sqsubseteq \exists P$	$A(x)$	$P(x,y)$	$P' \sqsubseteq P$	$P'(x,y)$
$P(x,\_)$	$\exists P' \sqsubseteq \exists P$	$P'(x,\_)$	$P(x,y)$	$P'^- \sqsubseteq P^-$	$P'(x,y)$
$P(x,\_)$	$\exists P'^- \sqsubseteq \exists P$	$P'(\_,x)$	$P(x,y)$	$P' \sqsubseteq P^-$	$P'(y,x)$
			$P(x,y)$	$P'^- \sqsubseteq P$	$P'(y,x)$

Here '$\_$' denotes an unbound variable, i.e., a variable that does not occur in any other literal of any of the queries.

Definition 2 (Derived Query). *Let $Q = L_1 \wedge \cdots \wedge L_m$ be a conjunctive query over a DL-Lite terminology. We write $Q \xrightarrow{r} Q'$ if $Q' = Q \vee Q[L_i/r(L_i,I)]$ for some literal L_i from Q. Let $\xrightarrow{r}{}^*$ denote the transitive closure of $\xrightarrow{r}$, then we call every Q' with $Q \xrightarrow{r}{}^* Q'$ a derived query of Q. Q' is called maximal if there is no Q'' such that $Q \xrightarrow{r}{}^* Q''$ and $Q' \xrightarrow{r}{}^* Q''$.*

Using the notion of a maximal derived query, we can establish the FOL-reducibility of *DL-Lite* by rephrasing the corresponding theorem from [11].

Theorem 1 (FOL-Reducibility of *DL-Lite* (from [11])). *Query answering in DL-Lite$_R$ is FOL-reducible. In particular, for every query Q with maximal derived query Q' and every DL-Lite$_R$ T-Box $\mathcal{T}$ and non-empty A-Box $\mathcal{A}$ we have $\mathcal{T} \cup \mathcal{A} \models Q$ if and only if $\mathcal{A} \models Q'$.*

Example. We illustrate the rewriting using the queries and the three ontological axioms from the motivating example. For the first query, the subclass relation between *politicianus* and *politician* triggers a rewriting leading to the new query $Q'(X) \Leftarrow politicianus(X), actor(X), hasoffice(X, president)$ which has to be united with the original query to form the new query $Q \vee Q'$. This query can be simplified to:

$$(politician(X) \vee politicianus(X)) \wedge$$
$$actor(X) \wedge hasoffice(X, president)$$

For the second query, the rewriting is slightly more complicated, but still can be computed in a single go through the query. Each of the axioms triggers a rewriting. Besides the rewriting already performed for the first query, the literal $agentcontrolledby(Y, Z)$ can be rewritten to $agentcontrols(Z, Y)$. Secondly, the literal $company(Z)$ can be rewritten to $acquired(Z, \_)$, after simplification leading to the following rewritten query:

$$(politician(X) \vee politicianus(X) \wedge spouse(X, Y)$$
$$\wedge(agentcontrolledby(Y, Z) \vee agentcontrols(Z, Y))$$
$$\wedge(company(Z) \vee acquired(Z, \_))$$

The resulting query can now directly be executed on the probabilistic database.

3.2 Correctness of Query Processing

Implementing TIP-OWL on top of probabilistic databases can now be done in the following way. The A-Box is stored in the probabilistic database. It is easy to see that the probabilistic semantics of TIP-OWL A-Boxes and tuple independent databases coincide. What remains to be done is to show that the idea of FOL-reducibility carries over to our probabilistic model. In particular, we have to show that a rewritten query has the same probability given a knowledge base with empty T-Box as the original query given a complete knowledge base. This result is established in the following corollary.

Corollary 1. *Let $\mathcal{TKB} = \langle \mathcal{T}, \mathcal{A}, \mathcal{P} \rangle$ be a TIP-OWL knowledge base and $\mathcal{TKB}' = \langle \emptyset, \mathcal{A}, \mathcal{P} \rangle$ the same knowledge base, but with an empty T-Box. Let further Q be a conjunctive query and Q' a union of conjunctive queries obtained by rewriting Q on the basis of $\mathcal{T}$, then*

$$P(Q|\mathcal{TKB}) = P(Q'|\mathcal{TKB}')$$

We can easily see that the semantics of queries over a TIP-OWL A-Box with empty T-Box directly corresponds to the tuple-independence semantics used in probabilistic databases [12], thus queries posed to correctly constructed probabilistic database have the same probability as a TIP-OWL query with empty T-Box. From Theorem 1 we get, that $P(Q|\mathcal{KB})$ does not change as $\mathcal{T}, \mathcal{A} \models Q$ if and only if $\mathcal{A} \models Q'$. Further, as $P(\mathcal{KB}|\mathcal{TKB})$ only depends on $\mathcal{A}$ this part also stays unchanged.

4   Benchmark Data for Probabilistic OBDA

We use two different datasets for our experimental evaluation. The first dataset is the ontology and knowledge base created by NELL, which presents a large real-world dataset consisting of uncertain data. Second, to assess the scalability of the approach, we created a modified version of the Lehigh University Benchmark (LUBM), a synthetic benchmark for OWL reasoners that generates datasets of various sizes.

4.1 NELL

NELL is an Open Information extraction system that extracts facts from text found in a large corpus of web pages. As a result, NELL generates triples like `wifeof(katie_holmes, cruise)`, called candidate beliefs, that are annotated with different levels of confidence in terms of a number in the range $(0, 1]$.

Within the context of our approach these candidate beliefs form the A-Box of our *TIP-OWL* knowledge base, while the confidences are interpreted as probabilities. NELL organizes extracted facts in a terminology consisting of concepts (called categories in NELL context) and roles (relations) and specifies domain and range restrictions, property symmetry, and disjointness of concepts and properties. We use the $DL\text{-}Lite_R$ fragment of this terminology as T-Box of our *TIP-OWL* knowledge base. We use the high confidence knowledge base of NELL (iteration 860) which contains only facts with a score of at least 0.75. It contains 2.3 million extracted facts about 1.8 million objects as compared to the full dataset with about 50 million. The T-Box, which is the same for all datasets, consists of 558 concepts, 1 255 properties, and 5 132 axioms (Table 1).

Table 1. Size of the different NELL and LUBM datasets.

Dataset	Assertions
NELL full	2 259 750
NELL filtered	467 943

Dataset	Assertions	% distinct
LUBM 1	717 250	54.77
LUBM 10	7 232 663	55.69
LUBM 100	71 698 666	55.66
LUBM 200	143 311 100	55.67
LUBM 500	361 432 844	55.12
LUBM 1000	719 097 512	55.32

To show the benefits and the scalability of our approach, we defined the following queries that are posed against the *TIP-OWL* version of NELL.

$Q_A(X) \Leftarrow person(X)$

$Q_B(X) \Leftarrow person(X), bornin(X, paris)$

$Q_C(X) \Leftarrow book(X), movie(X)$

$Q_D(Z) \Leftarrow hasParent(X, Y), hasparent(Y, Z)$

$Q_E(X) \Leftarrow actor(X), directordirectedmovie(X, Y), writerwrotebook(X, Z)$

$Q_F(X) \Leftarrow politician(X), actor(X), hasoffice(X, president)$

We use this dataset mainly to investigate the benefits of using background knowledge and reasoning on top of probabilistic data in terms of increased recall.

4.2 Probabilistic LUBM

The Lehigh University Benchmark (LUBM) [13] is a well known and widely used benchmark for OWL-based reasoning systems. Lutz et al. published a $DL\text{-}Lite_R$

version of LUBM [14]. Additionally, to restricting the expressivity of the ontology, they modified it to make it more suitable in an OBDA setting: First, they added multiple concept inclusions with existential restriction on the right hand side, second they extended the class hierarchy to be closer to real-world ontologies in its size.

We chose to extend LUBM over other synthetic benchmarks like SP^{2}Bench [15], BSBM [16], or FishMark [17]. SP^{2}Bench and BSBM only provide a very simple or no ontology, but rather focus on complex queries. FishMark contains an expressive ontology more suitable for our evaluation. However, it does not provide a generator for datasets of various sizes. Lutz et al.'s version of the LUBM ontology is of sufficient complexity to evaluate the scalability of reasoning in an OBDA setting, and it offers the possibility to generate datasets of different sizes.

We generated the benchmark dataset for probabilistic OBDA in two steps: 1. We extended the generator to create probabilistic ABoxes. 2. To increase the complexity of the probabilistic reasoning, we created redundancies in the dataset. In the first step we extended the implementation of the data generator to attach probabilities to every ABox axiom. Those probabilities are randomly distributed in (0,1]. We did not include a fixed percentage of certain axioms. The generator thus creates datasets of various size with probabilistic axioms. However, each axiom is contained exactly once, thereby trivializing the calculation of the final probabilities in a result set. To alleviate this, we used the option to change the seed that LUBM uses to determine the number of instances of departments, professors, students, etc. For every dataset size, we generated five ABoxes each with a different seed (0, 1, 42, 776, 141984). Combined, these five ABox serve as one probabilistic benchmark dataset. Note that our probabilistic version of LUBM is roughly five times larger than the normal LUBM of the same size, i.e. LUBM 1 contains only one university, whereas probabilistic LUBM 1 contains five different versions of that university, with different numbers of departments, professors, students, etc.

In the evaluation we used the original LUBM queries for which probabilities can be computed efficiently, i.e. queries 1, 3–6, and 10–14:

$$Q_1(X) \Leftarrow takesCourse(X, univ0\_dept0), type(X, graduateStudent)$$
$$Q_3(X, Y_1, Y_2, Y_3) \Leftarrow publicationAuthor(X, univ0\_asstProf0), type(X, publication)$$
$$Q_4(X) \Leftarrow worksFor(X, univ0\_dept0), name(X, Y_1), emailAddress(X, Y_2),$$
$$telephone(X, Y_3), type(X, professor)$$
$$Q_5(X) \Leftarrow memberOf(X, univ0\_dept0), type(X, person)$$
$$Q_6(Z) \Leftarrow type(X, student)$$
$$Q_{10}(X) \Leftarrow takesCourse(X, univ0\_graduateCourse0), type(X, student)$$
$$Q_{11}(X) \Leftarrow subOrgOf(X, univ0), type(X, researchGroup)$$
$$Q_{12}(X, Y) \Leftarrow worksFor(X, Y), type(X, chair), subOrgOf(Y, univ0),$$
$$type(Y, department)$$
$$Q_{13}(X) \Leftarrow hasAlumnus(univ0, X), type(X, person)$$
$$Q_{14}(X) \Leftarrow type(X, undergraduateStudent)$$

Q_{11} and Q_{12} require the reasoner to handle the transitive object property subOrgOf. However, transitive object properties are not allowed in $DL\text{-}Lite_R$. To circumvent this and still be able to use this query in the evaluation, we manually extended those queries to handle transitivity up to the maximum depth occurring in the data (in this case 2).

5   Experimental Evaluation

We evaluate our implementation of query processing for $TIP\text{-}OWL$ on the two datasets presented in the previous section. Our main goal is to show that our implementation scales to vers large A-Boxes and outperforms existing methods on safe queries.

5.1   Setting

Within our experiments we focus on answering the following two questions:

1. What are the benefits of exploiting the TBox by using it in the query rewriting process for a dataset like NELL?
2. How well does our algorithm scale with respect to different types of queries and subsets of NELL and a probabilistic LUBM, and compared to another system?

For answering the first question, we compare query results with and without query rewriting. We expect that rewriting the queries yields larger result sets. In particular, we expect that many interesting results are missed out when we ask the query directly without any expansion.

For answering the second question, we compare our implementation against the ProbLog system [7], which also uses the independent tuple semantics. While ProbLog does not support the complete expressivity of $DL\text{-}Lite^1$, it returns identical results for any safe conjunctive query over the dataset. For the comparison with ProbLog, we used a subset of the data consisting of about half a million facts. Additionally, to assess the general scalability of the approach, we run our implementation on different sizes of probabilistic LUBM, i.e. 1, 10, 100, 200, 500, and 1000 universities.

Experiments were run on a virtual machine with 4 cores (2.4 GHz) and 16 GB RAM running Ubuntu 14.10 Server. We used PostgreSQL 9.4 64 bit and ProbLog 2. We used to default settings of the database and did no special tuning apart from increasing the available RAM. The NELL dataset was loaded into a single table. The LUBM datasets use different tables for class, object, and data property assertions, one of each for different sizes of the benchmark. Query rewriting was done manually at this point, but we do not expect a significant impact on this step on the overall performance. As ProbLog always has to load all the data and does not provide persistent storage like a database system, we measured

---

[1] In particular, axioms of the form $A \sqsubseteq \exists R$ cannot be represented in ProbLog.

the time ProbLog takes to parse the file without a query and subtracted that amount from the query time.

An existing probabilistic database like MayBMS was not used, as we encountered serious issues with complex JOINs resulting from the query rewriting.

5.2   Results

NELL Dataset. Table 2 shows the results of comparing query answering with and without rewriting.

Table 2. Number of results with and without reasoning, and increase in query size (predicates)

	Plain		Rewritten	
	# res	# pred	# res	# pred
Q_A	5 405	1	319 986	148
Q_B	1	2	4	152
Q_C	352	2	414	12
Q_D	0	2	80	40
Q_E	0	3	1	11
Q_F	2	4	14	29

The number of results generally increases – sometimes dramatically (cf. Q_A) – and we can even find answers to Q_D which produced no results without rewriting. The large increase in results for Q_A is due to *person* being a very general concept of the NELL hierarchy and most instances are described using more specific concepts. Q_D has no answers without rewriting because the relation *personhasparent* is never used, but only its inverse *parentofperson.*

Exploiting the T-Box often changes the probabilities for an individual answer to a query as new evidence is added to the computation. For example the probability for *concept:person:sandy* being a *person* in Q_A or Q_B increase from 0.96875 to 1.0. The knowledge base only states that Sandy is a *person* with probability 0.96875. Through the rewriting step, the statement that Sandy graduated from State University with probability 1.0 is also included, resulting in her definitely being a person because of *graduatedfrom* having *person* as domain.

Tables 3 and 4 show the results of comparing *TIP-OWL* with ProbLog.

As expected our approach takes significantly more time loading the data as index structures have to be created and ProbLog only seems to do minimal preprocessing. The results in Table 4 show, however, that this effort is overcompensated by more efficient query answering.

Table 4 shows the time needed for answering queries over the full dataset and the reduced one. To accommodate for the fact that ProbLog always has to

Table 3. Dataset loading times (sec)

	full	filtered
ProbLog	24.393 sec	5.383 sec
SQL Loading	193.040 sec	12.163 sec
SQL Indexing	1 007.291 sec	47.427 sec

Table 4. Query performance in seconds, averaged over 10 runs

	Q_A	Q_B	Q_C	Q_D	Q_E	Q_F
ProbLog (filtered)	-	97.667	1.812	-	2.673	9.589
ProbLog (full)	-	-	-	-	-	-
Prob. SQL (filtered)	10.423	3.524	0.107	0.024	1.421	0.628
Prob. SQL (full)	8.846	5.488	0.097	0.011	0.888	0.617
SQL (full)	5.002	3.196	0.017	0.009	0.637	0.340

load the data anew, loading time has been subtracted from the query times for
ProbLog shown in Table 4.

The query response times clearly show that our database-driven approach is
more efficient for handling large datasets. ProbLog is not able to answer any of
the questions using the full dataset within a 30 min timeout. Also for the filtered
dataset, ProbLog fails for Q_A and Q_D with an out of memory error and not with
a timeout as for the full dataset. For the queries where both return answers,
our approach is between 15 and 30 times faster. We can observe that query
processing even becomes more efficient for the larger dataset. After analyzing
the generated query plans, we found that the query planner chooses a suboptimal
query plan for the smaller dataset. We suppose this is due to weaker statistics.
The overhead of the probabilistic SQL compared to the plain rewritten SQL
seems to be proportional to the number of computed answers. Q_A, Q_C and Q_D,
and Q_F, with a larger number of results, are twice to five times as slow as the
plain queries; Q_B and Q_E with very few results show almost no difference in
query time.

LUBM Dataset. Table 5 shows the results using the probabilistic LUBM
datasets. We only compared the performance of our implementation on different
size of the data. We ran the queries with a timeout of 60 min. ProbLog is not
able to handle even the smallest of those datasets.

The query response times show, that in general, the probabilistic reasoning
does not have a negative impact on scalability. Overall, the times increase linearly
in the size of the data. Query 1, which has a constant result that does not change
with the size of the dataset, also has a constant response time. When processing
Query 5, the database erroneously scans the complete table of data property
assertions, which takes most of the time for computing results. This could be

Table 5. Query response times (seconds) of our implementation on various sizes of the probabilistic LUBM dataset. A timeout (response time > 60 min) is denoted as "-".

	Q1	Q3	Q4	Q5	Q6	Q10	Q11	Q12	Q13	Q14
LUBM 1	< 0.1	< 0.1	1.5	2.7	0.7	0.3	0.2	0.8	0.3	0.3
LUBM 10	< 0.1	1.0	3.4	27.3	7.4	2.4	0.2	0.9	2.9	2.4
LUBM 100	< 0.1	32.2	50.5	350.9	74.2	33.2	0.3	2.0	76.2	33.2
LUBM 200	< 0.1	100.0	134.6	2 622.2	172.1	63.9	0.5	3.2	172.5	63.9
LUBM 500	< 0.1	201.8	440.2	-	508.0	192.0	1.1	6.9	612.1	1 941.9
LUBM 1000	< 0.1	643.8	904.7	-	874.7	365.1	1.6	232.3	1430.3	-

Table 6. Query mixes per hour (QMpH) for different dataset sizes. The number indicates how often the set of benchmark queries could be executed within one hour. The queries are executed in random order.

Dataset	LUBM 1	LUBM 10	LUBM 100	LUBM 200
QMpH	418.6	66.4	5.4	0.2
Optimum	514.3	75.0	5.8	1.0
%	81.4 %	88.5 %	93.1 %	20 %

alleviated by tuning the query planner, resulting in a better query execution plan. Query 13 and especially Query 14 produce a large number of results, thus they become I/O-bound for larger datasets, i.e. their performance is limited by disk speed, resulting in a large jump in the query time. The disks for our test virtual machine are attached via network, resulting in this large drop in performance.

To evaluate the scalability under a more realistic workload we tested how often the set of all ten queries can be executed within one hour (inspired by the BSBM benchmark). The queries are executed in random order. Table 6 shows the number of query mixes processed in an hour for different sizes of the LUBM dataset. Up to 70 million facts, the performance scales well and is close to the expected optimum based on the query times under ideal settings. For the smaller datasets, the response time are slightly farther away from the optimum due to a relatively higher overhead of establishing a database connection etc. Beyond 70 million facts, there is a huge drop in performance due to I/O-bound queries and the poor I/O performance of the virtual machine.

6   Related Work

Apart from ProbLog, two other systems for probabilistic reasoning with a similar semantic are Pronto [5] and Bundle [3]. They can handle probabilistic knowledge bases formulated in $\mathcal{SROIQ}$ and $\mathcal{SHOIN}(D)$, respectively. However, their main focus is not pOBDA, but probabilistic TBox reasoning (classification, satisfiability, ...), thus their performance in query answering is very limited. Both can only run simple instance checking for single individuals and classes. Probabilistic deductive databases [18] provide a similar solution, but to the best of our knowledge there is no system available and thus it is hard to estimate their scalability to large-scale knowledge bases.

Regarding the benchmark dataset, Klinov et al. [19] proposed a systematic approach to evaluate reasoning in probabilistic description logics which is, however, more geared towards complex TBoxes and not large-scale query answering. Lanti et al. [20] very recently published a dataset, based on real world data, specially tailored for benchmarking OBDA systems. They also provide a generator to scale the dataset in size. It will be interesting to analyze their dataset and also extend it for benchmarking probabilistic OBDA systems.

7 Conclusion and Future Work

In this paper we described a preliminary implementation of a probabilistic OBDA system for large-scale knowledge bases. It combines tractability for a certain class of queries with the benefits of ontology-based query rewriting. While making many simplifying assumptions the approach is well suited for large-scale knowledge bases with facts generated using machine learning techniques and provides a pragmatic alternative for theoretically more interesting but less feasible models as the one proposed in [3,5].

We used the NELL knowledge base as a real-world example and a probabilistic extension of LUBM to evaluate the system. We demonstrated that it scales well compared to another state-of-the-art system and is able to compute query answers in a reasonable amount of time for knowledge bases containing several million facts.

We plan to provide a stable implementation of the approach in the near future and address the problem of uncertain T-Box elements. Furthermore, we plan to develop a more thorough benchmark based on recent proposals for benchmarks specifically aimed at OBDA [20].

Acknowledgement. The authors want to thank Christian Meilicke for his ongoing support and fruitful discussions about the topic of this paper.

References

1. Artale, A., Calvanese, D., Kontchakov, R., Zakharyaschev, M.: The DL-Lite family and relations. J. Artif. Intell. Res. **36**, 1–69 (2009)
2. Lukasiewicz, T., Straccia, U.: Managing uncertainty and vagueness in description logics for the Semantic Web. Web Semant. **6**(4), 291–308 (2008)
3. Riguzzi, F., Bellodi, E., Lamma, E., Zese, R.: BUNDLE: a reasoner for probabilistic ontologies. In: Faber, W., Lembo, D. (eds.) RR 2013. LNCS, vol. 7994, pp. 183–197. Springer, Heidelberg (2013)
4. Riguzzi, F., Bellodi, E., Lamma, E., Zese, R.: Probabilistic description logics under the distribution semantics. Semant. Web-Interoperability, Usability, Applicability (2014, to appear)

5. Klinov, P., Parsia, B.: Pronto: a practical probabilistic description logic reasoner. In: Bobillo, F., Costa, P.C.G., d'Amato, C., Fanizzi, N., Laskey, K.B., Laskey, K.J., Lukasiewicz, T., Nickles, M., Pool, M. (eds.) URSW 2008-2010/UniDL 2010. LNCS, vol. 7123, pp. 59–79. Springer, Heidelberg (2013)

6. Niepert, M., Noessner, J., Stuckenschmidt, H.: Log-linear description logics. In: IJCAI International Joint Conference on Artificial Intelligence, pp. 2153–2158 (2011)

7. De Raedt, L., Kimmig, A., Toivonen, H.: ProbLog: a probabilistic prolog and its application in link discovery. In: IJCAI International Joint Conference on Artificial Intelligence, pp. 2468–2473 (2007)

8. Carlson, A., Betteridge, J., Kisiel, B.: Toward an architecture for never-ending language learning. In: Proceedings of the Conference on Artificial Intelligence, pp. 1306–1313 (2010)

9. Jung, J.C., Lutz, C.: Ontology-based access to probabilistic data with OWLQL. In: Cudré-Mauroux, P., Heflin, J., Sirin, E., Tudorache, T., Euzenat, J., Hauswirth, M., Parreira, J.X., Hendler, J., Schreiber, G., Bernstein, A., Blomqvist, E. (eds.) ISWC 2012, Part I. LNCS, vol. 7649, pp. 182–197. Springer, Heidelberg (2012)

10. Fader, A., Soderland, S., Etzioni, O.: Identifying relations for open information extraction. In: Proceedings of the Conference on Empirical Methods in Natural Language Processing, pp. 1535–1545 (2011)

11. Calvanese, D., Giacomo, G., Lembo, D., Lenzerini, M., Rosati, R.: Tractable reasoning and efficient query answering in description logics: the DL-Lite family. J. Autom. Reason. 39(3), 385–429 (2007)

12. Suciu, D., Olteanu, D., Ré, C., Koch, C.: Probabilistic Databases. Morgan & Claypool Publishers, San Rafael (2011)

13. Guo, Y., Pan, Z., Heflin, J.: LUBM: a benchmark for OWL knowledge base systems. Web Semant. 3, 158–182 (2005)

14. Lutz, C., Seylan, I., Toman, D., Wolter, F.: The combined approach to OBDA: taming role hierarchies using filters. In: Alani, H., Kagal, L., Fokoue, A., Biemann, C., Groth, P., et al. (eds.) ISWC 2013. LNCS, vol. 8218, pp. 314–330. Springer, Heidelberg (2013)

15. Schmidt, M., Hornung, T., Lausen, G., Pinkel, C.: SP2 Bench: a SPARQL performance benchmark. Data Eng. 25, 222–233 (2009)

16. Bizer, C., Schultz, A.: The Berlin SPARQL benchmark. Int. J. Semant. Web Inf. Syst. 5(2), 1–24 (2001)

17. Bail, S., Alkiviadous, S., Parsia, B., Workman, D., Van Harmelen, M., Goncalves, R.S., Garilao, C.: FishMark: a linked data application benchmark. In: CEUR Workshop Proceedings, vol. 943 (2012)

18. Lakshmanan, L.V.S., Sadri, F.: Probabilistic deductive databases. In: SLP, pp. 254–268, June 1994

19. Klinov, P., Parsia, B.: Optimization and evaluation of reasoning in probabilistic description logic: towards a systematic approach. In: Sheth, A.P., Staab, S., Dean, M., Paolucci, M., Maynard, D., Finin, T., Thirunarayan, K. (eds.) ISWC 2008. LNCS, vol. 5318, pp. 213–228. Springer, Heidelberg (2008)

20. Lanti, D., Rezk, M., Xiao, G., Calvanese, D.: The NPD benchmark : reality check for OBDA systems. In: Proceedings of the 18th International Conference on Extending Database Technology (2015)

Reasoning over Linear Probabilistic Knowledge Bases with Priorities

Nico Potyka[✉]

Department of Computer Science, Fern Universität in Hagen,
Hagen, Germany
nico.potyka@fernuni-hagen.de

Abstract. We consider the problem of reasoning over probabilistic knowledge bases with different priority levels. While we assume that the knowledge is consistent on each level, there can be inconsistencies between different levels. Examples arise naturally in hierarchical domains when general knowledge is overwritten with more specific information. We extend recent results on inconsistency-tolerant probabilistic reasoning to propose a solution for this problem.

1 Introduction

Often our evaluation of the likelihood of an event depends on the level of abstraction that we employ. For instance, you might agree that it is likely that a bird flies. However, a penguin is also a bird, but usually does not fly. It is intuitively clear that our beliefs about penguins are more specific and therefore should overwrite our beliefs about birds; but it can be difficult to automatically resolve conflicts betweens rules, in particular, if there are transitive dependencies.

Example 1. Let us consider a probabilistic version of an access control policy scenario from [4]. Suppose we have different files and different users and want to automatically deduce the probability that a user has access to a file. If the probability is 1, we might grant access immediately, otherwise we might send a confirmation request to the system administrator. If the probability is very low, say smaller than 0.1, we might want to send a warning in addition.

If no knowledge about the user and the file is available, the access probability should be 0. However, we want to have specialized rules for particular types of users and files. For instance, if we know that a user is an employee, we want to be less restrictive and increase the probability to 0.5. Of course there can be exceptions, for instance, if the file is confidential. On the other hand, this exception should not apply to executive managers. Finally, we never want to grant access if a user is blacklisted for some reason.

Obviously, we can make this example arbitrarily complex. The key problem is that if we ask for the probability that a user has access to a file, different rules may apply. How can our system decide, as autonomously as possible, which rules apply and which rules have to be excluded to avoid inconsistencies?

C. Beierle and A. Dekhtyar (Eds.): SUM 2015, LNAI 9310, pp. 121–136, 2015.
DOI: 10.1007/978-3-319-23540-0_9

This question is closely related to problems considered in belief merging [15] and non-monotonic reasoning [4,6,16] and several proposals have been made to deal with priorities in non-probabilistic logics [2,3,5,25]. There also exist several belief merging approaches for probabilistic logics when no priorities exist [1,7,13,26]. Whereas the goal in [13] is to consolidate the knowledge bases, the goal in [1,26] is to find a probability function that best captures all knowledge bases. In [7], a set of probability functions is considered, which is close to satisfying the inconsistent pieces of information (this will be made more precise at the end of the paper).

Our approach builds up on work in [22–24] and is related to ideas considered in [7] as will be discussed at the end of the paper. We suppose that our knowledge base consists of subsets of increasing priority and a set of integrity constraints that have to be maintained. Since we cannot assume that subsets of different priority are pairwise consistent, the overall knowledge base might be (and probably will be) inconsistent. That is, the knowledge base has no classical probabilistic models, i.e., there are no probability functions that satisfy all constraints in the knowledge base. We define two notions of priority models. *Strict priority models* are constructed by starting with the models of the integrity constraints. Then this set is successively decreased by selecting the best models with respect to a subset of our priority knowledge base starting with the subset of highest priority. This approach provides some nice guarantees for subsets with high priority, but subsets of low priority can be completely ignored. To overcome this problem, *weighted priority models* take all knowledge bases into account but weigh them with respect to their priority. We prove some interesting properties and illustrate both approaches by means of our access control policy example. We consider a general probabilistic framework, but illustrate the ideas by means of a relational probabilistic logic similar to those considered in [10,17].

The remainder of this paper is organized as follows: In Sect. 2, we explain our formal framework and discuss some basics from [22–24]. We then introduce and investigate priority knowledge bases, strict and weighted priority models in Sect. 3. In Sect. 4, we discuss related work and conclude in Sect. 5.

2   Linear Probabilistic Knowledge Bases and Generalized Models

To begin with, we describe a general framework to define our probabilistic knowledge bases. Let us assume that our knowledge can be represented by means of a set of *random variables* $\mathcal{X} = \{X_1, \ldots, X_n\}$. Each $X \in \mathcal{X}$ has a finite *domain* $dom(X)$. If $dom(X) = \{0, 1\}$, we call X a *Boolean* random variable. An *assignment* $(X_1 = x_1, \ldots, X_n = x_n)$ to $\mathcal{X}$ is sometimes abbreviated by $(x_1, \ldots, x_n)$ or just by x if the order of the variables in $\mathcal{X}$ is clear from the context or not important. If $\mathcal{Y} \subseteq \mathcal{X}$ and x is an assignment to $\mathcal{X}$, $x|_\mathcal{Y}$ denotes the assignment y to $\mathcal{Y}$ that is obtained from x by restricting to the variables in $\mathcal{Y}$. The set of all assignments to $\mathcal{X}$ is denoted by $\Omega_\mathcal{X}$ and is called the set of *possible worlds*.

Example 2. To model our access control policy, we consider a relational probabilistic language similar to [10,17]. We build up formulas over a finite set of typed predicate symbols, a finite set of typed individuals and an infinite set of (typed) variables. We allow the usual logical connectives, but do not allow quantifiers. Let us consider the types *User* and *File* and the predicates *grantAccess(User, File)*, *employee(User)*, *exec(User)*, *blacklisted(User)*, *confidential(File)*, where *exec* abbreviates *executive manager*. Let *alice* and *bob* be individuals of type *User* and let *file1*, *file2* be individuals of type *File*. We regard the 12 ground atoms *grantAccess (alice, file1)*, ..., *confidential(file2)* as Boolean random variables and our possible worlds correspond to truth assignments to the ground atoms.

Given a set of random variables $\mathcal{X}$, we denote by $\mathcal{P}_{\mathcal{X}}$ the set of all joint probability distributions over $\mathcal{X}$. If $P \in \mathcal{P}_{\mathcal{X}}$, $\mathcal{Y} \subseteq \mathcal{X}$, $\mathcal{Z} = \mathcal{X} \setminus \mathcal{Y}$, then the joint probability distribution $P_{\mathcal{Y}}$ over $\mathcal{Y}$ obtained from P by *marginalizing out* $\mathcal{Z}$ is $P_{\mathcal{Y}}(\mathcal{Y}) = \sum_z P(\mathcal{Y}, z)$, where the sum ranges over the variable assignments to $\mathcal{Z}$. Given a subset $\mathcal{Y} \subseteq \mathcal{X}$, a *linear probabilistic constraint function* l over $\mathcal{P}_{\mathcal{X}}$ with *scope* $\text{scope}(l) = \mathcal{Y}$ is a function $l : \mathcal{P}_{\mathcal{X}} \to \mathbb{R}$ that has the form

$$l(P) = \sum_y P_{\mathcal{Y}}(y) \, f_l(y),$$

where $f : \Omega_{\mathcal{Y}} \to \mathbb{R}$ is called the *feature function* of l. Roughly speaking, in probabilistic logics, constraint functions correspond to rules and feature functions indicate whether a world verifies or falsifies a rule, see, e.g., [10] for a detailed example. We say that $P \in \mathcal{P}_{\mathcal{X}}$ *satisfies* l iff $l(P) = 0$ and $l(P) = 0$ is called the *linear probabilistic constraint corresponding* to l. A *linear probabilistic knowledge base* over $\mathcal{P}_{\mathcal{X}}$ is a set $\mathcal{KB}$ consisting of linear probabilistic constraint functions over $\mathcal{P}_{\mathcal{X}}$. The *scope of* $\mathcal{KB}$ is the union of the scopes of the constraints in $\mathcal{KB}$, i.e., $\text{scope}(\mathcal{KB}) = \bigcup_{c \in \mathcal{KB}} \text{scope}(c)$. We say that $P \in \mathcal{P}_{\mathcal{X}}$ *satisfies* $\mathcal{KB}$ iff P satisfies all $l \in \mathcal{KB}$. The set

$$\text{Mod}(\mathcal{KB}) = \{P \in \mathcal{P}_{\mathcal{X}} \mid l(P) = 0 \text{ for all } l \in \mathcal{KB}\}$$

of all probability distributions satisfying $\mathcal{KB}$ is called the set of *models of* $\mathcal{KB}$. $\mathcal{KB}$ is called *consistent* if $\text{Mod}(\mathcal{KB}) \neq \emptyset$. Otherwise, $\mathcal{KB}$ is called *inconsistent*.

Remark 1. Note that each constraint function can as well be written as a sum over $\Omega_{\mathcal{X}}$:

$$l(P) = \sum_y P_{\mathcal{Y}}(y) \, f_l(y) = \sum_y \sum_z P(y, z) \, f_l(y) = \sum_{x \in \Omega_{\mathcal{X}}} P(x) \, f_l(x|y).$$

The second equation is obtained by putting in the definition of the marginal $P_{\mathcal{Y}}$, the third equation by using the fact that each two assignments y, z to $\mathcal{Y}, \mathcal{Z}$ correspond to an assignment x to $\mathcal{X} = \mathcal{Y} \uplus \mathcal{Z}$.

Example 3. In our running example, we represent rules by probabilistic conditionals $(\phi|\psi)[\rho]$, where the conclusion ϕ and the premise ψ are formulas in our language and $\rho \in [0,1]$ is a probability, c.f. [10,17]. For instance, the probabilistic conditional $(grantAccess(U,F) \mid confidential(F))[0]$ expresses intuitively that users usually do not have access to confidential files. If the premise ψ is tautological, we just omit it. For instance, $(blacklisted(U))[0.05]$ expresses that users are usually not blacklisted. We define the probability of a ground formula ϕ with respect to a joint probability distribution P to be

$$P(\phi) = \sum_{x \in \Omega} P(x)\, 1_{\{\phi\}}(x),$$

where the indicator function $1_{\{\phi\}}$ yields 1 iff ϕ evaluates to true under x in the usual sense (and 0 otherwise). P satisfies a ground conditional $(\phi|\psi)[\rho]$ iff

$$P(\phi \wedge \psi) = P(\psi) \cdot \rho.$$

Note that this definition coincides with conditional probability whenever $P(\psi) > 0$. P satisfies a general conditional $(\phi|\psi)[\rho]$ iff P satisfies each ground instance of $(\phi|\psi)[\rho]$. For instance, P satisfies $(blacklisted(U))[0.05]$, iff P satisfies its ground instances $(blacklisted(alice))[0.05]$ and $(blacklisted(bob))[0.05]$. To see that our conditionals indeed induce linear constraint functions, recall that P satisfies a ground conditional $(\phi|\psi)[\rho]$ iff

$$\begin{aligned}
0 &= P(\phi\psi) - P(\psi) \cdot \rho \;=\; P(\phi\psi) - (P(\phi\psi) + P(\overline{\phi}\psi)) \cdot \rho \\
&= P(\phi\psi) \cdot (1 - \rho) - P(\overline{\phi}\psi) \cdot \rho \\
&= \sum_{x \in \Omega} P(x)\, 1_{\{\phi\psi\}}(x) \cdot (1 - \rho) - \sum_{x \in \Omega} P(x)\, 1_{\{\overline{\phi}\psi\}}(x) \cdot \rho \\
&= \sum_{x \in \Omega} P(x) \cdot \underbrace{\left(1_{\{\phi\psi\}}(x) \cdot (1 - \rho) - 1_{\{\overline{\phi}\psi\}}(x) \cdot \rho\right)}_{:=f(\mathcal{Y})}.
\end{aligned}$$

From Remark 1, we see that this is a linear probabilistic constraint. The scope $\mathcal{Y}$ of the feature function f is the set of ground atoms appearing in $(\phi|\psi)[\rho]$ and the feature function is defined by $f(\mathcal{Y}) = 1_{\{\phi\psi\}}(x|y) \cdot (1 - \rho) - 1_{\{\overline{\phi}\psi\}}(x|y) \cdot \rho$. Note that the feature function yields $1 - \rho$ if the conditional is verified and $-\rho$ if the conditional is falsified. Correspondingly, each general conditional induces a set of constraints (one for each ground instance). For instance, $(blacklisted(U))[0.05]$ induces two constraints, one for $(blacklisted(alice))[0.05]$ and one for $(blacklisted(bob))[0.05]$. The scope of the first one is $blacklisted(alice)$, the scope of the second one is $blacklisted(bob)$. However, for the sake of clarity, we will usually just write the general conditional, but keep in mind that it represents several constraints.

To reason with probabilistic knowledge bases, we can use the probability distributions in $\mathrm{Mod}(\mathcal{KB})$ to compute (conditional) probabilities for arbitrary formulas. For instance, given a formula ϕ, the *probabilistic entailment problem* is to derive upper and lower bounds on $P(\phi)$ for $P \in \mathrm{Mod}(\mathcal{KB})$ [11,17,20]. Formally, we want to solve

$$\mathrm{opt}_{P \in \mathrm{Mod}(\mathcal{KB})}\, P(\phi),$$

where opt $\in \{\min, \max\}$. The *lower bound* l and the *upper bound* u on the probability of ϕ, is the result of the minimization an maximization problem, respectively.

Another way to reason with $\mathrm{Mod}(\mathcal{KB})$ is to find a unique probability function $P^* \in \mathrm{Mod}(\mathcal{KB})$ that optimizes some quality criterion like the entropy. Then one can use P^* directly to compute probabilities for formulas [10,12,21]. However, if $\mathcal{KB}$ is inconsistent, no such probability distribution exists and there is no way to infer reasonable information with these approaches.

To reason with inconsistent knowledge bases, we can replace $\mathrm{Mod}(\mathcal{KB})$ with a set of probability distributions which satisfy the knowledge base as best as possible [7,22]. The idea in [22] is to use those probability functions that *minimally violate* the knowledge base. To make this idea more precise, it is useful to represent linear constraint functions by matrices. To avoid ambiguity in this representation, we have to impose an ordering on the possible worlds and on the constraints. Let $N = |\Omega_\mathcal{X}|$ and consider an arbitrary but fixed order $x_1, \ldots, x_N$ of the worlds in $\Omega_\mathcal{X}$. Let l be a linear constraint over $\mathcal{P}_\mathcal{X}$. The *constraint matrix corresponding to* l is the $(1 \times N)$-matrix A_l which has the entry $f_l(x_j|y_i)$ at the j-th position for $1 \leq j \leq N$. Let $\mathcal{KB}$ be a linear probabilistic knowledge base over $\mathcal{P}_\mathcal{X}$, let $M = |\mathcal{KB}|$ and consider an arbitrary but fixed order $l_1, \ldots, l_M$ of the constraints in $\mathcal{KB}$. Then the *constraint matrix corresponding to* $\mathcal{KB}$ is the $(1 \times N)$-matrix

$$A_\mathcal{KB} = \begin{pmatrix} A_{l_1} \\ \ldots \\ A_{l_M} \end{pmatrix}.$$

To keep our notation simple, we identify probability functions P over $\Omega_\mathcal{X}$ with column vectors, whose i-th entry is the probability of the i-th world. Then P satisfies $\mathcal{KB}$ iff

$$A_\mathcal{KB} P = \begin{pmatrix} A_{l_1} P \\ \ldots \\ A_{l_M} P \end{pmatrix} = \begin{pmatrix} \sum_{x \in \Omega_\mathcal{X}} P(x)\, f_l(x|y_1) \\ \ldots \\ \sum_{x \in \Omega_\mathcal{X}} P(x)\, f_l(x|y_M) \end{pmatrix} = \begin{pmatrix} 0 \\ \ldots \\ 0 \end{pmatrix}.$$

Now given a knowledge base $\mathcal{KB}$ and some continuous vector norm $\|.\|$, we consider the following minimization problem:

$$\min_{P \in \mathcal{P}_\mathcal{X}} \|A_\mathcal{KB} P\| \tag{1}$$

The minimum exists; it is 0 if and only if $\mathcal{KB}$ is consistent [22]. In particular, in the latter case, the minimal solutions are just the models of $\mathcal{KB}$. Conversely, if $\mathcal{KB}$ is inconsistent, the minimal solutions minimally violate $\mathcal{KB}$ with respect to $\|.\|$. Therefore, the optimal solutions of (1) are called *generalized models of* $\mathcal{KB}$, see [22,23] for more details. Consistent probabilistic reasoning approaches can be generalized to inconsistency-tolerant probabilistic reasoning approaches by just replacing the models with the generalized models [23,24].

3 Linear Priority Knowledge Bases and Priority Models

Now let us get back to our initial problem. We have a knowledge base that contains rules with different levels of priority. Whereas we can assume that the knowledge base is consistent on each particular level, there can be conflicts between different levels. We could apply generalized reasoning approaches to the whole (inconsistent) knowledge base to deduce new information. However, the results will not necessarily reflect what we want. The reason is that we cannot define that some rules are more important than others. That is, instead of overwriting knowledge of low priority with knowledge of higher priority, we would merge the knowledge independently of the priority.

To overcome this problem, we will partition our knowledge base in subsets with different priority levels. In order to account for knowledge that has to be respected independently of the priority, we will also allow a set of integrity constraints that is guaranteed to be satisfied if it is consistent. We will call a Priority Knowledge Base *valid* iff the knowledge on each level is consistent with the integrity constraints.

Definition 1 (Linear Priority Knowledge Base, Validity). *Let X be a set of random variables. A* linear priority knowledge base over X *is a tuple $(\mathcal{KB}_1, \ldots, \mathcal{KB}_k, \mathcal{IC})$, where $\mathcal{KB}_1, \ldots, \mathcal{KB}_k, \mathcal{IC}$ are linear probabilistic knowledge bases over X. For $1 \leq i \leq k$, $\mathcal{KB}_i$ is called the* subset with priority i. *The elements in $\mathcal{IC}$ are called* integrity constraints. k *is called the* number of priority levels. *$(\mathcal{KB}_1, \ldots, \mathcal{KB}_k, \mathcal{IC})$ is called* valid *iff $\mathcal{KB}_i \cup \mathcal{IC}$ is consistent for $1 \leq i \leq k$.*

Remark 2. Note that validity implies that $\mathcal{IC}$ is consistent for otherwise $\mathcal{KB}_i \cup \mathcal{IC}$ is inconsistent for $1 \leq i \leq k$.

Example 4. Let us continue our running example and consider the priority knowledge base $\mathcal{KB} = (\mathcal{KB}_1, \mathcal{KB}_2, \mathcal{KB}_3, \mathcal{KB}_4, \mathcal{KB}_5, \mathcal{IC})$, where

$$\mathcal{KB}_1 = \{(grantAccess(U, F))[0], (blacklisted(U))[0.05]\}$$
$$\mathcal{KB}_2 = \{(grantAccess(U, F) \mid employee(U))[0.5],$$
$$(blacklisted(U) \mid employee(U))[0.01]\}$$
$$\mathcal{KB}_3 = \{(grantAccess(U, F) \mid confidential(F))[0]\}$$
$$\mathcal{KB}_4 = \{(grantAccess(U, F) \mid exec(U))[0.7],$$
$$(blacklisted(U) \mid exec(U))[0.001]\}$$
$$\mathcal{KB}_5 = \{(exec(alice))[1], (employee(bob))[1], (confidential(file1))[1]\}$$
$$\mathcal{IC} = \{(employee(U) \mid exec(U))[1], (grantAccess(U, F) \mid blacklisted(U)(F))[0]\}$$

On the first level, we define generic knowledge. If no knowledge is available, we do not want to grant access. Also, we make the assumption that it is rather unlikely that a user is blacklisted. On the second level, we increase the access probability and decrease the blacklist probability for employees. On level 3, we make an exception for confidential files. Afterwards, we further increase access probability and decrease blacklist probability for executive managers on level 4.

The last level contains domain knowledge. We know that *alice* is an executive manager, *bob* is an employee and *file1* is confidential. Our integrity constraints state that executive managers are employees and that we do not grant access to blacklisted users.

We will now make two proposals to define the models of priority knowledge bases.

3.1   Strict Priority Models

Our first approach is motivated by the desire to guarantee that all rules in the subset with the highest priority hold. After this goal is achieved, we look at the next subset successively. That is, we start with the models of our integrity constraints. Then we successively decrease this set by selecting the best models with respect to a subset of our priority knowledge base starting with the subset of highest priority. The following definition describes this approach precisely.

Definition 2 (Strict Priority Models). *Let* $\mathcal{KB} = (\mathcal{KB}_1, \ldots, \mathcal{KB}_k, \mathcal{IC})$ *be a linear priority knowledge base over* $\mathcal{X}$ *and let* $\|.\|$ *be some continuous vector norm. We let*

$$\mathrm{SPMod}_{\|.\|}^{k+1}(\mathcal{KB}) = \mathrm{Mod}(\mathcal{IC}) \quad and$$

$$\mathrm{SPMod}_{\|.\|}^{i}(\mathcal{KB}) = \arg \min_{P \in \mathrm{SPMod}_{\|.\|}^{i+1}(\mathcal{KB})} \|A_{\mathcal{KB}_i} P\| \ for \ i = k, \ldots, 1.$$

Let $\mathrm{SPMod}_{\|.\|}(\mathcal{KB}) = \mathrm{SPMod}_{\|.\|}^{1}(\mathcal{KB})$. *The elements in* $\mathrm{SPMod}_{\|.\|}(\mathcal{KB})$ *are called the* strict priority models *of* $\mathcal{KB}$.

Remark 3.   1.   To enhance readability, we usually omit the subscript $\|.\|$, but keep in mind that $\mathrm{SPMod}(\mathcal{KB})$ depends on the selected norm.
2.   Strict priority models are defined recursively. We let $\mathrm{SPMod}^{k+1}(\mathcal{KB})$ be the set of models of our integrity constraints $\mathcal{IC}$. Then we go backwards for $i = k, \ldots, 1$ and let $\mathrm{SPMod}^{i}(\mathcal{KB})$ be the set of probability distributions in $\mathrm{SPMod}^{i+1}(\mathcal{KB})$ that minimally violate the constraints in $\mathcal{KB}_i$.

Before looking at an example, we state some basic results of technical interest.

Lemma 1. *Let* $\mathcal{KB} = (\mathcal{KB}_1, \ldots, \mathcal{KB}_k, \mathcal{IC})$ *be a linear priority knowledge base over* $\mathcal{X}$ *and let* $\|.\|$ *be some continuous vector norm. If* $\mathcal{KB}$ *is valid, then*

1. $\mathrm{SPMod}^{i}(\mathcal{KB})$ *is non-empty, compact and convex for* $1 \leq i \leq k+1$.
2. $\emptyset \neq \mathrm{SPMod}(\mathcal{KB}) = \mathrm{SPMod}^{1}(\mathcal{KB}) \subseteq \cdots \subseteq \mathrm{SPMod}^{k}(\mathcal{KB})$.

Proof.   1.   First note that validity of $\mathcal{KB}$ implies that $\mathrm{Mod}(\mathcal{IC}) \neq \emptyset$. In particular, $\mathrm{Mod}(\mathcal{IC})$ is a subset of $\mathcal{P}_{\mathcal{X}}$ that is defined by linear equality constraints. Therefore, $\mathrm{SPMod}^{k+1}(\mathcal{KB}) = \mathrm{Mod}(\mathcal{IC})$ is also compact and convex. Now we proceed by induction and show that if $\mathrm{SPMod}^{i+1}(\mathcal{KB})$ is non-empty, compact and convex, so is $\mathrm{SPMod}^{i}(\mathcal{KB})$. Continuity of $\|.\|$ and compactness of

$\text{SPMod}^{i+1}(\mathcal{KB})$ imply that a minimum of $\arg\min_{P\in\text{SPMod}^{i+1}(\mathcal{KB})} \|A_{\mathcal{KB}_i}P\|$ exists and that the set of minima (that is, $\text{SPMod}^i(\mathcal{KB})$) is closed. As a subset of $\text{SPMod}^{i+1}(\mathcal{KB})$, $\text{SPMod}^i(\mathcal{KB})$ is also bounded and therefore compact. In particular, the objective function $f(x) = \|A_{\mathcal{KB}_i}x\|$ is convex since $\|.\|$ is convex (this follow from homogeneity and the triangular inequality for norms) and the composition of convex and linear functions is convex. This implies that $\text{SPMod}^i(\mathcal{KB})$ is also convex.

2. Non-emptiness follows from (1), the subset relationships follow from the definition. □

The practical importance of Lemma 1.1 is that it guarantees the existence of a minimum if we minimize some continuous function over $\text{SPMod}(\mathcal{KB})$ and that the minimum is unique if the function is also strictly convex. Lemma 1.2 is immediate, but is mentioned for emphasis. Given the strict priority models, Lemma 1.1 allows us to apply the usual reasoning approaches. For instance, we can compute upper and lower bounds on the probability of formulas [11,17,20] or select a best strict priority model to compute probability of formulas [10,12,21].

Example 5. Let us compute upper and lower bounds on some formulas for the knowledge base in Example 4. To keep things simple, we will only ask for the probability of ground formulas. More strictly speaking, given a ground formula ϕ, we want to solve

$$\text{opt}_{P\in\text{SPMod}(\mathcal{KB})} P(\phi),$$

where $\text{opt} \in \{\min, \max\}$. Like for the Probabilistic Entailment Problem, the *lower bound* l and the *upper bound* u on the probability of ϕ, is the result of the minimization an maximization problem, respectively. We write $\phi[l, u]$ to denote the result. If $l \approx u$, we sometimes just write $\phi[l]$ to enhance readability. For instance, we have the following rounded results when using the Euclidean norm to determine our strict priority models:

grantAccess(alice, file1)[0.7]	*grantAccess(bob, file1)*[0]
grantAccess(alice, file2)[0.7]	*grantAccess(bob, file2)*[0.5]
blacklisted(alice)[0.0001]	*blacklisted(bob)*[0.01].

Recall that *alice* is an executive manager and that *file1* is confidential. Note that the first query shows that the knowledge about executive managers in $\mathcal{KB}_4$ suppresses the knowledge about confidential files in $\mathcal{KB}_3$ as desired.

What can we say about strict priority models in general? The following proposition states that valid linear priority knowledge bases always have strict priority models and that these always satisfy our integrity constraints and the subset with highest priority.

Proposition 1 (Upmost Consistency). *Let* $\mathcal{KB} = (\mathcal{KB}_1, \ldots, \mathcal{KB}_k, \mathcal{IC})$ *be a linear priority knowledge base over* $\mathcal{X}$ *and let* $\|.\|$ *be some continuous vector norm. If* $\mathcal{KB}$ *is valid, then*

$$\emptyset \neq \text{SPMod}(\mathcal{KB}) \subseteq \text{Mod}(\mathcal{KB}_k \cup \mathcal{IC}). \tag{2}$$

Proof. Since $\mathrm{SPMod}(\mathcal{KB}) = \mathrm{SPMod}^1(\mathcal{KB})$, it follows that $\emptyset \neq \mathrm{SPMod}(\mathcal{KB}) \subseteq \mathrm{SPMod}^k(\mathcal{KB})$ from Lemma 1.2. Therefore, it suffices to show that $\mathrm{SPMod}^k(\mathcal{KB}) = \mathrm{Mod}(\mathcal{KB}_k \cup \mathcal{IC})$ to prove the claim. By validity, $\mathcal{KB}_k \cup \mathcal{IC}$ is consistent. Therefore, we have $\min_{P \in \mathrm{SPMod}^{k+1}(\mathcal{KB})} \|A_{\mathcal{KB}_k} P\| = 0$ and the minimal elements are models of $\mathcal{KB}_k$. In particular, they are also models of $\mathcal{IC}$ because we optimize over $\mathrm{SPMod}^{k+1}(\mathcal{KB})$. Hence, $\mathrm{SPMod}^k(\mathcal{KB}) \subseteq \mathrm{Mod}(\mathcal{KB}_k \cup \mathcal{IC})$. Conversely, if $P \in \mathrm{Mod}(\mathcal{KB}_k \cup \mathcal{IC})$, then $A_{\mathcal{KB}_k} P = 0$ and therefore $P \in \mathrm{SPMod}^k(\mathcal{KB})$. Hence, $\mathrm{SPMod}^k(\mathcal{KB}) = \mathrm{Mod}(\mathcal{KB}_k \cup \mathcal{IC})$, which completes the proof. $\square$

So strict priority models always satisfy the integrity constraints and the rules of highest priority. In fact, a slightly stronger property holds. If all knowledge bases from level l up to k are consistent, then the strict priority models are models of $\mathcal{KB}_l, \ldots, \mathcal{KB}_k$ and of $\mathcal{IC}$.

Proposition 2 (Upward Consistency). *Let $\mathcal{KB} = (\mathcal{KB}_1, \ldots, \mathcal{KB}_k, \mathcal{IC})$ be a linear priority knowledge base over $\mathcal{X}$ and let $\|.\|$ be some continuous vector norm. If $\mathcal{KB}$ is valid and $\bigcup_{i=l}^{k} \mathcal{KB}_i$ is consistent, then*

$$\mathrm{SPMod}(\mathcal{KB}) \subseteq \mathrm{Mod}(\bigcup_{i=l}^{k} \mathcal{KB}_i \cup \mathcal{IC}). \tag{3}$$

Proof. Like in the proof of Proposition 1, it suffices to show that $\mathrm{SPMod}^l(\mathcal{KB}) = \mathrm{Mod}(\bigcup_{i=l}^{k} \mathcal{KB}_i \cup \mathcal{IC})$. We prove the claim by induction on the difference $d = k - l$. For $d = 0$, i.e., $l = k$, we proved the claim in Proposition 1. Now suppose that the claim holds for all natural numbers lower than d and consider $\bigcup_{i=k-d-1}^{k} \mathcal{KB}_i$. Since $\bigcup_{i=k-d-1}^{k} \mathcal{KB}_i$ is consistent by assumption, so is $\bigcup_{i=k-d}^{k} \mathcal{KB}_i$ and therefore $\mathrm{SPMod}(\mathcal{KB})^{k-d} = \mathrm{Mod}(\bigcup_{i=k-d}^{k} \mathcal{KB}_i \cup \mathcal{IC})$ by our induction hypothesis. Now,

$$\mathrm{SPMod}^{k-d-1}(\mathcal{KB}) = \arg \min_{P \in \mathrm{SPMod}^{k-d}(\mathcal{KB})} \|A_{\mathcal{KB}_{k-d-1}} P\|$$

$$= \arg \min_{P \in \mathrm{Mod}(\bigcup_{i=k-d}^{k} \mathcal{KB}_i \cup \mathcal{IC})} \|A_{\mathcal{KB}_{k-d-1}} P\|.$$

Since, $\bigcup_{i=k-d-1}^{k} \mathcal{KB}_i$ is consistent by assumption, we can proceed like in the proof of Proposition 1 to show that $\min_{P \in \mathrm{Mod}(\bigcup_{i=k-d}^{k} \mathcal{KB}_i \cup \mathcal{IC})} \|A_{\mathcal{KB}_{k-d-1}} P\| = 0$ and to conclude from this that $\mathrm{SPMod}^{k-d-1}(\mathcal{KB}) = \mathrm{Mod}(\bigcup_{i=k-d-1}^{k} \mathcal{KB}_i \cup \mathcal{IC})$. $\square$

The assumption of Proposition 2 will usually be only satisfied for subsets with high priority. What can we say about subsets with low priority? Intuitively, there should be at least some guarantees for constraints that are independent of higher priority levels. Indeed, the following proposition states that if there is a subset of constraints $C \subseteq \mathcal{KB}_l$ on level l whose scope is disjunct from the scope of all knowledge bases with level greater than l, than C is still satisfiable on level $l + 1$.

Proposition 3 (Upward Independence). *Let $\mathcal{KB} = (\mathcal{KB}_1, \ldots, \mathcal{KB}_k, \mathcal{IC})$ be a valid linear priority knowledge base over $\mathcal{X}$ and let $\|.\|$ be some continuous vector norm. Let $C \subseteq \mathcal{KB}_l$ for some $l < k$ such that $\mathrm{scope}(C) \cap \mathrm{scope}(\bigcup_{p=l+1}^{k} \mathcal{KB}_p) = \emptyset$. Then*

$$\mathrm{Mod}(C) \cap \mathrm{SPMod}^{l+1}(\mathcal{KB}) \neq \emptyset. \tag{4}$$

Proof. First note that by validity, $\mathrm{Mod}(C) \neq \emptyset$ (for otherwise $\mathcal{KB}_l$ and hence $\mathcal{KB}_l \cup \mathcal{IC}$ were inconsistent) and $\mathrm{SPMod}^{l+1}(\mathcal{KB}) \neq \emptyset$. Let $\mathcal{Y} = \mathrm{scope}(C)$ and let $\mathcal{Z} = \mathcal{X} \setminus \mathcal{Y}$. Let $P^{(1)} \in \mathrm{Mod}(C)$ and let $P_{\mathcal{Y}}^{(1)}$ denote the probability distribution obtained from $P^{(1)}$ by marginalizing out $\mathcal{Z}$. Let $P^{(2)} \in \mathrm{SPMod}^{l+1}(\mathcal{KB})$ and let $P_{\mathcal{Z}}^{(2)}$ denote the probability distribution obtained from $P^{(2)}$ by marginalizing out $\mathcal{Y}$. Then $P(y, z) = P_{\mathcal{Y}}^{(1)}(y) P_{\mathcal{Z}}^{(2)}(z)$ is a probability distribution over $\mathcal{X}$ since for all assignments x to $\mathcal{X}$, we have $P(x) = P_{\mathcal{Y}}^{(1)}(y) P_{\mathcal{Z}}^{(2)}(z) \geq 0$ and $\sum_x P(x) = \sum_y \sum_z P(y, z) = (\sum_y P_{\mathcal{Y}}^{(1)}(y))(\sum_z P_{\mathcal{Z}}^{(2)}(z)) = 1$. Furthermore, for all $c^{(1)} \in \mathrm{SPMod}^{l+1}(\mathcal{KB})$ with scope $\mathrm{scope}(c^{(1)}) = \mathcal{Y}^{(1)}$, we have

$$c^{(1)}(P) = \sum_y \sum_z P_{\mathcal{Y}}^{(1)}(y) P_{\mathcal{Z}}^{(2)}(z)\, f_{c^{(1)}}(y|_{\mathcal{Y}^{(1)}})$$
$$= (\sum_z P_{\mathcal{Z}}^{(2)}(z))(\sum_y P_{\mathcal{Y}}^{(1)}(y)\, f_{c^{(1)}}(y|_{\mathcal{Y}^{(1)}})) = 0,$$

since $P^{(1)} \in \mathrm{Mod}(c^{(1)})$. Hence, $P \in \mathrm{Mod}(C)$. Analogously, it follows that for all $c^{(2)} \in \mathrm{SPMod}^{l+1}(\mathcal{KB})$, we have that $c^{(2)}(P) = c^{(2)}(P^{(2)})$. But this implies that also $P \in \mathrm{SPMod}^{l+1}(\mathcal{KB})$ (for $P^{(2)}$ and P yield the same objective value for all optimization problems) and therefore $\mathrm{Mod}(C) \cap \mathrm{SPMod}^{l+1}(\mathcal{KB}) \neq \emptyset$. $\square$

So if C is independent of subsets of higher priority, we know that there is at least one model of C in $\mathrm{SPMod}^{l+1}(\mathcal{KB})$. However, this does not mean that a model of C will be in $\mathrm{SPMod}^{l}(\mathcal{KB})$ because the probability distributions in $\mathrm{Mod}(C) \cap \mathrm{SPMod}^{l+1}(\mathcal{KB})$ might strongly violate the remaining constraints in $\mathcal{KB}_l \setminus C$. This case, however, is only possible if there are dependencies between C and $\mathcal{KB}_l \setminus C$ as explained in the following proposition. In fact, if there are no such dependencies, then even $\mathrm{SPMod}^{l}(\mathcal{KB}) \subseteq \mathrm{Mod}(C)$ holds, i.e., all $P \in \mathrm{SPMod}^{l}(\mathcal{KB})$ are models of C.

Proposition 4 (Level Independence). *Let $\mathcal{KB} = (\mathcal{KB}_1, \ldots, \mathcal{KB}_k, \mathcal{IC})$ be a valid linear priority knowledge base over $\mathcal{X}$ and let $\|.\|$ be some continuous vector norm. Let $C \subseteq \mathcal{KB}_l$ for some $l < k$ such that $\mathrm{Mod}(C) \cap \mathrm{SPMod}^{l+1}(\mathcal{KB}) \neq \emptyset$. If $\mathrm{scope}(C) \cap \mathrm{scope}(\mathcal{KB}_l \setminus C) = \emptyset$, then*

$$\mathrm{SPMod}^{l}(\mathcal{KB}) \subseteq \mathrm{Mod}(C). \tag{5}$$

Proof. For the sake of contradiction, assume that $P^{(l)} \in \mathrm{SPMod}^{l}(\mathcal{KB})$ and that $P^{(l)} \notin \mathrm{Mod}(C)$. Let $\mathcal{Y} = \mathrm{scope}(C)$ and let $\mathcal{Z} = \mathcal{X} \setminus \mathcal{Y}$. Then $\mathrm{scope}(\mathcal{KB}_l \setminus C) \subseteq \mathcal{Z}$ by assumption. Let $P^C \in (\mathrm{Mod}(C) \cap \mathrm{SPMod}^{l+1}(\mathcal{KB}))$ and let $P_{\mathcal{Y}}^C$ denote the probability distribution obtained from P^C by marginalizing out $\mathcal{Z}$. Let $P_{\mathcal{Z}}^{(l)}$ denote the probability distribution obtained from $P^{(l)}$ by marginalizing out $\mathcal{Y}$. Just like in the proof of Proposition 3, we can check that $P(y, z) = P^C(y) P_{\mathcal{Z}}^{(l)}(z)$ is a probability distribution over $\mathcal{X}$ that coincides with P^C for the constraints in C and that coincides with $P^{(l)}$ for the constraints in $\mathcal{KB}_l \setminus C$. That is, $c(P) = c(P^{(l)})$ for all $c \in \mathcal{KB}_l \setminus C$ and $c(P) = 0$ for all $c \in C$ since $P^C \in \mathrm{Mod}(C)$. This implies in particular that $P \in \mathrm{SPMod}^{l+1}(\mathcal{KB})$. But since $P^{(l)} \notin \mathrm{Mod}(C)$, there is a $c \in C$ such that $c(P^{(l)}) \neq 0$. But this means that $\|A_{\mathcal{KB}_l} P\| < \|A_{\mathcal{KB}_l} P^{(l)}\|$ contradicting $P^{(l)} \in \mathrm{SPMod}^l(\mathcal{KB})$ (for then $\|A_{\mathcal{KB}_l} P^{(l)}\| = \min_{P' \in \mathrm{SPMod}^{i+1}(\mathcal{KB})} \|A_{\mathcal{KB}_i} P'\|$). Hence, if $P^{(l)} \in \mathrm{SPMod}^l(\mathcal{KB})$, then $P^{(l)} \in \mathrm{Mod}(C)$ must also hold. $\qquad\square$

Remark 4. Note that by the subset relationships from Lemma 1.2, $\mathrm{SPMod}^l(\mathcal{KB}) \subseteq \mathrm{Mod}(C)$ implies that $\mathrm{SPMod}(\mathcal{KB}) \subseteq \mathrm{Mod}(C)$. That is, if the assumptions of Upward and Level Independence are satisfied for C, then each strict priority model of $\mathcal{KB}$ will also be a model of C.

3.2 Weighted Priority Models

Even though strict priority models have some nice properties, they cannot guarantee that subsets of low priority have any influence on the final outcome of $\mathrm{SPMod}(\mathcal{KB})$ unless they are consistent with or independent of the upper levels. In fact, in some extreme cases, $\mathrm{SPMod}^l(\mathcal{KB})$ might contain only a single distribution for some $l > 1$. Then $\mathrm{SPMod}^{l'}(\mathcal{KB}) = \mathrm{SPMod}^l(\mathcal{KB})$ whenever $1 \leq l' < l$. In order to allow that each subset of our priority knowledge base has some influence on the final outcome, let us consider another approach to define models of prioritized knowledge bases. Instead of considering the subsets successively based on their priorities, we consider them simultaneously but weigh them with respect to their priority.

Definition 3 (Weighted Priority Models). *Let* $\mathcal{KB} = (\mathcal{KB}_1, \ldots, \mathcal{KB}_k, \mathcal{IC})$ *be a linear priority knowledge base over* $\mathcal{X}$, *let* $\|.\|$ *be some continuous vector norm and let* $\mathrm{w} : \{1, 2, \ldots, k\} \to \mathbb{R}_{>0}$ *be some monotonically increasing weight function. We let*

$$\mathrm{WPMod}_{\|\cdot\|}^{\mathrm{w}}(\mathcal{KB}) = \arg \min_{P \in \mathrm{Mod}(\mathcal{IC})} \left\| \begin{pmatrix} \mathrm{w}(1) \cdot A_{\mathcal{KB}_1} \\ \cdots \\ \mathrm{w}(k) \cdot A_{\mathcal{KB}_k} \end{pmatrix} P \right\|$$

and call $\mathrm{WPMod}_{\|\cdot\|}^{\mathrm{w}}(\mathcal{KB})$ *the set of* weighted priority models *of* $\mathcal{KB}$.

Remark 5. 1. Again, we omit the superscript w and the subscript $\|.\|$ to enhance readability, but keep in mind that $\mathrm{SPMod}(\mathcal{KB})$ depends on both.
2. The symbol $\cdot$ denotes scalar multiplication. Hence, each row in $A_{\mathcal{KB}_i}$ is multiplied by $\mathrm{w}(i)$.

WPMod($\mathcal{KB}$) has the same nice properties like SPMod($\mathcal{KB}$). The claim follows from similar arguments like Lemma 1.1, so that we omit the proof.

Lemma 2. *Let $\mathcal{KB} = (\mathcal{KB}_1, \ldots, \mathcal{KB}_k, \mathcal{IC})$ be a linear priority knowledge base over $\mathcal{X}$ and let $\|.\|$ be some continuous vector norm. If $\mathcal{KB}$ is valid, then WPMod($\mathcal{KB}$) is non-empty, compact and convex.*

Hence, in particular, WPMod($\mathcal{KB}$) is always non-empty and by definition a subset of Mod($\mathcal{IC}$). We emphasize this as a counterpart to Propositions 1 and 2. Note that we can guarantee only that the integrity constraints are satisfied.

Proposition 5 (Integrity). *Let $\mathcal{KB} = (\mathcal{KB}_1, \ldots, \mathcal{KB}_k, \mathcal{IC})$ be a linear priority knowledge base over $\mathcal{X}$ and let $\|.\|$ be some continuous vector norm. If $\mathcal{KB}$ is valid, then*

$$\emptyset \neq \text{WPMod}(\mathcal{KB}) \subseteq \text{Mod}(\mathcal{IC}).$$

Example 6. Let us compute probability bounds like in Example 4, but this time using weighted priority models. We use again the Euclidean norm and the weight function $\text{w}(p) = 2 \cdot p$ (in theory, we might have used the identity function as well, but it caused numerical problems). This yields the following rounded results:

$grantAccess(alice, file1)[0.44]$ $grantAccess(bob, file1)[0.14]$

$grantAccess(alice, file2)[0.63]$ $grantAccess(bob, file2)[0.4]$

$blacklisted(alice)[0.005]$ $blacklisted(bob)[0.018].$

The fact that the access probability for *file2* is significantly lower for *alice* than for *bob* indicates that all levels have been taken into account. However, given the access probability of *alice* for *file1*, one might argue that the lower levels have too much influence (*alice* being an executive manager should weigh stronger than *file1* being confidential). To increase the weight of the upper priority levels, let us consider a non-linear weight function. We let $\text{w}(p) = 10^{p-1}$ for $1 \leq p \leq 5$. This yields the following rounded results:

$grantAccess(alice, file1)[0.693]$ $grantAccess(bob, file1)[0.005]$

$grantAccess(alice, file2)[0.7]$ $grantAccess(bob, file2)[0.5]$

$blacklisted(alice)[0.001]$ $blacklisted(bob)[0.01].$

There is still a minor decrease in the access probability of *alice* for *file1*, but overall the results are very close to what one might expect when looking at the priority knowledge base from Example 4.

There is probably no immediate counterpart to the independence properties of strict priority models if we do not make any restrictions on the weight function. In fact, the whole point of weighted priority models is to allow that each subset of the knowledge base influences the outcome, so that we should not expect strong independence properties between priority levels. We might prove some weaker independence properties which do not make use of the priorities, but we leave this for future work.

3.3 Implementations

Probabilistic Entailment with strict and with weighted priority models has been implemented in the Java library Log4KR[1]. The optimization problems are solved by OjAlgo[2]. You can find the source code and some source examples in the subdirectory

`edu.cs.ai.log4KR.structuredLogics.priorityReasoning`

of the corresponding directories. Note that numerical problems might cause odd results.

4   Related Work

If we consider a trivial priority knowledge base consisting only of a single subset with priority 1 and do not demand that this subset is consistent, both the strict and the weighted priority models correspond to the generalized models from [22–24]. In this sense, prioritized reasoning generalizes generalized reasoning approaches from [23, 24]. Generalized reasoning, in turn, generalizes common probabilistic reasoning [10–12, 17, 20, 21] in the sense that the generalized models are the usual probabilistic models if the knowledge base is consistent.

Daniel generalized probabilistic models in a similar way and called his generalization the *best candidates* [7]. To define best candidates, he identified linear constraint functions with the hyperplanes corresponding to their solution sets. Given a probability function P and a linear constraint c, he defined the *gap* between P and c as the Euclidean distance between P and the hyperplane corresponding to c. The best candidates can then be defined as the solution set

$$\arg\min_{P \in \mathcal{P}_{\mathcal{X}}} \prod_{c \in \mathcal{R}} h(\sqrt{2^n}\, \mathrm{gap}(P, c)),$$

where h is some strictly decreasing, (strictly) positive and continuous log-concave function such that $h(0) = 1$, see [7], Definition 13. The best candidates satisfy similar nice properties like the generalized models, namely they form a compact and convex set, which corresponds to the usual models if $\mathcal{R}$ is consistent. Daniel considered only reasoning with the best candidate having maximum entropy, but, in principle, the best candidates can also be applied to other probabilistic reasoning approaches.

Figure 1 illustrates the relationships between the different notions of models.

---

[1] https://www.fernuni-hagen.de/wbs/research/log4kr/index.html.
[2] http://ojalgo.org/.

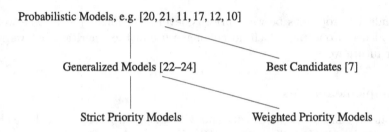

Fig. 1. Generalizations of Probabilistic Models: Generalized Models and Best Candidates generalize Probabilistic Models. Strict and Weighted Priority Models generalize Generalized Models (see Sect. 4 for details).

5   Conclusions

We proposed two methods to reason with linear probabilistic knowledge bases with priorities. *Strict priority models* provide some nice guarantees. The rules with highest priority are guaranteed to be satisfied; the same is true for rules with lower priority if they are consistent with or independent of higher priority levels. However, high priority rules can be so restrictive that low priority rules become meaningless. In such cases, *weighted priority models* can be more appropriate. If we make no restrictions on the weight function, they can only guarantee that the integrity constraints are satisfied. However, sometimes this is just what we want to guarantee that even low priority rules are taken into account.

Our results hold for linear probabilistic knowledge bases in general. These arise naturally from different probabilistic logics, see, e.g., [9,10,14,20] for some examples beyond our simple relational language. In fact, the results can be generalized to inequality constraints by just introducing slack variables as done in [8] for generalized models. Inequality constraints are, indeed, desirable to allow imprecise probabilities like in [9,17–19]. However, since the notation becomes more cumbersome, we did not consider inequality constraints here.

We also did not discuss computational aspects. However, note that we can apply similar ideas like in [23,24] to show that several interesting reasoning problems for priority knowledge bases can be solved by convex programming techniques. For instance, computing upper and lower bounds on the probability of formulas corresponds to a convex program. If we restrict to p-Norms, the problem remains convex if we allow conditional probabilities and becomes quadratic for $p = 2$ and linear for $p = 1, \infty$.

References

1. Adamcık, M.: Collective Reasoning under Uncertainty and Inconsistency. Ph.D. thesis, University of Manchester (2014)
2. Amgoud, L., Kaci, S.: An argumentation framework for merging conflicting knowledge bases. Int. J. Approximate Reasoning **45**(2), 321–340 (2007)

3. Benferhat, S., Dubois, D., Kaci, S., Prade, H.: Possibilistic merging and distance-based fusion of propositional information. Ann. Math. Artif. Intell. **34**(1–3), 217–252 (2002)
4. Bonatti, P.A., Faella, M., Sauro, L.: Adding default attributes to EL++. In: Proceedings of the Twenty-Fifth AAAI Conference on Artificial Intelligence, AAAI 2011 (2011)
5. Brewka, G.: Reasoning about priorities in default logic. In: AAAI 1994, pp. 940–945 (1994)
6. Britz, K., Heidema, J., Meyer, T.A.: Semantic preferential subsumption. In: KR 2008, pp. 476–484 (2008)
7. Daniel, L.: Paraconsistent Probabilistic Reasoning. Ph.D. thesis, L'École Nationale Supérieure des Mines de Paris (2009)
8. De Bona, G., Finger, M.: Measuring inconsistency in probabilistic logic: rationality postulates and dutch book interpretation. Artificial Intelligence (2015, to appear)
9. De Bona, G., Cozman, F.G., Finger, M.: Towards classifying propositional probabilistic logics. J. Appl. Logic **12**(3), 349–368 (2014)
10. Fisseler, J.: First-order probabilistic conditional logic and maximum entropy. Logic J. IGPL **20**(5), 796–830 (2012)
11. Jaumard, B., Hansen, P., Poggi, M.: Column generation methods for probabilistic logic. ORSA - J. Comput. **3**(2), 135–148 (1991)
12. Kern-Isberner, G.: Conditionals in Nonmonotonic Reasoning and Belief Revision. LNCS (LNAI), vol. 2087. Springer, Heidelberg (2001)
13. Kern-Isberner, G., Rödder, W.: Belief revision and information fusion in a probabilistic environment. In: Proceedings 16th International FLAIRS Conference, FLAIRS 2003, pp. 506–510. AAAI Press, Menlo Park (2003)
14. Kern-Isberner, G., Thimm, M.: Novel semantical approaches to relational probabilistic conditionals. In: Lin, F., Sattler, U., Truszczynski, M. (eds.) Proceedings Twelfth International Conference on the Principles of Knowledge Representation and Reasoning, KR 2010, pp. 382–391. AAAI Press (2010)
15. Konieczny, S., Pérez, R.P.: Logic based merging. J. Philos. Logic **40**(2), 239–270 (2011)
16. Kraus, S., Lehmann, D., Magidor, M.: Nonmonotonic reasoning, preferential models and cumulative logics. Artif. Intell. **44**(1), 167–207 (1990)
17. Lukasiewicz, T.: Probabilistic deduction with conditional constraints over basic events. J. Artif. Intell. Res. **10**, 380–391 (1999)
18. Lukasiewicz, T.: Expressive probabilistic description logics. Artif. Intell. **172**(6), 852–883 (2008)
19. Lutz, C., Schröder, L.: Probabilistic description logics for subjective uncertainty. In: Proceedings of KR 2010. AAAI Press (2010)
20. Nilsson, N.J.: Probabilistic logic. Artif. Intell. **28**, 71–88 (1986)
21. Paris, J.B., Vencovská, A.: On the applicability of maximum entropy to inexact reasoning. Int. J. Approximate Reasoning **3**(1), 1–34 (1989)
22. Potyka, N.: Linear programs for measuring inconsistency in probabilistic logics. In: Proceedings KR 2014. AAAI Press (2014)
23. Potyka, N., Thimm, M.: Consolidation of probabilistic knowledge bases by inconsistency minimization. In: Proceedings ECAI 2014, pp. 729–734. IOS Press (2014)

24. Potyka, N., Thimm, M.: Probabilistic reasoning with inconsistent beliefs using inconsistency measures. In: International Joint Conference on Artificial Intelligence 2015 (IJCAI 2015) (2015, to appear)
25. Qi, G., Liu, W., Bell, D.A.: Merging stratified knowledge bases under constraints. In: The Twenty-First National Conference on Artificial Intelligence and the Eighteenth Innovative Applications of Artificial Intelligence Conference, July 16–20, 2006, Boston, Massachusetts, USA, vol. 21, p. 281. AAAI Press, Menlo Park; MIT Press, Cambridge 1999 (2006)
26. Wilmers, G.: A foundational approach to generalising the maximum entropy inference process to the multi-agent context. Entropy 17(2), 594–645 (2015)

Intelligent Data Analytics

Evenness-Based Reasoning with Logical Proportions Applied to Classification

Myriam Bounhas[1,2]([⊠]), Henri Prade[3,4], and Gilles Richard[3]

[1] LARODEC Laboratory, ISG de Tunis, Le Bardo, Tunisia
myriam_bounhas@yahoo.fr
[2] Emirates College of Technology, Abu Dhabi, UAE
[3] IRIT, University of Toulouse, Toulouse, France
{prade,richard}@irit.fr
[4] QCIS, University of Technology, Sydney, Australia

Abstract. An approach to classification, based on a formal modeling of analogical proportions linking the features of 4 items, has been recently shown to be surprisingly successful on difficult benchmarks. Analogical proportions are homogeneous logical proportions. Homogeneous here refers both to the structure of their logical expression and to the specificity of their truth tables. In contrast, heterogeneous proportions express that there is an intruder among 4 truth values, which is forbidden to appear in a specific position. The 2 types of proportions are of an opposite nature. However heterogeneous proportions can also be considered as a basis for classification by considering that a new item can be added to a class only if its addition leaves the class as even as possible: the new item should rarely be an intruder with respect to any triple of items known to be in the class. Experiments show that this new evenness-based classifier gets good results on a number of representative benchmarks. Its accuracy is both compared to the ones of well-known classifiers and to previous analogy-based classifiers. A discussion investigates on what type of particular benchmarks the evenness-based classifiers outperform the analogical ones and when it is the opposite.

1 Introduction

It has been acknowledged for a long time that proportions play an important role in our perception and understanding of reality. Indeed proportions are a matter of comparisons expressed by differences or ratios that are equated to other differences or ratios. Two centuries ago, Gergonne [4,5] was the first to explicitly relate numerical (geometric) proportions to the ideas of interpolation and regression.

It is only in the last decade that *analogical* proportions, i.e., statements of the form A is to B as C is to D, where each capital letter refers to a situation described by a vector of feature values, have been formalized first in terms of subsets of properties that hold true in a given situation [6,15], and then in a logical manner [9]. Quite early, it was shown that a formal view of analogical

© Springer International Publishing Switzerland 2015
C. Beierle and A. Dekhtyar (Eds.): SUM 2015, LNAI 9310, pp. 139–154, 2015.
DOI: 10.1007/978-3-319-23540-0_10

proportions may be the basis of a new type of classifier that performs well on some difficult benchmarks [1,8]. This was confirmed by other implementations directly based on a logical view of analogical proportions [3].

Besides, it was shown that analogical proportions belong to a larger family of so-called logical proportions that relate a 4-tuple of Boolean variables [11], where the 8 logical proportions that are code-independent are of particular interest since their truth status remain unchanged if a property is encoded positively or negatively. These 8 logical proportions divide into 4 *homogeneous* proportions, which include the analogical proportion and 3 related proportions, and 4 *heterogeneous* proportions [14]. A heterogeneous proportion expresses the idea that there is an intruder among the 4 truth values, which is forbidden to appear in a specific position. Intuitively speaking, an item properly assigned to a class should not be (too much) an intruder in this class. It suggests that heterogeneous proportions may be also of interest as a basis for designing a new type of classifier. This is the topic of the paper.

The paper is organized as follows. The next section provides the necessary background on logical proportions, introducing the two types of proportions: the homogeneous ones and the heterogeneous ones, by especially emphasizing a code independency property. In Sect. 3, new results are established that single out these proportions in terms of permutations, of parity in the structure of their truth tables, and of unicity of the solution in an equation solving process. These particular features are meaningful when it comes to classification, and prepare the ground for Sect. 4 where heterogeneous proportions are extended to a logical formula defining evenness. This formula paves the way for Sect. 5 where an evenness measure is defined, quantifying the way a new item conforms with a set of existing items. This measure relies on a heterogeneous proportion: it is all the larger as the item is not an intruder with respect to any triple of items known to belong to the class. In Sect. 5, an evenness-based classifier is proposed, with two implementation options. Results on classical benchmarks are reported and analyzed in Sect. 6. They are also compared with both standard classifiers and analogical proportion-based classifiers. In spite of a very simple underlying formal framework, evenness-based classifiers appear to perform well on a number of benchmarks.

2   Logical Proportions: A Brief Overview

Logical proportions have been defined and studied in [13]: they are in some sense a Boolean counterpart to numerical proportions. They involve 4 items. Considering 2 Boolean variables a and b representing a given feature attached to 2 items A and B, $a \wedge b$ and $\overline{a} \wedge \overline{b}$ indicate that A and B behave similarly w.r.t. the given feature (they are called "similarity" indicators), $a \wedge \overline{b}$ and $\overline{a} \wedge b$ the fact that A and B behave differently (they are called "dissimilarity" indicators). When we have 4 items A, B, C, D, for comparing their respective behavior in a pairwise manner, we are led to consider logical equivalences between similarity, or dissimilarity indicators, such as $a \wedge b \equiv c \wedge d$ for instance.

Definition 1. *A logical proportion $T(a, b, c, d)$ is the conjunction of two equivalences between indicators for (a, b) on one side and indicators for (c, d) on the other side.*

For instance, $((\overline{a} \wedge \overline{b}) \equiv (c \wedge \overline{d})) \wedge ((\overline{a} \wedge b) \equiv (\overline{c} \wedge d))$ is a logical proportion. It has been established that there are 120 syntactically and semantically distinct logical equivalences. There are two ways for distinguishing remarkable subsets among the 120 proportions: either by investigating their structure, or by investigating their semantics (i.e. their truth table). In this section, we shall see that both investigations lead to the same conclusion: there are two groups of 4 proportions which stand out of the crowd.

A property which appears to be paramount in many reasoning tasks is code independency: there should be no distinction when encoding information positively or negatively. In other words, encoding truth (resp. falsity) with 1 or with 0 (resp. with 0 and 1) is just a matter of convention, and should not impact the final result. When dealing with logical proportions, this property is called *code independency* and can be expressed as

$$T(a, b, c, d) \implies T(\overline{a}, \overline{b}, \overline{c}, \overline{d})$$

From a structural viewpoint, remember that a proportion is built up with a pair of equivalences between indicators chosen among 16 equivalences. So, to ensure code independency, the only way to proceed is to first choose an equivalence then to pair it with its counterpart where every literal is negated: for instance $a \wedge b \equiv \overline{c} \wedge d$ should be paired with $\overline{a} \wedge \overline{b} \equiv c \wedge \overline{d}$ in order to get a code independent proportion. This simple reasoning shows that we have only $16/2 = 8$ code independent proportions whose logical expressions are given below.

$$\mathbf{A}: ((a \wedge \overline{b}) \equiv (c \wedge \overline{d})) \wedge ((\overline{a} \wedge b) \equiv (\overline{c} \wedge d))$$

$$\mathbf{R}: ((a \wedge \overline{b}) \equiv (\overline{c} \wedge d)) \wedge ((\overline{a} \wedge b) \equiv (c \wedge \overline{d}))$$

$$\mathbf{P}: ((a \wedge b) \equiv (c \wedge d)) \wedge ((\overline{a} \wedge \overline{b}) \equiv (\overline{c} \wedge \overline{d}))$$

$$\mathbf{I}: ((a \wedge b) \equiv (\overline{c} \wedge \overline{d})) \wedge ((\overline{a} \wedge \overline{b}) \equiv (c \wedge d))$$

$$\mathbf{H_1}: ((a \wedge \overline{b}) \equiv (c \wedge d)) \wedge ((\overline{a} \wedge b) \equiv (\overline{c} \wedge \overline{d}))$$

$$\mathbf{H_2}: ((\overline{a} \wedge b) \equiv (c \wedge d)) \wedge ((a \wedge \overline{b}) \equiv (\overline{c} \wedge \overline{d}))$$

$$\mathbf{H_3}: ((a \wedge b) \equiv (c \wedge \overline{d})) \wedge ((\overline{a} \wedge \overline{b}) \equiv (\overline{c} \wedge d))$$

$$\mathbf{H_4}: ((a \wedge b) \equiv (\overline{c} \wedge d)) \wedge ((\overline{a} \wedge \overline{b}) \equiv (c \wedge \overline{d}))$$

Only 4 among these proportions make use of similarity and dissimilarity indicators without mixing these types of indicators inside one equivalence: for this reason, these 4 proportions A, R, P, I are called *homogeneous proportions*. For instance, an informal reading of A would be: "*a* differs from *b* as *c* differs from

d and vice versa." This expresses the meaning of an analogical proportion, i.e., a statement of the form "a is to b as c is to d". In some sense, A is a qualitative form of comparison of differences, reminiscent to the concept of derivative where we study the ratio $\frac{f(a)-f(b)}{a-b}$ with $f(a) = c$ and $f(b) = d$, which is close to numerical proportions.

Obviously, we can permute the variables and check, for instance, if a given proportion still holds when permuting the 2 first variables. We denote p_{ij} the permutation of variable in position i with variable in position j. For instance p_{14} permutes the variables in extreme positions 1 and 4 while p_{23} permutes variables in mean positions. And $p_{12}(a) = b, p_{12}(b) = a, p_{12}(c) = c, p_{12}(d) = d$.

Definition 2. *A proportion T is stable w.r.t. permutation p_{ij} iff*

$$T(a, b, c, d) \implies T(p_{ij}(a), p_{ij}(b), p_{ij}(c), p_{ij}(d))$$

It can be checked that A is stable w.r.t. the extremes p_{14} or the means p_{23} permutations. This is observable on the truth tables (see Table 1, where only the patterns that make the logical proportion true appear).

In fact, R and P are closely related to A via permutations. Namely we have

$$A(a, b, c, d) \equiv P(a, d, c, b) \equiv R(a, b, d, c)$$

R and P enjoy other types of permutation stability which are easily deducible from the permutation properties of A. Besides, I is the only logical proportion that is stable w.r.t. any permutation of two of its variables. This last, remarkable result is proved in [12].

In fact, when d is fixed, exchanging the variables a, b, c amounts to move from one homogeneous proportion to another, with I remaining an exception. Thus A, R, P collectively maintain a form of exchangeability property with respect to a, b, c, while I ensures it by itself. These exchangeability properties are of particular interest when applying homogeneous logical proportions to classification.

The 4 remaining code independent logical proportions H_1, H_2, H_3, H_4 are called *heterogeneous proportions*: it is clear that they mix similarity and dissimilarity indicators inside each equivalence. The index i in H_i refers to a position inside the formula $H_i(a, b, c, d)$. The truth tables for code independent proportions are shown below where only the patterns leading to 1 (which make the logical proportion true) are given.

In the following section, we investigate new results, focusing on the heterogeneous proportions.

3   Specificity of Heterogeneous Proportions

In order to get a clear understanding of the heterogeneous proportions and to extract relevant properties, we now investigate their truth tables.

Table 1. Homogeneous/heterogeneous proportions valid patterns

A	R	P	I	H_1	H_2	H_3	H_4
0 0 0 0	0 0 0 0	0 0 0 0	1 1 0 0	1 1 1 0	1 1 1 0	1 1 1 0	1 1 0 1
1 1 1 1	1 1 1 1	1 1 1 1	0 0 1 1	0 0 0 1	0 0 0 1	0 0 0 1	0 0 1 0
0 0 1 1	0 0 1 1	1 0 0 1	1 0 0 1	1 1 0 1	1 1 0 1	1 0 1 1	1 0 1 1
1 1 0 0	1 1 0 0	0 1 1 0	0 1 1 0	0 0 1 0	0 0 1 0	0 1 0 0	0 1 0 0
0 1 0 1	0 1 1 0	0 1 0 1	0 1 0 1	1 0 1 1	0 1 1 1	0 1 1 1	0 1 1 1
1 0 1 0	1 0 0 1	1 0 1 0	1 0 1 0	0 1 0 0	1 0 0 0	1 0 0 0	1 0 0 0

3.1 Heterogeneity and Exchangeability

When observing the truth table of the heterogeneous proportions in Table 1, an obvious semantics appears: H_i holds when there are exactly 3 parameters with identical Boolean values (=1 for example) and the parameter in position i is one of these identical values.

Definition 3. *Given 4 Boolean values a, b, c, d in this order such that 3 of them are identical and the remaining one is different, the position $i \in [1, 4]$ of this remaining value is called the intruder position or the intruder for brevity.*

Then, H_i holds iff there is an intruder among the 4 values a, b, c, d and this intruder is not at position i. This suggests that H_i should be stable w.r.t. the permutations which do not affect position i. In fact a little bit more can be established:

Property 1. *Apart from I, H_i are the only logical proportions stable w.r.t any permutation which does not affect position i.*

The special case of I which satisfies any permutation has been already mentioned [13]. In Table 1, we can check that the H_i's are stable w.r.t. the permutations which do not affect position i. Showing that they are the only ones among the 120 logical properties to this permutation property requires a tedious checking procedure that cannot be summarized here.

Property 1 is quite satisfactory and confirms the informal semantics of H_i. It will be useful when applying heterogeneous proportions to classification. In the next subsection, we establish some results about the parity of the number of 1 or 0 in truth tables for heterogeneous proportions, which are contrasted with homogeneous proportions. This leads to a model of oddity and non oddity (or evenness) of a given value, among a set of 4 values.

3.2 Parity of the Number of 1 or 0 in Tables

Since logical proportions are Boolean formulas involving 4 variables, their truth tables have 16 rows, where only 6 lead to 1 (see [13] for a complete investigation).

One could ask if any truth table having 6 lines leading to 1 and 10 lines leading to 0 corresponds to a logical proportion. A simple numbering argument shows that this is not the case. On top of that, we can build classes of patterns which cannot be valid for any proportion:

Property 2. *A logical proportion can satisfy*
neither the class of valuation $\{0111, 1011, 1101, 1110\}$
nor the class $\{1000, 0100, 0010, 0001\}$.

Proof. An equivalence between indicators is of the form $l_1 \wedge l_2 \equiv l_3 \wedge l_4$. If this equivalence is valid for $\{0111, 1011\}$, it means that its truth value does not change when we switch the truth value of the 2 first literals from 0 to 1: there are only 2 indicators for a and b satisfying this requirement: $a \wedge b$ and $\overline{a} \wedge \overline{b}$. If this equivalence is still valid for $\{1101, 1110\}$, its truth value does not change when we switch the truth value of the 2 last literals from 0 to 1: there are only 2 indicators for c and d satisfying this requirement: $c \wedge d$ and $\overline{c} \wedge \overline{d}$. Then the equivalence $l_1 \wedge l_2 \equiv l_3 \wedge l_4$ is just $a \wedge b \equiv c \wedge d$, $a \wedge b \equiv \overline{c} \wedge \overline{d}$, $a \wedge b \equiv \overline{c} \wedge \overline{d}$ or $\overline{a} \wedge \overline{b} \equiv \overline{c} \wedge \overline{d}$. None of these equivalences satisfies the whole class $\{0111, 1011, 1101, 1110\}$. Same reasoning for the other class. □

Applying a similar reasoning, we can build other non satisfiable classes of patterns:

Property 3. *A logical proportion cannot satisfy a class of 4 patterns including 3 patterns of one of the previous classes appearing in Property 2 and where the 4th pattern is just the negation componentwise of the remaining pattern of the class.*

For instance, there is no logical proportion valid for $\{0111, 1011, 1101, 0001\}$ or for $\{0111, 0100, 1101, 1110\}$. This remark helps establishing the following result:

Property 4. *Heterogeneous proportions are the only proportions whose valid patterns have only an odd number of 1.*

Proof. From the truth tables, we observe that the only valid patterns for heterogeneous proportions have an odd number of 1. Let us now consider a proportion whose the 6 valid patterns carry odd number of 1. As there are exactly 8 patterns with an odd number of 1, and thanks to the previous property, this proportion includes necessarily 3 patterns from each of the previous classes. If the valid patterns in one class are obtained from the valid patterns from the other class just by negating all the variables, the proportion is code independent and then, it is a heterogeneous proportion. In the opposite case, it means that we have at least one pattern in the first class with no negated counterpart in the other class: for instance, $1110, 1101, 1011$ are valid but 0001 is not a valid pattern, leaving only $1000, 0100, 0010$ to complete the truth table of a logical proportion. Property 3 applies telling that there is no proportion valid for $1110, 1101, 1011, 1000$. □

A similar property holds for homogeneous proportions:

Property 5. *Homogeneous proportions are the only proportions whose valid patterns have only an even number of 1.*

Ultimately, we may consider that homogeneous proportions denote evenness while heterogeneous ones denote oddity. We now show the specificity of heterogeneous (and homogeneous) proportions from a reasoning point of view.

3.3 Inference and Univocal Proportions

There is a way to infer unknown properties of a partially known object D starting from the knowledge we have about its other specified properties, and assuming that a logical proportion T holds componentwise with three other objects A, B, C, also represented in terms of the same n Boolean features. This can be done via an induction principle that can be stated as follows (where J is a subset of $[1, n]$, and x_i denotes the truth value of feature i for object $X \in \{A, B, C, D\}$):

$$\frac{\forall i \in [1, n] \setminus J, T(a_i, b_i, c_i, d_i)}{\forall i \in J, T(a_i, b_i, c_i, d_i)}$$

This can be seen as a continuity principle assuming that if it is known that a proportion holds for some attributes, this proportion should still hold for the other attributes. It generalizes the inference principle used with the analogical proportion [12, 15] for prediction and classification purposes. From a strict logical viewpoint, this inference rule is unsound as there is no guarantee that the conclusion holds when the premisses hold. Nevertheless, specially when the ratio $\frac{|J|}{n}$ is close to 1, which means that proportions hold on a large number of attributes, it is natural to consider that such a proportion may also hold on the small number of remaining attributes.

This principle requires the unicity of the solution of equation $T(a, b, c, x) = 1$ where x is unknown, when it exists. Namely, given 3 Boolean values a, b, c, we want to determine for what logical proportion T the equation $T(a, b, c, x) = 1$ is solvable, and in such a case, if the solution is unique.

Definition 4. *A proportion T will be called 4-univocal iff, when the equation $T(a, b, c, x) = 1$ is solvable, the solution is unique. In a similar manner, one may define proportions that are 1, 2, or 3-univocal. T is univocal when it is i-univocal for every $i \in [1, 4]$.*

First of all, it is easy to see that there are always cases where the equation $T(a, b, c, x) = 1$ has no solution. Indeed, the triple a, b, c may take $2^3 = 8$ values, while any proportion T is true only for 6 distinct valuations, leaving at least 2 cases with no solution. For instance, when we deal with H_4, the equations $H_4(0, 0, 0, x)$ and $H_4(1, 1, 1, x)$ have no solution.

We have the following result:

Property 6 *The homogeneous and the heterogeneous proportions are the only proportions which are univocal.*

Proof: From the truth tables, we see that the 2 types of proportions satisfy the property. Now, a proportion which is not i-univocal is necessarily valid both for a pattern with an odd number of 1, and for a pattern with an even number of 1. Then, properties 4 and 5 exclude homogeneous and heterogeneous proportions. □

This shows how it would be possible to generate an object which is not an intruder with respect to three other objects.

4   Evenness via Heterogeneous Proportions

Since oddity, captured via heterogeneous proportions, is the opposite concept of evenness, there is a way to define evenness via heterogeneous proportions only. We investigate a way to do it in the next subsection.

4.1   A Proportion-Based Definition of Evenness

Let us recall the semantics of H_i: H_i holds iff there is an intruder among a, b, c, d and the parameter in position i is not this intruder. As a consequence, H_i implies that there is a majority of values among (a, b, c, d) and the value in position i conforms to the majority of values appearing among the 3 other positions (i.e. the set of values $\{a, b, c, d\}$ is more or less even). But the reverse implication does not hold since when the 4 parameters have identical value, $\forall i \in [1, 4]$, $H_i(a, b, c, d)=0$. Then, to have a concise Boolean definition for "there is a majority of values among the parameters a, b, c, d and the parameter in position i belongs to this majority of values", we need to consider the case where all the values are identical by using the following formula:

$$Even_i(a, b, c, d) =_{def} H_i(a, b, c, d) \vee Eq(a, b, c, d)$$

where $Eq(a, b, c, d) =_{def} (a = b) \wedge (a = c) \wedge (a = d)$. Thus, with $Even_i$ we take into account the special case where all the values are equal. The truth table of $Even_4$ is given in Table 2. It is clear that $Even_4$ holds only when the value of d belongs to a majority of the parameter's values. And does not hold in an opposite situation where there is no majority of values as it is the case for $Even_4(0011)$ or $Even_4(0110)$. From a practical perspective, when we have 3 Boolean values a, b, c, there is necessarily a value which constitutes a majority. In that case, when adding a new value d, $Even_4(a, b, c, d)$ can act as the flag indicating that d conforms to this majority, or if we prefer that adding d to the set a, b, c does not break the evenness of it. Note that $Even_4(a, b, c, d)$ does not depend on the ordering of a, b, c.

4.2   Dealing with Missing Values

Missing information is quite common in real life datasets and a way to extend the semantics of analogical proportion to deal with this issue has been deeply

Table 2. H_4, Eq and $Even_4$ truth values

	H_4	Eq	$Even_4$
0 0 0 0	0	1	1
0 0 0 1	0	0	0
0 0 1 0	1	0	1
0 0 1 1	0	0	0
0 1 0 0	1	0	1
0 1 0 1	0	0	0
0 1 1 0	0	0	0
0 1 1 1	1	0	1
1 0 0 0	1	0	1
1 0 0 1	0	0	0
1 0 1 0	0	0	0
1 0 1 1	1	0	1
1 1 0 0	0	0	0
1 1 0 1	1	0	1
1 1 1 0	0	0	0
1 1 1 1	0	1	1

investigated in [14] for instance. In fact, such an approach can be also applied here, as explained now.Let us focus on $Even_4$. Still keeping a logical approach and considering that ? denotes a missing value (i.e. an information is unknown), the idea is to extend the truth table of the $Even_4$ formula as follows: $Even_4(?,0,0,0) = Even_4(0,?,0,0) = Even_4(0,0,?,0) = 1$, $Even_4(?,1,1,1) = Even_4(1,?,1,1) = Even_4(1,1,?,1) = 1$, and $Even_4(x,y,z,t) = 0$ for any other pattern including at least a missing value ?. It is clear that, with the 6 first patterns, whatever the candidate value of the missing feature, the 4th argument belongs to the majority and cannot be an intruder. In all the remaining cases, where we have no certainty regarding the status of d, we adopt a cautious behaviour by considering that $Even_4$ does not hold. Note that, we are only dealing with Boolean data in this paper. An extension to numerical data is under work and we do not deal with dataset involving uncertain data in that context.

5   An Evenness-Based Classifier

Homogeneous proportions have been successfully used for classification purposes in the past ([1,3,10]). As classification is at most a matter of heterogeneity as a matter of homogeneity, it is quite natural to investigate the use of heterogeneous proportions for such tasks applying the following principle: *A new item can join a class if it does not appear odd in this class.* Obviously, to encode real life data, it is more realistic to use Boolean vectors instead of simple Boolean values. Thus, we need a way to measure evenness for Boolean vectors. Let us investigate how heterogeneous proportions thanks to the Boolean formula $Even_i$ can help us to design such a measure.

5.1   Evenness Measure for Boolean Vectors

To measure to what extent a vector $\overrightarrow{d}$ conforms to a set of three other vectors $\overrightarrow{a}, \overrightarrow{b}, \overrightarrow{c}$, it is quite natural to consider the number

$$m(\overrightarrow{a}, \overrightarrow{b}, \overrightarrow{c}, \overrightarrow{d}) =_{def} \Sigma_{i=1}^{n} Even_4(a_i, b_i, c_i, d_i) \in [0, n]$$

(for sake of the simplicity, we still keep the notation $Even_4$ for the formula and its truth value). When $m(\overrightarrow{a}, \overrightarrow{b}, \overrightarrow{c}, \overrightarrow{d}) = n$, there does not exist a feature where $\overrightarrow{d}$ behaves as an intruder. Then, it is acceptable that $\overrightarrow{d}$ joins the club of a, b, c (where a, b, c is a proper sampling). Clearly, the bigger $m(\overrightarrow{a}, \overrightarrow{b}, \overrightarrow{c}, \overrightarrow{d})$, the larger the number of features for which $\overrightarrow{d}$ conforms to the majority in $\overrightarrow{a}, \overrightarrow{b}, \overrightarrow{c}$, the better $\overrightarrow{d}$ conforms to vectors $\overrightarrow{a}, \overrightarrow{b}, \overrightarrow{c}$. Measuring to what extent a vector $\overrightarrow{d}$ conforms to a whole set S of vectors, can be done via the previous definition $m(\overrightarrow{a}, \overrightarrow{b}, \overrightarrow{c}, \overrightarrow{d})$. We compute $m(\overrightarrow{a}, \overrightarrow{b}, \overrightarrow{c}, \overrightarrow{d})$ for every distinct (to take into account exchangeability) triple $(\overrightarrow{a}, \overrightarrow{b}, \overrightarrow{c})$ in S^3, made of distinct elements, then we add all these numbers to get a clear evenness-measure as follows:

$$Even(S, \overrightarrow{d}) = \Sigma_{\overrightarrow{a}, \overrightarrow{b}, \overrightarrow{c} \in S} m(\overrightarrow{a}, \overrightarrow{b}, \overrightarrow{c}, \overrightarrow{d}).$$

The final result belongs to $[0, n \cdot \binom{|S|}{3}]$.

5.2   Algorithm

Given a class $\mathcal{C}$ having some homogeneity, adding to this class the new item $\overrightarrow{d}$ to be classified may perturb the homogeneity or evenness: in other words, $\mathcal{C} \cup \{\overrightarrow{d}\}$ may be less homogeneous or less even than $\mathcal{C}$ itself. Being able to measure this evenness leads to a quite natural classification principle: allocate to $\overrightarrow{d}$ the label corresponding to the class maximizing the evenness when $\overrightarrow{d}$ is added. The previous section provides a clear evenness measure of $\overrightarrow{d}$ w.r.t. a class $\mathcal{C}$ as $Even(\mathcal{C}, \overrightarrow{d})$. Obviously we have to take into account the relative size of the different classes $\mathcal{C}$. Then it would be fair to introduce a normalization factor:

- To consider the number of triples available as an increasing function of $|\mathcal{C}|$, we have to divide the actual evenness measure by $\binom{|\mathcal{C}|}{3}$ which, for large values of $|\mathcal{C}|$, has $|\mathcal{C}|^3$ as order of magnitude,
- To acknowledge the fact that, in case of equal evenness measure, we should favor the largest class: in some sense, we have to multiply the actual evenness measure by a factor proportional to $|\mathcal{C}|$, leading to a factor $\frac{1}{|\mathcal{C}|^2}$ instead of $\frac{1}{|\mathcal{C}|^3}$.

But, so it is relevant to consider the normalized version of *Even*, *norm-Even* as:

$$norm\text{-}Even(\mathcal{C}, \overrightarrow{d}) = \frac{1}{|\mathcal{C}|^2} Even(\mathcal{C}, \overrightarrow{d})$$

This leads to the following procedure which has been implemented in the Algorithm 1 below, where $\overrightarrow{d}$ is a new item to be classified:

1. For each class (or label) $\mathcal{C}$, compute $norm\text{-}Even(\mathcal{C}, \overrightarrow{d})$.
2. Allocate to $\overrightarrow{d}$ the label $argmax_{\mathcal{C}} \ norm\text{-}Even(\mathcal{C}, \overrightarrow{d})$

To optimize the computation, we have at least two options :

- In order to better control the meaning of $norm\text{-}Even(\mathcal{C}, \overrightarrow{d})$ (which has to be maximized), we may focus on triples for which $\overrightarrow{d}$ is an intruder for at most $n - l$ features, where $l = 0, 1, \cdots$. Instead of keeping all the triples, we can just choose a threshold $l \in [0, n]$, then consider $m(\overrightarrow{a}, \overrightarrow{b}, \overrightarrow{c}, \overrightarrow{d})$ only for the triples $(\overrightarrow{a}, \overrightarrow{b}, \overrightarrow{c})$ in $\mathcal{C}^3$ such that $m(\overrightarrow{a}, \overrightarrow{b}, \overrightarrow{c}, \overrightarrow{d}) \geq l$, i.e. we want *Even* to hold over at least l features. This leads to a modified definition of evenness measure as below:

$$norm\text{-}Even^l(\mathcal{C}, \overrightarrow{d}) = \frac{1}{|\mathcal{C}|^2} \Sigma_{\overrightarrow{a}, \overrightarrow{b}, \overrightarrow{c} \in S s.t. m(\overrightarrow{a}, \overrightarrow{b}, \overrightarrow{c}, \overrightarrow{d}) \geq l} \ m(\overrightarrow{a}, \overrightarrow{b}, \overrightarrow{c}, \overrightarrow{d})$$

- In terms of structural complexity, the initial function $norm\text{-}Even(\mathcal{C}, \overrightarrow{d})$ is in $\mathcal{O}(|\mathcal{C}|^3)$ just because we look for triples of elements. In order to reduce the complexity, we can filter the triples $(\overrightarrow{a}, \overrightarrow{b}, \overrightarrow{c})$ by choosing $\overrightarrow{c}$ to be a neighbour (in the sense of Hamming distance) of $\overrightarrow{d}$. So the normalization factor is now $\mathcal{C}$ instead of $|\mathcal{C}|^2$ which lead to a more tractable function $k\text{-}norm\text{-}Even(\mathcal{C}, \overrightarrow{d})$ which is square in the size of the sample set.

$$k\text{-}norm\text{-}Even(\mathcal{C}, \overrightarrow{d}) = \frac{1}{|\mathcal{C}|} \Sigma_{\overrightarrow{a}, \overrightarrow{b}, \overrightarrow{c} \in \mathcal{C}} m(\overrightarrow{a}, \overrightarrow{b}, \overrightarrow{c}, \overrightarrow{d})$$

where $\overrightarrow{c}$ is among the k nearest neighbours of $\overrightarrow{d}$.

Obviously, mixing the 2 optimizations leads to $k\text{-}norm\text{-}Even^l(\mathcal{C}, \overrightarrow{d})$:

$$k\text{-}norm\text{-}Even^l(\mathcal{C}, \overrightarrow{d}) = \frac{1}{|\mathcal{C}|} \Sigma_{\overrightarrow{a}, \overrightarrow{b}, \overrightarrow{c} \in S s.t. m(\overrightarrow{a}, \overrightarrow{b}, \overrightarrow{c}, \overrightarrow{d}) \geq l} \ m(\overrightarrow{a}, \overrightarrow{b}, \overrightarrow{c}, \overrightarrow{d})$$

where $\overrightarrow{c}$ is among the k nearest neighbours of $\overrightarrow{d}$. The previous procedure can now be described with the pseudo-code of Algorithm 1. Algorithm 1 can deal with missing values thanks to the extension of Sect. 4.2.

It is clear that l and k are parameters to be tuned and we experiment with different values in the next section. We have also implemented another way to predict labels for the new comer $\overrightarrow{d}$. We consider separately each $\overrightarrow{c}$ among

Algorithm 1. Evenness-based algorithm

Input: a training set TS of examples $z = (x, cl(x))$
a threshold $l \in [0, n]$ ▷ n is the number of attributes
an integer $k \geq 1$
a new item $\overrightarrow{d}$,
Partition TS into sets $\mathcal{C}$ of examples having the same label cl.   ▷ cl is the label of
the class $\mathcal{C}$
for each $\mathcal{C}$ **do**
    Compute $k\text{-}norm\text{-}Even^l(\mathcal{C}, \overrightarrow{d})$
end for   ▷ we get a list of integer
$cl(\overrightarrow{d}) = argmax_{label(\mathcal{C})}\{k\text{-}norm\text{-}Even^l(\mathcal{C}, \overrightarrow{d})\}$
return $cl(\overrightarrow{d})$

---

the k nearest neighbours, and combine $\overrightarrow{c}$ with pairs $(\overrightarrow{a}, \overrightarrow{b})$ in each subset $\mathcal{C}$, this leads to compute $k\text{-}norm\text{-}Even^l(\mathcal{C}, \overrightarrow{d})$ for each class and get a class label corresponding to this nearest neighbour $\overrightarrow{c}$ (by choosing the maximum value). Finally we apply a majority vote among all class labels obtained for different neighbours k. We refer to this procedure as Algorithm 2 (with vote).

6   Experimentations and Discussion

The experimental study is based on several data sets selected from the U.C.I. machine learning repository [7], focusing on classification problems with categorical attributes only:

- `Balance` and `Car` are multiple classes datasets.
- `TicTacToe`, `Voting`, `Spect`, `Monk1`, `Monk2` and `Monk3` datasets are binary class problems.
- `Monk3` has noise added (in the training set only).
- `Voting` and `Spect` datasets contain only binary attributes. `Voting` dataset contains missing attribute values.

For all categorical (non binary) attributes where the range of attribute values is finite, but greater than 2, we apply the following procedure to convert them into Boolean attributes. Considering an attribute domain $\{v_1, \cdots, v_m\}$, we can binarize it by means of the m properties "having or not value v_i". For instance, a trivalued attribute having candidate values v_1, v_2, v_3, can respectively be encoded as $100, 010, 001$. It means, in that case, that, e.g.,110 does not represent a value and will never appear in the dataset.

These data sets are described in Table 3.

Table 4 provides accuracies results for the two evenness-based classifiers obtained with a 10-fold cross validation and for two values of k and l (k being the number of nearest neighbors of d, l refers to the number of attributes j of d such that d_j belongs to a majority). The best results are in bold.

Table 3. Description of datasets

Datasets	Instances	Nominal Att.	Binary Att.	Classes
Balance	625	4	20	3
Car	743	6	21	4
TicTacToe	405	9	27	2
Voting	435	-	16	2
Spect Heart	267	-	22	2
Monk1	432	6	15	2
Monk2	432	6	15	2
Monk3	432	6	15	2

Table 4. Accuracy results for the two classifiers

Datasets		Without vote		With vote	
	k	$\ell = n$	$\ell = n - 1$	$\ell = n$	$\ell = n - 1$
Balance	5	83.06±3.43	88.02±1.11	85.43±1.58	**89.17 ± 2.15**
	11	84.16±2.64	86.88±2.11	86.7±2.6	88.82±2.23
Car	5	91.96±2.71	90.64±4.38	**92.73±3.44**	89.88±2.78
	11	92.23±2.89	91.31±4.46	92.33±3.2	90.56±2.54
TicTac	5	84.51±5.37	88.21±4.24	86.7±5.28	89.16±4.27
	11	85.97±5.31	**89.43±4.17**	86.72±5.01	89.16±4.12
Voting	5	95.17±2.66	95.16±3.03	95.38±2.1	95.38±2.1
	11	95.17±2.66	95.16±3.03	**95.83±2.06**	95.83±2.06
Spect	5	**84.34±3.98**	76.57±5.96	83.94±4.49	78.31±2.47
	11	**84.34±3.98**	77.63±4.32	83.17±4.94	77.53±2.84
Monk1	5	**100**	**100**	**100**	99.35±1.96
	11	**100**	**100**	**100**	99.56±1.3
Monk2	5	57.85±4.9	61.79±4.66	59.05±3.47	66.68±1.28
	11	62.02±4.44	64.11±4.12	65.29±3.18	**66.9±0.85**
Monk3	5	**100**	99.3±1.49	**100**	99.31±1.5
	11	**100**	99.3±1.49	**100**	99.07±1.56

We compare evenness-based classifiers to well-known classifers in Table 5, including SVM, k-Nearest Neighbors IBk for k=1, k=10 and JRip an optimized propositional rule learner. Accuracy results for SVM, IBk and JRip are obtained by applying the free implementation of Weka software. The column Analogy in Table 5 refers to the results obtained with [3] with the Boolean coding.

When we analyze results in Table 4, we can see that:

• In general, the best classification rates are obtained for $l = n$. This means that the classifier is likely to be more accurate when the classification is made on the

Table 5. Classification results of well-known classifiers

Datasets	SVM	IBk (k=1, k=10)	JRIP	Analogy	Algo2 (k=11,l=n)
Balance	90	83, 83	76	87	87
Car	91	92, 92	91	94	92
TicTac	100	98, 93	95	100	87
Voting	96	93, 92	95	78	**96**
Spect	81	75, 81	81	41	**83**
Monk1	75	100, 100	98	99	**100**
Monk2	67	44 , 64	73	99	65
Monk3	100	100, 99	100	99	**100**

basis of triples w.r.t. which d is not an intruder for *any* attributes. However, for some datasets such as Balance and Monk2, the classifier needs to consider more levels l when it is difficult to satisfy the constraint $m(\overrightarrow{a}, \overrightarrow{b}, \overrightarrow{c}, \overrightarrow{d}) \geq l$ for $l = n$ or even $l = n - 1$. Thus, we also tested smaller levels of l and for "Balance" data set, we get an accuracy equal to 90.13 ± 1.95 for $l = n - 3$.

• The classification success of the classifier for Balance, and Car (which have multiple classes) demonstrates its ability to deal with multiple class data sets.

• The two "evenness"-based classifiers (Algorithms 1 and 2), with or without vote, exhibit similar results, with the exception of "Balance" and "Monk2" where Algorithm 2 is *slightly* better.

• Table 5 highlights the fact that the proposed classifier performs more or less in the same way as the best known algorithms. Especially, the basic classifier, with large k works as well as any other classifiers for data sets Balance (for $l = n - 3$), Spect., Voting, Monk1 and Monk3 (for $l = n$). We can also see that "evenness"-based classifiers work as well as IBk for all datasets except Tic Tac Toe.

• If we compare the "evenness"-based classifiers with an analogical proportion-based classifier [3], we first noticed that both algorithms exhibit very good results for data sets Balance, Car, Monk1 and Monk3 when compared to the state of the art ML algorithms like IBK or SVM.

• However, "evenness"-based classifiers seems to be less efficient when classifying Monk2 and Tic Tac Toe data sets. Regarding Monk2, it is known that the underlying function ("having exactly two attributes with value 1") is more complicated than the functions underlying Monk1 and Monk3, and involves all the attributes (while in the two other functions only 3 attributes among 6 are involved in the discrete coding). We suspect that the existence of a large discontinuity in the classification of data (a nearest neighbor d of c will not generally be labeled with the same class $cl(c)$) may be too difficult to apprehend using heterogeneous proportions. For Tic Tac Toe, we also notice that all attributes are involved in the classification function. Moreover, this data set contains the largest number of attributes among all datasets which may require a larger amount of data for an accurate prediction.

• On the contrary, it is clear that "evenness"-based classifiers outperform the analogy-based classifier [3] for data sets Spect and Voting (see Table 5). From experiments, we notice that bad results for analogy-based classifier with Spect and Voting datasets seem to be due to the number of voters $(\overrightarrow{a}, \overrightarrow{b})$ which is equal to 0 for many examples to be classified. Regarding the analogy-based classifiers, when considering a particular item $\overrightarrow{d}$, and a neighbor $\overrightarrow{c} \in \mathcal{B}_H(\overrightarrow{d}, r)$ (where $\mathcal{B}_H(\overrightarrow{d}, r)$ denotes the Hamming ball with center $\overrightarrow{d}$ and radius r) the number of voters $(\overrightarrow{a}, \overrightarrow{b})$ is only a small subset of the set of pairs differing on r attributes. Due to the fact that two constraints have to be satisfied in the analogical proportion-based approach: the pairs $(\overrightarrow{a}, \overrightarrow{b})$ and $(\overrightarrow{c}, \overrightarrow{d})$ differ on the same attribute(s) and the associated class equation should be solvable, if only one attribute in the pair $(\overrightarrow{a}, \overrightarrow{b})$ is not satisfied, this pair will be discarded.

In order to reduce the effect of the first constraint in the analogy-based classifier [3], we reimplemented the analogy-based classifier for *numerical* data described in [2] on the datasets Spect and Voting (this algorithm seeks for only triples which form with d an analogy on a *maximum* number of attributes, and not necessarily on *all* attributes as the algorithm used in [3]). We obtained an accuracy respectively equal to 73.38±4.68 and 95.85±3.09 (using the function: A and $k = 11$). This accuracy improvement shows that, for some datasets whose attributes are highly dissimilar (the case of Spect for example), it is faithful to relax the constraint "the pairs $(\overrightarrow{a}, \overrightarrow{b})$ and $(\overrightarrow{c}, \overrightarrow{d})$ differ on the same attribute(s)" by satisfying the analogical proportion only on a *maximum* (as it is the case in [2] and in "evenness"-based classifiers) instead of *all* attributes.

Lastly, one may find that there is a flavor of conformal prediction [16] in our approach. Nevertheless, even if our evenness measure can also be considered as a conformity measure, the way we use it in our approach is quite different from the pure conformal framework. We do not make use of any p-value which is one of the added value underlying conformal predictors. The inherent complexity of conformal predictors, added to the one of our evenness-based classifier would likely lead to non tractable algorithms. However, this is an option that has to be investigated.

7 Conclusion

We have shown how specific are heterogeneous proportions among logical proportions, and their ability to express oddity. On this basis, we have described and experimented a new type of classifier directly based on the idea that a newcomer in a class should not appear too much as an intruder, using heterogeneous proportions. The results obtained on benchmarks are competitive with respect to well-established classifiers, as well as with analogy-based classifiers relying on homogeneous proportions.

References

1. Bayoudh, S., Miclet, L., Delhay, A.: Learning by analogy: a classification rule for binary and nominal data. In: Proceeding International Joint Conference on Artificial Intelligence IJCAI 2007, pp. 678–683 (2007)
2. Bounhas, M., Prade, H., Richard, G.: Analogical classification: handling numerical data. In: Straccia, U., Calì, A. (eds.) SUM 2014. LNCS, vol. 8720, pp. 66–79. Springer, Heidelberg (2014)
3. Bounhas, M., Prade, H., Richard, G.: Analogical classification: a new way to deal with examples. In: ECAI 2014–21st European Conference on Artificial Intelligence, 18–22 August 2014, Frontiers in Artificial Intelligence and Applications, vol. 263, pp. 135–140. IOS Press, Prague, Czech Republic (2014)
4. Gergonne, J.D.: Application de la méthode des moindres quarrés à l'interpolation des suites. Annales de Math. Pures et Appl **6**, 242–252 (1815)
5. Gergonne, J.D.: Théorie de la règle de trois. Annales de Math. Pures et Appl **7**, 117–122 (1816)
6. Lepage, Y.: Analogy and formal languages. In: Proceeding FG/MOL 2001, pp. 373–378 (2001). http://www.slt.atr.co.jp/lepage/pdf/dhdryl.pdf.gz
7. Mertz, J., Murphy, P.: Uci repository of machine learning databases (2000). http://www.ftp://ftp.ics.uci.edu/pub/machine-learning-databases
8. Miclet, L., Bayoudh, S., Delhay, A.: Analogical dissimilarity: definition, algorithms and two experiments in machine learning. JAIR **32**, 793–824 (2008)
9. Miclet, L., Prade, H.: Handling analogical proportions in classical logic and fuzzy logics settings. In: Sossai, C., Chemello, G. (eds.) ECSQARU 2009. LNCS, vol. 5590, pp. 638–650. Springer, Heidelberg (2009)
10. Moraes, R.M., Machado, L.S., Prade, H., Richard, G.: Classification based on homogeneous logical proportions. In: Bramer, M., Petridis, M. (eds.) Proceeding of AI-2013, The Thirty-third SGAI International Conference on Innovative Techniques and Applications of Artificial Intelligence, pp. 53–60. Springer, Cambridge, England, UK (2013)
11. Prade, H., Richard, G.: Reasoning with logical proportions. In: Lin, F.Z., Sattler, U., Truszczynski, M. (eds.) Proceeding 12th International Conference on Principles of Knowledge Representation and Reasoning, KR 2010, Toronto, 9–13 May 2010, pp. 545–555. AAAI Press (2010)
12. Prade, H., Richard, G.: Homogeneous logical proportions: Their uniqueness and their role in similarity-based prediction. In: Brewka, G., Eiter, T., McIlraith, S.A. (eds.) Proceeding 13th International Conference on Principles of Knowledge Representation and Reasoning (KR 2012), Roma, 10–14 June, pp. 402–412. AAAI Press (2012)
13. Prade, H., Richard, G.: From analogical proportion to logical proportions. Logica Universalis **7**(4), 441–505 (2013)
14. Prade, H., Richard, G.: Homogenous and Heterogeneous logical proportions. IfCoLog J. Logics Appl. **1**(1), 1–51 (2014)
15. Stroppa, N., Yvon, F.: Analogical learning and formal proportions: definitions and methodological issues. Technical report D004, ENST-Paris (2005)
16. Vovk, V., Gammerman, A., Saunders, C.: Machine-learning applications of algorithmic randomness. In: International Conference on Machine Learning, pp. 444–453 (1999)

Multivariate Cluster-Based Discretization for Bayesian Network Structure Learning

Ahmed Mabrouk[1], Christophe Gonzales[2]([✉]), Karine Jabet-Chevalier[1],
and Eric Chojnaki[1]

[1] Institut de Radioprotection et de Sûreté Nucléaire,
Cadarache, France
{Ahmed.Mabrouk,Karine.Chevalier-Jabet,Eric.Chojnaki}@irsn.fr
[2] Sorbonne Universités, UPMC Univ Paris 06, CNRS, UMR 7606, LIP6,
Paris, France
Christophe.Gonzales@lip6.fr

Abstract. While there exist many efficient algorithms in the literature
for learning Bayesian networks with discrete random variables, learning
when some variables are discrete and others are continuous is still an
issue. A common way to tackle this problem is to preprocess datasets
by first discretizing continuous variables and, then, resorting to classical
discrete variable-based learning algorithms. However, such a method is
inefficient because the conditional dependences/arcs learnt during the
learning phase bring valuable information that cannot be exploited by
the discretization algorithm, thereby preventing it to be fully effective
In this paper, we advocate to discretize while learning and we propose a
new multivariate discretization algorithm that takes into account all the
conditional dependences/arcs learnt so far. Unlike popular discretization
methods, ours does not rely on entropy but on clustering using an EM
scheme based on a Gaussian mixture model. Experiments show that our
method significantly outperforms the state-of-the-art algorithms.

Keywords: Multivariate discretization · Bayesian network learning

1 Introduction

For several decades, Bayesian networks (BN) have been successfully exploited
for dealing with uncertainties. However, while their learning and inference mechanisms are relatively well understood when they involve only discrete variables,
their coping with continuous variables is still often unsatisfactory. One actually
has to trade-off between expressiveness and computational complexity: on one
hand, conditional Gaussian models and their mixing with discrete variables are
computationally efficient but they definitely lack some expressiveness [12]; on the
other hand, mixtures of exponentials, bases or polynomials are very expressive
but at the expense of tractability [15,20]. In between lie discretization methods which, by converting continuous variables into discrete ones, can provide a
satisfactory trade-off between expressiveness and tractability.

© Springer International Publishing Switzerland 2015
C. Beierle and A. Dekhtyar (Eds.): SUM 2015, LNAI 9310, pp. 155–169, 2015.
DOI: 10.1007/978-3-319-23540-0_11

In many real-world applications, BNs are learnt from data and, when there exist continuous attributes, those are often discretized prior to learning, thereby opening the path to exploiting efficient discrete variable-based learning algorithms. However such an approach is doomed to be ineffective because the conditional dependences/arcs learnt during the learning phase bring valuable information that cannot be exploited by the discretization algorithm, thereby severely limiting its effectiveness. However, there exist surprisingly few papers on discretizing while learning, probably because it incurs substantial computational costs and it requires multivariate discretization instead of just a univariate one. In this direction, MDL and Bayesian scores used by search algorithms have been adapted to include multivariate discretizations taking into account the BN structure learnt so far [6,13]. But, to be naturally included into these scores, the latter heavily rely on entropy-related maximizations which, as we shall see, is not very well suited for BN learning. In [21], a non-linear dimensionality reduction process called GP-LVM combined with a Gaussian mixture model-based discretization is proposed for BN learning. Unfortunately, GP-LVM looses the random variable's semantics and the discretization does not rely on the BN structure. As a consequence, the method does not exploit all the useful information.

Unlike in BN learning, multivariate discretization has often been exploited in Machine Learning for supervised classification tasks [1,2,5,9,22]. But the goal is only to maximize the classification power w.r.t. one target variable. As such, only the individual correlations of each variable with the target are of interest and, thus, only bivariate discretization is needed. BN structure learning is fundamentally different because the complete set of conditional dependences between all sets of variables is of interest and multivariate discretization shall most often involve more than two variables. This makes these approaches not easily transferable to BN learning. In [11], the authors propose a general multivariate discretization relying on genetic algorithms to construct rulesets. However, the approach is very limited because it is designed to cope with only one target and the domain size of this variable needs to be small to keep the method tractable.

Discretizations have also been exploited in unsupervised learning (UL), but those are essentially univariate [4,8,16,17], which make them usable *per se* only as a preprocess prior to learning. However, BN learning can be related to UL in the sense that all the BN's variables can be thought of as targets whose discretized values are unobserved. This suggests that some key ideas underlying UL algorithms might be adapted for learning BN structures. Clustering is one such popular framework. In [14], for instance, multivariate discretization is performed by clustering but, unfortunately, independences between random variables are only considered given a latent variable. This limits considerably the range of applications of the method because numerous continuous variables require the latent one to have a large domain size in order to get good quality discretizations. This approach is therefore limited to small datasets and, by not exploiting the BN structure, it is best suited as a BN learning preprocess. Finally, by relying on entropy, its effectiveness for BN learning is certainly not optimal. However, here, we advocate to exploit clustering methods for discretization w.r.t. BN learning.

More precisely, we propose a new clustering-based approach for multivariate discretization that takes into account the conditional dependences among variables discovered during learning. By exploiting clustering rather than entropy, it avoids the shortcomings induced by the latter and, by taking into account the dependences between random variables, it significantly increases the quality of the discretization compared to state-of-the-art clustering approaches.

The rest of the paper is organized as follows. Section 2 recalls BN learning and discretizations. Then, in Sect. 3, we describe our approach and justify its correctness. Its effectiveness is highlighted through experiments in Sect. 4. Finally, some concluding remarks are given in Sect. 5.

2   Basics on BN Structure Learning and Discretization

Uppercase (resp. lowercase) letters X, Z, x, z, represent random variables and their instantiations respectively. Boldface letters represent sets.

Definition 1. *A (discrete) BN is a pair $(\mathcal{G}, \boldsymbol{\theta})$ where $\mathcal{G} = (\mathbf{X}, \mathbf{A})$ is a directed acyclic graph (DAG), $\mathbf{X} = \{X_1, ..., X_n\}$ represents a set of discrete random variables[1], $\mathbf{A}$ is a set of arcs, and $\boldsymbol{\theta} = \{P(X_i | \mathbf{Pa}(X_i))\}_{i=1}^n$ is the set of the conditional probability distributions (CPT) of the variables X_i in $\mathcal{G}$ given their parents $\mathbf{Pa}(X_i)$ in $\mathcal{G}$. The BN encodes the joint probability over $\mathbf{X}$ as:*

$$P(\mathbf{X}) = \prod_{i=1}^{n} P(X_i | \mathbf{Pa}(X_i)).$$ (1)

To avoid ambiguities between continuous variables and their discretized counterparts, letters, when superscripted by "o", e.g., $\mathring{X}, \mathring{x}$, represent variables and their instantiations prior to discretization, else they are discretized (for discrete variables, $X = \mathring{X}$ and $x = \mathring{x}$). In the rest of the paper, n always denotes the number of variables in the BN, and we assume that $\mathring{X}_1, \ldots, \mathring{X}_l$ are discrete whereas $\mathring{X}_{l+1}, \ldots, \mathring{X}_n$ are continuous. $\mathring{\mathcal{D}}$ and $\mathcal{D}$ denote the input databases before and after discretization respectively and are assumed to be complete, i.e., they do not contain any missing data. N refers to their number of records.

Given $\mathring{\mathcal{D}} = \{\mathring{\mathbf{x}}^{(1)}, \mathring{\mathbf{x}}^{(2)}, \ldots, \mathring{\mathbf{x}}^{(N)}\}$, BN learning consists of finding DAG $\mathcal{G}$ that most likely accounts for the observed data in $\mathring{\mathcal{D}}$. When all variables are discrete, i.e., $\mathcal{D} = \mathring{\mathcal{D}}$, there exist many efficient algorithms in the literature for solving this task. Those can be divided into 3 classes [10]: (i) the search-based approaches that look for the structure optimizing a score (BD, BDeu, BIC, AIC, K2, *etc.*); (ii) the constraint-based approaches that exploit statistical independence tests (χ^2, G^2, *etc.*) to find the best structure $\mathcal{G}$; (iii) the hybrid methods that exploit a combination of both. In the rest of the paper, we will focus on search-based approaches because our closest competitors, [6,13], belong to this class.

---

[1] By abuse of notation, we use interchangeably $X_i \in \mathbf{X}$ to denote a node in the BN and its corresponding random variable.

Basically, these algorithms start with a structure $\mathcal{G}_0$ (often empty). Then, at each step, they look in the neighborhood of the current structure for another structure, say $\mathcal{G}$, that increases the likelihood of structure $\mathcal{G}$ given observations $\mathcal{D}$, i.e., $P(\mathcal{G}|\mathcal{D})$. The neighborhood is often defined as the set of graphs that differ from the current one only by one atomic graphical modification (arc addition, arc deletion, arc reversal). $P(\mathcal{G}|\mathcal{D})$ is computed locally through the aforementioned scores, their differences stemming essentially from different *a priori* hypotheses. More precisely, assuming a uniform prior on all structures $\mathcal{G}$, we have that:

$$P(\mathcal{G}|\mathcal{D}) = \frac{P(\mathcal{D}|\mathcal{G})P(\mathcal{G})}{P(\mathcal{D})} \propto P(\mathcal{D}|\mathcal{G}) = \int_{\theta} P(\mathcal{D}|\mathcal{G},\theta)\pi(\theta|\mathcal{G})d\theta, \qquad (2)$$

where θ is the set of parameters of the CPTs of a (discrete) BN with structure $\mathcal{G}$. Different hypotheses on prior π and on θ result in the different scores (see, e.g., [18] for the hypotheses for the BIC score used later).

When database $\mathring{\mathcal{D}}$ contains continuous variables, those can be discretized. A discretization of a continuous variable $\mathring{X}$ is a function $f : \mathbb{R} \rightarrow \{0,\dots,g\}$ defined by an increasing sequence of g cut points $\{t_1, t_2, ..., t_g\}$ such that:

$$f(\mathring{x}) = \begin{cases} 0 & \text{if } \mathring{x} < t_1, \\ k & \text{if } t_k \leq \mathring{x} < t_{k+1}, \quad \text{for all } k \in \{1,\dots,g-1\} \\ g & \text{if } \mathring{x} \geq t_g. \end{cases}$$

Let $\mathcal{F}$ be a set of discretization functions, one for each continuous variable. Then, given $\mathcal{F}$, if $\mathcal{D}$ denotes the (unique) database resulting from the discretization of $\mathring{\mathcal{D}}$ by $\mathcal{F}$., Eq. (2) becomes:

$$P(\mathcal{G}|\mathring{\mathcal{D}},\mathcal{F}) \propto P(\mathring{\mathcal{D}}|\mathcal{G},\mathcal{F}) = P(\mathcal{D}|\mathring{\mathcal{D}},\mathcal{G},\mathcal{F})P(\mathring{\mathcal{D}}|\mathcal{G},\mathcal{F}) = P(\mathring{\mathcal{D}}|\mathcal{D},\mathcal{G},\mathcal{F})P(\mathcal{D}|\mathcal{G},\mathcal{F}),$$

Assuming that all databases $\mathring{\mathcal{D}}$ compatible with $\mathcal{D}$ given $\mathcal{F}$ are equiprobable, we thus have that:

$$P(\mathcal{G}|\mathring{\mathcal{D}},\mathcal{F}) \propto P(\mathcal{D}|\mathcal{G},\mathcal{F}) = \int_{\theta} P(\mathcal{D}|\mathcal{G},\mathcal{F},\theta)\pi(\theta|\mathcal{F},\mathcal{G})d\theta. \qquad (3)$$

BN structure learning therefore amounts to find structure $\mathcal{G}^*$ such that $\mathcal{G}^* = \text{Argmax}_{\mathcal{G}} \, P(\mathcal{G}|\mathcal{D},\mathcal{F})$. Note that $P(\mathcal{D}|\mathcal{G},\mathcal{F},\theta)$ corresponds to a classical score over discrete data. $\pi(\theta|\mathcal{F},\mathcal{G})$ is the prior over the parameters of the BN given $\mathcal{F}$. Equation (3) is precisely the one used when discretization is performed as a preprocess before learning.

When discretization is performed while learning, like in [6,13], both the structure and the discretization should be optimized simultaneously. In other words, the problem consists of computing $\text{Argmax}_{\mathcal{F},\mathcal{G}} \, P(\mathcal{G},\mathcal{F}|\mathring{\mathcal{D}})$, where finding the best discretization amounts to find the best set of cut points (including the best size for this set) for each continuous random variable. And we have that:

$$P(\mathcal{G},\mathcal{F}|\mathring{\mathcal{D}}) = P(\mathcal{G}|\mathcal{F},\mathring{\mathcal{D}})P(\mathcal{F}|\mathring{\mathcal{D}}) \propto P(\mathcal{F}|\mathring{\mathcal{D}})\int_{\theta} P(\mathcal{D}|\mathcal{G},\mathcal{F},\theta)\pi(\theta|\mathcal{F},\mathcal{G})d\theta. \quad (4)$$

Input: a database $\overset{\circ}{\mathcal{D}}$, an initial graph $\mathcal{G}$, a score function sc on discrete variables
Output: the structure $\mathcal{G}$ of the Bayesian network
1 **repeat**
2 | Find the best discretization $\mathcal{F}$ given $\mathcal{G}$
3 | $\{X_{l+1}, \dots, X_n\} \leftarrow$ discretize variables $\{\overset{\circ}{X}_{l+1}, \dots, \overset{\circ}{X}_n\}$ given $\mathcal{F}$
4 | $\mathcal{G} \leftarrow \mathcal{G}$'s neighbor that maximizes scoring function sc w.r.t. $\{X_1, \dots, X_n\}$
5 **until** $\mathcal{G}$ *maximizes the score*;

Algorithm 1. Our structure learning architecture.

As can be seen, the resulting equation combines the classical score on the discretized data (the integral) with a score $P(\mathcal{F}|\overset{\circ}{\mathcal{D}})$ for the discretization algorithm itself. The logarithm of latter corresponds to what [6,13] call $DL_\Lambda(\Lambda) + DL_{\overset{\circ}{\mathcal{D}} \to \mathcal{D}}(\overset{\circ}{\mathcal{D}}, \Lambda)$ and $\mathcal{S}_c(\Lambda; \overset{\circ}{\mathcal{D}})$ respectively.

3   A New Multivariate Discretization-Learning Algorithm

As mentioned earlier, we believe that taking into account the conditional dependences between random variables is important to provide high-quality discretizations. Our approach thus follows Eq. (4) and our goal is to compute $\text{Argmax}_{\mathcal{F},\mathcal{G}} P(\mathcal{G}, \mathcal{F}|\overset{\circ}{\mathcal{D}})$. Optimizing jointly over $\mathcal{F}$ and $\mathcal{G}$ is too computationally intensive a task to be usable in practice. Fortunately, we can approximate it efficiently through a gradient descent, alternating optimizations over $\mathcal{F}$ given a fixed structure $\mathcal{G}$ and optimizations over $\mathcal{G}$ given a fixed discretization $\mathcal{F}$. This suggests the BN structure learning method described as Algorithm 1.

Multivariate discretization is much more time consuming than univariate discretization. As such, Line 2 could thus incur a strong overhead to the learning algorithm because the discretization search space increases exponentially with the number of variables to discretize. To alleviate this problem without sacrificing too much in accuracy, we suggest a local search algorithm that iteratively fixes the discretizations of all the continuous variables but one and optimizes the discretization of the latter (given the other variables) until some stopping criterion is met. As such, discretizations being optimized one continuous variable at a time, the combinatorics and the computation time are significantly limited. Line 2 can thus be detailed as Algorithm 2.

Input: a database $\overset{\circ}{\mathcal{D}}$, a graph $\mathcal{G}$, a scoring function sc on discrete variables
Output: a discretization $\mathcal{F}$
1 **repeat**
2 | $i_0 \leftarrow$ Select an element in $\{l+1, \dots, n\}$
3 | Discretize $\overset{\circ}{X}_{i_0}$ given $\mathcal{G}$ and $\{X_1, \dots, X_{i_0-1}, X_{i_0+1}, \dots, X_n\}$
4 **until** *stopping condition*;

Algorithm 2. One-variable discretization architecture.

3.1 Discretization Criterion

To implement Algorithm 2, a discretization criterion to be optimized is needed. Basic ideas include trying to find cut points minimizing the discrepancy between the frequencies or the sizes of intervals $[t_k, t_{k+1})$. A more sophisticated approach consists of limiting as much as possible the quantity of information lost after discretization, or equivalently to maximize the quantity of information remaining after discretization. This naturally calls for maximizing an entropy. This is essentially what our closest competitors, [6,13], do.

But entropy may not be the most appropriate measure when dealing with BNs. Actually, consider a variable A with domain $\{a_1, a_2, a_3\}$. Then, it is possible that, for some BN, $P(A = a_1) = \frac{1}{6}$, $P(A = a_2) = \frac{1}{3}$ and $P(A = a_3) = \frac{1}{2}$. With a sufficiently large database $\mathcal{D}$, the frequencies of observations of a_1, a_2, a_3 in $\mathcal{D}$ would certainly lead to estimate $P(A) \approx [\frac{1}{6}, \frac{1}{3}, \frac{1}{2}]$. Now, assume that the observations in $\mathcal{D}$ are noisy, say with a Gaussian noise with an *infinitely small* variance, as in Fig. 1. Then, after discretization, we shall expect to have 3 intervals with respective frequencies $\frac{1}{6}$, $\frac{1}{3}$ and $\frac{1}{2}$, i.e., intervals similar to $(-\infty, t_1)$, $[t_1, t_2)$ and $[t_2, +\infty)$ of Fig. 1. However, w.r.t. entropy, the best discretization corresponds to intervals $[-\infty, s_1)$, $[s_1, s_2)$ and $[s_2, +\infty)$ of Fig. 1 whose frequencies are all approximately equal to $\frac{1}{3}$ (entropy is maximal for equiprobable intervals). Therefore, whatever the infinitesimal noise added to data in $\mathcal{D}$, an entropy-based discretization produces a discretized variable A with distribution $[\frac{1}{3}, \frac{1}{3}, \frac{1}{3}]$ instead of $[\frac{1}{6}, \frac{1}{3}, \frac{1}{2}]$. This suggests that entropy is probably not the best criterion for discretizing continuous variables for BN learning.

Figure 1 suggests that clustering would probably be more appropriate: here, one cluster/interval per Gaussian would provide a better discretization. *In this paper, we assume that, within every interval, each continuous random variable, say $\mathring{X}_{i_0}$, is distributed w.r.t. a truncated Gaussian.* Over its whole domain of definition, it is thus distributed as a mixture of truncated Gaussians, the weights of the latter being precisely the CPT of X_{i_0} in the discrete BN. In particular, if $\mathring{X}_{i_0}$ has some parents, there are as many mixtures as the product of the domain sizes of the parents. The parameters of such a discretization scheme are therefore: (i) a set of g cut points (to define $g+1$ intervals) and (ii) a mean and a variance for each interval (to define its Gaussian). Figure 1 actually illustrates the fact that the means of the Gaussians need not necessarily correspond to the middles of the intervals. For instance, the mean of the third Gaussian is a_3 whereas the

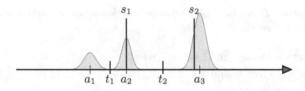

Fig. 1. Discretization: entropy v.s. clustering.

third interval, $[t_3, +\infty)$, has no finite middle. Here, even finite interval middles, like that of $[t_1, t_2)$, do not correspond to the means of the Gaussians.

For each continuous random variable $\overset{\circ}{X}_{i_0}$, this joint optimization problem is really hard due to the normalization requirements that the integrals of the truncated exponential of each interval must sum to 1 (which cannot be expressed using closed-form formulas). Therefore, to alleviate the discretization computational burden, we propose to approximate the computation of the cut points, means and variances using a two-step process: first, we approximate the density of the joint distribution of $\{X_1, \ldots, X_{i_0-1}, \overset{\circ}{X}_{i_0}, X_{i_0+1}, \ldots, X_n\}$ as a mixture of *untruncated* Gaussians and we determine by maximum likelihood the number of cut-points as well as the means and variances of the Gaussians. This can be easily done by an Expectation-Maximization (EM) approach. Then, in a second step, we compute the best cut points w.r.t. the Gaussians. As each Gaussian is associated with an interval, the parts of the Gaussian outside the interval can be considered as a loss of information and we will therefore look for cut points that minimize this loss. Now, let us delve into the details of the approach.

3.2   Discretization Exploiting the BN Structure

For the first discretization step of $\overset{\circ}{X}_{i_0}$, we estimate the number g of cut-points and the Gaussians' means and variances. Assume that structure $\mathcal{G}$ is fixed and that all the other variables are discrete. The density over all the variables, $p(\overset{\circ}{\mathbf{X}})$, is equal to $p(\overset{\circ}{X}_{i_0}|\mathbf{Pa}(\overset{\circ}{X}_{i_0})) \prod_{i \neq i_0} P(X_i|\mathbf{Pa}(X_i))$, where $p(\overset{\circ}{X}_{i_0}|\mathbf{Pa}(\overset{\circ}{X}_{i_0}))$ represents a mixture of Gaussians for each value of $\overset{\circ}{X}_{i_0}$'s parents (there are a finite number of values since all the variables but $\overset{\circ}{X}_{i_0}$ are discrete). $P(X_i|\mathbf{Pa}(X_i))$ should be the CPT of discrete variable X_i but, unfortunately, it is not well defined if $\overset{\circ}{X}_{i_0} \in \mathbf{Pa}(X_i)$ because, in this case, $\mathbf{Pa}(X_i)$ has infinitely many values. This is a serious issue since this CPT is used in the computation of $P(\mathcal{D}|\mathcal{G}, \mathcal{F}, \boldsymbol{\theta})$ of Eq. (4). Fortunately, this problem can be overcome by enforcing that $\overset{\circ}{X}_{i_0}$ has no child while guaranteeing that the density remains unchanged. Actually, in [19], an arc reversal operator is provided that, when applied, never alters the density/probability distribution. More precisely, when reverting arc $X \to Y$, Shachter showed that if all the parents of X are added to Y and all the parents of Y except X are added to X, then the resulting BN encodes the same distribution. As an example of these transformations, reversing

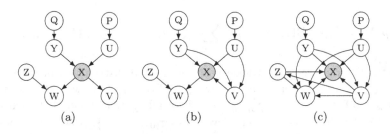

(a)   (b)   (c)

Fig. 2. Shachter's arc reversals.

arc $X \to V$ of Fig. 2.(a) results in Fig. 2.(b) and, then, reversing arc $X \to W$ results in Fig. 2.(c).

Therefore, to enforce that $\mathring{X}_{i_0}$ has no child, if $\{i_1, \ldots, i_c\}$ denotes the set of indices of the children variables of $\mathring{X}_{i_0}$, sorted by a topological order of $\mathcal{G}$, then, by reversing sequentially all the arcs $X_{i_j} \to \mathring{X}_{i_0}$, $j = 1, \ldots, c$, we get:

$$p(\mathring{\mathbf{X}}) = p(\mathring{X}_{i_0}|\mathbf{Pa}(\mathring{X}_{i_0})) \times \prod_{i \neq \{i_0, \ldots, i_c\}} P(X_i|\mathbf{Pa}(X_i)) \times \prod_{j=1}^{c} P(X_{i_j}|\mathbf{Pa}(X_{i_j})),$$

$$p(\mathring{\mathbf{X}}) = p(\mathring{X}_{i_0}|\mathbf{MB}(\mathring{X}_{i_0})) \times \prod_{i \neq \{i_0, \ldots, i_c\}} P(X_i|\mathbf{Pa}(X_i))$$

$$\times \prod_{j=1}^{c} P(X_{i_j}| \bigcup_{h=1}^{j} (\mathbf{Pa}(X_{i_h}) \backslash \{\mathring{X}_{i_0}\}) \cup \mathbf{Pa}(\mathring{X}_{i_0})),$$

where $\mathbf{MB}(\mathring{X}_{i_0})$ is the Markov blanket of $\mathring{X}_{i_0}$ in $\mathcal{G}$:

Definition 2. *The Markov blanket of any node in $\mathcal{G}$ is the set of its parents, its children and the other parents of its children.*

Note that, in the last expression of $p(\mathring{\mathbf{X}})$, only the first term involves $\mathring{X}_{i_0}$, hence all the other CPTs are well defined (they are finite CPTs). As a side effect, only $p(\mathring{X}_{i_0}|\mathbf{MB}(\mathring{X}_{i_0}))$ needs be taken into account to discretize $\mathring{X}_{i_0}$ since none of the other terms is related to $\mathring{X}_{i_0}$. *It shall be noted here that these arc reversals are applied only for determining the parameters of the discretization, i.e., the set of cut points, means and variances of the Gaussians, they are never used to learn the BN structure.* Now, let us see how the parameters of the mixture of Gaussians $p(\mathring{X}_{i_0}|\mathbf{MB}(\mathring{X}_{i_0}))$ maximizing the likelihood of dataset $\mathcal{D}$ can be easily estimated using an EM algorithm.

3.3 Parameter Estimation by an EM Algorithm

Let q_{i_0} represent the (finite) number of values of $\mathbf{MB}(\mathring{X}_{i_0})$. For simplicity, we will denote by $\{1, \ldots, q_{i_0}\}$ the set of values of the joint discrete random variable $\mathbf{MB}(\mathring{X}_{i_0})$. Let g denote the number of cut points in the discretization and let $\{\mathcal{N}(\mu_k, \sigma_k) : k \in \{0, \ldots, g\}\}$ be the corresponding set of Gaussians. Then:

$$p(\mathring{X}_{i_0} = \mathring{x}_{i_0}|\mathbf{MB}(\mathring{X}_{i_0}) = j) = \sum_{k=0}^{g} \pi_{jk} f(\mathring{x}_{i_0}|\boldsymbol{\theta}_k) \quad \forall j \in \{1, \ldots, q_{i_0}\},$$

where $f(\cdot|\boldsymbol{\theta}_k)$ represents the density of the normal distribution of parameters $\boldsymbol{\theta}_k = (\mu_k, \sigma_k)$, and π_{jk} represents the weights of the mixture (with the constraint that $\pi_{jk} \geq 0$ for all j, k and $\sum_{k=0}^{g} \pi_{jk} = 1$ for all j). Remember that each value of $\mathbf{MB}(\mathring{X}_{i_0})$ induces its own set of weights $\{\pi_{j0}, \ldots, \pi_{jg}\}$. Now, we propose to estimate parameters $\boldsymbol{\theta}_k$ from $\mathcal{D}$ by maximum likelihood. For this, EM is well-known to efficiently provide good approximations [3] (due to the mixture, direct maximum likelihood is actually hard to estimate). Assuming that data in $\mathring{\mathcal{D}}$ are i.i.d., the log-likelihood of $\mathring{\mathcal{D}}$ given $\Theta = \bigcup_{k=0}^{g}(\bigcup_{j=1}^{q_{i_0}}\{\pi_{jk}\} \cup \{\boldsymbol{\theta}_k\})$ is equal to:

$$\mathcal{L}(\mathring{\mathcal{D}}|\Theta) = \sum_{m=1}^{N} \log p(\mathring{X}_{i_0} = \mathring{x}_{i_0}^{(m)}|\mathbf{MB}(\mathring{x}_{i_0})^{(m)}, \Theta),$$

where $\mathring{x}_{i_0}^{(m)}$ represents the observed value of $\mathring{X}_{i_0}$ in the mth record of $\mathring{\mathcal{D}}$. Thus:

$$\mathcal{L}(\mathring{\mathcal{D}}|\Theta) = \sum_{j=1}^{q_{i_0}} \sum_{m:\mathbf{MB}(\mathring{x}_{i_0})^{(m)}=j} \log\left[\sum_{k=0}^{g} \pi_{jk}f(\mathring{x}_{i_0}^{(m)}|\boldsymbol{\theta}_k)\right]. \tag{5}$$

To solve $\text{Argmax}_\Theta \mathcal{L}(\mathring{\mathcal{D}}|\Theta)$, EM [3] iteratively alternates expectations (E-step) and maximizations (M-step) until convergence toward a local maximum which is guaranteed to correspond to the Argmax we look for due to the concavity of the log-likelihood function. In this paper, we just need to apply the standard EM, considering for weights π_{jk} only the records in the database that correspond to $\mathbf{MB}(\mathring{x}_{i_0})^{(m)} = j$. More precisely, for each record of $\mathring{\mathcal{D}}$, let $Z^{(m)}$ be a random variable whose domain is $\{0, \ldots, g\}$, and such that $Z^{(m)} = k$ if and only if observation $\mathring{x}_{i_0}^{(m)}$ has been generated from the kth Gaussian. Let $Q_m^t(Z^{(m)}) = P(Z^{(m)}|\mathring{x}_{i_0}^{(m)}, \Theta^t)$, i.e., $Q_m^t(Z^{(m)})$ represents the distribution that, at the tth step of the algorithm, $\mathring{x}_{i_0}^{(m)}$ is believed to have been generated by such and such Gaussian. Then, EM is described in Algorithm 3.

In the EM algorithm, only the M-step can be computationally intensive. Fortunately, here, we can derive in closed-form the optimal values of Line 4:

Proposition 1. *At the E-step, probability* $Q_m^{t+1}(k) = \dfrac{\pi_{jk}^t f(\mathring{x}_{i_0}^{(m)}|\boldsymbol{\theta}_k^t)}{\sum_{k'=0}^{g} \pi_{jk'}^t f(\mathring{x}_{i_0}^{(m)}|\boldsymbol{\theta}_{k'}^t)}$, *where* π_{jk}^t *and* $\boldsymbol{\theta}_k^t$ *are weights, means and variances in* Θ^t. *The optimal*

Input: a database $\mathring{\mathcal{D}}$, a number g of cut points
Output: an optimal set of parameters Θ
1 Select (randomly) an initial value Θ^0
2 **repeat**
    // E-step (expectation)
3    $Q_m^{t+1}(Z^{(m)}) \leftarrow P(Z^{(m)}|\mathring{x}_{i_0}^{(m)}, \Theta^t) \quad \forall\, m \in \{1, \ldots, N\}$
    // M-step (maximization)
4    $\Theta^{t+1} \leftarrow \text{Argmax}_\Theta \sum_{j=1}^{q_{i_0}} \sum_{m:\mathbf{MB}(\mathring{x}_{i_0})^{(m)}=j} \sum_{k=0}^{g} Q_m^{t+1}(k) \log\left[\dfrac{\pi_{jk}f(\mathring{x}_{i_0}^{(m)}|\boldsymbol{\theta}_k)}{Q_m^{t+1}(k)}\right]$
5 **until** *convergence*;

Algorithm 3. The EM algorithm.

parameters of the M-step are respectively:

$$\pi_{jk}^{t+1} = \frac{\sum_{m:\mathbf{MB}(\mathring{x}_{i_0})^{(m)}=j} Q_m^{t+1}(k)}{\sum_{m:\mathbf{MB}(\mathring{x}_{i_0})^{(m)}=j} \sum_{k'=0}^{g} Q_m^{t+1}(k')},$$

$$\mu_k^{t+1} = \frac{\sum_{m=1}^{N} Q_m^{t+1}(k)\mathring{x}_{i_0}^{(m)}}{\sum_{m=1}^{N} Q_m^{t+1}(k)} \qquad \sigma_k^{t+1} = \sqrt{\frac{\sum_{m=1}^{N} Q_m^{t+1}(k)(\mathring{x}_{i_0}^{(m)} - \mu_k^{t+1})^2}{\sum_{m=1}^{N} Q_m^{t+1}(k)}}.$$

Using Algorithm 3 with the formulas of Proposition 1, it is thus possible to determine the means and variances of the Gaussians. However, our ultimate goal is not to compute them but to exploit them to discretize variable $\mathring{X}_{i_0}$, i.e., to determine the best cut points $t_1, \ldots, t_g$. Let us see how this task can be performed.

3.4   Determination of the Cut Points

As mentioned at the end of Subsect. 3.1, each Gaussian $\mathcal{N}(\mu_k, \sigma_k)$ is associated with an interval $[t_k, t_{k+1})^2$ and the parts of the Gaussian outside the interval can be considered as a loss of information. The optimal set of cut points $\widehat{T} = \{\hat{t}_1, \ldots, \hat{t}_g\}$ is thus that which minimizes this loss. In other words, it is equal to:

$$\widehat{T} = \underset{\{t_1,\ldots,t_g\}}{\text{Argmin}} \sum_{k=1}^{g} \int_{t_k}^{+\infty} f(x|\boldsymbol{\theta}_{k-1})dx + \int_{-\infty}^{t_k} f(x|\boldsymbol{\theta}_k)dx,$$

where $\boldsymbol{\theta}_k$ represents pairs (μ_k, σ_k). As each Gaussian $\mathcal{N}(\mu_k, \sigma_k)$ is associated with interval $[t_k, t_{k+1})$, we can assume that $\hat{t}_k \in [\mu_{k-1}, \mu_k)$, for all k. Therefore:

$$\widehat{T} = \left\{ \underset{t_k \in [\mu_{k-1}, \mu_k)}{\text{Argmin}} \int_{t_k}^{+\infty} f(x|\boldsymbol{\theta}_{k-1})dx + \int_{-\infty}^{t_k} f(x|\boldsymbol{\theta}_k)dx : k \in \{1, \ldots, g\} \right\}. \quad (6)$$

All the $\hat{t}_k$ can thus be determined independently. In addition, as shown below, their values are the solution of a quadratic equation:

Proposition 2. *Let $u(t_k)$ represent the sum of the integrals in Eq. (6). Let α_k be a solution (if any) within interval (μ_{k-1}, μ_k) of the quadratic equation in t_k:*

$$t_k^2 \left(\frac{1}{\sigma_{k-1}^2} - \frac{1}{\sigma_k^2} \right) + 2t_k \left(\frac{\mu_k}{\sigma_k^2} - \frac{\mu_{k-1}}{\sigma_{k-1}^2} \right) + \left(\frac{\mu_{k-1}^2}{\sigma_{k-1}^2} - \frac{\mu_k^2}{\sigma_k^2} - 2log\frac{\sigma_k}{\sigma_{k-1}} \right) = 0. \quad (7)$$

Then $\hat{t}_k$ is, among $\{\mu_{k-1}, \mu_k, \alpha_k\}$, the element with the highest value of $u(\cdot)$ (which can be quickly approximated using a table of the Normal distribution).

---

[2] Without loss of generality, we consider here that the μ_k's resulting from the EM algorithm are sorted by increasing order.

Proof. Let $g(\cdot)$ and $h(\cdot)$ be two functions such that $\partial g(x)/\partial x = f(x|\boldsymbol{\theta}_{k-1})$ and $\partial h(x)/\partial x = f(x|\boldsymbol{\theta}_k)$. Then:

$$
\hat{t}_k = \operatorname*{Argmin}_{t_k \in [\mu_{k-1}, \mu_k)} u(t_k) = \operatorname*{Argmin}_{t_k \in [\mu_{k-1}, \mu_k)} \int_{t_k}^{+\infty} \frac{\partial g(x)}{\partial x} dx + \int_{-\infty}^{t_k} \frac{\partial h(x)}{\partial x} dx
$$

$$
= \operatorname*{Argmin}_{t_k \in [\mu_{k-1}, \mu_k)} -g(t_k) + h(t_k) + \lim_{t \to +\infty} [g(t) - h(-t)].
$$

Let us relax the optimization problem and try to find the Argmin over $\mathbb{R}$. Then the min is obtained when $\partial u(t_k)/\partial t_k = 0$ or, equivalently, when $\partial(-g(t_k) + h(t_k))/\partial t_k = -f(t_k|\boldsymbol{\theta}_{k-1}) + f(t_k|\boldsymbol{\theta}_k) = 0$. Since $f(\cdot|\boldsymbol{\theta})$ represents the density of the Normal distribution of parameters $\boldsymbol{\theta}$, this is equivalent to:

$$
-\frac{1}{\sqrt{2\pi}\sigma_{k-1}} \exp\left(-\frac{1}{2}\left(\frac{t_k - \mu_{k-1}}{\sigma_{k-1}}\right)^2\right) + \frac{1}{\sqrt{2\pi}\sigma_k} \exp\left(-\frac{1}{2}\left(\frac{t_k - \mu_k}{\sigma_k}\right)^2\right) = 0,
$$

or, equivalently:

$$
\frac{\sigma_k}{\sigma_{k-1}} = \frac{\exp\left[-\frac{1}{2}\left(\frac{t_k - \mu_k}{\sigma_k}\right)^2\right]}{\exp\left[-\frac{1}{2}\left(\frac{t_k - \mu_{k-1}}{\sigma_{k-1}}\right)^2\right]} = \exp\left[\frac{1}{2}\left(\frac{t_k - \mu_{k-1}}{\sigma_{k-1}}\right)^2 - \frac{1}{2}\left(\frac{t_k - \mu_k}{\sigma_k}\right)^2\right],
$$

which, by a log transformation, is equivalent to:

$$
2\log\frac{\sigma_k}{\sigma_{k-1}} = \frac{t_k^2}{\sigma_{k-1}^2} - \frac{2\mu_{k-1}t_k}{\sigma_{k-1}^2} + \frac{\mu_{k-1}^2}{\sigma_{k-1}^2} - \frac{t_k^2}{\sigma_k^2} + \frac{2\mu_k t_k}{\sigma_k^2} - \frac{\mu_k^2}{\sigma_k^2}.
$$

This corresponds precisely to Eq. (7). So, to summarize, if the optimal solution lies inside interval (μ_{k-1}, μ_k), then it satisfies Eq. (7). Otherwise, either $u(t_k)$ is strictly increasing or strictly decreasing within (μ_{k-1}, μ_k), which implies that the optimal solution for $\hat{t}_k$ is either μ_{k-1} or μ_k, which completes the proof. ∎

3.5 Score and Number of Cut Points

To complete the description of the algorithm, there remains to determine the number of cut points. Of course, the higher the number of cut points, the higher the likelihood but the lower the compactness of the representation. To reach of good trade-off, we simply propose to exploit the penalty functions included into the score used for the evaluation of different BN structures (see Line 5 of Algorithm 1). Here, we used the BIC score [18], which can be locally expressed as:

$$
BIC(\mathring{X}_{i_0}|\mathbf{MB}(\mathring{X}_{i_0})) = \mathcal{L}(\mathring{\mathcal{D}}|\boldsymbol{\Theta}) - \frac{|\boldsymbol{\Theta}|}{2}\log(N) \tag{8}
$$

where $\mathcal{L}(\mathring{\mathcal{D}}|\boldsymbol{\Theta})$ is the log-likelihood with the parameters estimated by EM, given the current structure $\mathcal{G}$. $|\boldsymbol{\Theta}|$ represents the number of parameters, i.e., $|\boldsymbol{\Theta}| = q_{i_0} \times g + 2 \times (g+1)$: the 1st and 2nd terms correspond to the number of parameters π_{jk} and of (μ_k, σ_k) needed to encode the conditional distributions (recall that there are $g+1$ Gaussians and q_{i_0} represents the domain size of $\mathbf{MB}(\mathring{X}_{i_0})$). Now, the best number of cut points is simply that which optimizes Eq. (8).

4   Experimentations

In this section, we highlight the effectiveness of our method, hereafter denoted MGCD (for Mixture of Gaussians Clustering-based Discretization), by comparing it with the algorithms provided in [6,17], hereafter called Ruichu and Friedman respectively. Step 4 of Algorithm 1 was performed using a simple *Tabu* search method. For the comparisons, three criteria have been taken into account: (i) the quality of the structure learnt by the algorithm (which strongly depends on that of the discretization); (ii) the computation time and (iii) the quality of the learnt CPT parameters, which has been evaluated by their prediction power on the values taken by some variables given observations.

For the first two criteria, we randomly generated discrete BNs following the guidelines given in [7]. Those contained from 10 to 30 nodes and from 12 to 56 arcs. Each node had at most 6 parents and its domain size was randomly chosen between 2 and 5. The CPTs of these BNs represented the π_{jk} of the preceding section. From these BNs, we generated continuous datasets containing from 1000 to 10000 records as follows: for each random variable X_i, we mapped its finite set of values into a set of consecutive intervals $\{[t_{k-1}, t_k)\}_{k=1}^{|X_i|}$ of arbitrary lengths. Then, we assigned a truncated Gaussian to each interval, the parameters (μ_k, σ_k) of which were randomly chosen. Finally, to generate a continuous record, we first generated a discrete record from the discrete BN using a logic sampling algorithm. Then, this record was mapped into a continuous one by sampling from the truncated exponentials. Overall, 350 continuous datasets were generated.

To compare them, the BN structures produced by Ruichu, Friedman and MGCD were converted into their Markov equivalence class, i.e., into a partially directed DAG (CPDAG). Such a transformation increases the quality of comparisons since two BNs encode the same distribution iff they belong to the same equivalence class. The CPDAGs were then compared w.r.t. their true and false positive rate metrics (TPR and FPR). TPR (resp. FPR) represents the percentage of arcs/edges belonging to the learnt CPDAG that also exist (resp. do not exist) in the original CPDAG. Both metrics describe how well the dependences

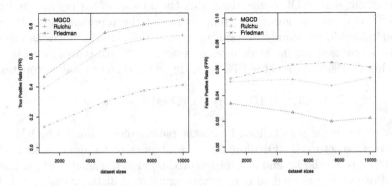

Fig. 3. Averages of the TPR (left) and FPR (right) metrics for BNs with 10 to 30 nodes in function of the sample sizes.

Table 1. Runtime ratio comparisons between discretization approaches.

Approaches / Dataset sizes	1000	5000	7500	10000
Friedman	2.762444	3.350782	3.404958	3.361540
Ruichu	0.8872389	1.1402535	1.1334637	1.1032982

Table 2. Prediction accuracy rates for discrete target variables in the Child and Sachs standard BNs (http://www.bnlearn.com/bnrepository/) w.r.t. the percentage of observed variables in the Markov blanket.

datasets	sizes	30 % Markov blanket			60 % Markov blanket			100 % Markov blanket		
		MGCD	Ruichu	Friedman	MGCD	Ruichu	Friedman	MGCD	Ruichu	Friedman
Child	1000	**60.90**	59.30	58.99	**62.76**	60.90	59.96	**67.53**	65.34	63.33
	2000	**61.59**	56.62	58.05	**62.41**	59.24	60.55	**67.29**	64.71	63.03
	5000	**64.88**	62.29	60.05	**66.07**	62.95	61.94	**69.42**	65.39	63.82
	10000	**65.81**	62.48	61.75	**67.44**	63.85	63.51	**70.49**	66.92	65.79
Sachs	1000	56.63	54.78	**57.59**	57.22	55.04	**58.74**	**65.67**	61.06	64.65
	2000	**56.96**	56.16	54.02	**59.72**	57.58	56.64	**65.80**	62.24	60.22
	5000	**57.69**	55.00	55.15	**59.80**	57.96	56.38	**65.51**	64.15	64.47
	10000	**60.35**	57.50	57.33	**61.67**	58.26	59.22	**70.04**	65.74	64.61

between variables are preserved by learning/discretization. Figure 3 shows the average TPR and FPR over the 350 generated databases. As can be seen, MGCD outperforms the others for all dataset sizes: MGCD's TPR is about 10 % higher than Ruichu and 40 % higher than Friedman, and MGCD's FPR is between 20 % and 40 % lower than the other methods. MGCD's performance w.r.t. Ruichu's can be explained by that fact that, unlike Ruichu's, it fully takes into account the conditional dependences between all the random variables. Its performance w.r.t. Friedman's can be explained by our choice of exploiting clustering rather than an entropy-based approach. Table 1 provides computation time ratios (other method's runtime / MGCD's runtime). As can be seen, our method slightly outperforms Ruichu's (but is 10 % better in terms of TPR and more than 20 % better in terms of FPR) and it significantly outperforms Friedman's (about 3 times faster) while at the same time being 40 % higher in terms of TPR.

Finally, we compared the discretizations w.r.t. the quality of the produced CPTs. To do so, we generated from two classical BNs, Child and Sachs, 100 continuous databases using the same process as above except that: (i) the distributions inside intervals were uniform instead of Gaussians (to penalize our approach since data do not fit its hypotheses), and (ii) some small sets of variables were kept discrete and served as multilabel targets. Databases were split into a learning (2/3) and a test (1/3) part. For each record in the latter, we computed the distribution (learnt by each of the 3 algorithms on the learning database) of each target given some observations on their Markov blanket and we estimated the value of the target by sampling it from the learnt distribution. The percentages of correct predictions are shown in Table 2. As we can see, our algorithm outperforms the other algorithms, especially Ruichu's, which fails to have correct predictions due to its univariate discretization not taking into account the conditional dependencies among random variables. Friedman's results are closer to ours but recall that it is about 3 times slower than ours.

5   Conclusion

We have proposed a new multivariate discretization algorithm designed for BN structure learning, taking into account the dependences among variables acquired during learning. Our experiments highlighted its efficiency and effectiveness compared to state-of-the-art algorithms, but more experiments are of course needed to better assess the strengths and the shortcoming of our proposed approach. For future work, we plan to improve our algorithm, notably by directly working with truncated Gaussians instead of the current approximation by mixture of Gaussians. But such an improvement is not trivial due to the fact that, in this case, no closed-form solution exists for determining the cut points.

Acknowledgment. This work was supported by the French Institute for Radioprotection and Nuclear Safety (IRSN), the Belgium's nuclear safety authorities (Bel V) and European project H2020-ICT-2014-1 #644425 Scissor.

References

1. Boullé, M.: MODL: a Bayes optimal discretization method for continuous attributes. Mach. Learn. **65**(1), 131–165 (2006)
2. Boullé, M.: Khiops: a statistical discretization method of continuous attributes. Mach. Learn. **55**(1), 53–69 (2004)
3. Dempster, A.P., Laird, N., Rubin, D.: Maximum likelihood from incomplete data via the EM algorithm. J. Roy. Stat. Soc. **39**(1), 1–38 (1977)
4. Dougherty, J., Kohavi, R., Sahami, M.: Supervised and unsupervised discretization of continuous features. In: proceeding of ICML 1995, pp. 194–202 (1995)
5. Fayyad, U., Irani, K.: Multi-interval discretization of continuous-valued attributes for classification learning. In: Proceeding of IJCAI 1993, pp. 1022–1029 (1993)
6. Friedman, N., Goldszmidt, M.: Discretizing continuous attributes while learning Bayesian networks. In: proceeding of ICML 1996, pp. 157–165 (1996)
7. Ide, J.S., Cozman, F.G., Ramos, F.T.: Generating random Bayesian networks with constraints on induced width. In: Proceeding of ECAI 2004, pp. 323–327 (2004)
8. Jiang, S., Li, X., Zheng, Q., Wang, L.: Approximate equal frequency discretization method. In: proceeding of GCIS 2009, pp. 514–518 (2009)
9. Kerber, R.: ChiMerge: Discretization of numeric attributes. In: Proceeding of AAAI 1992, pp. 123–128 (1992)
10. Koller, D., Friedman, N.: Probabilistic Graphical Models: Principles and Techniques. MIT Press (2009)
11. Kwedlo, W., Krętowski, M.: An evolutionary algorithm using multivariate discretization for decision rule induction. In: Żytkow, J.M., Rauch, J. (eds.) PKDD 1999. LNCS (LNAI), vol. 1704, pp. 392–397. Springer, Heidelberg (1999)

12. Lauritzen, S., Wermuth, N.: Graphical models for associations between variables, some of which are qualitative and some quantitative. Ann. Stat. **17**(1), 31–57 (1989)
13. Monti, S., Cooper, G.: A multivariate discretization method for learning Bayesian networks from mixed data. In: Proceeding of UAI 1998, pp. 404–413 (1998)
14. Monti, S., Cooper, G.: A latent variable model for multivariate discretization. In: proceeding of AIS 1999, pp. 249–254 (1999)
15. Moral, S., Rumí, R., Salmerón, A.: Mixtures of truncated exponentials in hybrid Bayesian networks. In: Benferhat, S., Besnard, P. (eds.) ECSQARU 2001. LNCS (LNAI), vol. 2143, pp. 156–167. Springer, Heidelberg (2001)
16. Ratanamahatana, C.: CloNI: Clustering of sqrt(n)-interval discretization. In: proc. of Int. Conf. on Data Mining & Comm Tech. (2003)
17. Ruichu, C., Zhifeng, H., Wen, W., Lijuan, W.: Regularized Gaussian mixture model based discretization for gene expression data association mining. Appl. Intell. **39**(3), 607–613 (2013)
18. Schwarz, G.: Estimating the dimension of a model. Ann. Stat. **6**(2), 461–464 (1978)
19. Shachter, R.: Evaluating influence diagrams. Oper. Res. **34**(6), 871–882 (1986)
20. Shenoy, P., West, J.: Inference in hybrid Bayesian networks using mixtures of polynomials. Int. J. Approximate Reasoning **52**(5), 641–657 (2011)
21. Song, D., Ek, C., Huebner, K., Kragic, D.: Multivariate discretization for Bayesian network structure learning in robot grasping. In: proceeding of ICRA 2011, pp. 1944–1950 (2011)
22. Zighed, D., Rabaséda, S., Rakotomalala, R.: FUSINTER: a method for discretization of continuous attributes. Int. J. Uncertainty Fuzziness Knowl. Based Syst. **6**(03), 307–326 (1998)

Modeling and Forecasting Time Series of Compositional Data: A Generalized Dirichlet Power Steady Model

Mohamad Mehdi[1]([✉]), Elise Epaillard[3], Nizar Bouguila[2],
and Jamal Bentahar[2]

[1] Engineering and Computer Science, Concordia University, Montreal, Canada
mo_mehdi@encs.concordia.ca
[2] Concordia Institute for Information Systems Engineering, Concordia University,
Montreal, Canada
{bouguila,bentahar}@ciise.concordia.ca
[3] ELectrical and Computer Engineering, Concordia University, Montreal, Canada
e_epail@encs.concordia.ca

Abstract. This paper presents GDPSM a power steady model (PSM) based on generalized Dirichlet observations for modeling and predicting compositional time series. The model's unobserved states evolve according to the generalized Dirichlet conjugate prior distributions. The observations' distribution is transformed into a set of Beta distributions each of which is re-parametrized as a unidimensional Dirichlet in its exponential form. We demonstrate that dividing the modeling problem into multiple smaller problems leads to more accurate predictions. We evaluate this model with the web service selection application. Specifically, we analyze the proportions of the quality classes that are assigned to the web services interactions. Our model is compared with another PSM that assumes Dirichlet observations. The experiments show promising results in terms of precision errors and standardized residuals.

Keywords: Time series · State space models · Generalized Dirichlet

1 Introduction

Time series of continuous proportions or compositional data, have been analyzed and modeled using various approaches [1,5]. This kind of series presents itself in domains varying from economics (e.g., yearly gross domestic product), to chemistry (e.g., chemical compositions), to political sciences (e.g., vote and seat shares). Generally, time series of compositional data are multivariate and denoted by time-dependent vectors of proportions that sum to one. To model such data, one might resort to standard techniques such as the multivariate autoregressive integrated moving average (ARIMA) [17] and Kalman filters [9]. However, due to the positive nature of the components of compositional data and their sum to one constraint, these techniques are not applicable [1].

© Springer International Publishing Switzerland 2015
C. Beierle and A. Dekhtyar (Eds.): SUM 2015, LNAI 9310, pp. 170–185, 2015.
DOI: 10.1007/978-3-319-23540-0_12

Various approaches have been proposed to deal with the positivity and dependence of the compositional data's components. For instance, Aitchison proposed the mapping of the data from the positive simplex $\mathbb{S}^d = \{(s_1, \ldots, s_d), s.t. \sum_{i=1}^{d} s_i < 1\}$, to the d-dimensional real space $\mathbb{R}^d$ [1]. Specifically, he suggested the additive and multiplicative logistic transforms. Inspired by Aitchison's proposals, the authors in [5] employed the multivariate ARIMA to model compositional time series transformed using the above additive logistic transform. The practicality of this transform has been shown via a public opinion polls application. However, one limitation of such approach is dealing with zero values of s_i which yield $y_i = \pm\infty$. In the same line of research, [12] used the same transform with multivariate dynamic linear models. To circumvent the zero-infinity issue, looking for a replacement for the additive transformation might be the answer. For instance, [19] proposed an alternative approach that employs a hyperspherical transform. This was intended to overcome the positivity and unit-sum constraints of compositional data. It also promised to solve the problems that arise with cases that have zero-valued components. This approach is based on modeling each component of the time series by the available data instances. The time series are first mapped through a non-linear dimensionality reduction approach onto a hypersphere. As such, the dimension d of a time series is reduced to $d-1$. Furthermore, [2] suggested the Box-Cox transformation which is a general form of the additive logistic transformation. Afterwards, the authors proposed a regression model, framed in a dynamic Bayesian structure, to model compositional time series.

Additionally, forecasting is another major part of the body of time series literature. [7] provides a review for time series forecasting models including exponential smoothing methods, ARIMA, state space and structural models, Kalman filters, and autoregressive conditional heterscedastic models. A stochastic extension to traditional autoregressive moving average (ARMA) time series models was proposed in [16]. State space models consist of observation and state processes that may be non-linear and non-Gaussian. The main usage of such models is to deduce the properties of the states given the knowledge from the observations. It is noteworthy to mention that all ARMA and ARIMA may be written as state space models. In the case of linear processes, Kalman filters are used to solve the corresponding state space models. The authors in [8] developed a Dirichlet state space model to forecast compositional time series. They also propose an estimation approach of the trends, covariates, and interventions in time series. A motor vehicle production data set that consists of the number of vehicle production in Japan, the United States, and the rest of the world during the years 1947 to 1987.

Motivation: As mentioned earlier, a wide range of real applications in varied domains involve compositional time series. The majority of these applications handle series that consist of yearly, quarterly, or monthly proportions. However, with the plethora of online data, some compositional time series may arise on a daily or even hourly basis. For example, the geographic distribution of the users of social media websites may be measured on an hourly basis for various business related functions. Therefore, given the large amount of available data, the need to understand this data, and the benefits in turning it into actionable insights, building a modeling and forecasting model for compositional time series becomes of unprecedented significance.

Contributions: In this paper, we build upon and extend the literature of compositional time series forecasting by the following contributions. (1) We propose to model and forecast compositional time series based on a novel PSM in which the observations are assumed to follow a generalized Dirichlet (GD) distribution. (2) We transform the GD distribution of d dimensions to d Beta distributions which, in turn, are transformed to d unidimensional Dirichlet distributions in their exponential form. This approach partitions the modeling and forecasting of $(d + 1)$-dimensional time series into d smaller problems with fewer parameters to learn. (3) We evaluate our model with a new application, web service selection, in comparison to outdated ones used in the literature. (4) We compare our model's forecasting performance to that of the Dirichlet-based power steady model (DPSM) proposed in [8]. We show the merits of our model via standardized residuals and mean squared error (MSE) of the predictions.

The rest of the paper is organized as follows. Section 2 describes the GD distribution and the transformations that it undergoes to be represented by multiple Dirichlet distributions. Section 3 highlights the characteristics of state space models and the details of the proposed time series model based on the GDPSM are described in Sect. 4. The experimental evaluation of the proposed model using various simulated data are presented and discussed in Sect. 5. Section 6 concludes the paper by summarizing its main contributions and suggesting directions for future work.

2   Generalized Dirichlet Formulation

Let $\boldsymbol{X} = (X_1, \ldots, X_{d+1})$ denote a vector of proportions that follows a d-dimensional GD distribution with the parameters vector $\boldsymbol{\alpha} = (\alpha_1, \beta_1, \ldots, \alpha_d, \beta_d)$. The probability distribution function of $\boldsymbol{X}$ is given by:

$$p(X_1, ..., X_d) = \prod_{l=1}^{d} \frac{\Gamma(\alpha_l + \beta_l)}{\Gamma(\alpha_l) + \Gamma(\beta_l)} X_l^{\alpha_l - 1} \left(1 - \sum_{j=1}^{l} X_j\right)^{\gamma_l}, \tag{1}$$

for $\sum_{l=1}^{d} X_d < 1$ and $0 < X_l < 1$, for $l = 1, ..., d$, where $\alpha_l > 0$, $\beta_l > 0$, $\gamma_l = \beta_l - \alpha_{l+1} - \beta_{l+1}$, for $l = 1, ..., d - 1$, and $\gamma_d = \beta_d - 1$. Also, note that $\Gamma(x) = \int_0^{\infty} t^{x-1} e^{-t} dt$. Since $\boldsymbol{X}$ follows a generalized Dirichlet and is completely neutral, it can be transformed to d independent Beta distributions [3,20]. Let $\boldsymbol{Y} = (Y_1, \ldots, Y_d)$ be the result of the following transformation:

$$Y_j = \begin{cases} X_j, & \text{if } j = 1, \\ \frac{X_j}{1 - X_1 - \cdots - X_{j-1}}, & \text{if } 2 \leq j \leq d. \end{cases} \tag{2}$$

The parameters vector $\boldsymbol{\alpha}$ can be estimated by considering that each of the Y_j has a Beta distribution with parameters α_j and β_j. Therefore, the joint probability distribution of $\boldsymbol{Y}$ can be written as follows:

$$p(\boldsymbol{Y}|\boldsymbol{\alpha}) = \prod_{l=1}^{d} B(\alpha_l, \beta_l)^{-1} Y_l^{\alpha_l-1}(1-Y_l)^{\beta_l-1}, \tag{3}$$

where $B(\alpha_l, \beta_l) = \frac{\Gamma(\alpha_l)\Gamma(\beta_l)}{\Gamma(\alpha_l+\beta_l)}$. The Beta distribution belongs to the exponential family in which each density is given by the following:

$$p(\boldsymbol{Y}|\theta) = H(\boldsymbol{Y})exp\left(\sum_{s=1}^{S} \eta_s(\theta)T_l(\boldsymbol{Y}) + \Phi(\theta)\right), \tag{4}$$

where $\eta_s(\theta)$ are called the natural parameters, $T_l(\boldsymbol{Y})$ are the sufficient statistics, $H(\boldsymbol{Y})$ is a base measure, and $\Phi(\theta)$ is referred to as the log-partition function. Equation (3) can thus be written as an exponential density:

$$p(\boldsymbol{Y}|\boldsymbol{\alpha}) = \exp\left[\sum_{l=1}^{d} \log(B(\alpha_l, \beta_l)^{-1}) + (\alpha_l - 1)\log Y_l + (\beta_l - 1)\log(1-Y_l)\right]$$

$$= \prod_{l=1}^{d} \frac{1}{Y_l(1-Y_l)} \exp\left[\sum_{l=1}^{d} \log(\frac{\Gamma(\alpha_l + \beta_l)}{\Gamma(\alpha_l)\Gamma(\beta_l)})\right.$$

$$\left. + \alpha_l \log Y_l + \sum_{l=d+1}^{2d} \beta_l \log(1-Y_l)\right]. \tag{5}$$

Let $S = 2d$, then we have:

$$H(\boldsymbol{Y}) = \prod_{l=1}^{d} \frac{1}{Y_l(1-Y_l)}, \tag{6}$$

$$T(Y_l) = \begin{cases} \log Y_l & \text{for } l = 1, \ldots, d, \\ \log(1-Y_l) & \text{for } l = d+1, \ldots, 2d, \end{cases} \tag{7}$$

$$\eta_l(\theta) = \begin{cases} \alpha_l & \text{for } l = 1, \ldots, d, \\ \beta_l & \text{for } l = d+1, \ldots, 2d, \end{cases} \tag{8}$$

$$\Phi(\theta) = \sum_{l=1}^{d} \log\left(\frac{\Gamma(\alpha_l + \beta_l)}{\Gamma(\alpha_l)\Gamma(\beta_l)}\right). \tag{9}$$

In the case of exponential density functions, a conjugate prior on θ is of the following form [13]:

$$\pi(\theta) \propto \exp\left(\sum_{s=1}^{S} \rho_s\eta_s(\theta) + \kappa\Phi(\theta)\right), \tag{10}$$

where $(\rho_1, \ldots, \rho_S)$ and κ are the prior's hyperparameters. Therefore, the conjugate prior family to d-dimensional GD distributions transformed to d independent Beta written in their exponential form (Eq. (5)) is given by:

$$\pi(\theta) \propto \exp\left[\sum_{l=1}^{d} \rho_l \alpha_l + \sum_{l=d+1}^{2d} \rho_l \beta_l + \kappa \sum_{l=1}^{d} \log\left(\frac{\Gamma(\alpha_l + \beta_l)}{\Gamma(\alpha_l)\Gamma(\beta_l)}\right)\right]. \quad (11)$$

The d Beta distributions that generate $\boldsymbol{Y}$, are also simplified unidimensional Dirichlet distributions. In $K+1$ dimensions, the Dirichlet density of a vector of proportions, $\boldsymbol{Y} = (Y_1, \ldots, Y_{K+1})$, is given by:

$$p(\boldsymbol{Y}|\boldsymbol{\alpha}) = \frac{\prod_{j=1}^{K+1} \Gamma(\alpha_j)}{\Gamma(\sum_{j=1}^{K+1} \alpha_j)} \prod_{j=1}^{K+1} Y_j^{\alpha_j - 1}, \quad (12)$$

where $\boldsymbol{\alpha} = (\alpha_1, \ldots, \alpha_{K+1})$ is the parameters vector, $\sum_{j=1}^{K+1} Y_j = 1$ and $0 < Y_j < 1$. This distribution can also be depicted by $Y \sim Dir(\boldsymbol{\alpha})$. In the exponential form, the density (12) becomes:

$$p(\boldsymbol{X}|\boldsymbol{\theta}) = \exp\left[\log\left(\Gamma\left(\sum_{l=1}^{K+1} \alpha_l\right) - \sum_{l=1}^{K+1} \log(\Gamma(\alpha_l))\right.\right.$$
$$\left.\left. + \sum_{l=1}^{K+1} \alpha_l \log(X_l) - \sum_{l=1}^{K+1} \log(X_l)\right]. \quad (13)$$

In [8], Eq. (13) is re-parametrized to separate the effects of its mean, $\boldsymbol{\theta} = \left(\frac{\alpha_1}{\sum_{j=1}^{K+1} \alpha_j}, \ldots, \frac{\alpha_{K+1}}{\sum_{j=1}^{K+1} \alpha_j}\right)$, and spread $\tau = \sum_{j=1}^{K+1} \alpha_j$. As a result, we get:

$$p(\boldsymbol{Z}|\boldsymbol{\theta}, \tau) = \exp\left[\tau \boldsymbol{Z}^T \boldsymbol{\theta} + \tau \frac{\sum_{j=1}^{K+1} W_j}{K+1} - \log\left(\frac{\prod_{j=1}^{K+1} \Gamma(\theta_j \tau)}{\Gamma(\sum_{j=1}^{K+1} \theta_j \tau)}\right)\right], \quad (14)$$

where $\boldsymbol{W} = \log(\boldsymbol{X})$ and $\boldsymbol{Z} = \boldsymbol{W} - \frac{\sum_{l=1}^{K+1} W_l}{K+1}$. In case $K=2$, the Beta distribution of each Y_j can be written in the exponential form of a unidimensional Dirichlet as follows:

$$p(Y_j|\boldsymbol{\theta}) = \exp\left[\log\left(\frac{\Gamma(\alpha_1 + \alpha_2)}{\Gamma(\alpha_1)\Gamma(\alpha_2)}\right) + \alpha_2 \log(1 - Y_j)\right.$$
$$\left. + \alpha_1 \log(Y_j) - \log(Y_j) - \log(1 - Y_j)\right]. \quad (15)$$

Following the same re-parametrization, Eq. (15) becomes:

$$p(\boldsymbol{Z}'|\boldsymbol{\theta}', \tau') = \exp\left[\tau'(\boldsymbol{Z}'^T \boldsymbol{\theta}' + W_j') - \log\left(\frac{\prod_{l=1}^{K} \Gamma(\theta_l' \tau')}{\Gamma(\sum_{l=1}^{K} \theta_l' \tau')}\right)\right], \quad (16)$$

where $\boldsymbol{\theta}' = (\theta_1' = \frac{\alpha_1}{\tau'}, \theta_2' = \frac{\alpha_2}{\tau'})$, $\tau' = \alpha_1 + \alpha_2$, $W_j' = \frac{\log(Y_j)+\log(1-Y_j)}{2}$, and $\boldsymbol{Z}_j' = (\log(Y_j) - W_j', \log(1 - Y_j) - W_j')$. Given these parameters, we represent the distribution of $\boldsymbol{Y}$ by $\boldsymbol{Y} \sim DirBeta(\tau'\boldsymbol{\theta}')$. A conjugate prior family to the Dirichlet distributions in their exponential form is:

$$p(\boldsymbol{\theta}'|\sigma, \kappa, \tau') \propto \exp\left[\sigma\left[\tau'\boldsymbol{\kappa}^T\boldsymbol{\theta}' - \log\left(\frac{\Gamma(\tau'\theta_1')\Gamma(\tau'\theta_2')}{\Gamma(\tau'\theta_1' + \tau'\theta_2')}\right)\right]\right]. \tag{17}$$

3   State Space Models

Dynamic linear models (DLM) can be represented in what is called a state space form. This representation consists in identifying the change of an observed variable (aka observation vector) in terms of another unobserved variable (aka state vector). The authors in [10] proposed a steady model for DLM that is only defined for normal distributions and is equivalent to an ARIMA(0,1,1) model. However, [14] generalized this model by redefining it across non-Gaussian distributions. More specifically, it was generalized to cases where the conditional probability of the observations given the states follows an exponential family distribution. The generalized model, also known as the PSM, is defined by:

$$\begin{cases} (x_t|y^t) \sim PR(\omega), \\ p(x_{t+1}|y^t) \propto p(x_t|y^t)^k, \end{cases} \tag{18}$$

where $0 < k < 1$ and $PR(\omega)$ is the conjugate prior for the exponential family distribution of $p(y_t|x_t)$. This model was first developed to be applied to univariate observations. However, [15] generalized the PSM of a time series to handle multivariate processes. Specifically, a symmetric multivariate PSM in which the process evolution is defined based on the density of the parameter vector is proposed. This generalization was also introduced as part of a Bayesian forecasting framework. However, this model undergoes some limitations when the observations follow a Dirichlet distribution [8], which are mostly due to the fact that the PSM estimates both the dispersion and location of the distribution at the same time. This problem can be solved by using the re-parametrized form of the Dirichlet distribution (Eq. (13)) which allows the separation of the dispersion τ and the location θ.

4   Generalized Dirichlet Power Steady Model (GDPSM)

Given a time series of proportions denoted by $X : \{\boldsymbol{X}_t = (X_{t1}, \ldots, X_{td+1})\}$, where $t = 1, \ldots, T$, we first assume that each vector in this time series follows a GD distribution $\boldsymbol{X}_t \sim GD(\alpha_{t1}, \ldots, \alpha_{td}, \beta_{t1}, \ldots, \beta_{td})$. Afterward, we apply the geometric transformation denoted by Eq. (2) on each of these vectors [3,4]. Using this transformation, $\boldsymbol{X}_t$ is transformed to $\boldsymbol{W}_t$ that follows a Beta distribution with parameters $(\alpha_{tl}, \beta_{tl})$ which define the GD distribution of $\boldsymbol{X}_t$, where $1 \leq l \leq d$. Since Beta distributions are special cases of Dirichlet distributions, we finally model the time series by $T \times d$ unidimensional Dirichlet distributions, $X_{tl} = Dir(\alpha_{tl_1}, \alpha_{tl_2})$. Each of the $T \times d$ distributions is then re-parametrized as per Eqs. (15) and (16). Subsequently, we build $T \times d$ state space models, each of which is based on an unobserved state $\boldsymbol{\theta}'$, to model each of the observations $(W_{11}, \ldots, W_{1d}, \ldots, W_{T1}, \ldots, W_{Td})$. These observations are denoted by:

$$(W_{tj} | \boldsymbol{\theta}'_t, \tau'_t) \sim DirBeta(\tau'_t \boldsymbol{\theta}'_t), \tag{19}$$

where $1 \leq j \leq d$ and $\boldsymbol{\theta}'_t$ follows the PSM given by Eq. (18). In other words, the conditional probability of $\boldsymbol{\theta}'_{t+1}$ given the observations $W_{1j}, \ldots, W_{tj}$, is defined as:

$$p(\boldsymbol{\theta}'_{t+1} | \boldsymbol{W}_t) \propto p(\boldsymbol{\theta}'_t | \boldsymbol{W}_t)^\gamma \quad \text{where } 0 < \gamma < 1. \tag{20}$$

Equation (20) reveals an interesting property of the $(\boldsymbol{\theta}'_{t+1} | \boldsymbol{W}_t)$ and $(\boldsymbol{\theta}'_t | \boldsymbol{W}_t)$ distributions; their modes are equal, but the dispersion of the former is greater.

4.1　Time Series Model

The GD time series model is defined by two steps similar to those of a Gaussian Kalman filter: a prediction and an update step. In the prediction step, $p(\boldsymbol{\theta}'_{t+1} | \boldsymbol{W}_t)$ is computed using Eq. (20). Both sides of this equation follow the conjugate prior given by Eq. (17), each with different parameters. Formally, this is given by:

$$p(\boldsymbol{\theta}'_{t+1} | \boldsymbol{W}_t) \sim \exp \left[\sigma_{t+1|t} \left[\tau'_t \boldsymbol{\kappa}^T_{t+1|t} \boldsymbol{\theta}'_{t+1} - \log \left(\frac{\Gamma(\tau'_t \theta'_{t+1,1}) \Gamma(\tau'_t \theta'_{t+1,2})}{\Gamma(\tau'_t \theta'_{t+1,1} + \tau'_t \theta'_{t+1,2})} \right) \right] \right], \tag{21}$$

$$p(\boldsymbol{\theta}'_t | \boldsymbol{W}_t) \sim \exp \left[\sigma_{t|t} \left[\tau'_t \boldsymbol{\kappa}^T_{t|t} \boldsymbol{\theta}_t' - \log \left(\frac{\Gamma(\tau'_t \theta'_{t1}) \Gamma(\tau'_t \theta'_{t2})}{\Gamma(\tau'_t \theta'_{t1} + \tau'_t \theta'_{t2})} \right) \right] \right]. \tag{22}$$

The prediction step consists of Eq. (24), which is a known fact and (25), which is derived by taking the log of Eq. (20), and the two equations above:

$$\sigma_{t+1|t} \left[\tau'_t \boldsymbol{\kappa}^T_{t+1|t} \boldsymbol{\theta}' - \log \left(\frac{\Gamma(\tau'_t \theta'_1) \Gamma(\tau'_t \theta'_2)}{\Gamma(\tau'_t \theta'_1 + \tau'_t \theta'_2)} \right) \right]$$
$$= \gamma \sigma_{t|t} \left[\tau'_t \boldsymbol{\kappa}^T_{t|t} \boldsymbol{\theta}' - \log \left(\frac{\Gamma(\tau'_t \theta'_1) \Gamma(\tau'_t \theta'_2)}{\Gamma(\tau'_t \theta'_1 + \tau'_t \theta'_2)} \right) \right], \tag{23}$$

knowing that:

$$\boldsymbol{\kappa}_{t+1|t} = \boldsymbol{\kappa}_{t|t}, \tag{24}$$

therefore,

$$\sigma_{t+1|t} = \gamma \sigma_{t|t}. \tag{25}$$

γ is a model parameter, such that $0 < \gamma < 1$. As for the update step, we need to compute $p(\boldsymbol{\theta}'_{t+1} | \boldsymbol{W}_{t+1})$ which, according to Bayes' theorem, can be written as:

$$p(\boldsymbol{\theta}'_{t+1} | \boldsymbol{W}_{t+1}) = p(\boldsymbol{W}_{t+1} | \boldsymbol{\theta}'_{t+1}) \times p(\boldsymbol{\theta}'_{t+1}), \tag{26}$$

where $p(\boldsymbol{W}_{t+1}|\boldsymbol{\theta}'_{t+1})$ is the data likelihood that follows, in this case, the GD reformulated as $DirBeta(\tau'_t\boldsymbol{\theta}'_t)$ and given by Eq. (16). $p(\boldsymbol{\theta}'_{t+1})$ is the prior and is given by Eq. (17). Therefore, applying the log to both sides of Eq. (26) yields the following:

$$\sigma_{t+1|t+1} = \sigma_{t+1|t} + 1, \tag{27}$$

$$\kappa_{t+1|t+1} = \left(1 - \frac{1}{\sigma_{t+1|t+1}}\right)\kappa_{t+1|t} + \frac{1}{\sigma_{t+1|t+1}}z_{t+1}. \tag{28}$$

4.2 Model Evaluation

We evaluate our model by the standardized residuals, the mean squared error (MSE) of the predictions, and the correlations between the residuals at lag 0. We compare our results with those of *DPSM*. The standardized residuals are computed as follows [8]:

$$R_t = \frac{\boldsymbol{Z}_t - E[\boldsymbol{Z}_t|D_{t-1}])}{var[\boldsymbol{Z}_t|D_{t-1}]}, \tag{29}$$

where D_{t-1} denotes all the observations available at time $(t-1)$. $E[\boldsymbol{Z}_t|D_{t-1}]$ and $var[\boldsymbol{Z}_t|D_{t-1}]$ are the respective posterior mean and variance of the prediction density, formally given by:

$$p(\boldsymbol{Z}_{t+1}|\boldsymbol{W}_t) = \int p(\boldsymbol{Z}_{t+1}|\boldsymbol{\theta}_{t+1})p(\boldsymbol{\theta}_{t+1}|\boldsymbol{W}_t)d\boldsymbol{\theta}_{t+1}. \tag{30}$$

Since there is no direct solution for this density, we use the approximation proposed in [18] and used in [8]. Given the density $p(\boldsymbol{Z}_{t+1}|\boldsymbol{W}_t)$, its mean is approximated as follows:

$$E[p(\boldsymbol{Z}_{t+1}|\boldsymbol{W}_t)] = E[p(\boldsymbol{Z}_{t+1}|\boldsymbol{\Theta})], \tag{31}$$

where $\boldsymbol{\Theta} = (\sigma, \kappa, \tau')$. According to [6,8], if a variable follows the Dirichlet distribution in Eq. (16) with the conjugate prior given by Eq. (17), then the following equality holds:

$$E[p(\boldsymbol{Z}_{t+1}|\boldsymbol{\Theta})] = E[p(\boldsymbol{Z}_{t+1}|\sigma, \kappa, \tau')] = \kappa. \tag{32}$$

Therefore, the posterior mean of the density in Eq. (30) is equal to κ. The posterior variance also lacks an exact solution and is solved in [18] by approximating each term of the following:

$$var[p(\boldsymbol{Z}_{t+1}|D_{t-1})] = E[p(\boldsymbol{Z}_{t+1}|D_{t-1})^2] - (E[p(\boldsymbol{Z}_{t+1}|D_{t-1})])^2. \tag{33}$$

Furthermore, we compute the correlations at lag 0 between the residuals of each pair of dimensions in the analyzed time series. These correlations are additional indicators of the model quality; the weaker the correlations the better the model. Stronger correlations imply that further modeling is necessary to better fit the time series [8].

5   Experiments

Application: Web Service Selection. Business applications are increasingly being deployed as autonomous web applications that are published and used on the web (Web Services). The abundance of web services that provide similar functionalities creates a competitive market while rendering the selection of services that best meet the consumers requirements a challenging task. A common solution to this problem considers the trustworthiness of services as a selection criterion based on the outcomes of various quality of service (QoS) metrics, including response time, throughput, reliability, availability, security, and cost. Therefore, the quality of a web service changes continuously during its lifetime. Therefore, we assume that a component for service performance monitoring already exists [21]. A web service consumer can then evaluate and store, after each interaction with any web service, the values of multiple QoS metrics. Then, each vector of QoS metrics' values are classified into a priori defined quality classes [11]. Afterwards, we count based on a predefined time interval, the number of interactions with a web service that belong to each of the defined quality classes.

The main objective is then to model the QoS-based behavior of web services and predict their future performance to assist the web service selection process. To evaluate our GD time series model, we run different simulations with synthetic data due to the unavailability of real QoS data sets. We are aware of two real data sets; QWS and WS-Dream. The former includes the averages over time of multiple QoS metrics' measurements of $2,507$ web services monitored over a six-day period. As such, the data set includes one quality for each of the monitored web services. The latter reports the response time, http code, and http message of 100 web services over a large number of invocations from 150 computers distributed in more than 20 countries. However, the time of each invocation is not available which makes it hard to build a realistic time-series model for each of the monitored web services. Therefore, we validate our approach with multiple simulated data that embed time-variant processes.

5.1   Simulation 1: Trigonometric Functions

We evaluate our model with the outcomes of a web service's transactions that are classified into D quality classes such as *Very Good*, *Good*, and *Average*. We make the assumption that the proportions $\mathcal{C} = \{C_1; ...; C_D\}$ of each class during a specific period of time, follow a latent model that we specify using trigonometric functions. At each time step, the number of transactions are counted and assigned to their corresponding quality class. Their overall evolution is modeled by oscillating functions as a web service performance is not constant. We propose to use trigonometric functions as they are easy to handle and to generate with various settings (mean, amplitude, frequency). These functions can be expressed in the form $C_i'(t) = F_i + \gamma_i \, trig_i(f_i t) + \nu_{it}, \quad i = 1, \ldots, D$, where $trig_i(f_i t)$ is either the *cosine* or *sine* function of frequency f_i randomly taken

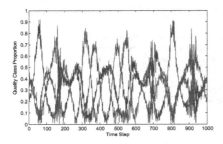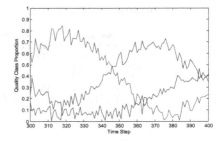

Fig. 1. Sample data (left) with zoom (right)

in the range $[0.0001, 0.009]$. These values keep the functions variations at a reasonable level, see Fig. 1. γ_i is a scaling factor controlling the amplitude of the number of transactions within a given class, A_i is a translation coefficient that controls the average number of transactions of a given class per time step, and ν_t is a white Gaussian noise. t represents the time steps and D the number or quality classes (equal to 3 or 6 here). As ν_{it} is unbounded, the functions C_i' can sporadically go below 0. These rare occurrences are individually handled by assigning a low random value to the sample, within the predefined range $[10, 50]$. In our experiments, we fixed $A_i = 1200$ and $\gamma_i = 1000$ for all i's, and the Signal-to-Noise Ratio has been set to 20. These values can be adapted if a more realistic model is needed without impact on the overall performance of the method presented here. All values are rounded in order to get integers which represent the number of transactions over a given period of time for a given quality class. The proportion vectors are finally obtained by normalizing the C_i' functions, $C_i(t) = \frac{C_i'(t)}{\sum_{d=1}^{D} C_d'(t)}$, where the function C_i represents the proportions of requests that have been processed in *Good, Average, Poor,...* standing by the web service among the total number of requests sent during a given period of time. The algorithm takes these proportions $C_i(t), t = 1, \ldots, 1000$, as input data, of which the first 20 samples are only used for training purpose and the 980 remaining samples are used as testing data. In the first experiment, we compare *GDPSM* and *DPSM* with the data obtained from Simulation 1 for 5 different values of $\gamma = \{0.001\ 0.250\ 0.500\ 0.750\ 0.999\}$, averaged over 10 runs, for the cases of 3 and 6 quality classes.

Three Quality Classes Results. Figure 1 displays the data simulated in one of the 10 runs. The standardized residuals computed by Eq. (29) for *GDPSM* and *DPSM* averaged over each of the $d - 1$ dimensions, are displayed in Table 1 (left). For all values of γ, our model's residuals are slightly smaller than the ones given by the *DPSM*. The correlations at lag 0 between the residuals of the first two dimensions computed by *DPSM* and *GDPSM* are -0.4529 and -0.0092, respectively. This shows that our model explains the time series better than *DPSM*.

Table 1. Standardized residuals (left) and MSE (right) for 3-dimensional data

γ	Dimension 1		Dimension 2	
	DPSM	GDPSM	DPSM	GDPSM
0.001	0.665	0.632	0.647	0.639
0.250	0.666	0.631	0.653	0.645
0.500	0.673	0.637	0.663	0.658
0.750	0.670	0.667	0.696	0.698
0.999	0.855	0.838	0.899	0.900

γ	Dimension 1		Dimension 2	
	DPSM	GDPSM	DPSM	GDPSM
0.001	0.166	**0.067**	0.190	**0.144**
0.250	0.134	**0.054**	0.154	**0.117**
0.500	0.116	**0.048**	0.132	**0.101**
0.750	0.119	**0.050**	0.132	**0.101**
0.999	0.558	**0.276**	0.587	**0.445**

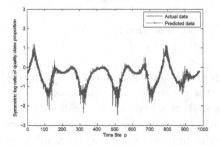 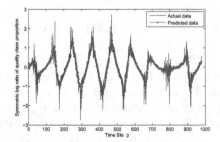

Fig. 2. Actual versus predicted data for the first (left) and second (right) dimensions

Furthermore, Table 1 (right) shows the MSE of both *GDPSM* and *DPSM* which demonstrate that our model yields more accurate predictions than *DPSM* for both dimensions. To visualize the prediction performance of our model, we display $\boldsymbol{Z}$, the symmetric log ratio of the quality class proportions after being transformed using Eq. (2) (actual data) versus the predicted data ($E[p(\boldsymbol{Z}_{t+1}|\boldsymbol{W}_t)]$) in Fig. 2. This figure demonstrates that our model is capable of predicting the time series and providing a smoother distribution than that of the actual ones. The latter is actually due to the fact that we are using a noisy signal. The prediction mostly fits the functional part of the model.

Six Quality Classes Results. We rerun the same experiment with another set of 10 different simulated 6-dimensional data, each of which is represented by the trigonometric function defined earlier. This aims to further validate the efficiency of partitioning the time series model into d simpler problems to solve and thus lead to lower prediction errors. The average of the standardized residuals of *GDPSM* and *DPSM* over the 10 simulated data are displayed in Table 2. For clarity, we only present the results for $\gamma = 0.001$ which raised the best performance for the 3-dimensional residuals (see Table 1). It is noteworthy to mention that other values of γ give equivalent results with the exception of $\gamma = 0.999$ that leads to significantly degraded results. This mostly confirms what has been observed in [8]. The correlations at lag 0 between each pair of dimensions are given in Table 3.

Our model shows better performance due to the overall smaller correlations. Table 4 illustrates the out-performance of *GDPSM* in comparison to *DPSM* in terms of goodness-of-fit. The MSE of *GDPSM*'s predictions for all the dimensions

Table 2. Standardized residuals for *GDPSM* and *DPSM* with 6-dimensional data

	Dimension				
	1	2	3	4	5
DPSM	0.642	0.625	0.669	0.705	0.693
GDPSM	0.555	0.541	0.620	0.675	0.674

Table 3. Residuals correlations at lag 0

DPSM	GDPSM
1	1
-0.242 1	-0.040 1
-0.197 -0.193 1	-0.027 -0.032 1
-0.144 -0.200 -0.160 1	-0.005 -0.014 -0.017 1
-0.195 -0.172 -0.196 -0.182 1	0.001 0.001 0.001 0.0003 1

Table 4. MSE for *GDPSM* and *DPSM* with 6-dimensional data

γ	Dimension 1		Dimension 2		Dimension 3		Dimension 4		Dimension 5	
	DPSM	GDPSM	DPSM	GDPSM	DPSM	GDPSM	DPSM	GDPSM	DPSM	GDPSM
0.001	0.238	0.075	0.226	0.072	0.221	0.071	0.215	0.074	0.230	0.135
0.250	0.194	0.061	0.185	0.059	0.179	0.057	0.174	0.060	0.186	0.110
0.500	0.169	0.054	0.162	0.052	0.155	0.050	0.149	0.051	0.161	0.095
0.750	0.178	0.057	0.170	0.055	0.156	0.050	0.146	0.052	0.165	0.096
0.999	0.730	0.241	0.714	0.246	0.735	0.254	0.719	0.281	0.677	0.427

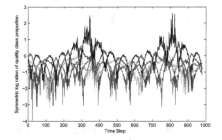 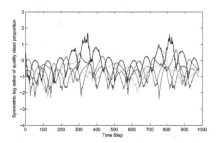

Fig. 3. Actual (left) versus predicted (right) 6-dimensional function-based data

are two to three times smaller than those of *DPSM*. Figure 3 reports the actual and predicted data.

5.2 Simulation 2: Random Data

In this simulation, we test our model with randomly generated 3 and 6 dimensional data. The quality class of a web service interactions do vary according

Table 5. Standardized residuals (left) and MSE (right) for 3-dimensional random data

γ	Dimension 1		Dimension 2		γ	Dimension 1		Dimension 2	
	DPSM	GDPSM	DPSM	GDPSM		DPSM	GDPSM	DPSM	GDPSM
0.001	0.798	0.792	0.801	0.810	0.001	0.126	0.071	0.126	0.097
0.250	0.799	0.792	0.800	0.810	0.250	0.101	0.056	0.102	0.078
0.500	0.800	0.792	0.799	0.810	0.500	0.084	0.047	0.085	0.065
0.750	0.804	0.794	0.802	0.810	0.750	0.072	0.040	0.073	0.055
0.999	0.808	0.798	0.804	0.804	0.999	0.064	0.036	0.064	0.049

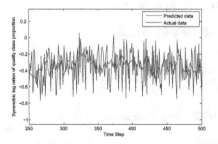 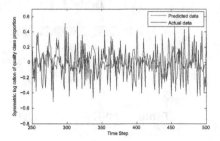

Fig. 4. Actual (left) versus Predicted (right) 3-dimensional random data

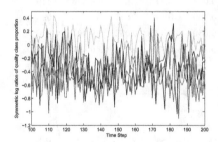 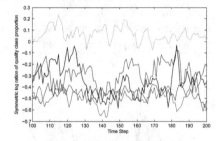

Fig. 5. Actual (left) versus Predicted (right) 6-dimensional random data

to the time they occurred. In Simulation 1, we showed that *GDPSM* is capable of modeling and forecasting time series generated from noisy time-varying functions. However, it is equally essential for the proposed model to perform well with random data to prove its robustness. Similar to Simulation 1, we compute the standardized residuals, the residuals correlations, and the MSE of the predictions.

Three Quality Classes Results. Table 5 displays the *DPSM* and *GDPSM* standardized residuals (left) and MSE (right) of predictions. Figure 4 shows the actual and predicted data. The correlations between the residuals of the two dimensions as computed by *DPSM* and *GDPSM* are −0.4862 and 0.0116, respectively (Fig. 5).

Table 6. Standardized residuals for *GDPSM* and *DPSM* with 6-dimensional random data

	Dimension				
	1	2	3	4	5
DPSM	0.801	0.816	0.803	0.794	0.789
GDPSM	0.802	0.810	0.811	0.795	0.799

Table 7. Residuals correlations at lag 0

DPSM	GDPSM
1	1
-0.067 1	-0.051 1
-0.011 -0.070 1	-0.101 -0.107 1
-0.300 -0.278 -0.299 1	-0.103 -0.103 -0.242
-0.169 -0.201 -0.199 -0.309 1	0.015 0.0174 0.026 0.034 1

Table 8. MSE for *GDPSM* and *DPSM* with 6-dimensional random data

γ	Dimension 1		Dimension 2		Dimension 3		Dimension 4		Dimension 5	
	DPSM	GDPSM	DPSM	GDPSM	DPSM	GDPSM	DPSM	GDPSM	DPSM	GDPSM
0.001	0.137	0.046	0.203	0.075	0.175	0.077	0.494	0.203	0.314	0.158
0.250	0.111	0.037	0.162	0.060	0.140	0.062	0.394	0.162	0.252	0.127
0.500	0.093	0.031	0.135	0.050	0.117	0.051	0.327	0.135	0.210	0.105
0.750	0.079	0.027	0.116	0.043	0.100	0.044	0.280	0.115	0.180	0.090
0.999	0.069	0.023	0.102	0.038	0.088	0.038	0.245	0.101	0.158	0.080

Six Quality Classes Results. We repeat the same experiment above with 6-dimensional simulated random data. It is noteworthy to mention that we select 6 as the higher number of dimensions since it would not realistically make sense to classify a web service quality into more than 6 classes (Tables 6, 7 and 8).

6   Conclusion

This paper presents a power steady model that is based on observations that follow a generalized Dirichlet distribution. This model is optimized by dividing the problem of the model parameters estimation into multiple smaller problems. As such, the resulting model consists of multiple power steady models that depend on Dirichlet distributed observations. We evaluate the proposed approach by applying it to the web service selection problem where the time series consist of the proportions of quality classes to which a web service was assigned over a period of time. These time series are simulated using two different mechanisms; either generated from trigonometric functions or from random distributions. The experimental results show that our model performs better than a single Dirichlet

PSM in terms of standardized residuals and goodness-of-fit of the predictions. Evaluating this model with real compositional time series is left for a future work after collecting the values of various QoS metrics of multiple web services for a three-month period.

References

1. Aitchison, J.: The statistical analysis of compositional data. J. R. Stat. Soc. Ser. B (Methodol.) **44**(2), 139–177 (1982)
2. Bhaumik, A., Dey, D.K., Ravishanker, N.: Joint statistical meetings - Bayesian statistical science - time series analysis of compositional data using a dynamic linear model approach (1999)
3. Bouguila, N., Ziou, D.: High-dimensional unsupervised selection and estimation of a finite generalized dirichlet mixture model based on minimum message length. IEEE Trans. Pattern Anal. Mach. Intell. **29**(10), 1716–1731 (2007)
4. Boutemedjet, S., Bouguila, N., Ziou, D.: A hybrid feature extraction selection approach for high-dimensional non-Gaussian data clustering. IEEE Trans. Pattern Anal. Mach. Intell. **31**(8), 1429–1443 (2009)
5. Brunsdon, T.M., Smith, T.: The time series analysis of compositional data. J. Off. Stat. **14**(3), 237–253 (1998)
6. Diaconis, P., Ylvisaker, D.: Conjugate priors for exponential families. Ann. Stat. **7**(2), 269–281 (1979)
7. Gooijer, J.G.D., Hyndman, R.J.: 25 years of time series forecasting. Int. J. Forecast. **22**(3), 443–473 (2006)
8. Grunwald, G.K., Raftery, A.E., Guttorp, P.: Time series of continuous proportions. J. R. Stat. Soc. Ser. B **55**, 103–116 (1993)
9. Hamilton, J.D.: Time Series Analysis. Princeton University Press, Princenton (1994)
10. Harrison, P.J., Stevens, C.F.: Bayesian forecasting. J. R. Stat. Soc. Ser. B (Methodol.) **38**(3), 205–247 (1976)
11. Mehdi, M., Bouguila, N., Bentahar, J.: Probabilistic approach for QoS-aware recommender system for trustworthy web service selection. Appl. Intell. **41**(2), 503–524 (2014)
12. Quintana, J.M., West, M.: Time series analysis of compositional data. In: Bernando, J.M., DeGroot, M.H., Lindley, D.V., Smith, A.F.M. (eds.) Bayesian Statistics 3, pp. 747–756. Oxford University Press, Oxford (1988)
13. Robert, C.P.: The Bayesian Choice: From Decision-Theoretic Foundations to Computational Implementation, 2nd edn. Springer, New York (2007)
14. Smith, J.Q.: A generalization of the Bayesian steady forecasting model. J. R. Stat. Soc. Ser. B (Methodol.) **41**(3), 375–387 (1979)
15. Smith, J.Q.: The multiparameter steady model. J. R. Stat. Soc.Ser. B (Methodol.) **43**(2), 256–260 (1981)
16. Thiesson, B., Chickering, D.M., Heckerman, D., Meek, C.: Arma time-series modeling with graphical models. In: Proceedings of the 20th Conference on Uncertainty in Artificial Intelligence, pp. 552–560. AUAI Press (2004)
17. Tiao, G.C., Box, G.E.P.: Modeling multiple times series with applications. J. Am. Stat. Assoc. **76**(376), 802–816 (1981)
18. Tierney, L., Kadane, J.B.: Accurate approximations for posterior moments and marginal densities. J. Am. Stat. Assoc. **81**(393), 82–86 (1986)

19. Wang, H., Liu, Q., Mok, H.M.K., Fu, L., Tse, W.M.: A hyperspherical transformation forecasting model for compositional data. Eur. J. Oper. Res. **179**(2), 459–468 (2007)
20. Wong, T.T.: Parameter estimation for generalized Dirichlet distributions from the sample estimates of the first and the second moments of random variables. Comput. Stat. Data Anal. **54**(7), 1756–1765 (2010)
21. Zeng, L., Lei, H., Chang, H.: Monitoring the QoS for web services. In: Krämer, B.J., Lin, K.-J., Narasimhan, P. (eds.) ICSOC 2007. LNCS, vol. 4749, pp. 132–144. Springer, Heidelberg (2007)

Linguistic and Graphical Explanation of a Cluster-Based Data Structure

Grégory Smits[(✉)] and Olivier Pivert

IRISA - University of Rennes 1, UMR 6074, Lannion, France
{gregory.smits,olivier.pivert}@irisa.fr

Abstract. On the one hand, clustering methods are of a particular interest to automatically identify the inner structure of a data set. On the other hand, fuzzy partitions are particularly suitable to define a subjective and domain dependent vocabulary that may then be used to personalize an information system. To make the translation of raw data into knowledge easier, we propose in this paper to generate personalized linguistic and graphical explanations of a cluster-based data structure.

1 Introduction

Initiated by the U.S. government in the late 60's, the data publishing phenomenon has then emerged later in Europe, between 2000 and 2010. Published data now constitute an essential source of knowledge for various professions and its management raises many interesting challenges. Indeed, a raw data set cannot be directly managed and interpreted by a final user, who is generally a domain expert but not often a computer scientist specialized in data and knowledge management. Making data talk and giving end users clear explanations of the data is e.g. the role of communication managers and journalists as well. The fact that making editorial content from raw data is now considered as a journalistic area in its own right, namely data journalism, shows how crucial this data is. This is why efficient methods and intuitive tools have to be developed to help domain experts, data journalists [1–3] and communication managers in our case, make the most of these open data sets.

As a first step in the process of translating raw data into knowledge, we propose in this paper to generate a graphical visualization and linguistic explanations of the data inner structure. Whereas most of the existing approaches to data summarization generate linguistic and graphical explanations directly from raw data, the approach presented in this paper is guided by a cluster-based structure of the data. Clustering methods are indeed of a particular interest to automatically discover and to summarize the structure of a data set. Even if grouping objects by similarity is a natural cognitive process and leads to a data structure that is more understandable than the one obtained with other data models (relational, ontological, hierarchical, etc.), interpreting clusters may be abstruse for unexperienced users. This is why we propose to generate linguistic explanations of such a cluster-based data structure. A way to make these explanations even more valuable and to speed up the appropriation of a new data set

© Springer International Publishing Switzerland 2015
C. Beierle and A. Dekhtyar (Eds.): SUM 2015, LNAI 9310, pp. 186–200, 2015.
DOI: 10.1007/978-3-319-23540-0_13

is to personalize the knowledge discovery process through the use of the expert's vocabulary. Experts usually possess their own vocabulary to describe data and properties, and a vocabulary is obviously composed of subjective and sometimes imprecise linguistic terms.

In our approach, instead of producing a flat set of linguistic explanations, the idea is to exhibit the structural properties of the different clusters using linguistic and graphical explanations. These explanations thus describe the typical properties of each cluster, which by definition means to focus on properties shared exclusively by objects of the same group. Thus, the goal of these explanations is to help users determine the discriminative properties of the most typical elements of each cluster. An example of such an explanation is: "most of the elements from cluster C_1 satisfy P^a" whereas "most of the elements of cluster C_2 satisfy P^b", where P^a and P^b are conjunctions of terms taken from the expert vocabulary. Moreover, each explanation is associated with a truth degree and a score related to the typicality of the subset covered by the explanation.

The rest of the paper is organized as follows. Section 2 recalls the basic notions involved in this approach, namely: clustering and fuzzy-set-based vocabulary. Section 3 introduces the notion of a typicality-based explanation of a clustering and explains how these explanations may be efficiently computed. Section 4 then shows how these explanations are presented to the expert, first linguistically and then graphically. Finally, before concluding, Sect. 5 puts our approach in perspective with respect to existing works in the domain of linguistic summaries and the personalization of knowledge management systems.

2   Preliminaries

2.1   Cluster-Based Data Structure

An open data set is generally a tabular structure, which contains the description of n items $\{x_1, x_2, ..., x_n\}$. Each item, say x, is described by m attributes $A_1, A_2, ..., A_m$, respectively defined on domain $\mathcal{D}_j$, $j = 1..m$, where $x.A$ denotes the value taken by the item x on attribute A.

The data underlying structure, defined by subgroups of similar data, is automatically obtained using a clustering algorithm [4]. In the considered task, three requirements must be taken into account when selecting an appropriate algorithm: it must be scalable in order to be able to process large data sets and if possible incremental so as to manage data evolution in an efficient way. It must also be able to automatically determine the appropriate number of clusters: the aim being to identify the underlying data structure, it is not justified to assume that the data expert knows how many clusters should be identified. Third, each cluster must be associated with a representative, called its centre, corresponding to the most typical object of the cluster. With its linear/incremental way of managing the data and the automatic detection of the number of clusters, the *l-fcmed-select* algorithm introduced in [5] fulfills our requirements. However, to obtain a data structure easier to explain and understand, one will use a crisp (i.e. non fuzzy) version of the algorithm.

Thus, the data set is structured into k clusters: $\{C_1, C_2, ..., C_k\}$, each of them possessing a centre denoted by $M_l, l = 1..k$. During this structuration process, each item is assigned to one cluster, $items(C)$ denotes the set of items assigned to cluster C and $cluster(x)$ denotes the cluster to which item x has been assigned.

An item is assigned to a cluster because of its similarity wrt. the other items of the group, whose quantification requires a ressemblance measure r, and its dissimilarity wrt. items of the other groups, that relies on a dissimilarity measure, here denoted by d. The distance d between two items is computed as follows: $d(x_1, x_2) = 1/m \sum_{i=1}^{m} d_i(x_1.A_i, x_2.A_i)$, where d_i is $d_i(a, b) = \frac{|a-b|}{max(a,b)}$ if A_i is of a numerical type, whereas for categorical values $d_i(a, b) = 1$ if $a \neq b$, $d_i(a, b) = 0$. The distance used for numerical attributes has been chosen to as to take into account the stretch of the attribute's definition domain. The resemblance measure r is simply based on the distance $r(x_1, x_2) = 1 - d(x_1, x_2)$. The combination of these two measures of intra-similarity, denoted by $intrasim(x)$, and inter-dissimilarity, $interdis(x)$, may be interpreted as a degree of typicality [6] that we denote by $typ(x)$. Internal similarity, external similarity and typicality of an object x are defined as follow:

$$intrasim(x) = \frac{1}{|cluster(x)|} \sum_{y \in cluster(x)} r(x, y), \tag{1}$$

$$interdis(x) = \frac{1}{|D| - |cluster(x)|} \sum_{y \notin cluster(x)} d(x, y), \tag{2}$$

$$typ(x) = \top(intrasim(x), interdis(x)), \tag{3}$$

where $\top$ is an aggregation operator that may be pessimistic ($\top = min$), optimistic ($\top = max$) or be an operator of compromise (mean, weighted-mean, OWA, etc.). The weighted-mean $typ(x) = \frac{\alpha \times intrasim(x) + \beta \times interdis(x))}{\alpha + \beta}$ is used in this work with weights defined so as to favor intra-similarity ($\alpha = .6, \beta = .4$).

From a cluster C and a typicality measure typ, a fuzzy set denoted by FC is defined to represent the typicality of its members:

$$FC = \{typ(x_1)/x_1, ..., typ(x_l)/x_l\}. \tag{4}$$

This approach leads to a fuzzy interpretation of the clusters, but differs from fuzzy clustering techniques: instead of having items belonging gradually to different clusters, an item belongs to one cluster and possesses a certain degree of typicality. We consider this data structure easier to understand and explain.

2.2 Vocabulary

Generally, a lexicon explaining, more or less roughly, the meaning of each attribute is published with the data set. Based on this lexicon, a subjective and domain dependent vocabulary $\mathcal{V}$ is defined through linguistic variables, associating each attribute of interest with a set of linguistic labels and a strong fuzzy partition [7]: formally, for attribute A_j, $j = 1..m$, a_j denotes the number of associated modalities and $V_j = \{v_{j1}, \ldots v_{ja_j}\}$ their associated fuzzy sets. The strong partition property imposes that $\forall j = 1..m$, $\forall x \in \mathcal{D}_j$, $\sum_{k=1}^{a_j} \mu_{v_{jk}}(x.A_j) = 1$. It is also imposed that an item cannot satisfy more than 2 modalities on each dimension.

Figure 1 and Table 1 illustrate partitions defined respectively on a numerical and a categorical domain.

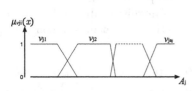

Fig. 1. Numerical partition

	$\mathcal{D}_i$			
V_i	$e_1 \in \mathcal{D}_i$	$e_2 \in \mathcal{D}_i$	...	$e_h \in \mathcal{D}_i$
v_{i1}	$\mu_{v_{i1}}(e_1)$	$\mu_{v_{i1}}(e_2)$	...	$\mu_{v_{i1}}(e_h)$
v_{i2}	$\mu_{v_{i2}}(e_1)$	$\mu_{v_{i2}}(e_2)$	...	$\mu_{v_{i2}}(e_h)$
...	...	...	...	...
v_{ia_i}	$\mu_{v_{ia_i}}(e_1)$	$\mu_{v_{ia_i}}(e_2)$	...	$\mu_{v_{ia_i}}(e_h)$

Table 1. Categorical partition

An item $x \in \mathcal{D}$ can then be rewritten as a vector of $\sum_{k=1}^{m} a_k$ membership degrees $\langle \mu_{v_{11}}(x.A_1), \ldots, \mu_{v_{1a_1}}(x.A_1), \ldots, \mu_{v_{m1}}(x.A_m), \ldots, \mu_{v_{ma_m}}(x.A_m) \rangle$. Due to the partition properties, the above vector has at most $2m$ nonzero components.

Figure 2 graphically presents a toy example of a clustering on a very simple two-dimensional data set, and also shows how these domains are rewritten by two fuzzy partitions. Table 2 details the items assignment in the two clusters as well as their rewriting in terms of the vocabulary.

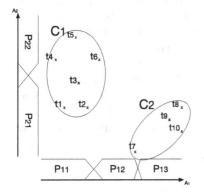

Fig. 2. Clusters and a vocabulary

t	(x, y)	$\mu_{P_{11}}$	$\mu_{P_{12}}$	$\mu_{P_{13}}$	$\mu_{P_{21}}$	$\mu_{P_{22}}$	$typ(t)$
t_3	$(3, 6)$	1	0	0	.9	.1	.65
t_6	$(3, 10)$	.3	.7	0	0	1	.634
t_2	$(4, 4)$	.9	.1	0	1	0	.628
t_4	$(1, 8)$	1	0	0	0	1	.627
t_1	$(2, 4)$	1	0	0	1	0	.614
t_5	$(5, 8)$	1	0	0	0	1	.585
t_{10}	$(12, 1)$	0	0	1	1	0	.679
t_9	$(12, 2)$	0	0	1	1	0	.677
t_8	$(11, 3)$	0	0	1	1	0	.674
t_7	$(8, 0)$	0	.9	.1	1	0	.514

Table 2. Tuples ordered by typicality

3   Explaining Clusters Using the Expert Vocabulary

3.1   Item/Set Explanations and Typicality-Cut of a Cluster

The data set is structured into groups according to the intra-similarity of their members and the inter-dissimilarity with members of other groups, as a result of the clustering algorithm. Explanations of this structure are in our approach also guided by these two measures and more precisely by their aggregation as a measure of typicality (cf Eq. (3)).

Thus, explanations have to focus on the properties that make a group of elements typical. An explanation is composed of linguistic labels taken from the expert vocabulary and expressed in one conjunctive way, at most a conjunct per attribute.

Definition 1. *Let x be an item whose satisfaction vector wrt. a vocabulary $\mathcal{V}$ is $\langle \mu_{v_{11}}(x.A_1), \ldots, \mu_{v_{1a_1}}(x.A_1), \ldots, \mu_{v_{m1}}(x.A_m), \ldots, \mu_{v_{ma_m}}(x.A_m) \rangle$. An* **expla-nation** *$E$ of an item x is a conjunction of $v_{ij}, i = 1..m, j = 1..a_i$ such that $\forall v_{ij}, \mu_{v_{ij}}(x.A_j) > 0$, and $\forall v_{ij}, v_{kl} \in E, i \neq k$. $\mu_E(x)$ is classically interpreted as the degree of satisfaction/validity of the explanation E for the item x (the min t-norm is used in our case to aggregate the $\mu_{v_{ij}}(x)$'s st. $v_{ij} \in E$).*

Definition 2. *A conjunction of labels E taken from $\mathcal{V}$ is an* **explanation of a set** *S iff. E is an explanation of all $x \in S$ and the sigma-count [8] is used to quantify the validity of E regarding S:*

$$\mu_E(S) = \Sigma_E^{count}(s) = \frac{\sum_{x \in S} \mu_E(x)}{|S|}. \tag{5}$$

A cluster of size n contains $2^n - 1$ nonempty subsets and for each of them an exponential number of explanations, wrt. the number of dimensions on which each item is described, may be envisaged. It would obviously be inefficient and not informative to consider all these possible combinations of subsets and expla-nations. The overall objective being to explain the typicality-based inner struc-ture of the data set, explanations our approach provides are also guided by the typicality of the clusters' items. As illustrated by Fig. 3, the idea is to find expla-nations for nested α-cuts of the clusters, and more precisely their fuzzy versions representing the typicality of their members (Sect. 2.1). Let FC be the fuzzy interpretation of a cluster C. In the same way as the α-cut applied to a classical fuzzy set E, the α-cut of FC, denoted by FC^α, returns a crisp set containing the items belonging to C whose typicality is at least equal to α:

$$FC^\alpha = \{x \in C, typ(x) \geq \alpha\}.$$

The goal of the explanation process is to associate explanations to α-cuts of the different clusters, i.e. to α-cuts. For an explanation E, one looks for the

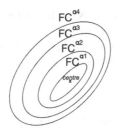

Fig. 3. Typicality-cuts of a cluster FC, where $\alpha 1 > \alpha 2 > \alpha 3 > \alpha 4 > 0$

largest subset (α-cut) S of a cluster that E explains, this largest subset S being called the maximal set explained by E.

Definition 3. *Let FC be the fuzzy set representing the typicality of the elements assigned to cluster C, and E an explanation of FC^{α}, i.e. the α-cut of FC. FC^{α} is said to be* **the maximal set explained by E** *iff. $\nexists \alpha' < \alpha$ such that E is also an explanation of $FC^{\alpha'}$.*

3.2   Computation of the Explanations and Their Maximal Sets

As a first step of the clustering explanation process, one generates all the possible pairs ⟨*explanation, maximal explained α-cut*⟩ for each cluster. By definition of an explanation (Definitions 1 and 2) and by the fact that one wants to associate each explanation with its maximal explained set (Definition 3), the candidate explanations to explore are all the possible explanations of the cluster's centre.

As detailed by Algorithm 1, in order to identify the explanations and the associated maximal α-cuts of a cluster C_i (or more precisely its fuzzy interpretation FC_i), one considers as candidate explanations all the possible conjunctions of modalities satisfied by M_i (line 1), M_i being the centre of C_i, taking care that each conjunction involves at most one modality for each dimension. Then, to determine the maximal subset of C_i that is explained by each candidate explanation, α-cuts from C_i are treated in a decreasing order of their typicality degree (loop line 5). If an explanation E is valid for a given α-cut (FC^{α}) but is no more verified when considering the next α'-cut, $\alpha' < \alpha$ (line 9), then the pair ⟨E, α-cut⟩ is stored as a validated explanation and its maximal associated set is FC^{α} (line 11). Moreover, E and other conjunctions containing E are removed from the set of candidate explanations (line 12). This process is iterated until there is no more candidate explanation or the whole cluster is explained.

This algorithm has only a linear data complexity wrt. the cardinality of the cluster to explain and an exponential data complexity with respect to the number of dimensions on which each item is described. This is not a real limitation in practice as this number is generally rather low.

Data: α-cuts of a cluster C_i in a decreasing order of α:
$$FC_i^{\alpha^1} \subseteq FC_i^{\alpha^2} \subseteq ... \subseteq FC_i^{\alpha^n} = C_i\}$$
Result: set of pairs $\langle explanation, explained\ subset\rangle$
1: $candidateExp \leftarrow allExp(M_i)$
2: $validExp \leftarrow \emptyset$
3: $curS = \{FC_i^{\alpha 1}\}$
4: $j = 2$
5: **while** $candidateExp \neq \emptyset$ **and** $j \leq n$ **do**
6:     $S^{temp} = curS \cup \{FC_i^{\alpha^j}\}$
7:     **for** $E \in candidateExp$ **do**
8:         **if** E not an explanation of S^{temp} **then**
9:             $validExp \leftarrow validExp \cup \{\langle E, curS\rangle\}$
10:            $candidateExp \leftarrow candidateExp/\{E' \subseteq E\}$
11:        **end if**
12:    **end for**
13:    $curS = S^{temp}$
14: **end while**

Algorithm 1: Explanations of a cluster

Table 3. Rolling out of Algorithm 1 on C_1

α-cuts of $C_1 \downarrow typicality$	Candidate explanations	Validated explanations
$S^1 = FC_1^{.65} = \{t_3\}$	$\{P_{11}, P_{21}, P_{22}, P_{11} \wedge P_{21},$ $P_{11} \wedge P_{22}\}$	$\emptyset$
$S^2 = FC_1^{.634} = S^1 \cup \{t_6\}$	$\{P_{11}, P_{22}, P_{11} \wedge P_{22}\}$	$\{(P_{21}, S^1), (P_{11} \wedge P_{21}, S^1)\}$
$S^3 = FC_1^{.628} = S^2 \cup \{t_2\}$	$\{P_{11}\}$	$\{(P_{22}, S^2), (P_{11} \wedge P_{22}, S^2)\}$
$S^3 = FC_1^{.627} = S^3 \cup \{t_6\}$	$\{P_{11}\}$	
$S^3 = FC_1^{.614} = S^3 \cup \{t_1\}$	$\{P_{11}\}$	
$S^3 = FC_1^{.585} = S^3 \cup \{t_5\}$	$\{P_{11}\}$	$\{(P_{11}, S^3)\}$
α-cuts of $C_2 \downarrow typicality$	Candidate explanations	Validated explanations
$S^{1'} = FC_2^{.679} = \{t_{10}\}$	$\{P_{13}, P_{21}, P_{13} \wedge P_{21}\}$	$\emptyset$
$S^{1'} = FC_2^{.677} = S^{1'} \cup \{t_9\}$	$\{P_{13}, P_{21}, P_{13} \wedge P_{21}\}$	$\emptyset$
$S^{1'} = FC_2^{.674} = S^{1'} \cup \{t_8\}$	$\{P_{13}, P_{21}, P_{13} \wedge P_{21}\}$	$\emptyset$
$S^{1'} = FC_2^{.514} = S^{1'} \cup \{t_7\}$	$\{P_{13}, P_{21}, P_{13} \wedge P_{21}\}$	$\{(P_{13}, S^{1'}), (P_{21}, S^{1'}), (P_{13} \wedge P_{21}, S^{1'})\}$

Table 3 shows the result produced by Algorithm 1 when applied to the clusters from Table 2.

3.3   Representativity of an Explanation

From Algorithm 1, one obtains a set of explanations for different α-cuts of each cluster (it is worth recalling that α is a typicality threshold in our case). To be informative, an explanation should also make it possible to distinguish between a cluster, or a subset of this cluster, and other clusters. Thus, a degree of representativity may be attached to an explanation in order to quantify how much this explanation discriminates a given set of items from subsets of other clusters.

To quantify the representativity of an explanation E for a given set S, one checks the extent to which E is also valid for subsets of the other clusters having at least the same level of typicality as S. As said in the introduction, we consider that, in our context of a cluster-based data structure, explanations must help users quickly identify the discriminative properties of the most typical elements of each cluster.

Let E be a validated explanation for the α-cut of FC_i. The representativity of E with respect to FC_i^α, denoted by $\tau_E(FC_i^\alpha)$, quantifies the extent to which E explains FC_i^α and does not explain any another $FC_j^{\alpha'}, j \neq i, \alpha' \geq \alpha$, hence:

$$\tau_E(FC_i^\alpha) = min(\mu_E(FC_i^\alpha), 1 - max_{FC_j^{\alpha'}, j \neq i, \alpha' \geq \alpha} \mu_E(FC_j^{\alpha'})). \qquad (6)$$

Example 1. Table 3 gives the explanations found for different subsets of the clusters illustrated in Fig. 2. For each of these validated explanations, one computes its representativity degree. To illustrate the way the representativity of an explanation is quantified, let us take the example of the explanation P_{21} associated with the subset $S^1 = \{t_3\}$ of cluster C_1. The validity of P_{21} for S^1 is $\mu_{P_{21}}(S^2) = 0.9$ and its typicality level is 0.65. To determine the representativity of P_{21} for S^1, one checks the validity of P_{21} for α'-cuts of C_2, $\alpha' \geq 0.65$: as $FC_2^{.65} = \{t_{10}, t_9, t_8\}$ and $\mu_{P_{21}}(FC_2^{.65}) = 1$, then $\tau_{P_{21}}(FC_1^{.65}) = 0$ and $\tau_{P_{21}}(FC_2^{.65}) = 1 - 0.9 = 0.1$. Table 4 gives the representativity degree of the explanations enumerated in Table 3.

Table 4. Representativity of the explanations

Explanations	Representativity
P_{21} for $S^1 = \{t_3\}$	0
$P_{11} \wedge P_{21}$ for $S^1 = \{t_3\}$	0.9
$P_{11} \wedge P_{22}$ for $S^2 = \{t_3, t_6\}$	0.55
P_{22} for $S^2 = \{t_3, t_6\}$	0.55
P_{11} for $S^3 = C_1$	0.87
P_{13} for $S^{1'} = C_2$	0.775
P_{21} for $S^{1'} = C_2$	0.1
$P_{13} \wedge P_{21}$ for $S^{1'} = C_2$	0.775

4   Explanations Visualization

4.1   Explanations Ranking

The process described in Sect. 3 generates for each cluster a set of explanations. An explanation E is related to a subset S of the concerned cluster C, and more precisely an α-cut of its fuzzy version representing the typicality of its members. Four indicators may be attached to each pair (E, S), namely:

- $\frac{|S|}{|C|}$ the proportion of elements from C in S,
- α the typicality threshold of the α-cut S, $\alpha = min_{x \in S} typ(s)$,
- $\mu_E(S)$ the validity of E for the subset S,
- $\tau_E(S)$ the representativity degree of the explanation E relatively to S.

We consider that the most informative explanations are those discriminating the most typical elements of each cluster. So as to rank-order the explanations according to the data structure, the representativity of the explanations ($\tau_E(S)$) and the typicality of the elements (α) it explains are aggregated to obtain a score that is attached to each explanation:

$$score(E, FC^\alpha) = \alpha \times \tau_E(S).$$

For each cluster, explanations are ordered according to their scores. The two other properties, namely the proportion ($\frac{|S|}{|C|}$) and the validity of the explanation ($\mu_E(S)$), are linguistically and graphically presented to the user (Subsects. 4.2 and 4.3).

4.2   Linguistic Explanation

A first way to help users understand the data structure is to translate the explanations associated with each cluster into linguistic statements. One uses the classical protoform $Q\,y's\,are\,E$ to generate these linguistic explanations. In our case, Q is replaced by a linguistic label representing the proportion $\frac{|S|}{|C|}$ and E is the conjunction of properties taken from the expert's vocabulary that explains S.

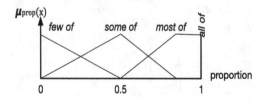

Fig. 4. Linguistic variable to describe a proportion

Thus, in addition to the domain-specific vocabulary (Sect. 2.2), a linguistic variable is used to linguistically express the proportion of items explained by E (Fig. 4). This partition is predefined using common sense default values, but it may be revised by the user. When a proportion falls in the transition area between two modalities of the partition, for the sake of clarity, only one linguistic explanation is generated using the most representative modality. However, the satisfaction degree is used to quantify the truth value of the linguistic explanation [9]:

$$\Psi(Q \text{ elements of } C \text{ are } E) = \mu_Q\left(\frac{1}{|C|}\sum_{x \in S}\mu_E(x)\right). \tag{7}$$

Example 2. Explanations found in Example 1 are ordered according to their score (Subsect. 4.1) and translated into linguistic explanations (Table 5):

Table 5. Linguistic version of the explanations

Explanation	Q	Score	Truth	Linguistic translation
$P_{11} \wedge P_{21}$ for $S^1 = \{t_3\}$	Few	0.65	0.15	Few of the elements from C_1 are P_{11} and P_{21}
P_{11} for $S^3 = C_1$	All	0.585	1	All the elements from C_1 are P_{11}
$P_{11} \wedge P_{22}$ for $S^2 = \{t_3, t_6\}$	Some	0.55	0.07	Some of the elements from C_1 are P_{11} and P_{22}
P_{22} for $S^2 = \{t_3, t_6\}$	Some	0.55	1	Some of the elements from C_1 are P_{22}
P_{13} for $S^{1'} = C_2$	All	0.514	0.8	All of the elements from C_2 are P_{13}
$P_{13} \wedge P_{21}$ for $S^{1'} = C_2$	All	0.514	1	All of the elements from C_2 are P_{13} and P_{21}
P_{21} for $S^{1'} = C_2$	All	0.1	0.8	All of the elements from C_2 are P_{21}

4.3 Graphical Visualization

As a picture speaks a million words, a graphical representation of these explanations is also generated using the D3.js javascript library (http://d3.js). Each explanation is represented by a bubble whose radius tells us about the proportion of items from the cluster that is covered by the explanation, and the darkness of the bubble is proportional to the score of the explanation. This way, it is easy to identify at first sight big dark bubbles that correspond to the most informative and global explanations. To better analyze the discriminative properties of each cluster, two views of these explanations are proposed. The first one gathers all the found explanations (Fig. 5), whereas the second one splits them according to the cluster they describe (Fig. 6).

Figures 5 and 6 graphically show the explanations found for the toy clustering illustrated in Fig. 2 [1]. This visualization, and especially the split view by cluster, tells us a lot about the cluster-based structure, using terms of the vocabulary. The size of the bubble for cluster C_2 indicates the homogeneity of its items regarding their rewriting in terms of the vocabulary. It indeed shows that all the items from C_2 satisfy the properties P^{13} and P^{21}, and according to the shade of gray of the bubbles one may determine that individually P^{13} (bottom left bubble) is a more discriminative explanation than P^{21} (top left bubble) for C_2.

The proposed explanation process of a cluster-based data structure has then been applied to the well know iris dataset available on the UCI machine learning

---

[1] The animated graphical explanations of the toy dataset may be found at the following url http://gsmits.iutlan.univ-rennes1.fr/toyExample.html.

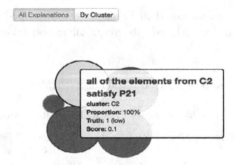

Fig. 5. Gathering of all the explanations for the toy dataset

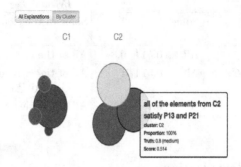

Fig. 6. Explanations by cluster for the toy dataset

repository (https://archive.ics.uci.edu/ml/datasets/Iris)[2]. Figure 7 illustrates the explanations found for the three clusters found by the clustering algorithm applied to the iris dataset. The small but dark bubbles associated with cluster C_3 show that the most typical iris of this type (Iris-virginica) possesses discriminative properties, e.g. "petal length is long and petal width is large", but also that these explanations covers a small proportion of the cluster. On the contrary, cluster C_1 (for iris-setosa) is entirely covered by two explanations but that are not so discriminative (the two big light bubbles for the explanations "sepal length is short and petal length is short" and "sepal length is short and petal width is narrow"). Cluster C_2 containing iris of type versicolor is mainly explained by the properties 'petal length is medium and petal width is medium', properties that are shared by most of the items from C_2.

Interesting conclusions may be quickly drawn from such a graphical visualization of the explanations found for a cluster-based data structure. E.g., from a visual explanation composed of small bubbles only, one may conclude that the properties of the cluster centres are not shared by the remainder of their respective cluster, thus that the clusters are of a low compactness. A visual explanation

---

[2] The animated graphical explanations of the iris dataset may be found at the following url http://gsmits.iutlan.univ-rennes1.fr/iris.html.

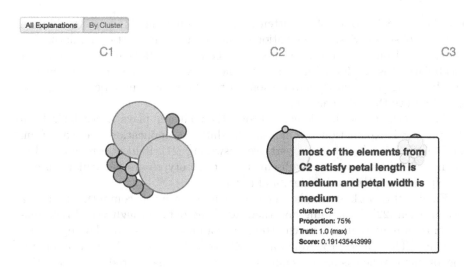

Fig. 7. Explanations by cluster for the iris dataset

composed of light bubbles only may result from an inadequate vocabulary wrt. the data structure. This inadequacy may e.g. come from too generic terms of the vocabulary that cover a too large part of the definition domain. Perspectives for future works concern the interpretation of such visualization to assess the cluster validity and vocabulary adequacy wrt. to the data structure as it is done numerically in [10,11].

Finally, the application of the proposed approach to these two small datasets confirms the low complexity of our explanation process. The time needed to rewrite the items in the terms of the expert vocabulary and to generate the explanations, the linguistic as well as the graphical explanations, is really low (around 0.2 s for the iris data set).

5   Related Works

Helping users or experts understand a data set by summarizing it is not a new research area as many interesting investigations have been conducted on this problem [12]. Existing approaches mainly differ on the nature of the considered summaries. Whereas linguistic protoforms constitute the most commonly used explanation structure [13–15], Raschia and Mouaddib introduced in [16] a hierarchical summarizing process and in [17] Yager shows how the OWA operator may be used as a generic tool to generate many interesting statistics about a data set.

This task of data summarization has then been adapted to particular types of data as time series [18] or logs, web logs for instance [19]. However, to the best of our knowledge, only a few of the existing summarizing approaches take the data structure into account [16]. Whereas clusters are used in our approach as a data model, groups of rewritten items are formed in [16] so as to obtain a hierarchy of summaries. Aforementioned approaches to linguistic summarization are all based on a confrontation between the data or its structure and a user vocabulary. It is considered in this work that the vocabulary is in adequacy wrt. the clustering [10] even if modifications of the vocabulary may be envisaged so as to improve this adequacy [11].

The typicality of the elements inside their cluster plays a crucial role in the proposed explanation mechanism. In different application contexts, from machine learning [6,20] to cooperative answering [21], it has been shown that the notions of typicality and subsequently of prototypes are particularly useful to help users explore and understand a data set.

Finally, this work also makes a connection between soft computing and data visualization [22]. Explaining and then exploring data through visual representation is an issue that gets much attention from researchers [23], developers and designers. The gallery available on the d3.js website is an illustration of this phenomenon. Whereas data visualization generally aims at producing attractive and synthesized views of raw data, we produce a graphical representation of knowledge extracted from the data. Knowledge extraction from Open Data sets and its visualization is a crucial issue for helping data journalists and communication managers make the most of published data [24,25].

6   Conclusion

The Open Data initiative has changed the work methods of many journalists and communication managers that have to rapidly transform raw data into editorial contents or presentations. To help them in this task, we propose a novel summarization strategy guided by the cluster-based structure of the data sets. The underlying hypotheses of this work are that, first it is easier to understand the main trends of a data set when groups of somewhat similar items are formed and, then when synthetic explanations are produced. Finally, as a first step toward the understanding of a new dataset, it appears natural to produce discriminative explanations of the typical items of each group. It is in this sense that the proposed approach generates first synthetic linguistic explanations of the data structure and then an additional graphical visualization of these explanations. An interesting property of this explanation process is its low complexity and the fact that only a small number of explanations is generated. However, these explanations only concern the *typical* items of each cluster. This is why we are currently working on an incremental explanation process and its visualization that will make it possible to identify both typical and atypical properties.

References

1. Garrison, B.: Computer-Assisted Rreporting. Psychology Press, Hove (1998)
2. Parasie, S., Dagiral, E.: Data-driven journalism and the public good: "computer-assisted-reporters" and "programmer-journalists" in chicago. New Media & Society (2012) 1461444812463345

3. Léchenet, A.: Global database investigations: The role of the computer-assisted reporter. Reuters institure for the study of journalism (2014)
4. Jain, A., Murty, M., Flynn, P.: Data clustering: a review. ACM Comput. Surv. **31**(3), 264–323 (1999)
5. Damez, M., Lesot, M.-J., Revault d'Allonnes, A.: Dynamic credit-card fraud profiling. In: Torra, V., Narukawa, Y., López, B., Villaret, M. (eds.) MDAI 2012. LNCS, vol. 7647, pp. 234–245. Springer, Heidelberg (2012)
6. Lesot, M.-J., Rifqi, M., Bouchon-Meunier, B.: Fuzzy prototypes: from a cognitive view to a machine learning principle. In: Bustince, H., Herrera, F., Montero, J. (eds.) Fuzzy Sets and Their Extensions: Representation, Aggregation and Models, vol. 220, pp. 431–452. Springer, Heidelberg (2008)
7. Guillaume, S., Charnomordic, B.: Generating an interpretable family of fuzzy partitions from data. IEEE Trans. Fuzzy Syst. **12**(3), 324–335 (2004)
8. Zadeh, L.A.: Fuzzy logic. Computer **21**(4), 83–93 (1988)
9. Zadeh, L.A.: Toward a theory of fuzzy information granulation and its centrality in human reasoning and fuzzy logic. Fuzzy Sets Syst. **90**(2), 111–127 (1997)
10. Lesot, M.J., Smits, G., Pivert, O., et al.: Adequacy of a user-defined vocabulary to the data structure. In: FUZZ-IEEE, pp. 1–8. (2013)
11. Smits, G., Pivert, O., Lesot, M.-J.: A vocabulary revision method based on modality splitting. In: Laurent, A., Strauss, O., Bouchon-Meunier, B., Yager, R.R. (eds.) Information Processing and Management of Uncertainty in Knowledge-Based Systems. Communications in Computer and Information Science, vol. 444, pp. 140–149. Springer, Heidelberg (2014)
12. Bouchon-Meunier, B., Moyse, G.: Fuzzy linguistic summaries: where are we, where can we go? In: 2012 IEEE Conference on Computational Intelligence for Financial Engineering & Economics (CIFEr), pp. 1–8. IEEE (2012)
13. Yager, R.R.: A new approach to the summarization of data. Inf. Sci. **28**(1), 69–86 (1982)
14. Dubois, D., Prade, H., Rannou, E.: User-driven summarization of data based on gradual rules. In: Proceedings of the Sixth IEEE International Conference on Fuzzy Systems, vol. 2, pp. 839–844. IEEE (1997)
15. Kacprzyk, J., Zadrożny, S.: Linguistic database summaries and their protoforms: towards natural language based knowledge discovery tools. Inf. Sci. **173**(4), 281–304 (2005)
16. Raschia, G., Mouaddib, N.: Saintetiq: a fuzzy set-based approach to database summarization. Fuzzy Sets Syst. **129**(2), 137–162 (2002)
17. Yager, R.R.: A human directed approach for data summarization. In: 2006 IEEE International Conference on Fuzzy Systems, pp. 707–712. IEEE (2006)
18. Castillo-Ortega, R., Marín, N., Sánchez, D.: Time series comparison using linguistic fuzzy techniques. In: Hüllermeier, E., Kruse, R., Hoffmann, F. (eds.) IPMU 2010. LNCS, vol. 6178, pp. 330–339. Springer, Heidelberg (2010)
19. Zadrozny, S., Kacprzyk, J.: Summarizing the contents of web server logs: a fuzzy linguistic approach. In: IEEE International Fuzzy Systems Conference, FUZZ-IEEE 2007, pp. 1–6. IEEE (2007)
20. Rifqi, M., Monties, S.: Fuzzy prototypes for fuzzy data mining. In: Pons, O., Vila, M.A., Kacprzyk, J. (eds.) Knowledge Management in Fuzzy Databases, pp. 275–286. Springer, Heidelberg (2000)
21. Pivert, O., Smits, G., Jaudoin, H.: Finding similar objects in relational databases — an association-based fuzzy approach. In: Larsen, H.L., Martin-Bautista, M.J., Vila, M.A., Andreasen, T., Christiansen, H. (eds.) FQAS 2013. LNCS, vol. 8132, pp. 425–436. Springer, Heidelberg (2013)

22. Steele, J., Iliinsky, N.: Designing Data Visualizations: Representing Informational Relationships. O'Reilly Media Inc., Sebastopol (2011)
23. Fayyad, U.M., Wierse, A., Grinstein, G.G.: Information Visualization in Data Mining and Knowledge Discovery. Morgan Kaufmann, San Francisco (2002)
24. Léchenet, A.: Global database investigations: The role of the computer-assisted reporter (2014)
25. Gray, J., Chambers, L., Bounegru, L.: The Data Journalism Handbook. O'Reilly Media Inc., Sebastopol (2012)

Possibility Theory, Belief Functions and Transformations

Probability-Possibility Transformations: Application to Credal Networks

Salem Benferhat[1,2], Amélie Levray[1,2], and Karim Tabia[1,2(✉)]

[1] Université Lille Nord de France, 59000 Lille, France
[2] UArtois, CRIL UMR CNRS 8188, 62300 Lens, France
{benferhat,levray,tabia}@cril.fr
http://www.cril.fr

Abstract. This paper deals with belief graphical models and probability-possibility transformations. It first analyzes some properties of transforming a credal network into a possibilistic one. In particular, we are interested in satisfying some properties of probability-possibility transformations like dominance and order preservation. The second part of the paper deals with using probability-possibility transformations in order to perform MAP inference in credal networks. This problem is known for its high computational complexity in comparison with MAP inference in Bayesian and possibilistic networks. The paper provides preliminary experimental results comparing our approach with both exact and approximate inference in credal networks.

Keywords: Credal networks · Possibilistic networks · Probability-possibility transformations

1 Introduction

Belief graphical models such as Bayesian [4], credal [3] and possibilistic networks [2] are powerful tools for representing and reasoning with uncertain information. Bayesian networks allow to compactly encode a probability distribution thanks to the conditional independence relationships existing between the variables. Credal networks, based on the theory of credal sets, generalize Bayesian networks in order to allow some flexibility regarding the model parameters. Indeed, credal networks are often seen as probabilistic graphical models with relaxed parameters. They are for instance used in robustness analysis and for encoding incomplete and ill-known knowledge and reasoning with the knowledge of groups of experts. Possibilistic networks are the counterparts of Bayesian networks based on possibility theory [7,17].

Many uncertainty frameworks exist, some of which are generalizations of some other ones. For instance, imprecise probability theory [10,15] is a generalization of probability theory while possibility theory [7,17] is an alternative non additive uncertainty theory particularly suited for handling incomplete, qualitative and partial information. In order to cast the information encoded within

© Springer International Publishing Switzerland 2015
C. Beierle and A. Dekhtyar (Eds.): SUM 2015, LNAI 9310, pp. 203–219, 2015.
DOI: 10.1007/978-3-319-23540-0_14

one setting into another uncertainty framework, transformations are used. They are mechanical transformations satisfying some desirable properties like consistency and order preservation. Many works are done for instance for transforming probability measures into possibilistic ones [9,17]. However, in the context of belief graphical models and knowledge bases, only few works addressed some related issues [1,13]. Transformations can be useful in various contexts such as (i) using the existing tools (e.g. algorithms and software) developed in one setting rather than developing everything from scratch for the other setting or (ii) exploiting information provided in different uncertainty languages as it is often the case in some multiple expert applications. In this paper, we are mainly interested in probability-possibility transformations for computational complexity purposes. More precisely, in this preliminary work, our objective is to exploit probability-possibility transformations to efficiently perform MAP inference in credal networks where this task is NP^{PP}-hard in the general case [12]. The main contributions of the paper are:

- Proposing and analyzing a transformation allowing to turn a credal network into a possibilistic network.
- Proposing a kind of approximate approach for MAP inference in credal networks by means of probability-possibility transformations.
- Providing preliminary experimental studies showing that MAP inference could efficiently be carried out using our approach with a high accuracy rate.

2   A Brief Refresher on Credal and Possibilistic Networks

This section briefly presents the belief graphical models dealt with in this paper.

2.1   Bayesian Networks

Bayesian networks ($\mathcal{BN}$) are well-known probabilistic graphical models [4]. They allow to compactly represent a probability distribution over a set of variables of interest. A $\mathcal{BN}$ is specified by:

- A *graphical component* with nodes and edges forming a directed acyclic graph (DAG). Each node represents a variable A_i of the modeled problem and the edges encode independence relationships among variables.
- A *quantitative component*, where each variable A_i is associated with a local probability distribution $p(A_i|par(A_i))$ for A_i in the context of its parents $par(A_i)$.

The joint probability distribution encoded by a Bayesian network is computed using the following chain rule:

$$P(A_1, .., A_n) = \prod_{i=1}^{n} P(A_i|par(A_i)) \tag{1}$$

2.2   Credal Networks

Credal networks are probabilistic graphical models based on imprecise probabilities. Imprecise probability theory [10,15] generalizes probability theory to
encode imprecise and ill-known information. A key notion in this theory is the
one of credal set.

Definition 1 (Credal set). *A credal set is a convex set of probability distributions.*

Intuitively, if K is a convex set of probability measures, then *mixing* any two
distributions p_1 and p_2 from K will result in a distribution p belonging to K.
Mixing here means linearly combining a set of distributions p_1 ... p_k as follows:
$p = \sum_{i=1}^{k}(a_i * p_i)$ where $\sum_{i=1}^{k} a_i = 1$. A credal set is often interpreted as a
set of imprecise beliefs in the sense that the true uncertainty model (probability
measure) is in this set but there is no way to determine it exactly due to lack
of knowledge. In order to characterize a credal set, one can use a (finite[1]) set of
extreme points (edges of the polytope representing the credal set), probability
intervals or linear constraints.

Interval-based probability distributions (IPD for short) are a very natural and
common way to specify imprecise and ill-known information. In an IPD IP, every
interpretation $\omega_i \in \Omega$ is associated with a probability interval $IP(\omega_i) = [l_i, u_i]$
where l_i (resp. u_i) denotes the lower (resp. upper) bound of the probability of ω_i.

Definition 2 (Interval-based probability distribution). *Let Ω be the set
of possible worlds. An interval-based probability distribution IP is a function that
maps every interpretation $\omega_i \in \Omega$ to a closed interval $[l_i, u_i] \subseteq [0, 1]$.*

An IPD should satisfy the following constraints in order to ensure that the
underlying credal set is not empty and every lower/upper probability bound is
reachable.

$$\sum_{\omega_i \in \Omega} l_i \leq 1 \leq \sum_{\omega_i \in \Omega} u_i$$

$$\forall \omega_i \in \Omega, l_i + \sum_{\omega_j \neq i \in \Omega} u_j \geq 1 \ and \ u_i + \sum_{\omega_j \neq i \in \Omega} l_j \leq 1$$

In order to give a formal semantics for IPDs, let us first define the concept of
compatible probability distribution.

Definition 3 (Compatible probability distribution). *A probability distri-
bution p over Ω is said compatible with IP iff $\forall \omega_i \in \Omega$, $p(\omega_i) \in IP(\omega_i)$.*

---

[1] It is important to note that the number of extreme points can reach $N!$ where N is
the number of interpretations [16].

Note that while a standard probability distribution p induces a complete order over the set of possible worlds Ω, an IPD IP may induce a partial order since some interpretations may be incomparable in case of overlapping intervals. In this paper, a credal set K_i associated with a variable A_i having an interval-based probability distribution IP denotes the closed convex set of (standard) probability distributions p that are compatible with IP. Let us now introduce probabilistic graphical models based on credal sets, called credal networks [3,12].

Definition 4 (Credal network). *A credal network* $\mathcal{CN} = <G, K>$ *is a probabilistic graphical model where*

- *$G = <V, E>$ is a directed acyclic graph (DAG) encoding conditional independence relationships where $V = \{A_1, A_2, .., A_n\}$ is the set of variables of interest (D_i denotes the domain of variable A_i) and E is the set of edges of G.*
- *$K = \{K_1, K_2, .., K_n\}$ is a collection of local credal sets, each K_i is associated with the variable A_i in the context of its parents $par(A_i)$.*

Such credal networks are called separately specified credal networks as the only constraints on probabilities are specified in local tables for each variable in the context of its parents. Note that in practice, in local tables, one can either specify a set of extreme points characterizing the credal set as in JavaBayes[2] software or directly local IPDs.

A credal network $\mathcal{CN}$ can be seen as a set of Bayesian networks $\mathcal{BN}$s, each encoding a joint probability distribution. In this paper, we deal only with discrete variables and semantics associated with a $\mathcal{CN}$ is a set of compatible $\mathcal{BN}$s, defined as follows:

Definition 5 (Compatible Bayesian network). *Let $\mathcal{CN} = <G, K>$ be a credal network and $\mathcal{BN} = <G, CPT>$ be a Bayesian networks. $\mathcal{BN}$ is said compatible with $\mathcal{CN}$ iff*

1. *$\mathcal{BN}$ and $\mathcal{CN}$ have exactly the same structure (hence they encode the same conditional independence relations).*
2. *For each variable A_i, $\forall a_i \in D_i, p_{\mathcal{BN}}(a_i|par(a_i)) \in K_i(a_i|par(a_i))$.*

According to this semantics, a credal network $\mathcal{CN}$ encodes a set of joint probability distributions, called extensions and denoted $ext(\mathcal{CN})$, where each joint distribution $p \in ext(\mathcal{CN})$ is encoded by a compatible Bayesian network. Given an extension $ext(\mathcal{CN})$, one can compute a joint IPD (interval-based joint probability distribution) as follows:

$$\underline{P}(a_1 a_2 .. a_n) = \min_{p \in ext(\mathcal{CN})} (p(a_1 a_2 .. a_n)) \tag{2}$$

$$\overline{P}(a_1 a_2 .. a_n) = \max_{p \in ext(\mathcal{CN})} (p(a_1 a_2 .. a_n)) \tag{3}$$

---

[2] http://www.cs.cmu.edu/~javabayes/Home/.

In Eqs. 2 and 3, $p(a_1 a_2..a_n)$ is computed with the well-known chain rule in Bayesian networks (see Eq. 1). Note that the vertices of $ext(\mathcal{CN})$ can be obtained by considering only the set of vertices of the local credal sets K_i associated with the variables [3]. As for marginalization and conditioning, they are defined as follows:

Let $K(A_1..A_n)$ be a credal set over the set of variables $A = \{A_1..A_n\}$. Let X and Y be two disjoint subsets of A such that $X \cup Y = A$. Then,

$$K(X) = CH(\{\sum_Y p(X,Y) \ with \ p(X,Y) \in K(A_1..A_n)\}) \qquad (4)$$

where CH is the convex hull operator. As for conditioning, let e be an evidence, then

$$K(A_1..A_n|e) = CH(\{p(A_1..A_n|e) \ with \ p(A_1..A_n) \in K(A_1..A_n) \ and \ p(e) > 0\}) \qquad (5)$$

2.3 Possibilistic Networks

A possibilistic network $\mathcal{PN} = <G, \Theta>$ is specified by:

(i) *A graphical component* G consisting of a directed acyclic graph (DAG) where vertices represent the variables and edges represent direct *dependence* relationships between variables.

(ii) *A numerical component* Θ allowing to weight the uncertainty relative to each variable using local possibility tables. The possibilistic component consists in a set of local possibility tables $\theta_i = \pi(A_i|par(A_i))$ for each variable A_i in the context of its parents $par(A_i)$ in the network $\mathcal{PN}$.

Note that all the local possibility distributions θ_i must be normalized, namely $\forall i = 1..n$, for each parent context $par(a_i)$, $\max_{a_i \in D_i}(\pi(a_i \mid par(a_i))) = 1$.

In the possibilistic setting, the joint possibility distribution is factorized using the following possibilistic counterpart of the chain rule:

$$\pi(a_1, a_2, .., a_n) = \otimes_{i=1}^n (\pi(a_i|par(a_i))). \qquad (6)$$

where $\otimes$ denotes the product or the min-based operator depending on the quantitative or the qualitative interpretation of the possibilistic scale [7].

3 Probability-Possibility Transformations

3.1 Form Probability Distributions to Possibilistic Ones

Many probability-possibility transformations exist [6,9,17]. Most of the works address desirable properties and propose some transformations that satisfy such properties. Among these transformations, the optimal transformation (OT) [6] defines a consistency condition requiring that

1. the obtained possibility distribution π dominates the original probability distribution p (namely, $\phi \subseteq \Omega, P(\phi) \leq \Pi(\phi)$).
2. the obtained possibility distribution π preserves the order of elementary worlds encoded in p (namely, $\forall (\omega_i, \omega_j) \in \Omega^2, p(\omega_i) > p(\omega_j) \Rightarrow \pi(\omega_i) > \pi(\omega_j)$ and $p(\omega_i) = p(\omega_j) \Rightarrow \pi(\omega_i) = \pi(\omega_j)$).

The optimal transformation (OT) transforms p into π as follows:

$$\pi_i = \sum_{j/p_j \leq p_i} p_j, \tag{7}$$

where π_i (resp. p_i) denotes $\pi(\omega_i)$ (resp. $p(\omega_i)$). The transformation of Eq. 7 guarantees that the obtained possibility distribution π is the most specific[3] (hence most informative) one that is consistent and preserving the order of interpretations.

The author in [14] addressed the commutativity of transformations with respect to some operations but the aim was to show that the obtained distributions are not identical. Some of these issues were also dealt with in the context of fuzzy interval analysis [8]. In [1], we dealt with some issues about probability-possibility transformations especially those regarding reasoning tasks and graphical models. In particular, we showed that:

- there is no transformation that can preserve the order of arbitrary events through some reasoning operations like marginalization.
- for the independence of events and variables, we showed that there is no transformation that preserves the independence relations,
- when the uncertain information is encoded by means of graphical models, we showed that no transformation can preserve the order of interpretations and events.

In this paper, we deal with some of these issues in the context of credal networks.

3.2 From Interval-Based Probability Distributions to Possibilistic Ones

When transforming uncertain information expressed by means of probability intervals to a possibility distribution, there is to the best of our knowledge only one work [11] where the authors learn possibility distributions from empirical data by transforming confidence intervals to possibility distributions. The starting point of this transformation is to consider an IPD as a means of encoding a partial order $\mathcal{M}$ over Ω. Indeed, contrary to precise probability distributions which encode complete order relations over Ω, interval-based ones encode partial orders in the form $\omega_i <_{IP} \omega_j$ in case where $u_i < l_j$. Let $\mathcal{M}$ be the partial order encoded by an IPD IP and let $\mathcal{C}$ be the set of linear extensions (complete orders) that are compatible with the partial order $\mathcal{M}$. The transformation proposed in [11] proceeds as follows:

---

[3] Let π' and π'' be two possibility distributions, π' is more specific than π'' iff $\forall \omega_i \in \Omega$, $\pi'(\omega_i) \leq \pi''(\omega_i)$.

– For every linear extension $C_l \in C$ and for each $\omega_i \in \Omega$, compute:

$$\pi^{C_l}(\omega_i) = \max_{p_1..p_n} \left(\sum_{p_j \leq p_i} p_i \right) \tag{8}$$

subject to the following constraints (in order to explore only compatible probability distributions satisfying the current linear extension C_l):

$$\begin{cases} p_i \in [l_i, u_i] \\ \sum_{i=1..n} p_i = 1 \\ p_1..p_n \ satisfy \ the \ linear \ extension \ C_l \end{cases}$$

– Build the distribution π that dominates all the distributions π^{C_l} as follows: $\forall \omega_i \in \Omega$,

$$\pi(\omega_i) = \max_{C_l \in C}(\pi^{C_l}(\omega_i)) \tag{9}$$

The motivation of using Eq. 9 is to guarantee that the obtained possibility distribution π dominates the probability intervals IP. This transformation tries on one hand to preserve the order of interpretations induced by IP and the dominance principle requiring that $\forall \phi \subseteq \Omega, P(\phi) \leq \Pi(\phi)$ on the other hand.

There are two main drawbacks with the transformation of Eqs. 8 and 9:

– The first issue is about the computational complexity of such transformation. Applied directly, this latter can consider in the worst case $N!$ linear extensions where N is the number of possible worlds. The authors proposed in [11] an algorithm allowing to achieve some improvements during this transformation but it is still very costly when one considers variables having domains exceeding a dozen values, which is common in many applications.
– The second concern lies in the fact that this transformation does not guarantee that the obtained distribution is optimal is terms of specificity. Indeed, it was shown in [5] that the transformation of Eq. 9 results in a loss of information as it is not the most specific one dominating the considered IPD. The authors in [5] suggest that any upper generalized R-cumulative distribution $\overline{F}$ built from one linear extension $C_l \in C$ can be viewed as a possibility distribution and it also dominates all the probability distributions that are compatible with the IPD. Let C_l be a linear extension compatible with the partial order M induced by an IPD. Let $\phi_1, \phi_2..\phi_n$ be subsets of Ω such that $\phi_i = \{\omega_j | \omega_j \leq_{C_l} \omega_i\}$. The upper cumulative distribution $\overline{F}$ built from one linear extension C_l is as follows (see [5] for more details):

$$\overline{F}(\phi_i) = \min\left(\sum_{\omega_j \in \phi_i} u_j, 1 - \sum_{\omega_j \notin \phi_i} l_j \right) \tag{10}$$

The obtained cumulative distribution $\overline{F}$ is a possibility distribution dominating IP and it is such that $\overline{P}(\phi_i) = \Pi(A_i)$.

Regarding the commutativity of transformations with respect to change operations like marginalization and conditioning used to answer MAP queries, since probability distributions are special cases of IPDs, it can be expected that for the commutativity issue, the transformations exhibit the same properties. This is the focus of the next section.

4   Commutativity of Interval-Based Probability-Possibility Transformations with Respect to Marginalization and Conditioning

This section checks whether the interval-based probability-possibility transformations are commutative with respect to two major change operations that are marginalization and conditioning. Namely, the question dealt with here is: Given an IPD IP, do we get exactly the same results when (i) we first transform IP into a possibility distribution π then apply the change operation in the possibilistic setting and when (ii) we first apply the change operation in the interval-based setting then transform the result into a possibility distribution. Proposition 1 provides the answer for marginalization:

Proposition 1. *Let TR be an interval-based probability-possibility transformation[4]. Then there exists an IPD IP, two events $\phi \subseteq \Omega$, $\psi \subseteq \Omega$ with $\phi \neq \psi$, and $\pi = TR(IP)$ such that $\overline{P}(\phi) < \underline{P}(\psi)$ but $\Pi(\phi) > \Pi(\psi)$.*

Proposition 1 asserts that no interval-based probability-possibility transformation can guarantee the preservation of the order of events as shown in the following example.

Example 1. Let IP be an IPD of Table 1 where $\Omega = \{\omega_1, \omega_2, \omega_3, \omega_4\}$ and $\pi = TR(IP)$. In this example, α_1, α_2 and α_3 are possibility degrees such that $1 > \alpha_1 > \alpha_2 > \alpha_3$ in order to satisfy the preference preservation principle. Now, let ϕ and ψ be two events such that $\phi = \{\omega_1\}$ and $\psi = \{\omega_2, \omega_3\}$. We have $\Pi(\phi) = 1 > \Pi(\psi) = \max(\alpha_1, \alpha_2)$ while $\overline{P}(\phi) = .4 < \underline{P}(\psi) = .6$.

As shown in Example 1, the strict order of events is not preserved by TR because of the different behavior of the additivity axiom in the probabilistic setting and the maxitivity axiom of the possibilistic setting used by the marginalization operation.

As a consequence of Proposition 1, we have the following Lemma:

---

[4] In the rest of this paper, TR denotes an interval-based probability-possibility transformation satisfying the following principles:

- *Dominance:* The possibility distribution π obtained from the IPD IP by TR dominates every probability distribution p compatible with IP, namely $\forall \phi \subseteq \Omega$, $\pi(\phi) \geq p(\phi)$.
- *Order preservation:* Given two interpretations $\omega_i \in \Omega$ and $\omega_j \in \Omega$, $\pi(\omega_i) < \pi(\omega_j)$ iff $\overline{p}(\omega_i) < \underline{p}(\omega_j)$.

Table 1. Example showing the loss of the order of events.

ω_i	$IP(\omega_i)$	$\pi(\omega_i)$
ω_1	[.36, .4]	1
ω_2	[.35, .35]	α_1
ω_3	[.25, .25]	α_2
ω_4	[0, .04]	α_3

Lemma 1. *Let TR be an interval-based probability-possibility transformation. Then there exists an IPD IP over $\Omega = \{\omega_1, \omega_2, .., \omega_n\}$ and a partition $\Omega' = \{W_1, W_2...W_k\}$ of Ω with $k < n$. Let $\pi = TR(IP)$, IP' is obtained by marginalizing IP on Ω' according to Eq. 4 and π' is obtained by marginalizing π on Ω' in the possibilistic setting. Then there may exist an event $W_i \in \Omega'$ such that*

$$\pi(W_i) \neq \pi'(W_i).$$

Proof (Proof sketch). The proof follows from Proposition 1 since if the order of events is not preserved then the underlying marginalized distributions must be different.

Let us now check the commutativity issue with respect to conditioning. For standard probability distributions, we have the following finding [1]:

Proposition 2. *Let p be a probability distribution over Ω and let $\phi \subseteq \Omega$ be an evidence. Let TR be a probability-possibility transformation, p' be a probability distribution obtained by conditioning p by ϕ, $\pi'' = TR(p')$ and π' is the possibility distribution obtained by conditioning $\pi = TR(p)$ by ϕ. Then, $\forall \omega_i, \omega_j \in \Omega$,*

$$\pi'(\omega_i) < \pi'(\omega_j) \text{ iff } \pi''(\omega_i) < \pi''(\omega_j).$$

Note that Proposition 2 is valid in both the product and the min-based possibilistic settings and it states that the order of interpretations is not affected by the order of applying the transformation and the conditioning operation. For IPDs, the following proposition states that the partial order encoded by IP after conditioning is preserved in the (complete) order induced by π after conditioning on the same evidence.

Proposition 3. *Let IP be an IPD over Ω and let $\phi \subseteq \Omega$ be an evidence. Let TR be an interval-based probability-possibility transformation, $IP' = IP(.|\phi)$ be a posterior probability distribution obtained by conditioning IP by ϕ, $\pi'' = TR(IP')$ and $\pi' = \pi(.|\phi)$ is the possibility distribution obtained by conditioning $\pi = TR(IP)$ by ϕ. Then,*

$$\forall \omega_i, \omega_j \in \Omega, \pi'(\omega_i) < \pi'(\omega_j) \text{ iff } \pi''(\omega_i) < \pi''(\omega_j).$$

Proof (Proof sketch). The idea of the proof is that since conditioning in both the probabilistic and possibilistic settings consists in discarding the worlds that are not models of the evidence ϕ (by assigning them a 0 probability/possibility degree) then renormalizing the obtained distribution. Hence, the order of interpretations that are models of ϕ is not affected by the order of application of transformation/conditioning operations.

Let us now see how one can use probability-possibility transformations to perform some inference queries in credal networks.

5 A Probability-Possibility Transformation Based Approach for Inference in Credal Networks

In [13] a natural transformation of Bayesian networks into possibilistic networks is proposed using the existing probability-possibility transformations such as OT.

5.1 From Credal Networks to Possibilistic Networks

A straightforward way to transform a credal network into a possibilistic network is as follows:

Definition 6 (Credal-possibilistic network transformation). *Let $\mathcal{CN}$ be a credal network, $\mathcal{PN}_{CN}$ is a possibilistic network obtained from $\mathcal{CN}$ and defined by:*

- *A graphical component G which is the same graph as the credal network hence $\mathcal{PN}_{CN}$ encodes the same independence relations as $\mathcal{CN}$.*
- *A collection of local possibility tables π_i obtained by transforming local credal sets K_i with TR, a transformation from interval-based probability distribution into possibilistic ones.*

The advantage of transforming a graphical model using Definition 6 is to preserve the independence relationships while transforming only local tables.

Example 2. Let $\mathcal{CN}$ be the credal network of Fig. 1 over two binary variables A and B. Using the transformation of Eq. 9, the credal network $\mathcal{CN}$ of Fig. 1 will be transformed to the possibilistic network $\mathcal{PN}$ of Fig. 2.

In the following, we address two main questions: (i) Does the distribution $\pi_{\mathcal{PN}}$ dominate $IP_{\mathcal{CN}}$ (the joint interval-based distribution encoded by $\mathcal{CN}$)? and (ii) Is the partial order of interpretations induced by $IP_{\mathcal{CN}}$ preserved by the transformation TR?

Regarding the first question, the two following propositions provide the answer. For elementary worlds $\omega_i \in \Omega$, Proposition 4 ensures that the computed possibility distribution dominates the corresponding probability degrees in case where the credal network $\mathcal{CN}$ is a Bayesian network (namely, all the intervals in $\mathcal{CN}$ are singletons).

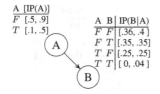

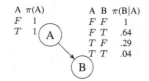

Fig. 1. Example of a credal network $\mathcal{CN}$.

Fig. 2. The possibilistic network $\mathcal{PN}_{CN}$ obtained from the credal network $\mathcal{CN}$ of Fig. 1.

Proposition 4. *Let TR be a probability-possibility transformation. Let $\mathcal{BN}$ be a standard Bayesian network and let $p_{\mathcal{BN}}$ be the underlying joint probability distribution encoded by $\mathcal{BN}$. Let $\mathcal{PN}$ be a possibilistic network such that $\mathcal{PN} = TR(\mathcal{BN})$ and $\pi_{\mathcal{PN}}$ be the joint possibility distribution encoded by $\mathcal{PN}$. Then $\forall \omega_i \in \Omega$,*

$$\pi_{\mathcal{PN}}(\omega_i) \geq P_{\mathcal{BN}}(\omega_i).$$

Proof (Proof sketch). Let $\omega_i = a_1 a_2 ... a_n$ be an instantiation of the network variables A_1, $A_2 ... A_n$. We have in the product-based possibilistic setting, for every variable value a_i in its parents context $par(a_i)$, $p_{\mathcal{BN}}(a_i|par(a_i)) \leq \pi_{\mathcal{PN}}(a_i|par(a_i))$, guaranteed by the transformation TR. Then $\prod_{i=1}^{n}(p_{\mathcal{BN}}(a_i|par(a_i))) \leq \prod_{i=1}^{n}(\pi_{\mathcal{PN}}(a_i|par(a_i)))$. The proof follows similarly for min-based possibilistic networks.

Now, regarding arbitrary events $\phi \subseteq \Omega$, the issue is still open. If we use the optimal transformation OT, the following proposition states that the obtained joint possibility distribution does not guarantee to dominate the joint probability distribution.

Proposition 5. *Let OT be the optimal probability-possibility transformation. There may exist a standard Bayesian network $\mathcal{BN}$ encoding a joint probability distribution denoted $p_{\mathcal{BN}}$. Let $\mathcal{PN}$ be a possibilistic network such that $\mathcal{PN} = OT(\mathcal{BN})$ and $\pi_{\mathcal{PN}}$ be the joint possibility distribution encoded by $\mathcal{PN}$. Then there may exist an event $\phi \subseteq \Omega$ such that*

$$\Pi_{\mathcal{PN}}(\phi) \not\geq P_{\mathcal{BN}}(\phi)$$

The following counter-example shows that $\Pi_{\mathcal{PN}}(\phi) \geq P_{\mathcal{BN}}(\phi)$ is not guaranteed when using the optimal transformation OT.

Example 3. Let $\mathcal{BN}$ be the Bayesian network of Fig. 3 over two variables A and B having the domains $D_A = \{a_1, a_2\}$ and $D_B = \{b_1, b_2, b_3\}$ respectively.

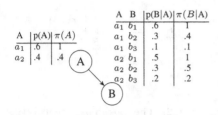

A B	p(B|A)	π(B|A)
$a_1\ b_1$	.6	1
$a_1\ b_2$	.3	.4
$a_1\ b_3$	.1	.1
$a_2\ b_1$	.5	1
$a_2\ b_2$	.3	.5
$a_2\ b_3$	.2	.2

A	p(A)	π(A)
a_1	.6	1
a_2	.4	.4

A B	p(A,B)	π(A,B)
$a_1\ b_1$	.36	1
$a_1\ b_2$	.18	.4
$a_1\ b_3$	.06	.1
$a_2\ b_1$	.2	.4
$a_2\ b_2$	.12	.2
$a_2\ b_3$	.08	.08

Fig. 3. Example of a Bayesian network $\mathcal{BN}$ and the possibilistic network $\mathcal{PN}$ obtained from $\mathcal{BN}$ using the optimal transformation OT.

Fig. 4. Joint probability and possibility distributions encoded by the networks $\mathcal{BN}$ and $\mathcal{PN}$ of Fig. 3.

The joint distributions encoded by the networks $\mathcal{BN}$ and $\mathcal{PN}$ are given in Fig. 4. From Fig. 4, one can compute $P(b_3) = .06 + .08 = .14 > \Pi(b_3) = \max(.1, .08) = .1$.

Example 3 clearly shows that the transformation OT does not guarantee that when transforming a Bayesian network to a possibilistic network, the underlying joint possibility distribution dominates the corresponding probability distribution.

Now, how about the order of interpretations encoded by a credal network when it is transformed into a possibilistic network? The following proposition answers this question. Recall that the objective here is to check if the order of interpretations induced by $IP_{\mathcal{CN}}$ (the joint IPD encoded by the credal network $\mathcal{CN}$) is preserved in the obtained joint possibility distribution $\pi_{\mathcal{PN}}$ encoded by the possibilistic network $\mathcal{PN}$.

Proposition 6. *Let TR be a transformation from credal networks to possibilistic ones. Then there exists a credal network $\mathcal{CN}$ and two interpretations $\omega_i \in \Omega$ and $\omega_j \in \Omega$ such that*

$$\overline{p}_{CN}(\omega_i) < \underline{p}_{CN}(\omega_j) \text{ but } \pi_{PN}(\omega_i) \geq \pi_{PN}(\omega_j).$$

where $\underline{p}_{CN}$ and $\overline{p}_{CN}$ denote lower and upper bounds induced by $\mathcal{CN}$ and π_{PN} denotes the joint possibility distribution induced by $\mathcal{PN}$ using the transforamtion of Definition 6. The following gives a counter-example.

Example 4. Let $\mathcal{CN}$ be the credal network of Fig. 5 over two disconnected variables A and B. Note that the IPD $IP(A)$ in $\mathcal{CN}$ is a permutation[5] of the IPD of B. Hence, the transformation of $IP(A)$ and $IP(B)$ by TR gives $\pi(A)$ and $\pi(B)$ where $\pi(B)$ is also a permutation of $\pi(A)$. In this example, since TR is assumed to preserve the partial order of interpretations, we have $1 > \alpha_1 > \alpha_2 > \alpha_3$. The probability and possibility degrees of interpretations $a_1 b_3$ and $a_2 b_2$ are

---

[5] The permutation property of probability-possibility transformations is discussed in [14].

$\underline{p}(a_1b_3) = 0.36 * 0.24 = 0.0864$ and $\overline{p}(a_2b_2) = 0.26 * 0.26 = 0.0676$. Clearly, $\underline{p}(a_1b_3) > \overline{p}(a_2b_2)$. Now, $\pi(a_1b_3) = \min(\alpha_2, 1)$ and $\pi(a_2b_2) = \min(\alpha_1, \alpha_1)$ then, $\pi(a_1b_3) < \pi(a_2b_2)$. It is clear that the relative order of interpretations is reversed whatever is the used transformation in the ordinal setting. In the same way, in the product-based possibilistic setting, the relative order of interpretations can not be preserved by any transformation.

A	$IP(A)$	$\pi(A)$
a_1	$[.36, .4]$	1
a_2	$[.26, .26]$	α_1
a_3	$[.24, .24]$	α_2
a_4	$[.1, .14]$	α_3

$\left(A\right)$ $\left(B\right)$

B	$IP(B)$	$\pi(B)$
b_1	$[.1, .14]$	α_3
b_2	$[.26, .26]$	α_1
b_3	$[.24, .24]$	α_2
b_4	$[.36, .4]$	1

Fig. 5. Counter-example for Proposition 6.

Up to now, the findings of this paper are rather negative but transformations from credal networks into possibilistic ones can be very helpful for certain types of queries in credal networks as it is shown in the following sections.

5.2 MAP Inference in Credal Networks Through Credal-Possibilistic Network Transformation

Inference in probabilistic graphical models generally consists in computing the probability of an event. In credal networks, this equivalently comes down to computing lower or upper probabilities of an event of interest. Let $A = \{A_1, A_2...A_n\}$ be the set of variables of the probabilistic model. Let $\mathcal{O} \subseteq A$ be the set of observed variables and let o be an instantiation of observation variables $\mathcal{O}$. Let also $\mathcal{Q} \subseteq A$ be the set of query variables and let q be instantiation of the query variables. There are three main query types when reasoning with belief graphical models:

- Computing the probability of an event q of interest (Pr) given an evidence o.
- Computing the most plausible explanation (MPE). Given an observation o of some variables, the objective is to compute the most probable instantiation q of all the remaining (unobserved) variables $\mathcal{Q}$. Note that here $\mathcal{O} \cup \mathcal{Q} = A$ and $\mathcal{Q} \cap \mathcal{O} = \emptyset$.
- Computing the maximum a posteriori (MAP). Given some observations o of the values of some variables $\mathcal{O}$, the objective is to compute the most probable instantiation q of the query variables $\mathcal{Q}$. In MAP queries, $\mathcal{Q} \cap \mathcal{O} = \emptyset$. Note that MPE queries are special cases of MAP ones.

In credal networks, the inference problem equivalently comes down to compute either the lower or the upper bound of an event of interest. As for MPE and MAP queries, there are different criteria to choose the *most probable* instantiation of query variables given the observations. The commonly used criterion in credal networks is the one of *interval-dominance* and refers to non-dominated instantiations of query variables.

Definition 7 (Interval-dominance). *An instantiation q_i of query variables $\mathcal{Q}$ dominates another instantiation q_j iff $\underline{P}(q_i|o) > \overline{P}(q_j|o)$ where o is an instantiation of observation variables $\mathcal{O}$.*

The following table summarizes complexity results of inference in Bayesian and credal networks [12].

	Query	Polytree	Bounded treewidth	Multiply-connected
Bayesian Networks	Pr	Polynomial	Polynomial	PP-Complete
	MPE	Polynomial	Polynomial	NP-Complete
	MAP	NP-Complete	NP-Complete	NP^{PP}-Complete
Credal Networks	Pr	NP-Complete	NP-Complete	NP^{PP}-Complete
	MPE	Polynomial	Polynomial	NP-Complete
	MAP	Σ_2^P-Complete	Σ_2^P-Complete	NP^{PP}-Hard

It is obvious that even in polytrees, MAP inference is a hard task. In practice, the size of networks and the set of extreme points of local credal sets is often large. This motivates approximate inference approaches where the goal is to provide bounds of the real bounds of probabilities. In this work, we provide a kind of approximate inference approach for MAP inference in $\mathcal{CN}$s by transforming the credal network $\mathcal{CN}$ into a possibilistic network $\mathcal{PN}$ used to answer the queries. Note that the complexity of inference in possibilistic networks is similar to inference in Bayesian networks.

6   Experimental Studies

The objective of this section is to empirically evaluate the accuracy of performing MAP inference in credal networks by transforming them into possibilistic networks. In order to evaluate our approach, we carried out a set of experimentations on the well-known and publicly available credal networks benchmark[6]. This latter contains a set of credal networks with different topologies and parameters in *.bif* format. Table 2 gives some details on the networks used in our experimentations. Table 2 shows that the number of variables in the used networks varies from 6 up to 37. As for variable domains, their sizes vary between 2 and 8. In this preliminary study, we are interested only in MAP queries where given some observed variables, the task is to find the most probable values of some other non observed variables, called query variables. In this experimentation, we report results where the number of observed variables and the observed values are randomly chosen. The queries concern only one variable chosen randomly. Now, in order to compare the results of MAP inference in credal networks and their possibilistic counterpart, each query Q is submitted to a credal network

---

[6] http://ipg.idsia.ch/software/.

$\mathcal{CN}$ then to the corresponding possibilistic network $\mathcal{PN}$ obtained from $\mathcal{CN}$. The results are compared through the accuracy criterion defined as follows:

$$accuracy(Q_1, Q_2...Q_n) = \frac{1}{n} \sum_{i:1..n} \frac{|CN_{MAP}(Q_i) \cap PN_{MAP}(Q_i)|}{|CN_{MAP}(Q_i) \cup PN_{MAP}(Q_i)|}, \quad (11)$$

where $CN_{MAP}(Q_i)$ (resp. $PN_{MAP}(Q_i)$) denotes the results of the query Q_i submitted to the network $\mathcal{CN}$ (resp. $\mathcal{PN}$). This criterion evaluates the coincidence between the results of $\mathcal{CN}$ to the MAP queries and the ones of $\mathcal{PN}$.

Table 2. Credal networks used in the experimentations.

Networks	Topology	# Nodes	max domain size
Alarm	Multiply-connected	37	4
Insurance	Multiply-connected	27	5
Poly	Polytree	10, 20, 30	4
Multi	Multiply-connected	6, 10, 20	8

In Table 3, we provide the accuracy (see Eq. 11) of MAP inference achieved through our credal-possibilistic network transformation approach with respect to the results of credal networks. More precisely, the column *Exact* vs *Appr* provides the accuracy of an approximate inference algorithm in credal networks achieved with the GL2U software[7] on each network category. The column *Exact* vs *CD* (resp. *Exact* vs *MD*) provides the accuracy of possibilistic networks obtained by our credal-possibilistic network transformation where local tables are transformed using the cumulative distribution of Eq. 10 (resp. Masson and Denoeux's transformation [11] considering all the linear extensions). Note also that we evaluated only product-based possibilistic networks and the experiments are performed on a few dozen requests on a laptop.

Table 3. Credal networks used in the experimentations.

Networks	Exact vs Appr	Exact vs CD	Exact vs MD
Alarm	75 %	100 %	timeout
Insurance	50 %	100 %	timeout
Poly	83 %	100 %	timeout
Multi	90 %	38 %	48 %

The results of Table 3 clearly show that, on one hand, the credal-possibilistic network transformation based approach can ensure a high accuracy rate and, on the other hand, the results are often better than those obtained with an approximate approach.

---

[7] http://people.idsia.ch/~sun/gl2u-ii.htm.

7   Conclusions

This paper dealt with probability-possibility transformations in the context of credal networks. We first analyzed some issues related to the commutativity of transformations with respect to marginalization and conditioning, two main change operations used for MAP inference. We then proposed an approach allowing to perform MAP inference in credal networks with a lower computational costs. Finally, we provided experimental studies showing the efficiency of the proposed approach in terms of accuracy. Future works will deal with extensive experimental studies as well as using our credal-possibilistic network transformation based approach for achieving classification with credal networks in real applications.

Acknowledgements. This work is done with the support of a CNRS funded project PEPS FaSciDo 2015 - MAPPOS.

References

1. Benferhat, S., Levray, A., Tabia, K.: On the analysis of probability-possibility transformations: changing operations and graphical models. In: ECSQARU 2015, Compiegne, France, 15–17 July 2015
2. Borgelt, C., Kruse, R.: Learning possibilistic graphical models from data. IEEE Trans. Fuzzy Syst. **11**(2), 159–172 (2003)
3. Cozman, F.G.: Credal networks. Artif. Intell. **120**(2), 199–233 (2000)
4. Darwiche, A.: Modeling and Reasoning with Bayesian Networks. Cambridge University Press, New York (2009)
5. Destercke, S., Dubois, D., Chojnacki, E.: Transforming probability intervals into other uncertainty models. In: EUSFLAT 2007 Proceedings, vol. 2, pp. 367–373. Universitas Ostraviensis, Ostrava, Czech Republic (2007)
6. Dubois, D., Foulloy, L., Mauris, G., Prade, H.: Probability-possibility transformations, triangular fuzzy sets, and probabilistic inequalities. Reliable Comput. **10**(4), 273–297 (2004)
7. Dubois, D., Prade, H.: Possibility Theory: An Approach to Computerized Processing of Uncertainty. Plenum Press, New York (1988)
8. Dubois, D., Prade, H.: Random sets and fuzzy interval analysis. Fuzzy Sets Syst. **42**(1), 87–101 (1991)
9. Klir, G.J., Geer, J.F.: Information-preserving probability-possibility transformations: recent developments. In: Lowen, R., Roubens, M. (eds.) Fuzzy Logic, pp. 417–428. Kluwer Academic Publishers, Dordrecht (1993)
10. Levi, I.: The Enterprise of Knowledge: An Essay on Knowledge, Credal Probability, and Chance/Isaac Levi. MIT Press, Cambridge, Mass (1980)
11. Masson, M.-H., Denoeux, T.: Inferring a possibility distribution from empirical data. Fuzzy Sets Syst. **157**(3), 319–340 (2006)
12. Mauá, D., de Campos, C.P., Benavoli, A., Antonucci, A.: Probabilistic inference in credal networks: new complexity results. J. Artif. Intell. Res. (JAIR) **50**, 603–637 (2014)

13. Slimen, Y.B., Ayachi, R., Amor, N.B.: Probability-possibility transformation: application to Bayesian and possibilistic networks. In: Masulli, F. (ed.) WILF 2013. LNCS, vol. 8256, pp. 122–130. Springer, Heidelberg (2013)
14. Sudkamp, T.: On probability-possibility transformations. Fuzzy Sets Syst. **51**, 73–81 (1992)
15. Walley, P.: Towards a unified theory of imprecise probability. Int. J. Approx. Reason. **24**(23), 125–148 (2000)
16. Wallner, A.: Extreme points of coherent probabilities in finite spaces. Int. J. Approx. Reason. **44**(3), 339–357 (2007)
17. Zadeh, L.A.: Fuzzy sets as a basis for a theory of possibility. Fuzzy Sets Syst. **100**, 9–34 (1999)

Planning in Partially Observable Domains with Fuzzy Epistemic States and Probabilistic Dynamics

Nicolas Drougard[1(✉)], Didier Dubois[2], Jean-Loup Farges[1], and Florent Teichteil-Königsbuch[1]

[1] Onera – The French Aerospace Lab, Toulouse, France
nicolas.drougard@onera.fr
[2] IRIT, CNRS and Université de Toulouse, Toulouse, France

Abstract. A new translation from Partially Observable MDP into Fully Observable MDP is described here. Unlike the classical translation, the resulting problem state space is finite, making MDP solvers able to solve this simplified version of the initial partially observable problem: this approach encodes agent beliefs with possibility distributions over states, leading to an MDP whose state space is a finite set of epistemic states. After a short description of the POMDP framework as well as notions of Possibility Theory, the translation is described in a formal manner with semantic arguments. Then actual computations of this transformation are detailed, in order to highly benefit from the factored structure of the initial POMDP in the final MDP size reduction and structure. Finally size reduction and tractability of the resulting MDP is illustrated on a simple POMDP problem.

1 Introduction

It is claimed that Partially Observable Markov Decision Processes (POMDPs) [17] finely models an agent acting under uncertainty in a partially hidden environment. However, solving a POMDP, *i.e.* the computation of an optimal strategy for the agent, is a really difficult task: the problem is PSPACE-complete [12]. Classical approaches try to solve this problem using Dynamic Programming [3], or via approximate computation. These include for instance heuristic search [18] and Monte Carlo approaches [16].

The approach proposed here simplifies a POMDP problem before solving it. The transformation described leads to a fully observable MDP on a finite number of epistemic states, *i.e.* a problem modeling an agent acting under uncertainty in a fully observable environment [13]. As such a finite state space MDP problem is P-complete [12] this transformation qualifies as a simplification, and any MDP solver can return a policy for this translated POMDP.

Most of the POMDP algorithms draw upon the *agent belief* during the process, defined as the probability of the actual system state knowing all the system observations and agent actions from the beginning. This belief is updated

© Springer International Publishing Switzerland 2015
C. Beierle and A. Dekhtyar (Eds.): SUM 2015, LNAI 9310, pp. 220–233, 2015.
DOI: 10.1007/978-3-319-23540-0_15

at each time step using Bayes' rule and the new observation. The initial belief, or *prior* probability distribution over the system states, is part of the definition of the POMDP. However in practice, the initial system state can be unknown: for instance, in a robotic exploration context, the initial location of the agent, or initial presence of an entity in the scene. Defining the process with a uniform probability distribution as initial belief (*e.g.* over all locations or over entity presence) is a subjectivist answer [5], *i.e.* all probabilities are the same because no event is more plausible than another: it corresponds to equal betting rates. However the subsequent belief updates will eventually mix up frequentist probability distributions defining the POMDP with the initial belief which is a subjective probability, and it does not always make sense.

More than only a simplification of the initial POMDP problem, the theoretical framework used here for the belief representation formally models an agent's knowledge about the system state: the proposed translation defines beliefs as *possibility distributions* over system states $s \in S$: these kinds of distributions are denoted by π (counterpart of probability notation $\mathbf{p}$) and represent fuzzy sets of system states, as the indicator (characteristic) function of this set. Recall that the indicator function of a classical set $A \subseteq S$ is $\mathbb{1}_A(s) = 1$ if $s \in A$ and 0 otherwise. Values of a fuzzy set indicator function π are chosen in a finite and totally ordered scale $\mathcal{L} = \{1 = l_1, l_2, \ldots, 0\}$ with $l_1 > l_2 > \ldots > 0$ *i.e.* $\pi : S \rightarrow \mathcal{L}$. If $s \in S$ is such that $\pi(s) = l_i$, s is in the fuzzy set described by π, with degree l_i. Possibilistic beliefs used in this work will represent fuzzy sets of possible states. If the current possibilistic belief coincides with the distribution $\pi(s) = 1 \; \forall s \in S$, all system states are totally possible, and it models therefore a total ignorance about the current system state: qualitative possibilistic beliefs can model agent initial ignorance. The perfect knowledge of the current state, say $\tilde{s} \in S$, is encoded by a possibility distribution equal to the classical indicator function of the singleton $\pi(s) = \mathbb{1}_{\{s = \tilde{s}\}}(s)$. Between these two extrema, the current knowledge of the system is described by a set of entirely possible states, $\{s \in S \text{ s.t. } \pi(s) = 1\}$, and successive sets of less plausible ones $\{s \in S \text{ s.t. } \pi(s) = l_i\}$ down to the set of impossible states $\{s \in S \text{ s.t. } \pi(s) = 0\}$.

The major originality of this work comes from the finiteness of the scale $\mathcal{L}$: the number of possible beliefs about the system state is, as well, finite (smaller than $\#(\mathcal{L}^S) = (\#\mathcal{L})^{\#S}$), while the set of all probability distributions over S is infinite. The translation described here leads then to an MDP whose finite state space is the set of possible possibilistic beliefs, or *epistemic states*.

In addition to POMDP simplification and knowledge modelling, this qualitative possibilistic framework offers some interesting properties: the possibilistic counterpart of Bayes' rule leads to a special belief behaviour. Indeed the agent can possibly change their mind radically and rapidly, and under some conditions the increased specificity of the belief is enforced, *i.e.* the knowledge about the current state is non decreasing with time steps [6]. Finally, in order to fully define the resulting MDP, the translation has to attach a reward function to its states: as the new (epistemic) state of the problem is a possibility distribution, a dual measure, called necessity, can be computed from it. Defined as the Choquet integral using the necessity measure, the reward of an epistemic state is a pessimistic evaluation of the actual reward.

However the number of possibilistic belief distributions, or *fuzzy epistemic states*, grows exponentially with the number of initial POMDP system states. The so called simplification of the problem does not transform the PSPACE POMDP problem into a polynomial one: as the new state space size is exponential in the previous one, the resulting problem is EXPTIME. The proposed translation tries to generate as few epistemic states as possible taking carefully into account potential factorized structures of the initial POMDP.

The first section is devoted to the presentation of the Markov Decision Processes, the main concern of this paper. Tools from Possibility Theory are also defined to make this paper self-contained. Follows a section describing the first contribution of this work, which is the translation itself, presented in a formal way. As the resulting state space of the built MDP is too big to make this problem tractable without factorization tricks in practice, the next section details the proper way to preprocess its attributes. Finally, the last section illustrates the relevance of this approach with a simple robotic mission problem.

2   Background

The work developed in this paper remains in the classical MDP and POMDP frameworks, which are recalled in this section: possibilistic material necessary to build the promised translation are then presented.

2.1   Markov Decision Processes

A Markov Decision Process (MDP) [1] is a well suited framework for sequential decision making under uncertainty, when the agent involved has a full knowledge of the actual system state. Such a process is formally defined by a 4-tuple $\langle \mathcal{S}, \mathcal{A}, T, r \rangle$ where $\mathcal{S}$ is a finite set of system states $s \in \mathcal{S}$. The finite set $\mathcal{A}$ consists of all actions $a \in \mathcal{A}$ available for the agent. The Markov dynamics of the system is described by the transition function $T : \mathcal{S} \times \mathcal{A} \times \mathcal{S} \rightarrow [0, 1]$. This function is defined as the transition probability distribution of the system states: if action $a \in \mathcal{A}$ is chosen by the agent, and the current system state is $s \in \mathcal{S}$, the next state $s' \in \mathcal{S}$ is reached with probability $T(s, a, s') = \mathbf{p}(s' \mid s, a)$. Finally, a reward function $r : \mathcal{S} \times \mathcal{A} \rightarrow \mathbb{R}$ is defined to model the goal of the agent. Indeed, solving an infinite horizon MDP problem consists in computing a *strategy*, *i.e.* a function d defined on $\mathcal{S}$ and whose values are actions $a \in \mathcal{A}$, maximizing the expected discounted total reward: $\mathbb{E}\left[\sum_{t=0}^{+\infty} \gamma^t \cdot r\left(s_t, d_t\right)\right]$ where $d_t = d(s_t) \in \mathcal{A}$ and $0 < \gamma < 1$ is the discount factor.

A Partially Observable MDP (POMDP) [17] makes a step further in the modeling flexibility, allowing the agent not to know which system state is the current one. The formal definition of a POMDP is the 7-tuple $\langle \mathcal{S}, \mathcal{A}, T, \Omega, O, r, b_0 \rangle$, where the system state $\mathcal{S}$, the set of actions $\mathcal{A}$, the transition function T and the reward function r remain the same as for the MDP definition. In this model, the current system state $s \in \mathcal{S}$ cannot be used as available information for the agent: the agent knowledge about the actual system state comes from observations $o \in \Omega$, where Ω is a finite set. The observation function $O : \mathcal{S} \times \mathcal{A} \times \Omega \rightarrow [0, 1]$ gives for each action $a \in \mathcal{A}$ and reached system state $s' \in \mathcal{S}$, the probability distribution over possible observations $o' \in \Omega$: $O(s', a, o') = \mathbf{p}(o' \mid s', a)$. Finally, the initial belief $b_0 : \mathcal{S} \rightarrow [0, 1]$ is the *prior* probability distribution over the state space $\mathcal{S}$: $b_0(s) = \mathbf{p}(s_0 = s)$, $\forall s \in \mathcal{S}$.

At a given time step $t > 0$, the agent belief is defined as the probability of the tth system state s_t conditioned on all the past actions and observations, and with the prior b_0, i.e. $b_t(s) = \mathbf{p}_{s_0 \sim b_0}(s_t = s \mid a_0, o_1, \ldots, a_{t-1}, o_t)$. It can be easily recursively computed using Bayes' rule: at time step t, if the belief is b_t, the chosen action $a \in \mathcal{A}$ and the new observation $o' \in \Omega$, the next belief is $b_{t+1}(s') \propto O(s', a, o') \cdot \sum_{s \in \mathcal{S}} T(s, a, s') \cdot b_t(s)$. Successive beliefs are computed from observations perceived by the agent, and are then available during the process. Let us denote by $\mathbb{P}_\mathcal{S}$ the infinite set of probability distributions over $\mathcal{S}$: seen as an MDP whose states are probabilistic beliefs, an optimal strategy for the infinite horizon POMDP is looked for among strategies $d : \mathbb{P}_\mathcal{S} \rightarrow \mathcal{A}$ such that successive actions $d_t = d(b_t) \in \mathcal{A}$ maximize the expected discounted total reward, which can be rewritten

$$\mathbb{E}\left[\sum_{t=0}^{+\infty} \gamma^t \cdot r(s_t, d_t)\right] = \mathbb{E}\left[\sum_{t=0}^{+\infty} \gamma^t \cdot r(b_t, d_t)\right], \tag{1}$$

defining $r(b_t, a) = \sum_{s \in \mathcal{S}} r(s, a) \cdot b_t(s)$ as the reward of belief b_t. As the focused problem (POMDP) has been formally defined, possibilistic tools are now presented in the next section.

2.2 Possibility Theory

In our context, distributions defined in the Possibility Theory framework are valued in a totally ordered scale $\mathcal{L} = \{1 = l_1, l_2, \ldots, 0\} \subseteq [0, 1]$ with $l_1 > l_2 > \ldots > 0$. A possibility measure Π defined on a finite set $\mathcal{S}$ is a fuzzy measure valued in $\mathcal{L}$, such that $\forall A, B \subset \mathcal{S}$, $\Pi(A \cup B) = \max\{\Pi(A), \Pi(B)\}$, $\Pi(\emptyset) = 0$ and $\Pi(\mathcal{S}) = 1$. It follows that this measure is entirely defined by the associated possibility distribution, i.e. the measure of the singletons: $\forall s \in \mathcal{S}$, $\pi(s) = \Pi(\{s\})$. Properties of this measure lead to the possibilistic normalization: $\max_{s \in \mathcal{S}} \pi(s) = \Pi(\mathcal{S}) = 1$. If $\overline{s}, \underline{s} \in \mathcal{S}$ are such that $\pi(\overline{s}) < \pi(\underline{s})$, it means that $\overline{s}$ is less plausible than $\underline{s}$. States with possibility degree 0, i.e. states $s \in \mathcal{S}$ such that $\pi(s) = 0$, are impossible (same meaning as $\mathbf{p}(s) = 0$), and those such that $\pi(s) = 1$ are entirely possible (but not necessarily the most probable ones).

After the introduction of a possibility measure over a set Ω, the joint possibility measure on $\mathcal{S} \times \Omega$ is defined in a qualitative way: $\forall A \subset \mathcal{S}$, $\forall B \subset \Omega$

$$\Pi(A, B) = \min\{\Pi(A \mid B), \Pi(B)\} = \min\{\Pi(B \mid A), \Pi(A)\}. \tag{2}$$

Note the similarities between Possibility and Probability Theory, replacing max by + and min by ×. Moreover, Possibility Theory has its own conditioning [9]:

$$\Pi(A \mid B) = \begin{cases} 1 & \text{if } \Pi(A, B) = \Pi(B) \\ \Pi(A, B) & \text{otherwise} \end{cases} \tag{3}$$

which is nothing more than the least specific measure fulfilling the condition described by Eq. 2. It can also be seen more easily as the joint measure normalized in a possibilistic manner.

These tools from Qualitative Possibility Theory are enough to define the announced translation. The next section is then devoted to the building of an MDP with fuzzy epistemic states from a POMDP.

3   A Hybrid POMDP

As claimed by Zadeh, "most information/intelligent systems will be of hybrid type" [19]: the idea developed here is to use a granulated representation of the agent knowledge using possibilistic beliefs instead of probabilistic beliefs in the POMDP framework. The first advantage of this granulation is that strategy computations are performed by reasoning on a finite set of possibilistic beliefs (called then epistemic states): the set of all possibility distributions defined over $\mathcal{S}$, denoted by $\Pi_{\mathcal{S}}$ is the set $\mathcal{L}^{\mathcal{S}}$ without non-normalized functions, and then

$$\#\Pi_{\mathcal{S}} = \#\mathcal{L}^{\#\mathcal{S}} - (\#\mathcal{L} - 1)^{\#\mathcal{S}}, \tag{4}$$

while the set of probability distributions over $\mathcal{S}$ is infinite. First, such beliefs are formally defined, as well as their own updates.

3.1   Possibilistic Belief

Consider that possibility distributions similar to those used to define the initial POMDP are available: a transition distribution, giving the possibility degree of reaching $s' \in \mathcal{S}$ from $s \in \mathcal{S}$ using action $a \in \mathcal{A}$, $\pi(s' \mid s, a) \in \mathcal{L}$; as well as an observation one, giving the possibility degree of observing $o' \in \Omega$, in a system state $s' \in \mathcal{S}$ after the use of $a \in \mathcal{A}$, $\pi(o' \mid s', a) \in \mathcal{L}$. Indeed, this work is devoted to two kinds of practical problems. On the one hand real problems modeled as POMDPs are often intractable: our granulated approach is in this case a simplification of the initial POMDP, and possibility distributions are computed from the POMDP probability distributions, using a possibility-probability transformation [10]. On the other hand, some problems lead to POMDPs with partially defined probability distributions: some estimated probabilities have no strong guarantees. A more faithful representation is given with possibility distributions modeling the inherent imprecision, defining transition and observation possibility distributions.

Let $b_0^\pi : \mathcal{S} \to \mathcal{L}$ be an initial possibilistic belief, normalized as any possibility distribution: $\max_{s \in \mathcal{S}} b^\pi(s) = 1$. As in the probabilistic case, possibilistic belief can be defined recursively using the possibilistic belief update [6], derived from Bayes' rule based on the conditioning (3): at time step t, if the possibilistic belief is b_t^π, action $a \in \mathcal{A}$ and observation $o' \in \Omega$ specify the next belief

$$b_{t+1}^\pi(s') = u(b_t^\pi, a, o')(s') = \begin{cases} 1 \text{ if } \pi(o', s' \mid b_t^\pi, a) = \max_{\tilde{s} \in \mathcal{S}} \pi(o', \tilde{s} \mid b_t^\pi, a) \\ \pi(o', s' \mid b_t^\pi, a) \text{ otherwise} \end{cases} \tag{5}$$

where the joint possibility distribution over $\Omega \times \mathcal{S}$ $\pi\left(o', s' \mid b_t^\pi, a\right)$ is equal to $\max_{s \in \mathcal{S}} \min\left\{\pi\left(o' \mid s', a\right), \pi\left(s' \mid s, a\right), b_t^\pi(s)\right\}$. Note that keeping a qualitative view for the belief update, $i.e.$ using the min operator to compute joint possibility distributions as defined in Eq. 2, allows to reason on a finite set of beliefs, as no new values are created: the classical product is used in the quantitative part of Possibility Theory, but is not considered in this work. Moreover, the use of a qualitative belief update has already been used in planning [7].

3.2   Setting Up Transition Functions

If the agent selects action $a \in \mathcal{A}$ in epistemic state $b^\pi \in \Pi_\mathcal{S}$, the next epistemic state depends only on the next observation, as highlighted by possibilistic belief update (5). The probability distribution over observations conditioned on the reached state is part of the POMDP definition via the observation function O. The probability distribution over observations conditioned on the previous state is obtained using transition function T: $\mathbf{p}\left(o' \mid s, a\right) = \sum_{s' \in \mathcal{S}} O(s', a, o') \cdot T(s, a, s')$. This distribution and the possibilistic belief b^π about the system state, can lead to an approximated probability distribution over the next observations. Indeed, a probability distribution over the system state, $\overline{b^\pi} \in \mathbb{P}_\mathcal{S}$, can be derived from b^π: a proper way to construct $\overline{b^\pi} \in \mathbb{P}_\mathcal{S}$ is the use of the pignistic transformation [8], minimizing the arbitrariness in the translation into probability distribution: numbering system states with the order induced by distribution b^π, $1 = b^\pi(s_1) \geqslant b^\pi(s_2) \geqslant \ldots \geqslant b^\pi(s_{\#\mathcal{S}+1}) = 0$, with $s_{\#\mathcal{S}+1}$ an artificial state such that $b^\pi(s_{\#\mathcal{S}+1}) = 0$ introduced to simplify the formula,

$$\overline{b^\pi}(s_i) = \sum_{j=i}^{\#\mathcal{S}} \frac{b^\pi(s_j) - b^\pi(s_{j+1})}{j}. \tag{6}$$

Note that this probability distribution corresponds to the center of mass of the probability distributions family induced by the possibility measure defined by distribution b^π [10], and respects the Laplace principle of Insufficient Reason (ignorance leads to uniform probability). Then an approximate distribution over $o' \in \Omega$ is defined as

$$\mathbf{p}\left(o' \mid b^\pi, a\right) = \sum_{s \in \mathcal{S}} \mathbf{p}\left(o' \mid s, a\right) \cdot \overline{b^\pi}(s). \tag{7}$$

Finally, summing over concerned observations, the transition probability distribution over epistemic states is defined as

$$\tilde{T}(b^\pi, a, (b^\pi)') = \mathbf{p}\left((b^\pi)' \mid b^\pi, a\right) = \sum_{o' \mid u(b^\pi, a, o') = (b^\pi)'} \mathbf{p}\left(o' \mid b^\pi, a\right). \tag{8}$$

3.3   Reward Aggregation

After the transition function, it remains to assign a reward to each epistemic state: in the classical probabilistic translation, the reward assigned to a belief b is the reward expectation according to the probability distribution b: $\sum_{s \in \mathcal{S}} r(s, a) \cdot b(s)$. Here, the agent knowledge is represented with a possibility distribution b^π, which is less informative than a probability one: it accumulates uncertainty due to possibilistic discretization and due to possible agent ignorance. A way to define a reward pessimistic about these uncertainties for each epistemic state b^π is to aggregate the reward on states using the dual measure of the possibility distribution b^π, and the *Choquet integral*.

The dual measure of a possibility measure $\Pi : 2^{\mathcal{S}} \to \mathcal{L}$ is called *necessity measure* and is denoted by $\mathcal{N}$. This measure is defined by $\forall A \subseteq \mathcal{S}$, $\mathcal{N}(A) = 1 - \Pi(\overline{A})$ where $\overline{A}$ is the complementary set of A: $\overline{A} = \mathcal{S} \setminus A$. Recall notation $\mathcal{L} = \{l_1 = 1, l_2, l_3, \ldots, 0\}$. For a given action $a \in \mathcal{A}$, reward values, $\{r(s, a) \mid s \in \mathcal{S}\}$ are denoted by $\{r_1, r_2, \ldots, r_k\}$ with $r_1 > r_2 > \ldots > r_k$, and $k \leqslant \#\mathcal{S}$. An artificial value $r_{k+1} = 0$ is also introduced to simplify the formulae.

The discrete Choquet integral of the reward function with respect to the necessity measure $\mathcal{N}$ is defined, and then simplified, as follows:

$$Ch(r, \mathcal{N}) = \sum_{i=1}^{k} (r_i - r_{i+1}) \cdot \mathcal{N}(\{r(s) \geqslant r_i\}) = \sum_{i=1}^{\#\mathcal{L}-1} (l_i - l_{i+1}) \cdot \min_{\substack{s \in \mathcal{S} \text{ s.t.} \\ \pi(s) \geqslant l_i}} r(s). \quad (9)$$

More on possibilistic Choquet integrals can be found in [4]. This reward aggregation using the necessity measure leads to a pessimistic estimation of the reward: as an example, the reward $\min_{s \in \mathcal{S}} r(s, a)$ is assigned in case of total ignorance. Note that, if the necessity measure $\mathcal{N}$ is replaced by a probability measure $\mathbb{P}$, Choquet integral coincides with the expected reward based on $\mathbb{P}$.

3.4 MDP with Epistemic States

This section summarizes the complete translation using the main equations of the previous sections. This translation takes for input a POMDP: $\langle \mathcal{S}, \mathcal{A}, T, \Omega, O, r, b_0 \rangle$ and returns an epistemic state-based MDP: $\langle \tilde{\mathcal{S}}, \mathcal{A}, \tilde{T}, \tilde{r} \rangle$. The state space is $\tilde{\mathcal{S}} = \Pi_{\mathcal{S}}$. The (approximate) transition functions are $\tilde{T}$, such that $\forall (b^\pi, \tilde{b}^\pi) \in \Pi_{\mathcal{S}}^2$, $\forall a \in \mathcal{A}$, $\tilde{T}(b^\pi, a, (b^\pi)') = \mathbf{p}((b^\pi)' \mid b^\pi, a)$ defined with Eqs. 7 and 8. The reward of a belief b^π is $\tilde{r}(a, b^\pi) = Ch(r(a, .), \mathcal{N}_{b^\pi})$, defined with Eq. 9 and where $\mathcal{N}_{b^\pi}$ is the necessity measure computed from b^π. Finally, as in the probabilistic framework (see Eq. 1), the criterion of this MDP is the expected total reward: $\mathbb{E}_{(b^\pi) \sim \tilde{T}} \left[\sum_{t=0}^{+\infty} \gamma^t \cdot \tilde{r}(b_t^\pi, d_t) \right]$.

While the resulting state space is finite, only really small POMDP problems can be solved with this translation without computation tricks. Indeed, $\Pi_{\mathcal{S}}$ grows exponentially with the number of system states (see Eq. 4), which makes the problem intractable even for state of the art MDP solvers.

Purely possibilistic counterparts of the (PO)MDPs, called Qualitative Possibilistic (PO)MDPs, have been already defined [14] and efficiently used for planning under uncertainty problems [7]. These π-(PO)MDPs are quite different from the model exposed in this paper. For instance, they do not use quantitative data as probabilities or rewards. Dynamics is described in a purely qualitative possibilistic way. Frequentist information about the problem cannot be encoded: these frameworks are indeed dedicated to situations where the probabilistic dynamic of the studied system is lacking. Moreover, possible values of the reward function are chosen among the degrees of the qualitative possibilistic scale. A commensurability assumption between reward and possibility degrees, i.e. a meaning of why they share the same scale, is needed to use the criteria proposed in these frameworks. Our model bypass these demands: a real number is assigned to each possibilistic belief (epistemic state), using the Choquet integral, instead of a qualitative utility degree: it represents the reward got by the agent when reaching this belief (in an MDP fashion) as detailed in Sect. 3.3. Moreover, the dynamics of our process is described with probability distributions: approximate probabilistic transition functions between current and next beliefs, or epistemic states, are given in Sect. 3.2. Finally, our model can be solved by any MDP solver in practice: it eventually becomes a classical probabilistic fully observable MDP whose state space is the finite set Π_S. Here, the term hybrid is used because only the beliefs about the current state are defined as possibility distributions, and all variables keep a probabilistic dynamics: the agent reasons based on a possibilistic analysis of the system state (the possibilistic belief, or epistemic state), and transition probability distributions are defined between these epistemic states.

4   Benefit from Factorization

This section carefully derives a tractable MDP problem from a factored POMDP [2]: some factorization and computational tricks are described here to reduce its size and to make it factorized. First, the definition of a factored POMDP is briefly outlined, followed by some notations about variable dependences helpful for describing how distributions are dealt with. Next, a classification of the state variables is made to strongly adapt computations according to the nature of the system state. The way possibility distributions are defined is presented, and the description of the use of the possibilistic Bayes' rule in practice ends this section.

4.1   Factored POMDPs

Partially Observable Markov Decision Processes can be defined in a factorized way. The state space is described with Boolean variables of the set $\mathbb{S} = \{s_1, \ldots, s_m\}$: $\mathcal{S} = s_1 \times \ldots \times s_m$. The notation $\mathbb{S}' = \{s'_1, \ldots, s'_m\}$ is also used. The set of Boolean observation variables $\mathbb{O} = \{o_1, \ldots, o_n\}$ describes also the observation space $\Omega = o_1 \times \ldots \times o_n$. For simplicity, and as state $s \in \mathcal{S}$ and observation $o \in \Omega$ notations are no longer reused in this paper, only variables are denoted with these letters from now: $s_j \in \mathbb{S}$ and $o_i \in \mathbb{O}$.

The factorized description continues defining, $\forall j \in \{1, \ldots, m\}$ and $\forall a \in \mathcal{A}$, a transition function $T_j^a(\mathbb{S}, s'_j) = \mathbf{p}\left(s'_j \mid \mathbb{S}, a\right)$, about the state variable s'_j. One observation function is also given for each observation variable: $O_i^a(\mathbb{S}', o'_i) = \mathbf{p}\left(o'_i \mid \mathbb{S}', a\right)$, $\forall i \in \{1, \ldots, n\}$ and $\forall a \in \mathcal{A}$. It is here understood that $\mathbb{S}'$ are independent conditional on $\mathbb{S}$ and the action a, and that $\{o_i'\}_{i=1}^n$ are independent conditional on $\mathbb{S}'$ and a.

4.2   Notations and Observation Functions

Transitions of the final MDP make it more handy if each variable depends on only few previous variables: the procedure to exploit such simplifications brought by the structure of the initial POMDP during the translation, needs the following notations. In practice, for each $i \in \{1, \ldots, n\}$ not all state variables influence observation variable o_i'; similarly, for each $j \in \{1, \ldots, m\}$, not all current state variables influence the next state variable s_j': observation variable o_i' depends on some state variables which are called *parents* of o_i' as they appears as "parents nodes" in a Bayesian network illustrating dependencies of the process, and denoted by $\mathcal{P}(o_i') = \{s_j' \in \mathbb{S}'$ s.t. o_i' depends on $s_j'\}$. Likewise, probability distributions of the next state variable s_j' depend on some current state variables, denoted by $\mathcal{P}(s_j') = \{s_k \in \mathbb{S}$ s.t. s_j' depends on $s_k\}$. It leads to the following rewriting of probability distributions: $T_j^a(\mathcal{P}(s_j'), s_j') = \mathbf{p}\left(s_j' \mid \mathcal{P}(s_j'), a\right)$ and $O_i^a(\mathcal{P}(o_j'), o_i') = \mathbf{p}\left(o_i' \mid \mathcal{P}(o_j'), a\right)$. Finally, the following subset of $\mathbb{S}$ is useful to specify observation dynamics: $\mathcal{Q}(o_i') = \{s_k \in \mathbb{S}$ s.t. $\exists s_j' \in \mathcal{P}(o_i')$ s.t. $s_k \in \mathcal{P}(s_j')\} = \cup_{s_j' \in \mathcal{P}(o_i')} \mathcal{P}(s_j') \subseteq \mathbb{S}$. Probability distributions of variables $\mathcal{P}(o_i')$ also benefit from previous rewritings: thanks to state variables independences, $\forall i = 1, \ldots, n$,

$$\mathbf{p}\left(\mathcal{P}(o_i') \mid \mathbb{S}, a\right) = \prod_{s_j' \in \mathcal{P}(o_i')} T_j^a(\mathcal{P}(s_j'), s_j') = \mathbf{p}\left(\mathcal{P}(o_i') \mid \mathcal{Q}(o_i'), a\right). \qquad (10)$$

The observation probability distributions knowing previous state variables are

$$\forall i = 1, \ldots, n, \quad \mathbf{p}\left(o_i' \mid \mathcal{Q}(o_i'), a\right) = \sum_{v \in 2^{\mathcal{P}(o_i')}} \mathbf{p}\left(o_i' \mid v, a\right) \cdot \mathbf{p}\left(v \mid \mathcal{Q}(o_i'), a\right). \quad (11)$$

Therefore a possibilistic belief defined on $2^{\mathcal{Q}(o_i')}$ is enough to get the approximate probability distribution of an observation variable: such an epistemic state, leads to a probability distribution $\overline{b^\pi}$ over $2^{\mathcal{Q}(o_i')}$ via the pignistic transformation (6). The approximate probability distribution of the ith observation variable, factorized counterpart of Eq. 7, is: $\forall i = 1, \ldots, n$,

$$\mathbf{p}\left(o_i' \mid b^\pi, a\right) = \sum_{v \in 2^{\mathcal{Q}(o_i')}} \mathbf{p}\left(o_i' \mid v, a\right) \cdot \overline{b^\pi}(v). \qquad (12)$$

4.3   State Variable Classification

State variables $s \in \mathbb{S}$ do not play the same role in the process: as already studied in the literature [11], some variables can be visible for the agent, and this *mixed-observability* leads to important computational simplifications. Moreover, some variables do not affect observation variables, and this structure reduces the final MDP complexity.

- A state variable s_j is said to be **visible**, if $\exists o_i \in \mathbb{O}$, observation variable, such that $\mathcal{P}(o_i') = \{s_j'\}$ and $\forall a \in \mathcal{A}$, $\mathbf{p}\left(o_i' \mid s_j', a\right) = \mathbb{1}_{\{o_i'=s_j'\}}$ *i.e.* if $o_i' = s_j'$ surely. The set of visible state variables is denoted by $\mathbb{S}_v = \{s_{v,1}, s_{v,2}, \ldots, s_{v,m_v}\}$. The observation variables corresponding to the visible state variables can be removed from the set of observation variables: the number of observation variables becomes $\tilde{n} = n - m_v$.
- **Inferred hidden variables** are simply $\cup_{i=1}^{\tilde{n}} \mathcal{P}(o_i')$, *i.e.* all hidden variables influencing the (remaining) observation variables. The set of inferred hidden variables is $\mathbb{S}_h = \{s_{h,1}, s_{h,2}, \ldots, s_{h,m_h}\}$ and it contains possibly visible variables.
- **Non-inferred hidden variables** or **fully hidden variables**, denoted by $\mathbb{S}_f$, consists of hidden state variables which do not influence any observation, *i.e.* all the remaining state variables. The fully hidden variables are denoted by $s_{f,1}, s_{f,2}, \ldots, s_{f,m_f}$, and the corresponding set is $\mathbb{S}_f$.

The classification allows to avoid some computations for visible variables: if $s_v \in \mathbb{S}_v$, and o_v is the associated observation, computations of the distribution over $\mathcal{P}(o_v')$, Eq. 10, and of the distribution over o_v', Eq. 11, are unnecessary: the distribution over s_v' ($= o_v'$) needed is simply given by $T^a\left(\mathcal{P}(s_v'), s_v'\right)$. The counterpart of Eq. 12 is then simply

$$\mathbf{p}\left(s_v' \mid b^\pi, a\right) = \sum_{2^{\mathcal{P}(s_v')}} T^a\left(\mathcal{P}(s_v'), s_v'\right) \cdot \overline{b^\pi}(\mathcal{P}(s_v')) \tag{13}$$

where $\overline{b^\pi}$ is the probability distribution over $2^{\mathcal{P}(s_v')}$ extracted from the possibilistic belief over the same space, using pignistic transformation (6).

4.4   Beliefs Process Definition and Handling

This section is meant to define marginal beliefs instead of a global one, in order to benefit from the structure of the initial POMDP. Possibilistic belief distributions have different definitions according to which class of state variables they concern.

As visible state variables are directly observed, there is no uncertainty over these variables. Two epistemic states (possibilistic belief distribution) are possible for visible state variable $s_{v,j}'$: $b_{v,T}'(s_{v,j}') = \mathbb{1}_{\{s_{v,j}'=\top\}}$ and $b_{v,F}'(s_{v,j}') = \mathbb{1}_{\{s_{v,j}'=\bot\}}$. As a consequence, one Boolean variable $\beta_{v,j}' \in \{\top, \bot\}$ per visible state variables is enough to represent this belief distribution in practice: if $s_{v,j}' = \top$, then the next belief is $b' = b_{v,T}'$ represented by the belief variable assignment $\beta_{v,j}' = \top$, otherwise the next belief is $b' = b_{v,F}'$, and $\beta_{v,j}' = \bot$.

For each $i \in 1, \ldots, \tilde{n}$, each inferred hidden variable constituting $\mathcal{P}(o_i')$ is an input of the same possibilistic belief distribution: the non-normalized belief is

$$\forall i = 1, \ldots, \tilde{n}, \quad \tilde{b}'(\mathcal{P}(o_i')) = \max_{v \in 2^{\mathcal{Q}(o_i')}} \min\left\{\pi\left(o_i', \mathcal{P}(o_i') \mid v, a\right), b(v)\right\}, \tag{14}$$

where the joint possibility distributions over $o_i' \times \mathcal{P}(o_i')$ are $\pi\left(o_i', \mathcal{P}(o_i') \mid \mathcal{Q}(o_i'), a\right)$
$= \min\left\{\pi\left(o_i' \mid \mathcal{P}(o_i'), a\right), \min_{s_j' \in \mathcal{P}(o_i')} \pi\left(s_j' \mid \mathcal{P}(s_j'), a\right)\right\}$. The   possibilistic

normalization, $\forall w \in 2^{\mathcal{P}(o'_i)}$, $b'(w) = \begin{cases} 1 \text{ if } w \in \text{argmax}_{v \in 2^{\mathcal{P}(o'_i)}} \tilde{b}'(v); \\ \tilde{b}'(w) \text{ otherwise.} \end{cases}$ finalizes

this rewriting of the belief update (5). In practice, if $l = \#\mathcal{L}$, and $p_i = \#\mathcal{P}(o'_i)$, the number of belief states is $l^{2^{p_i}} - (l-1)^{2^{p_i}}$, and then the number of belief variables is $n_{h,i} = \lceil \log_2(l^{2^{p_i}} - (l-1)^{2^{p_i}}) \rceil$. A belief variable of an inferred hidden state variable is denoted by β_h. As well, for each $j \in 1, \dots, m_f$, the non-normalized belief defined on fully hidden variable $s_{f,j}$ is

$$\tilde{b}'(s'_{f,j}) = \max_{v \in 2^{\mathcal{P}(s'_{f,j})}} \min \{ \pi (s'_{f,j} \mid v,a), b(v) \}, \tag{15}$$

which leads to the actual new belief b' after the possibilistic normalization. As each fully hidden variable is considered independently from the others, the number of belief variables is $n_f = \lceil \log_2(l^2 - (l-1)^2) \rceil = \lceil \log_2(2l-1) \rceil$. A belief variable of a fully hidden state variable is denoted by β_f.

Finally the actual global epistemic state $b'(\mathbb{S}')$ is upper bounded by the less informative belief $(b^\pi)'(\mathbb{S}') = \min \left\{ \min_{j=1}^{m_v} b'(s'_{v,j}), \min_{i=1}^{\tilde{n}} b'(\mathcal{P}(o'_i)), \min_{k=1}^{m_f} b'(s'_{f,k}) \right\}$. The latter is considered as the agent belief to make the final MDP factorized. Note that the belief over the inferred hidden variables (14) and the distribution over observation variables (12), need a belief distribution over $\mathcal{Q}(o'_i) \subseteq \mathbb{S}$. Likewise, the belief over the fully hidden state variables (15) needs a belief distribution over variables $\mathcal{P}(s'_{f,j}) \subseteq \mathbb{S}$. Moreover, approximate probability distributions over visible state variables (13) need a belief distribution over $\mathcal{P}(s'_{v,i}) \subseteq \mathbb{S}$. These beliefs can be computed by marginalizing $b^\pi(\mathbb{S})$ using the max operator.

5   Solving a POMDP with a Discrete MDP Solver

A practical version of the factored MDP achieved in the previous section is described here. A concrete POMDP problem and the resulting MDP illustrate then the state space size reduction of our detailed possibilistic translation.

5.1   Resulting Factored MDP

A belief update depends only on the next observation (see Eq. 5): the transition of a belief is then deterministic conditional on the next observation. A simple trick is used to keep this determinism in the final MDP: a *flipflop* Boolean variable is introduced, changing its state at each step, denoted by f. It artificially divides a classical time step of the POMDP into two phases. During the first phase, called *the observation generation phase*, non-identity transition functions are the probability distributions over observation variables (12) and visible state variables (13). During the second phase, called *the belief update phase*, non-identity transition functions are the deterministic transitions of the belief variables: variables β_v are updated knowing the value of the corresponding visible variable s_v; variables $\beta_h^1, \dots, \beta_h^{n_{h,i}}$ are updated knowing the value of the observation variables o_i, and using update (14); finally, variables $\beta_f^1, \dots, \beta_f^{n_f}$ are updated using update

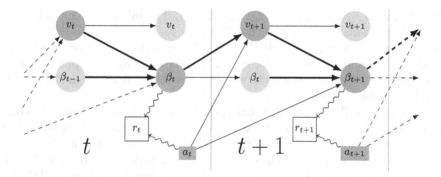

Fig. 1. ID of the resulting MDP: thickest arrows are non-identity transitions.

(15). The state space is then defined as: $\mathcal{S} = f \times s_v^1 \times \ldots \times s_v^{m_v} \times o^1 \times \ldots \times$ $o^{\tilde{n}} \times \beta_v^1 \times \ldots \times \beta_v^{m_v} \times \beta_h^1 \times \ldots \times \beta_h^{\tilde{n}} \times \beta_f^1 \times \ldots \times \beta_f^{m_f}$, where $\forall i = 1, \ldots, \tilde{n}$, β_h^i represents Boolean variables $\beta_h^{1,i}, \ldots, \beta_h^{n_{h,i},i}$, and $\forall k = 1, \ldots, m_f$, β_f^j represents Boolean variables $\beta_f^{1,j}, \ldots, \beta_f^{n_f,j}$. Figure 1 is the Influence Diagram (ID) of the resulting MDP where β_t represents all belief variables, and v_t the visible variables: the flipflop variable f, observations and visible state variables. The resulting MDP is a factored MDP thanks to the flipflop trick.

5.2   For a Concrete POMDP

A problem inspired by the RockSample problem [18] is described in this section to illustrate the factorized possibilistic discretization of the agent belief, from a factored POMDP: a rover is navigating in a place described by a finite number of locations $l_1, \ldots, l_n$, and where m rocks stand. Some of these m rocks are of interest in the scientific mission of the rover, and it has to sample them. However, sampling a rock is an expensive operation. The rover is thus equipped with a long range sensor making it capable of estimating if the rock has to be sampled. Finally operating time of the rover is limited, but its battery level is available.

Variables of this problem can now be set, and classified as in Sect. 4.3: as the battery level is directly observable by the agent (the rover), the set of visible state variables consists of the Boolean variables encoding it: $\mathbb{S}_v = \{B_1, B_2, \ldots, B_k\}$. The agent knows the different locations of the rocks, however the nature of each rock is to be estimated. The set of inferred hidden state variables consists of m Boolean variables R_i encoding the nature of the ith rock, $\top$ for "scientifically good" and $\bot$ otherwise: $\mathbb{S}_h = \{R_1, R_2, \ldots, R_m\}$. When the ith rock is observed using the sensor, it returns a noisy observation of the rock in $\{\top, \bot\}$, modeled by the Boolean variable O_i: the set of observation variables is then $\mathbb{O} = \{O_1, O_2, \ldots, O_m\}$. Finally, no localization equipment is provided: the agent estimates its location from its initial information, and its actions. Each location of the rover is formally described by a variable L_j, which equals $\top$ if the rover is at the jth location, and $\bot$ otherwise. The set of fully hidden variables consists thus of these n variables: $\mathbb{S}_f = \{L_1, L_2, \ldots, L_n\}$. The initial location is known, leading to a deterministic initial belief: $b_0^\pi(\mathbb{S}_h) = 1$ if $L_1 = \top$ and $L_j = \bot \; \forall j \neq 1$, and 0 otherwise. However the initial nature of each rock is not known. Instead of a uniform probability distribution, Possibility Theory allows to represent initial ignorance about the nature of rocks with the possibility distribution $b_0^\pi(\mathbb{S}_h) = 1$, for each variable assignment.

Classical POMDP solvers are based on probabilistic beliefs over the state space defined by $\mathbb{S}_h$, $\mathbb{S}_f$ and even $\mathbb{S}_v$ if Mixed-Observability [11] is not taken into account. The approach presented in this paper leads to an MDP with a finite space of epistemic states. Finally, the factorization tricks lead to a reduction of the state space size: with a flat translation of this POMDP, $\lceil \log_2(\#\mathcal{L}^{2^{n+m+k}} - (\#\mathcal{L} - 1)^{2^{n+m+k}}) \rceil$ Boolean variables are necessary. Taking advantage of the POMDP structure, the resulting state space is encoded with $1 + 2k + m + (m+n)\lceil \log_2(2\#\mathcal{L} - 1) \rceil$ Boolean variables: the flipflop variable, the visible variables and associated beliefs variables, the observation variables, and the belief variables associated to the fully hidden and inferred hidden variables. Moreover, the dynamics of the resulting MDP is factored, and lots of transitions are deterministic, thanks to the flipflop variable trick. These simplifying structures are beneficial to the MDP solvers, leading to faster computations.

6   Conclusion

This paper describes a hybrid translation of a POMDP into a finite state space MDP one. Qualitative Possibility Theory is used to maintain an epistemic state during the process: the belief space has a granulated representation, instead of a continuous one as in the classical translation. The resulting MDP is entirely defined computing transition and reward functions over these epistemic states. Definitions of these functions use respectively the pignistic transformation, used to recover a probability distribution from an epistemic state, and the Choquet integral with respect to the necessity, making the agent pessimistic about its ignorance. A practical way to implement this translation is then described: with these computation tricks, a factored POMDP leads to a factored and tractable MDP. This promising approach will be tested on the POMDPs of the IPPC competition [15] in a future work: the provided problem descriptions are indeed in the form of the factored POMDPs introduced in Sect. 4.

References

1. Bellman, R.: A Markovian decision process. Indiana Univ. Math. J. **6**, 679–684 (1957)
2. Boutilier, C., Poole, D.: Computing optimal policies for partially observable decision processes using compact representations. In: Proceedings of the 13th National Conference on Artificial Intelligence, AAAI 1996, Portland, Oregon, vol. 2, pp. 1168–1175 (1996). http://www.aaai.org/Library/AAAI/1996/aaai96-173.php

3. Cassandra, A., Littman, M.L., Zhang, N.L.: Incremental pruning: a simple, fast, exact method for partially observable Markov decision processes. In: Proceedings of the 13th Conference on Uncertainty in Artificial Intelligence, pp. 54–61. Morgan Kaufmann Publishers (1997)
4. De Cooman, G.: Integration and conditioning in numerical possibility theory. Ann. Math. Artif. Intell. **32**(1–4), 87–123 (2001)
5. De Finetti, B.: Theory of Probability: A Critical Introductory Treatment. Wiley Series in Probability and Mathematical Statistics. Wiley, New York (1974)
6. Drougard, N., Teichteil-Königsbuch, F., Farges, J.L., Dubois, D.: Qualitative possibilistic mixed-observable MDPs. In: Proceedings of 29th Conference on Uncertainty in Artificial Intelligence, UAI 2013, pp. 192–201. AUAI Press, Oregon (2013)
7. Drougard, N., Teichteil-Königsbuch, F., Farges, J., Dubois, D.: Structured possibilistic planning using decision diagrams. In: Proceedings of 28th AAAI Conference on Artificial Intelligence, Québec City, Canada, pp. 2257–2263 (2014). http://www.aaai.org/ocs/index.php/AAAI/AAAI14/paper/view/8553
8. Dubois, D.: Possibility theory and statistical reasoning. Comput. Stat. Data Anal. **51**, 47–69 (2006)
9. Dubois, D., Prade, H.: The logical view of conditioning and its application to possibility and evidence theories. Int. J. Approximate Reasoning **4**(1), 23–46 (1990). http://www.sciencedirect.com/science/article/pii/0888613X9090007O
10. Dubois, D., Prade, H., Sandri, S.: On possibility/probability transformations. In: Proceedings of the 4th IFSA Conference, pp. 103–112. Kluwer Academic Publiction (1993)
11. Ong, S., Png, S., Hsu, D., Lee, W.: Planning under uncertainty for robotic tasks with mixed observability. Int. J. Rob. Res. **29**(8), 1053–1068 (2010)
12. Papadimitriou, C., Tsitsiklis, J.N.: The complexity of Markov decision processes. Math. Oper. Res. **12**(3), 441–450 (1987)
13. Puterman, M.L.: Markov Decision Processes: Discrete Stochastic Dynamic Programming, 1st edn. Wiley, New York (1994)
14. Sabbadin, R.: A possibilistic model for qualitative sequential decision problems under uncertainty in partially observable environments. In: Proceedings of the 15th Conference on Uncertainty in Artificial Intelligence, UAI 1999. Morgan Kaufmann Publishers Inc., San Francisco (1999)
15. Sanner, S.: Probabilistic track of the 2011 international planning competition (2011). http://users.cecs.anu.edu.au/~ssanner/IPPC_2011
16. Silver, D., Veness, J.: Monte-carlo planning in large POMDPs. In: Advances in Neural Information Processing Systems, Vancouver, Canada, vol. 23, pp. 2164–2172 (2010)
17. Smallwood, R.D., Sondik, E.J.: The optimal control of partially observable Markov processes over a finite horizon, vol. 21. INFORMS (1973)
18. Smith, T., Simmons, R.: Heuristic search value iteration for POMDPs. In: Proceedings of the 20th Conference on Uncertainty in Artificial Intelligence, UAI 2004, pp. 520–527. AUAI Press, Arlington (2004)
19. Zadeh, L.A.: Some reflections on soft computing, granular computing and their roles in the conception, design and utilization of information/intelligent systems. Soft Comput. **2**(1), 23–25 (1998)

Propagation of Belief Functions in Singly-Connected Hybrid Directed Evidential Networks

Wafa Laâmari[⊠] and Boutheina Ben Yaghlane

LARODEC Laboratory - Institut Supérieur de Gestion de Tunis, Tunis, Tunisia
wafa.lamari@gmail.com

Abstract. Directed evidential networks (DEVNs) can be seen, at present, as an extremely powerful graphical tool for representing and reasoning with uncertain knowledge in the framework of evidence theory.

The main purpose of this paper is twofold. Firstly, it introduces hybrid directed evidential networks which generalize the standard DEVNs. Secondly, it presents an algorithm for performing inference over singly-connected hybrid evidential networks.

1 Introduction

Over the few last decades, many different theories of uncertainty management have been introduced, such as probability theory [5], fuzzy set theory [11], and Dempster-Shafer belief function theory [6]. The belief function theory has proven to be prevalent and suitable for managing uncertainties in many domains. The invention of evidential networks has represented an important stepping stone in the development of approaches for knowledge representation and reasoning under the theory of belief functions.

Several evidential networks have been reported in the literature [1,7,10]. Xu and Smets have presented Evidential Networks with Conditional belief functions (ENCs) [10]. Unlike conditional probabilities which are specified per child node in probabilistic graphical models, conditional belief functions encoding the independence relations among the variables in these evidential networks are specified per edge[1]. In spite of their remarkable power for representation of uncertain knowledge and evidential inference, ENCs have many inherent limitations. One of their limitations is the fact that the algorithm used for the propagation of belief functions in these networks is limited to graphs having only binary relations between the variables.

Directed EVidential Networks with conditional belief functions (DEVNs) have been proposed as an alternative framework to ENCs [1]. One of the most

---

[1] In Bayesian networks, conditional probabilities are specified per child node, i.e. for each node given all its parents. However, in ENCs, if there are two edges going from nodes A and B to their common child node C, then two conditional belief function distributions have to be defined: a distribution for C given A and another one for C conditionally to B.

© Springer International Publishing Switzerland 2015
C. Beierle and A. Dekhtyar (Eds.): SUM 2015, LNAI 9310, pp. 234–248, 2015.
DOI: 10.1007/978-3-319-23540-0_16

powerful features of DEVNs is their ability to model not only binary relations between variables but also n-ary relations (i.e. relations for any number of nodes). Compared to ENCs, DEVNs are more flexible for the representation of conditional relations. In fact, DEVNs can be weighted by conditional belief functions specified either per edge (i.e. per single parent), like in ENCs, or per child node (i.e. for all parents) like in Bayesian networks. Despite the great flexibility of DEVNs in modeling conditional relations between the variables, algorithms for belief propagation in singly-connected DEVNs were limited to networks with conditionals specified per edge [2]. To fill this gap, we presented, in a recent paper, a new algorithm for reasoning in singly-connected DEVNs where conditional beliefs are specified per child node [4].

Quantifying a DEVN amounts to assessing belief function distributions for each of the network's variables conditional on their direct predecessors in the directed graph. In many domains, information is available to this end from domain experts. In some cases, an expert may be able to provide a conditional belief function distribution for a variable given all its parents. But in some other cases, he may feel confident providing estimates of belief functions for a variable given each of its parents separately. For instance, a medical domain's expert may easily report the belief function distribution of a particular disease given all its symptoms. If we consider also a situation in which late trains and broken alarm clocks are valid excuses or causes for being late, an expert may easily estimate the belief function of being late knowing that the alarm clock did not go off and that trains were delayed. Now, if we consider an other situation in which anaemia can cause feeling cold. Knowing that being in Russia in winter can also be a cause of feeling cold, an expert may be more confident to provide two belief function distributions for feeling cold given each of its causes separately (i.e. feeling cold given anaemia and feeling cold given being in Russia in winter) than one belief function distribution of feeling cold given the two causes together. Since the choice of the quantification manner made by an expert may vary depending on the variable and its parents, and assuming that the expert have to use the same manner for specifying all the conditional distributions in a DEVN (i.e. either all the distributions are specified given all the parents (per child node) or all of them are defined per single parent (per edge)), an evidential model that would allow combining the two manners for quantifying the network has been lacking so far.

With the purpose of coupling conditionals specified per edge with conditionals specified per child node in the same network, we introduce hybrid DEVNs (HDEVNs). The motivation to develop HDEVNs stems from the desire to provide more freedom and flexibility. This flexibility regards the quantification of the evidential graphical models with numerical belief functions: HDEVNs can cope with both conditionals specified per one parent and conditionals specified for all parents.

The remainder of the paper is organized as follows. Section 2 reminds some formal background on belief function theory. In Sect. 3, we recall the DEVNs. Section 4 first introduces the proposed HDEVNs for dealing with conditionals

specified per edge and conditionals specified per child node and then presents
our algorithm for the belief propagation in singly-connected HDEVNs. Section 5
illustrates the HDEVN and the proposed algorithm for inference over it.

2   Belief Function Theory: Theoretical Background

The belief function theory is a rich and flexible framework for handling incomplete and uncertain knowledge [3,6,9]. In this section, we briefly introduce the
theory of belief functions with an emphasis on its basic concepts.

2.1   Basics of Belief Function Theory

In belief function theory, the *frame of discernment* of a variable N_i denoted by
Θ_{N_i}, is defined to be a finite non empty set of all its possible elementary values.
These elementary values are exhaustive and mutually exclusive.

Each subset of Θ_{N_i} belongs to its *power set* which is denoted by $2^{\Theta_{N_i}}$ and
formally defined as: $2^{\Theta_{N_i}} = \{S : S \subseteq \Theta_{N_i}\}$.

A *basic belief assignment* (bba), referred also to as a *mass function* is a
mapping $m^{N_i} \colon 2^{\Theta_{N_i}} \to [0, 1]$ verifying:

$$\sum_{S \subseteq \Theta_{N_i}} m^{N_i}(S) = 1 \tag{1}$$

where $m^{N_i}(S)$, called *basic belief mass*, is considered to be the part of belief that
supports exactly the proposition S without supporting any strict subset of S.

Total ignorance is represented by the mass function $m^{N_i}(\Theta_{N_i}) = 1$ which is
called a *vacuous belief*.

A *plausibility function* associated with a bba m^{N_i} assigns to every subset S of
Θ_{N_i} the sum of basic belief masses of the subsets Q of Θ_{N_i} which are compatible
with S. Given a bba m^{N_i}, the plausibility function, denoted by pl^{N_i}, is defined,
for $S \subseteq \Theta_{N_i}$, as follows:

$$pl^{N_i}(S) = \sum_{Q \cap S \neq \emptyset} m^{N_i}(Q) \tag{2}$$

The bba m^{N_i} can be recovered from pl^{N_i} as follows:

$$m^{N_i}(S) = \sum_{Q \subseteq S} (-1)^{|S-Q|}(1 - pl^{N_i}(\neg Q)) \tag{3}$$

2.2   Operations on Belief Functions

We consider two variables X and Y associated with the frames of discernment
Θ_X and Θ_Y, respectively, and two finite sets of variables A and B associated with
the frames Θ_A and Θ_B, respectively. The frames Θ_A and Θ_B are the Cartesian
products of the frames associated with the variables the sets A and B include,
respectively.

A bba m^{AB} defined on the product space $\Theta_{AB} = \Theta_A \times \Theta_B$ can be marginalized on Θ_A by transferring each mass $m^{AB}(S)$ for $S \subseteq \Theta_{AB}$ to its projection on Θ_A. The *marginalization* of m^{AB} on Θ_A is defined as follows:

$$m^{AB\downarrow A}(S') = m^A(S') = \sum_{S \subseteq (\Theta_{AB}), S^{\downarrow A} = S'} m^{AB}(S) \qquad (4)$$

where $S^{\downarrow A}$ represents the projection of $S \subseteq \Theta_{AB}$ to Θ_A.

In general, it is not possible to retrieve the original bba m^{AB} from its marginal $m^{AB\downarrow A}$ on Θ_A. However, the least informative bba, such that its projection on Θ_A is $m^{AB\downarrow A}$ can be computed. This defines the *vacuous extension* of m^A to the product space Θ_{AB} which is computed as follows:

$$m^{A\uparrow AB}(S') = m^{AB}(S') = \begin{cases} m^A(S) \text{ if } S' = S \times \Theta_B, S \subseteq \Theta_A \\ 0 \qquad \text{otherwise} \end{cases} \qquad (5)$$

Two bba's m^A and m^B defined on the spaces Θ_A and Θ_B, respectively, can be combined to produce a single bba m^{AB} defined $\forall S \subseteq \Theta_{AB}$, as follows:

$$(m^A \otimes m^B)(S) = m^{AB}(S) = \sum_{S_1 \cap S_2 = S} m^{A\uparrow AB}(S_1) \times m^{B\uparrow AB}(S_2) \qquad (6)$$

where both $m^{A\uparrow AB}$ and $m^{B\uparrow AB}$ are computed using the Eq. (5).

Now, let us consider a set of conditional plausibility functions $\{pl^X[y'](x) : y' \in \Theta_Y, x \subseteq \Theta_X\}$ which quantifies the plausibility of a subset x of Θ_X when we know which element y' of Θ_Y holds. The *Disjunctive Rule of Combination* (DRC) has been derived by Smets [8] to build the plausibility function $pl^X[y](x)$ for any $x \subseteq \Theta_X$ conditionally to any subset $y \subseteq \Theta_Y$ as follows:

$$pl^X[y](x) = 1 - \prod_{y' \in y}(1 - pl^X[y'](x)) \qquad (7)$$

Smets has also derived the *Generalized Bayesian Theorem* (GBT) as a dual function of the DRC to build the conditional plausibility function $pl^Y[x](y)$ for any subset y of Θ_Y given any subset $x \subseteq \Theta_X$ as follows:

$$pl^Y[x](y) = pl^X[y](x) = 1 - \prod_{y' \in y}(1 - pl^X[y'](x)) \qquad (8)$$

3 Directed Evidential Networks with Conditional Belief Functions

A directed evidential network with conditional belief functions (DEVN) is defined by a directed acyclic graph (DAG) $G = (N, E)$ where $N = \{N_1, \ldots, N_n\}$ denotes a finite set of nodes and $E = \{E_1, \ldots, E_y\}$ denotes a set of edges. Each node N_i in G represents a random variable and takes its values on a frame of discernment Θ_{N_i}. Each root node N_i in G is associated with an a priori mass function m^{N_i} satisfying the axiom given by the Eq. (1). Local conditional distributions can be defined in DEVNs in two different manners:

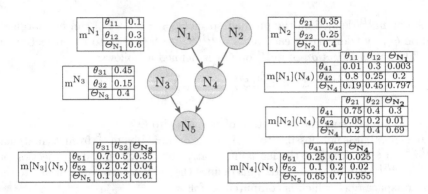

Fig. 1. A DEVN with conditionals per edge

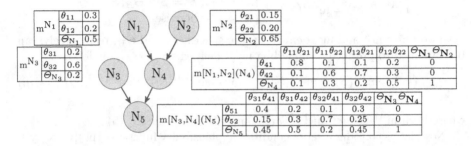

Fig. 2. A DEVN with conditionals per child node

(i) **Per edge:** each edge going from a node N_k to a node N_i in G is associated with a conditional mass function $m[N_k](N_i)$[2] over N_i given N_k. By adopting this manner for specifying the conditionals, we get a DEVN weighted with conditionals defined **per single parent** and close to conditionals in ENCs. A DEVN with conditional distributions defined per edge is illustrated in Fig. 1. Each variable N_i $(i = 1, \ldots, 5)$ takes its values on the frame Θ_{N_i} $= \{\theta_{i1}, \theta_{i2}\}$.

(ii) **Per child node:** each child node N_i is associated with a conditional mass function $m[Pa(N_i)](N_i)$ over N_i given all its parent nodes $Pa(N_i)$. When we adopt this manner for specifying conditionals, we get a DEVN weighted with conditionals defined per child node **in the context of all the parents**. Figure 2 shows a DEVN with conditionals specified per child node.

3.1 Reasoning in Singly-Connected DEVNs with Conditional Belief Functions Defined per Edge

The DRC and the GBT, proposed by Smets for dealing with conditionals specified for a variable conditionally to another one, provide the necessary tools

---

[2] The notations $m[N_k](N_i)$ and $m^{N_i}[N_k]$ used throughout this paper correspond to the classical notation $m(N_i|N_k)$.

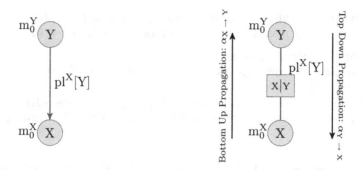

Fig. 3. DRC and GBT for belief propagation in DEVNs with conditionals specified per edge

for belief propagation in singly-connected evidential networks with conditionals specified per edge. A simple DEVN **D**, shown on the left side of Fig. 3, illustrates the application of the DRC and the GBT for reasoning in these evidential networks. **D** consists of two nodes: a parent node Y and a child node X which are associated with the a priori bba's m_0^Y and m_0^X, defined over Θ_Y and Θ_X, respectively. The edge (Y, X) is weighted by a set of conditional plausibility functions $pl^X[y'](x)$ defined for $x \subseteq \Theta_X$ conditionally to $y' \in \Theta_Y$.

The DRC can be applied for top down propagation to compute the message $\alpha_{Y \to X}$ that the parent node Y sends to its child node X for each $x \subseteq \Theta_X$. $\alpha_{Y \to X}$ is defined as follows [8]:

$$\alpha_{Y \to X} = pl^X(x) = \sum_{y \subseteq \Theta_Y} m_0^Y(y) \times pl^X[y](x) \tag{9}$$

where $pl^X[y](x)$ is given by Eq. (7).

Similarly, the GBT can be used for bottom up propagation to compute the message $\alpha_{X \to Y}$ that the parent node Y receives from its child node X for any $y \subseteq \Theta_Y$. $\alpha_{X \to Y}$ is defined as follows:

$$\alpha_{X \to Y} = pl^Y(y) = \sum_{x \subseteq \Theta_X} m_0^X(x) \times pl^Y[x](y) \tag{10}$$

where $pl^Y[x](y)$ is given by Eq. (8).

The algorithm for inference in singly-connected directed evidential networks with conditionals specified per edge is based on a local propagation down and up the network, using (9) or (10), depending on the direction in which a message is circulated between two neighboring nodes. To illustrate this algorithm, which is detailed in [2], let us consider again the DEVN **D** of Fig. 3. Reasoning with conditionals specified per edge in this network consists in two phases as follows:

(1) **Top Down Propagation**
   (a) Compute the message $pl_{Y \to X}$ that Y sends to X: $pl_{Y \to X} = \alpha_{Y \to X}$ (using (9))

(b) Compute the bba $m_{Y \to X}$ corresponding to $pl_{Y \to X}$ (using (3))
(c) Compute the mass function distribution m^X of X: $m^X = (m_{Y \to X} \otimes m_0^X)$

(2) **Bottom Up Propagation**
  (a) Compute the message $pl_{X \to Y}$ that X sends to Y: $pl_{X \to Y} = \alpha_{X \to Y}$ (using (10))
  (b) Compute the bba $m_{X \to Y}$ corresponding to $pl_{X \to Y}$ (using (3))
  (c) Compute the mass function distribution m^Y of Y: $m^Y = (m_{X \to Y} \otimes m_0^Y)$

3.2 Reasoning in Singly-Connected DEVNs With Conditional Belief Functions Defined per Child Node for All the Parents

In [4], we have explained how it is still possible to use the DRC and the GBT to perform the top down propagation and the bottom up propagation, respectively, in singly-connected DEVNs weighted by conditionals defined for each child node given all its parent nodes. The left side of Fig. 4 shows a singly-connected DEVN **D** with conditionals defined per child node: X is a child node associated with the conditional plausibility function $pl^X[P_1, \ldots, P_n]$, and $Y = \{P_1, \ldots, P_n\}$ is the set of its parent nodes, where each parent node $P_i \in Y$ has an a priori bba $m_0^{P_i}$.

Our propagation algorithm first transforms the initial evidential network with conditionals defined per child node given all the parents **D** into a tree structure **D"** (see the right side of Fig. 4) [4]. The tree is obtained by merging all the parent nodes P_i of each child node X in the DEVN in a single joint node Y. Going through a transformation of the initial network into a tree structure allows to exploit the GBT and the DRC in order to perform the top down propagation and the bottom up propagation. As shown on the right side of Fig. 4, all the parent nodes P_i of X are merged in a single node $Y = \{P_1, \ldots, P_n\}$ in **D"**. All the mass distributions $m_0^{P_i}$ of the single parent nodes of X are extended to the joint space Θ_Y using (4), then they are combined using (6) to produce the mass distribution m_0^Y of Y. The resulting combined mass function m_0^Y is stored in the table associated with the node Y. Once we have computed the distribution of the composed parent node Y in **D"**, we can make the propagation up and down the tree using the DRC and GBT through a conditional node X|Y (i.e. $X|P_1, \ldots, P_n$) linking the child node X with the new parent node Y.

Reasoning over the DEVN **D** of Fig. 4 proceeds in two phases as follows:

(1) **Top Down Propagation**
  (a) Compute for each parent node $P_i \in Y$ the message $m_{P_i \to Y}$ that P_i sends to Y: $m_{P_i \to Y} = m_0^{P_i \uparrow Y}$ (using (5))
  (b) Compute the mass function distribution m_0^Y of Y: $m_0^Y = (m_{P_1 \to Y} \otimes .. \otimes m_{P_n \to Y})$
  (c) Compute the message $pl_{Y \to X}$ that Y sends to X: $pl_{Y \to X} = \alpha_{Y \to X}$ (using (9))
  (d) Compute the bba $m_{Y \to X}$ corresponding to $pl_{Y \to X}$ (using (3))
  (e) Compute the mass function distribution m^X of X: $m^X = (m_{Y \to X} \otimes m_0^X)$

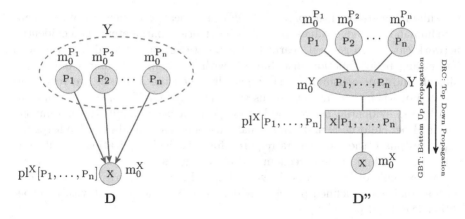

Fig. 4. DRC and GBT for inference in singly-connected DEVN with conditionals specified per child node

(2) **Bottom Up Propagation**
   (a) Compute the message $pl_{X \to Y}$ that X sends to Y: $pl_{X \to Y} = \alpha_{X \to Y}$ (using (10))
   (b) Compute the bba $m_{X \to Y}$ corresponding to $pl_{X \to Y}$ (using (3))
   (c) Compute for each parent node P_i in Y the message $m_{Y \to P_i}$ that Y sends to P_i: $m_{Y \to P_i} = ((\otimes m_{P_j \to Y}) \otimes m_{X \to Y})^{\downarrow P_i}$, where $P_j \in Y - P_i$
   (d) Compute for each parent node P_i in Y the mass function distribution m^{P_i}: $m^{P_i} = m_0^{P_i} \otimes m_{Y \to P_i}$

The construction algorithm of the tree structure D" corresponding to any singly-connected DEVN D with conditionals specified per child node, and the belief propagation algorithm over it can be found in [4].

4 Hybrid Directed Evidential Networks for Handling Conditionals Specified per Edge and Conditionals Specified per Child Node

The following section introduces our hybrid directed evidential network with conditional belief functions (HDEVN) and details how to deal with both conditionals specified per edge and conditionals specified per child node.

4.1 Knowledge Representation in Singly-Connected HDEVNs

A hybrid directed evidential network with conditional belief functions is a DEVN in which some conditionals are defined per edge and some others are specified per child node given all the parents. HDEVNs appear to be better suited to capture human knowledge values: Indeed, the numerical parameters of an evidential network can be, in general, learned from a data set or obtained by an expert. Estimating belief function distributions with the help of human experts is a difficult and time consuming task, especially when the problems are very complicated or when there are numerous variables involved. Thus, the development of a hybrid evidential network able to cope with both conditional parameters specified per child node and conditional parameters specified per edge helps the expert estimate the parameters required for a typical application in a flexible way and gives him greater freedom and more choices.

An example of HDEVN is shown in Fig. 5. In this network N_5 is associated with a conditional defined per child node, while N_6 is associated with two conditionals defined per edge.

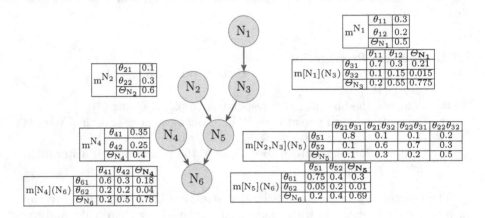

Fig. 5. A hybrid directed evidential network with conditional belief functions

4.2   Reasoning in Singly-Connected HDEVNs

The algorithm for reasoning in singly-connected HDEVNs includes a linkage between the principle of reasoning in singly-connected DEVNs with conditionals specified per edge proposed in [2] and presented in Sect. 3.1 and the one for reasoning in singly-connected DEVNs with conditionals specified per child node presented in Sect. 3.2.

In a singly-connected DEVN with conditionals specified per edge, the belief function propagation algorithm acts directly on the initial graph by applying the DRC and the GBT for inferring beliefs between each child node and each of its parent nodes through the conditional node between them. For coping with conditionals specified given all the parents, we have to go through a transformation of the initial network into a tree structure. This transformation allows to exploit the DRC and the GBT for the propagation up and down through the tree [4].

The inference algorithm that we propose in this paper for reasoning in singly-connected HDEVNs is principally based on an extended version of our tree construction's method proposed in [4]. This extension allows to consider the two kinds of conditionals.

Since in a HDEVN there are child nodes with conditionals specified per single parent and child nodes associated with conditionals defined for all the parent nodes, the corresponding tree is constructed so that at a time a child node N_i in the HDEVN is selected and its corresponding subtree is built depending on the following two cases:

- **Case 1:** if N_i is associated with conditional distributions specified per edge, then the corresponding subtree is obtained by introducing a conditional node between N_i and each of its parent nodes $Pa(N_i)$ (as shown on the right side of Fig. 3). Each conditional node connecting N_i to one of its parent nodes $N_k \in Pa(N_i)$ is weighted by the conditional distribution $m[N_k](N_i)$.
- **Case 2:** if N_i is associated with a conditional belief function specified for all its parents $Pa(N_i)$, then the corresponding subtree is obtained by merging all the parent nodes in a single node, representing all the predecessors of N_i, then by introducing a conditional node connecting N_i to the single parent node (as shown on the right side of Fig. 4). The conditional node is weighted by the conditional distribution $m[Pa(N_i)](N_i)$.

The tree construction's method ends when all the child nodes in the HDEVN are selected. This method is formally presented in the following section.

Construction of the Tree Structure for Reasoning in Singly-Connected HDEVNs: Let $H = (N, E)$ denote a singly-connected HDEVN and let $C \subseteq N$ be the set of child nodes with only one parent in H, $F_{Parents} \subseteq N$ be the set of child nodes having more than one parent and associated with conditionals given all their parents in H and $F_{Edge} \subseteq N$ be the set of child nodes having more than one parent and for which conditionals are specified per edge. $Pa(N_i)$ denotes the parents of a node N_i in H. The construction process of the tree structure H" corresponding to H is formally described by Algorithm 1.

Belief Propagation in Singly-Connected HDEVNs: The propagation algorithm in singly-connected HDEVNs is based on the propagation principles described in Sect. 3.1 where conditional distributions are defined per one parent and in Sect. 3.2 where conditional distributions are defined for all the parents.

The belief propagation algorithm is based on a message passing through the graph H". The message passing algorithm (MPA) involves principally two phases: the top down propagation phase and the bottom up propagation phase.

Algorithm 1. Construct a Tree Structure H" from a HDEVN H

Require: H=(N,E)
Ensure: H"=(N",E")
  Initialization
  $\eta \leftarrow C \cup F_{Edge}$;
  $\zeta \leftarrow F_{Parents}$;
  $\vartheta \leftarrow \emptyset$; /* ϑ denotes the set of non conditional nodes in H"*/
  $\beta \leftarrow \emptyset$; /* β denotes the set of conditional nodes in H"*/
  $E" \leftarrow \emptyset$; /* E"denotes the set of undirected edges in H"*/
  $N" \leftarrow \emptyset$;/* N" denotes the set of nodes in H"*/
  while $|\zeta| \geq 1$ **do**
    Pick a candidate variable $c \in \zeta$
    $P \leftarrow Pa(c)$
    $n1 \leftarrow P$
    $\vartheta \leftarrow \vartheta \cup \{n1\}$
    while $|P| \geq 1$ **do**
      $n2 \leftarrow p$ where $p \in P$
      $\vartheta \leftarrow \vartheta \cup \{n2\}$
      $E" \leftarrow E" \cup \{(n2,n1)\}$
      $P \leftarrow P - \{n2\}$
    end while
    $n3 \leftarrow \{c|Pa(c)\}$
    $n4 \leftarrow \{c\}$
    $\vartheta \leftarrow \vartheta \cup \{n4\}$
    $\beta \leftarrow \beta \cup \{n3\}$
    $E" \leftarrow E" \cup \{(n1,n3),(n3,n4)\}$
    $\zeta \leftarrow \zeta - \{c\}$
  end while
  while $|\eta| \geq 1$ **do**
    Pick a candidate variable $c \in \eta$
    $P \leftarrow Pa(c)$
    $n1 \leftarrow \{c\}$
    $\vartheta \leftarrow \vartheta \cup \{n1\}$
    while $|P| \geq 1$ **do**
      $n2 \leftarrow p$ where $p \in P$
      $\vartheta \leftarrow \vartheta \cup \{n2\}$
      $n3 \leftarrow \{c|p\}$
      $\beta \leftarrow \beta \cup \{n3\}$
      $E" \leftarrow E" \cup \{(n2,n3),(n3,n1)\}$
      $P \leftarrow P - \{n2\}$
    end while
    $\eta \leftarrow \eta - \{c\}$
  end while
  $N" \leftarrow \vartheta \cup \beta$

Before running the MPA, an initialization phase is applied during which each conditional distribution in H is associated with the corresponding conditional node in H" and each a priori distribution in H is associated with the corresponding non-conditional node in H". A vacuous belief is associated with each non-conditional node in H" having no a corresponding a priori distribution in H.

The MPA consists in two phases:

(1) **A Top down Propagation Phase** applied by passing messages inwards, starting from the roots towards the leaves. During this phase, each non-root node N_i:
   (a) receives:
      (i) either a message from each of its parent nodes (**Case 1**)[3]. Each message is computed using the steps (a) and (b) of the top down propagation method described in Sect. 3.1,
      (ii) or one message from all its parent nodes (**Case 2**)[4]. This message is computed using the steps (a), (b), (c) and (d) of the top down propagation method described in Sect. 3.2
   (b) updates its distribution by combining its own a priori one with the message(s) received from its parent(s) using (6).

(2) **A Bottom up Propagation Phase** applied by distributing messages away from the leaves, until reaching the roots. Each non-leaf node N_j:
   (a) receives a message from each of its child nodes (**Case 3**)[5]. Each message is computed using the steps (a) and (b) of the bottom up propagation method described in Sect. 3.1
   (b) receives a message from each of its child nodes (**Case 4**)[6]. This message is computed using the steps (a), (b) and (c) of the bottom up propagation method described in Sect. 3.2
   (c) updates its distribution by combining its a priori distribution with the message(s) received from its parent(s) and its child node(s) using (6).

5   Illustration

Let us consider the simple HDEVN, the a priori mass functions and the conditional plausibility functions corresponding to the mass functions of Fig. 5. The mass functions $m_0^{N_3}$, $m_0^{N_5}$ and $m_0^{N_6}$ correspond to the vacuous beliefs.

---

[3] **Case 1** occurs when the child node N_i is associated with conditionals specified per single parent.

[4] **Case 2** occurs when the child node N_i has one conditional distribution defined for all its parents.

[5] **Case 3** occurs when N_j has one or more child nodes associated with conditionals specified per single parent.

[6] **Case 4** occurs when N_j has one or more child nodes associated with conditionals specified for all the parents.

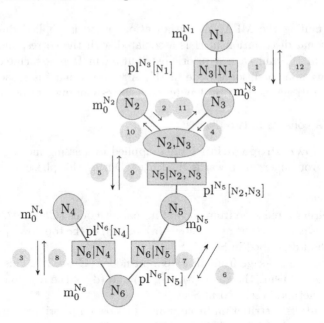

Fig. 6. The tree structure H" corresponding to the HDEVN of Fig. 5

The tree H" is built from the HDEVN H of Fig. 5 using Algorithm 1. The messages of the inward pass (i.e. the top down propagation phase) and the outward pass (i.e. the bottom up propagation phase) performed on H" are generated in the order shown in Fig. 6. These messages are computed as follows:

Message	Computation	Rule	Message's Value
1	$1 = m_{N_1 \rightarrow N_3}$	(9)	$m_{N_1 \rightarrow N_3}(\{n_{31}\}) = 0.375;$ $m_{N_1 \rightarrow N_3}(\{n_{32}\}) = 0.0675;$ $m_{N_1 \rightarrow N_3}(\{\Theta_{N_3}\}) = 0.5575$
2	$2 = m_{N_2 \rightarrow \{N_2,N_3\}} = m_0^{N_2}$		$m_{N_2 \rightarrow \{N_2,N_3\}}(\{n_{21}\}) = 0.1;$ $m_{N_2 \rightarrow \{N_2,N_3\}}(\{n_{22}\}) = 0.3;$ $m_{N_2 \rightarrow \{N_2,N_3\}}(\Theta_{N_2}) = 0.6$
3	$3 = m_{N_4 \rightarrow N_6}$	(9)	$m_{N_4 \rightarrow N_6}(\{n_{61}\}) = 0.357;$ $m_{N_4 \rightarrow N_6}(\{n_{62}\}) = 0.136;$ $m_{N_4 \rightarrow N_6}(\Theta_{N_6}) = 0.507$
4	$4 = m_{N_3 \rightarrow \{N_2,N_3\}} = (m_0^{N_3} \otimes 1)$	(6)	$m_{N_3 \rightarrow \{N_2,N_3\}}(\{n_{31}\}) = 0.375;$ $m_{N_3 \rightarrow \{N_2,N_3\}}(\{n_{32}\}) = 0.0675;$ $m_{N_3 \rightarrow \{N_2,N_3\}}(\Theta_{N_3}) = 0.5575$

Message	Computation	Rule	Message's Value
5	$2^{\uparrow\{N_2,N_3\}} = m^{N_2\uparrow\{N_2,N_3\}}$	(5)	$m^{N_2\uparrow\{N_2,N_3\}}(\{(n_{21},n_{31}),(n_{21},n_{32})\})$ $=0.1$; $m^{N_2\uparrow\{N_2,N_3\}}(\{(n_{22},n_{31}),(n_{22},n_{32})\})$ $=0.3$; $m^{N_2\uparrow\{N_2,N_3\}}(\{(n_{21},n_{31}),(n_{22},n_{31}),$ $(n_{21},n_{32}),(n_{21},n_{32})\})$ $= 0.6$;
	$4^{\uparrow\{N_2,N_3\}} = m^{N_3\uparrow\{N_2,N_3\}}$	(5)	$m^{N_3\uparrow\{N_2,N_3\}}(\{(n_{31},n_{21}),(n_{31},n_{22})\})$ $=0.375$; $m^{N_3\uparrow\{N_2,N_3\}}(\{(n_{32},n_{21}),(n_{32},n_{22})\})$ $=0.0675$; $m^{N_3\uparrow\{N_2,N_3\}}(\{(n_{31},n_{21}),(n_{32},n_{21}),$ $(n_{31},n_{22}),(n_{31},n_{22})\})= 0.5575$
	$m^{N_2,N_3} = 2^{\uparrow\{N_2,N_3\}} \otimes 4^{\uparrow\{N_2,N_3\}}$	(6)	
	$5 = m_{\{N_2,N_3\}\to N_5}$	(9)	$m_{\{N_2,N_3\}\to N_5}(\{n_{51}\}) = 0.0731$; $m_{\{N_2,N_3\}\to N_5}(\{n_{52}\}) = 0.1584$; $m_{\{N_2,N_3\}\to N_5}(\Theta_{N_5}) = 0.7685$
6	$m_{N_5} = (m_0^{N_5} \otimes 5)$	(6)	$m_{N_5}(\{n_{51}\}) = 0.0731$; $m_{N_5}(\{n_{52}\}) = 0.1584$; $m_{N_5}(\Theta_{N_5}) = 0.7685$
	$6 = m_{N_5\to N_6}$	(9)	$m_{N_5\to N_6}(\{n_{61}\}) = 0.3487$ $m_{N_5\to N_6}(\{n_{62}\}) = 0.0430$; $m_{N_5\to N_6}(\Theta_{N_6}) = 0.6083$
7	$m_{N_6} = (m_0^{N_6} \otimes 3 \otimes 6)$	(6)	$m_{N_6}(\{n_{61}\}) = 0.5532$; $m_{N_6}(\{n_{62}\}) = 0.1178$; $m_{N_6}(\Theta_{N_6}) = 0.329$
	$7 = m_{N_6\to N_5}$	(10)	$m_{N_6\to N_5}(\{n_{51}\}) = 0$; $m_{N_6\to N_5}(\{n_{52}\}) = 0$; $m_{N_6\to N_5}(\Theta_{N_5}) = 1$
8	$8 = m_{N_6\to N_4}$	(10)	$m_{N_6\to N_4}(\{n_{41}\}) = 0$; $m_{N_6\to N_4}(\{n_{42}\}) = 0$; $m_{N_6\to N_4}(\Theta_{N_4}) = 1$
	$m_{N_4} = (m_0^{N_4} \otimes 8)$	(6)	$m_{N_4}(\{n_{41}\}) = 0.35$; $m_{N_4}(\{n_{42}\}) = 0.25$; $m_{N_4}(\Theta_{N_4}) = 0.4$
9	$m_{N_5} = (m_0^{N_5} \otimes 7 \otimes 5)$	(6)	$m_{N_5}(\{n_{51}\}) = 0.0731$; $m_{N_5}(\{n_{52}\}) = 0.1584$; $m_{N_5}(\Theta_{N_5}) = 0.7685$
	$9 = m_{N_5\to N_2,N_3}$	(10)	
10	$10 = m_{\{N_2,N_3\}\to N_2}$	(5)	$m_{\{N_2,N_3\}\to N_2}(\{n_{21}\}) = 0$; $m_{\{N_2,N_3\}\to N_2}(\{n_{22}\}) = 0$; $m_{\{N_2,N_3\}\to N_2}(\Theta_{N_2}) = 1$
11	$11 = m_{\{N_2,N_3\}\to N_3}$	(5)	$m_{\{N_2,N_3\}\to N_3}(\{n_{31}\}) = 0$; $m_{\{N_2,N_3\}\to N_3}(\{n_{32}\}) = 0$; $m_{\{N_2,N_3\}\to N_3}(\Theta_{N_3}) = 1$

Message	Computation	Rule	Message's Value
12	$m_{N_3} = (m_0^{N_3} \otimes 11 \otimes 1)$	(6)	$m_{N_3}(\{n_{31}\}) = 0.375;$ $m_{N_3}(\{n_{32}\}) = 0.0675;$ $m_{N_3}(\Theta_{N_3}) = 0.5575$
	$12 = m_{N_3 \to N_1}$	(10)	$m_{N_3 \to N_1}(\{n_{11}\}) = 0;$ $m_{N_3 \to N_1}(\{n_{12}\}) = 0;$ $m_{N_3 \to N_1}(\Theta_{N_1}) = 1$
	$m_{N_1} = (m_0^{N_1} \otimes 12)$	(6)	$m_{N_1}(\{n_{11}\}) = 0.3;$ $m_{N_1}(\{n_{12}\}) = 0.2;$ $m_{N_1}(\Theta_{N_1}) = 0.5$

6 Conclusion and Future Work

An extension of the DEVNs, called HDEVNs, was proposed in this paper to deal with both conditionals specified per child node and those specified per edge. An inference algorithm for belief propagation in singly-connected HDEVNs was also proposed in this paper, based on the DRC and the GBT which can be used to deal with both conditionals specified per edge and conditionals defined per child node. In future works, we propose to deal with multiply-connected HDEVNs.

References

1. Ben Yaghlane, B., Mellouli, K.: Inference in directed evidential networks based on the transferable belief model. IJAR **48**(2), 399–418 (2008)
2. Ban Yaghlane, B., Mellouli, K.: Updating directed belief networks. In: Hunter, A., Parsons, S. (eds.) ECSQARU 1999. LNCS (LNAI), vol. 1638, pp. 43–54. Springer, Heidelberg (1999)
3. Dempster, A.P.: Upper and lower probabilities induced by a multivalued mapping. Ann. Math. Stat. **38**, 325–339 (1967)
4. Laâmari, W., Ben Yaghlane, B.: Reasoning in singly-connected directed evidential networks with conditional beliefs. In: Likas, A., Blekas, K., Kalles, D. (eds.) SETN 2014. LNCS, vol. 8445, pp. 221–236. Springer, Heidelberg (2014)
5. Pearl, J.: Probabilistic Reasoning in Intelligent Systems: Networks of Plausible Inference. Morgan Kaufmann, San Mateo (1988)
6. Shafer, G.: A Mathematical Theory of Evidence. Princeton University Press, Princeton (1976)
7. Shenoy, P.P.: Valuation networks and conditional independence. In: Uncertainty in Artificial Intelligence, pp. 191–199 (1993)
8. Smets, Ph.: Belief function: the disjunctive rule of combination and the generalized Bayesian theorem. Int. J. Approx. Reasoning **9**, 1–35 (1993)
9. Smets, Ph., Kennes, R.: The transferable belief model. Artif. Intell. **66**, 191–234 (1994)
10. Xu, H., Smets, Ph.: Evidential reasoning with conditional belief functions. In: Heckerman, D., et al. (eds.) Proceedings of Uncertainty in Artificial Intelligence (UAI 1994), pp. 598–606. Morgan Kaufmann, San Mateo (1994)
11. Zadeh, L.A.: Fuzzy sets. Inf. Control **8**, 338–353 (1965)

Uncertain Logical Gates in Possibilistic Networks. An Application to Human Geography

Didier Dubois[1], Giovanni Fusco[2], Henri Prade[1], and Andrea Tettamanzi[3(✉)]

[1] IRIT – CNRS, 118, route de Narbonne, Toulouse, France
{dubois,prade}@irit.fr
[2] Univ. Nice Sophia Antipolis/CNRS, ESPACE, UMR7300, Nice, France
fusco@unice.fr
[3] Univ. Nice Sophia Antipolis/CNRS, I3S, UMR7271, Sophia Antipolis, France
andrea.tettamanzi@unice.fr

Abstract. Possibilistic networks offer a qualitative approach for modeling epistemic uncertainty. Their practical implementation requires the specification of conditional possibility tables, as in the case of Bayesian networks for probabilities. This paper presents the possibilistic counterparts of the noisy probabilistic connectives (and, or, max, min, . . .). Their interest is illustrated on an example taken from a human geography modeling problem. The difference of behaviors in some cases of some possibilistic connectives, with respect to their probabilistic analogs, is discussed in details.

1 Introduction

Bayesian networks [11] can be built in two ways: statistical and subjective. In the first case, a supposedly large dataset involving a number of variables is available, and the Bayesian network is obtained by some machine learning procedure. The probability tables thus obtained have a frequentist flavor, and the simplest network possible is searched for. On the contrary, Bayesian networks can be specified using expert knowledge. In this case, the structure of a network relating the variables is first given, often relying on causal connections between variables and conditional independence relations the expert is aware of. Then probability tables must be filled by the expert. They consist, for each variable in the network, of conditional probabilities for that variable, conditioned on each configuration of its parent variables. Note that, even if causal relations as perceived by the expert are instrumental in building a simple and interpretable network, the joint probability distribution obtained by combining the probability tables no longer accounts for causality. Another difficulty arises for causality-based Bayes networks: if variables are not binary and/or the number of parent variables is more than two, the task of eliciting numerical probability tables becomes tedious, if not impossible to fulfill. Indeed, the number of probability values to be supplied increases exponentially with the number of parent variables.

To alleviate the elicitation task, the notion of noisy logical gate (or connective) has been introduced, based on the assumption of independent causal

© Springer International Publishing Switzerland 2015
C. Beierle and A. Dekhtyar (Eds.): SUM 2015, LNAI 9310, pp. 249–263, 2015.
DOI: 10.1007/978-3-319-23540-0_17

influences that can be combined. As a result, one small conditional probability table is elicited per parent variable, and the probability table of each variable given its parents is obtained by combining these small tables by a so-called noisy connective [6,10], which may include a so-called leakage factor summarizing the causal effect of variables not explicitly present in the network.

While the notion of noisy connective solves the combinatorial problem of collecting many probability values to a large extent, the issue remains that people cannot always provide precise probability assessments. Let alone the fact that the probability scale is too fine-grained for human perception of belief or frequencies, some conditional probability values may be ill-known or plainly unknown to the experts. The usual Bayesian recommendation in the latter case is to use uniform distributions, but it is well-known that these do not properly model ignorance. Alternatively, one may use imprecise probability networks (called credal networks) [12], qualitative Bayesian networks [14] or possibilistic networks [3]. While the two first options extend probabilistic networks to ill-known parameters (with an interval-based approach for the former and an ordinal approach for the latter), possibilistic networks represent a more drastic departure from probabilistic networks. In their qualitative version, possibilistic networks can be defined on a finite chain of possibility values and do not refer to numerical values. This feature may make the collection of expert information on conditional tables easier than requiring precise numbers obeying the laws of probability.

In this paper, we propose possibilistic counterparts of noisy connectives of probabilistic networks. As possibilistic uncertainty is merely epistemic and due to a lack of information, we shall speak of *uncertain connectives*. After recalling probabilistic networks with noisy gates, we present the corresponding approach for possibilistic networks and present various uncertain gates, especially the AND, OR, MAX, and MIN functions.[1] Finally, the approach, including algorithmic issues, is illustrated on a belief network stemming from an application to human geography.

2   Probabilistic Networks with Independent Causal Influences

Consider a set of independent variables $X_1, \ldots, X_n$ that influence the value of a variable Y. In the ideal case, there is a deterministic function f such that $Y = f(X_1, X_2, \ldots, X_n)$. In order to account for uncertainty, one may assume the existence of intermediary variables $Z_1, \ldots, Z_n$, such that Z_i expresses the fact that X_i will have a causal influence on Y, and to what extent (Z_i has the same domain as Y). It is assumed that the relation between X_i and Z_i is

---

[1] The idea of possibilistic uncertain gates was first considered empirically by [13] directly in the setting of possibilistic logic, at a time where possibilistic networks had not yet been introduced. It seems that the question of possibilistic uncertain gates has not been reconsidered ever since, if we except a recent study in the broader setting of imprecise probabilities [1] and a preliminary outline in French by the authors [4].

probabilistic and that X_i is independent of other variables given Z_i. Besides, we consider the deterministic function as affected by the auxiliary variables Z_i only. In other words, we get a probabilistic network such that

$$P(Y, Z_1, \ldots, Z_n, X_1, X_2, \ldots, X_n) = P(Y, Z_1, \ldots, Z_n) \cdot \prod_{i=1}^{n} P(Z_i \mid X_i), \quad (1)$$

where $P(Y, Z_1, \ldots, Z_n) = 1$ if $Y = f(Z_1, Z_2, \ldots, Z_n)$ and 0 otherwise. This is called a noisy function. In particular, notice that the dependence tables between Y and $X_1, \ldots, X_n$ can now be obtained by combining simple conditional probability distributions pertaining to single factors:

$$P(y \mid x_1, \ldots, x_n) = \sum_{z_1, \ldots, z_n : y = f(z_1, \ldots, z_n)} \prod_{i=1}^{n} P(z_i \mid x_i). \quad (2)$$

This is the assumption of *independence of causal influence* (ICI) [6]. In the case of Boolean variables, it is assumed that $P(z_i = 0 \mid x_i = 0) = 1$ (no cause, no effect), while $P(z_i = 0 \mid x_i = 1)$ can be positive (the effect may or may not appear when the cause is present).

Canonical ICI models are obtained by means of specific choice of the function f. For instance, if all variables are Boolean, f will be a logical connective. In this case, we speak of noisy OR ($f = \vee$), noisy AND ($f = \wedge$); if the range of the Z_i's and Y is a totally ordered set, usual gates are the noisy MAX ($f = \max$), or MIN ($f = \min$).

The approach may be further refined by allowing f to summarize the potential effect of external variables not taken into account: this is the leaky model. Now, Y also depends on a leak variable Z_ℓ not explicitly related to specifically identified causes, i.e., $Y = f(Z_1, Z_2, \ldots, Z_n, Z_\ell)$. The domain of Z_ℓ is supposed to be the range of f, i.e., the domain of Y and this variable is independent of the other ones. Hence, the leakage model may be written as:

$$P(Y, Z_1, \ldots, Z_n, Z_\ell, X_1, \ldots, X_n) = P(Y, Z_1, \ldots, Z_n) \cdot P(Z_\ell) \cdot \prod_{i=1}^{n} P(Z_i \mid X_i),$$

so that

$$P(y \mid x_1, \ldots, x_n) = \sum_{z_1, \ldots, z_n, z_\ell : y = f(z_1, \ldots, z_n, z_\ell)} P(z_\ell) \cdot \prod_{i=1}^{n} P(z_i \mid x_i). \quad (3)$$

For instance, in the case of Boolean variables, $P(y = 1 \mid x_1 = 0, \ldots, x_n = 0)$ may be positive due to such external causes.

We will now turn to the question whether the same kind of ICI approach can be used to elicit possibilistic networks as well.

3   Canonical Possibilistic Networks

Possibility theory [7,16] is based on maxitive set functions associated to possibility distributions. Formally, given a universe of discourse U, a possibility distribution $\pi : U \to [0,1]$ pertains to a variable X ranging on U and represents the available (incomplete) information about the more or less possible values of X, assumed to be single-valued. Thus, $\pi(u) = 0$ means that $X = u$ is impossible. The consistency of information is expressed by the normalization of π : $\exists u \in U, \pi(u) = 1$, namely, at least one value is fully possible for X. Distinct values u and u' may be simultaneously possible at degree 1. A state of complete ignorance is represented by the distribution $\pi_?(u) = 1, \forall u \in U$. A possibility measure of an event $A \subseteq U$ is defined by

$$\Pi(A) = \sup_{u \in A} \pi(u).$$

Possibility measures are maxitive, i.e.,

$$\forall A, \forall B, \Pi(A \cup B) = \max(\Pi(A), \Pi(B)).$$

The underlying assumption is that the agent focuses on most plausible values, neglecting other ones. A dual measure of necessity $N(A) = 1 - \Pi(U \setminus A)$ expresses the certainty of event A as the impossibility of non-A.

A possibilistic network [2,3] has the same structure as a Bayesian network. The joint possibility for n variables linked by an acyclic directed graph is defined by

$$\pi(x_1, \ldots, x_n) = *_{i=1,\ldots,n} \pi(x_i \mid pa(X_i)),$$

where x_i is an instantiation of the variable X_i, and $pa(X_i)$ an instantiation of the parent variables of X_i. The operation $*$ is the minimum (in the qualitative case) or the product (in the numerical case).

Deterministic models $Y = f(X_1, \ldots, X_n)$ are defined as in the probabilistic case:

$$\pi(y \mid x_1, \ldots, x_n) = \begin{cases} 1 & \text{if } y = f(x_1, \ldots x_n); \\ 0 & \text{otherwise.} \end{cases} \tag{4}$$

Let us define possibilistic models with *independent causal influences* (ICI). We use a deterministic function $Y = f(Z_1, \ldots, Z_n)$ with n intermediary causal variables Z_i, as for the probabilistic models. Now, $\pi(y \mid x_1, \ldots, x_n)$ is of the form:

$$\pi(y \mid z_1, \ldots, z_n) * \pi(z_1, \ldots, z_n \mid x_1, \ldots, x_n),$$

where $\pi(y \mid z_1, \ldots, z_n)$ obeys Eq. 4. Again, each variable Z_i only depends (in an uncertain way) on the variable X_i. Thus, we have $\pi(z_1, \ldots, z_n \mid x_1, \ldots, x_n) = *_{i=1,\ldots,n} \pi(z_i \mid x_i)$. This leads to the equality

$$\pi(y \mid x_1, \ldots, x_n) = \max_{z_1, \ldots, z_n : y = f(z_1, \ldots, z_n)} *_{i=1,\ldots,n} \pi(z_i \mid x_i), \tag{5}$$

whose similarity with Eq. 2 is striking. Notice that, when $* = \min$, Eq. 5 boils down to applying the extension principle [16] to function f, assuming fuzzy-valued inputs $F_1, \ldots, F_n$, where the membership function of F_i is defined by $\mu_{F_i}(z_i) = \pi(z_i \mid x_i)$.

In case we suppose that y depends also in an uncertain way on other causes summarized by a leak variable Z_ℓ, then the counterpart of Eq. 3 reads:

$$\pi(y \mid x_1, \ldots, x_n) = \max_{z_1, \ldots, z_n, z_\ell : y = f(z_1, \ldots, z_n, z_\ell)} *_{i=1, \ldots, n} \pi(z_i \mid x_i) * \pi(z_\ell). \quad (6)$$

In the following, we provide a detailed analysis of possibilistic counterparts of noisy gates.

3.1 Uncertain OR and AND Gates

The variables are assumed to be Boolean (i.e., $Y = y$ or $\neg y$, etc.). The uncertain OR (counterpart of the probabilistic "noisy OR") assumes that $X_i = x_i$ for at least one variable X_i represents a sufficient cause for getting $Y = y$, and $Z_i = z_i$ indicates that $X_i = x_i$ has caused $Y = y$. This gives $f(Z_1, \ldots, Z_n) = \bigvee_{i=1}^n Z_i$. The uncertainty indicates that the causes may fail to produce their effects. $Z_i = \neg z_i$ indicates that $X_i = x_i$ did not cause $Y = y$ due to the presence of some inhibitor that prevents the effect from taking place. We assume it is more possible that $X_i = x_i$ causes $Y = y$ than the opposite (otherwise one could not say that $X_i = x_i$ is sufficient for causing $Y = y$). Then we must define $\pi(z_i \mid x_i) = 1$ and $\pi(\neg z_i \mid x_i) = \kappa_i < 1$. Besides, $\pi(z_i \mid \neg x_i) = 0$, since when X_i is absent, it does not cause y. Hence the causal elementary possibility table:

$\pi(Z_i \mid X_i)$	x_i	$\neg x_i$
z_i	1	0
$\neg z_i$	κ_i	1

Note that in the case of a probabilistic network, $\pi(z_i \mid x_i)$ is replaced by $1 - \kappa_i$ in the above table. We can then obtain the table of the conditional possibility distribution $\pi(Y \mid X_1, \ldots, X_n)$ by means of Eq. 5:

$$\pi(y \mid X_1, \ldots, X_n) = \max_{z_1, \ldots, z_n : z_1 \vee \ldots \vee z_n = 1} *_{i=1}^n \pi(z_i \mid X_i)$$
$$= \max_{i=1}^n \pi(z_i \mid X_i) * (*_{j \neq i} \max(\pi(z_j \mid X_j)\pi(\neg z_j \mid X_j)));$$

$$\pi(\neg y \mid X_1, \ldots, X_n) = \max_{z_1, \ldots, z_n : z_1 \vee \ldots \vee z_n = 0} *_{i=1}^n \pi(z_i \mid X_i)$$
$$= \pi(\neg z_1 \mid X_1) * \ldots * \pi(\neg z_n \mid X_n).$$

Let us denote by $\mathbf{x}$ a configuration of $(X_1, \ldots, X_n)$, and let $I_+(\mathbf{x}) = \{i : X_i = x_i\}$ and $I_-(\mathbf{x}) = \{i : X_i = \neg x_i\}$. Then we get:

- $\pi(\neg y \mid \mathbf{x}) = *_{i=1,\ldots,n} \pi(\neg z_i \mid X_i = \mathbf{x}_i) = *_{i \in I_+(\mathbf{x})} \kappa_i$;
- $\pi(y \mid \mathbf{x}) = 1$ when $\mathbf{x} \neq (\neg x_1, \ldots, \neg x_n)$;
- $\pi(\neg y \mid \neg x_1, \ldots, \neg x_n) = 1$, $\pi(y \mid \neg x_1, \ldots, \neg x_n) = 0$: $\neg y$ (no effect) can be obtained for sure only if all the causes are absent.

For $n = 2$, this gives the conditional tables:

$\pi(y \mid X_1 X_2)$	x_1	$\neg x_1$	$\pi(\neg y \mid X_1 X_2)$	x_1	$\neg x_1$
x_2	1	1	x_2	$\kappa_1 * \kappa_2$	κ_2
$\neg x_2$	1	0	$\neg x_2$	κ_1	1

More generally, if there are n causes, we have to provide the values of n parameters κ_i.

For the uncertain OR with leak, we now assume that $f(Z_1, \ldots, Z_n) = \bigvee_{i=1}^n Z_i \vee Z_\ell$, where Z_ℓ is an unknown external cause. We assign $\pi(z_\ell) = \kappa_\ell < 1$ considering that z_ℓ is not a usual cause. We thus obtain

- $\pi(\neg y \mid \mathbf{x}) = *_{i=1,\ldots,n}\pi(\neg z_i \mid X_i = \mathbf{x}_i) * \pi(\neg z_\ell) = *_{i \in I_+(\mathbf{x})}\kappa_i;$
- $\pi(y \mid \mathbf{x}) = 1$, if $\mathbf{x} \neq (\neg x_1, \ldots, \neg x_n);$
- $\pi(\neg y \mid \neg x_1, \ldots, \neg x_n) = 1;$
- $\pi(y \mid \neg x_1, \ldots, \neg x_n) = \kappa_\ell$ (even if the causes x_i are absent, there is still a possibility for having $Y = y$, namely if the external cause is present).

Indeed, we get (letting $\neg x = \neg x_1, \ldots, \neg x_n$),

$$\pi(y \mid \neg x_1, \ldots, \neg x_n) = \max(\pi(y \mid \neg x, z_\ell) * \pi(z_\ell), \pi(y \mid \neg x, \neg z_\ell) * \pi(\neg z_\ell))$$
$$= \max(1 * \kappa_\ell, 0 * 1) = \kappa_\ell.$$

For $n = 2$, the conditional table becomes:

$\pi(y \mid X_1 X_2)$	x_1	$\neg x_1$		$\pi(\neg y \mid X_1 X_2)$	x_1	$\neg x_1$
x_2	1	1		x_2	$\kappa_1 * \kappa_2$	κ_2
$\neg x_2$	1	κ_ℓ		$\neg x_2$	κ_1	1

The only 0 entry has been replaced by the leakage coefficient. For n causes, we have now to provide the values of $n + 1$ parameters κ_i.

The uncertain AND (counterpart of the probabilistic "noisy AND") uses the same local conditional tables but it assumes that $X_i = x_i$ represents a *necessary* cause for $Y = y$. We again build the conditional possibility table $\pi(Y \mid X_1, \ldots, X_n)$ by means of Eq. 5 with $f(Z_1, \ldots, Z_n) = \bigwedge_{i=1}^n Z_i$. Thus, we find

- $\pi(\neg y \mid x_1, \ldots, x_n) = \max_{z_1, \ldots, z_n : \neg y = z_1 \wedge \ldots \wedge z_n} *_{i=1}^n \pi(z_i \mid x_i) = \max_{i=1}^n \pi(\neg z_i \mid x_i) = \max_{i=1}^n \kappa_i;$
- $\pi(y \mid x_1, \ldots, x_n) = 1;$
- $\pi(\neg y \mid \mathbf{x}) = 1, \pi(y \mid \mathbf{x}) = 0$ if $\mathbf{x} \neq (x_1, \ldots, x_n)$ (if at least one of the causes is absent, the effect is necessarily absent).

For $n = 2$, Eq. 5 yields the conditional tables:

$\pi(y \mid X_1 X_2)$	x_1	$\neg x_1$		$\pi(\neg y \mid X_1 X_2)$	x_1	$\neg x_1$
x_2	1	0		x_2	$\max(\kappa_1, \kappa_2)$	1
$\neg x_2$	0	0		$\neg x_2$	1	1

More generally, if there are n causes, we have to assess n values for the parameters κ_i. The case of the uncertain AND with leak corresponds to the possibility $\pi(z_L) = \kappa_L < 1$ that an external factor $Z_L = z_L$ causes $Y = y$ independently of the values of the X_i. Namely $f(Z_1, \ldots, Z_n, Z_L) = (\bigwedge_{i=1}^n Z_i) \vee Z_L$. For $n = 2$, Eq. 5 then gives the conditional tables:

$\pi(y \mid X_1 X_2)$	x_1	$\neg x_1$		$\pi(\neg y \mid X_1 X_2)$	x_1	$\neg x_1$
x_2	1	κ_L		x_2	$\max(\kappa_1, \kappa_2)$	1
$\neg x_2$	κ_L	κ_L		$\neg x_2$	1	1

3.2   Comparison with Probabilistic Gates

It is interesting to compare the possibilistic and probabilistic tables. Consider those of the noisy OR [6], where $\kappa_i = P(\neg z_i \mid x_i)$:

$P(y \mid X_1 X_2)$	x_1	$\neg x_1$		$P(\neg y \mid X_1 X_2)$	x_1	$\neg x_1$
x_2	$1 - \kappa_1 \kappa_2$	$1 - \kappa_2$		x_2	$\kappa_1 \kappa_2$	κ_2
$\neg x_2$	$1 - \kappa_1$	0		$\neg x_2$	κ_1	1

There is an important difference between the behaviors of uncertain and noisy OR if $* = \min$. In the possibilistic tables, we see (using the associated necessity measure N, and Boolean notations for the instantiations of x_1 and x_2) that $N(y \mid 11) = \max(N(y \mid 10), N(y \mid 01))$ while $P(y \mid 11) > \max(P(y \mid 10), P(y \mid 01))$, so that the presence of two causes does not reinforce the certainty of the effect wrt the presence of the most influential cause. Hence qualitative possibility networks will be less expressive than probabilistic networks. If $* = $ product, $N(y \mid 11) = 1 - \kappa_1 \kappa_2 > \max(N(y \mid 10), N(y \mid 01))$ as with the probability case.

Another major difference will occur in case the effects of causes are not frequent, as when $P(\neg z_i \mid x_i) = \kappa_i > 0.5, i = 1, 2$. Then it may happen that $P(y \mid x_1 x_2) = 1 - \kappa_1 \kappa_2 > 0.5$, that is the presence of the two causes makes the effect frequent. Then a possibilistic rendering of this case must be such that $\pi(\neg z_i \mid x_i) = 1 > \pi(z_i \mid x_i) = \lambda_i$ (say). However, there is no way of observing this reversal effect, since $\pi(y \mid x_1 x_2) = \max(\lambda_1 * \lambda_2, \lambda_1, \lambda_2) = \max(\lambda_1, \lambda_2) < 1$. Hence $\pi(\neg y \mid x_1 x_2) = 1$ and $N(y \mid x_1 x_2) = 0$. In other words, using the uncertain OR, two causes that are individually insufficient to make an effect plausible are still insufficient to make it plausible if joined together. Note that this fact reminds of the property of closure under conjunction for necessity measures in possibility theory $(N(y_1) > 0$ and $N(y_2) > 0$ imply $N(y_1 \wedge y_2) > 0)$ which fail to hold in probability theory.

One way to address this problem is to define the global conditional possibility tables $\pi(Y \mid X_1, X_2)$ enforcing $\pi(y \mid x_1 x_2) > \pi(\neg y \mid x_1 x_2)$ even if $\pi(y \mid x_1) < \pi(\neg y \mid x_1)$ and $\pi(y \mid x_2) < \pi(\neg y \mid x_2)$, which is perfectly compatible with possibility theory. We will outline a solution of this kind in the next section for the uncertain MAX, which is a generalization of the uncertain OR. However, one cannot build the global table from the marginal ones using an uncertain OR.

3.3   Uncertain MAX and MIN Gates

The uncertain MAX is a multiple-valued extension of the uncertain OR, where the output variable (hence the variables Z_i) is valued on a finite, totally ordered, severity or intensity scale $L = \{0 < 1 < \cdots < m\}$. We assume that $Y = \max(Z_1, \ldots, Z_n)$. $Z_i = z_i \in L$ represents the fact that X_i alone has increased the value of Y at level z_i. The conditional possibility distributions $\pi(y \mid x_i)$ are supposed to be given. We can then compute the conditional tables, as

$$\pi(y \mid x_1, \ldots, x_n) = \max_{z_1, \ldots, z_n : y = \max(z_1, \ldots, z_n)} *_{i=1}^{n} \pi(z_i \mid x_i)$$

$$= \max_{i=1}^{n} \pi(Z_i = y \mid x_i) * (*_{j \neq i} \Pi(Z_j \leq y \mid x_j)).$$

In a causal setting, we assume that $y = 0$ is a normal state, and $y > 0$ is more or less abnormal, $y = m$ being fully abnormal. Suppose that the domain of X_i is L as well. It is natural to assume that:

- if $X_i = j$ then $Z_i = j$, which means $\Pi(Z_i = j \mid X_i = j) = 1$;
- $\Pi(Z_i > j \mid X_i = j) = 0$ (a cause having a weak intensity cannot induce an effect with strong severity);
- $0 < \Pi(Z_i < j \mid X_i = j) < 1$ (a cause having strong intensity may sometimes only induce an effect with weak severity, or may even have no effect at all);
- An effect with severity weaker than the intensity of a cause is all the less plausible as the effect is weak. This leads to suppose the following inequalities:

$$0 < \pi(Z_i = 0 \mid X_i = j) < \pi(Z_i = 1 \mid X_i = j) < \ldots < \pi(Z_i = j \mid X_i = j) = 1.$$

This leads to state the left-hand side table below (for 3 levels of strength 0, 1, 2).

$\pi(Z_i \mid X_i)$	$X_i = 2$	$X_i = 1$	$X_i = 0$
$Z_i = 2$	1	0	0
$Z_i = 1$	κ_i^{12}	1	0
$Z_i = 0$	κ_i^{02}	κ_i^{01}	1

$\pi(Z_i \mid X_i)$	$X_i = 2$	$X_i = 0$
$Z_i = 2$	1	0
$Z_i = 1$	κ_i^{12}	0
$Z_i = 0$	κ_i^{02}	1

where $0 < \kappa_i^{02} < \kappa_i^{12} < 1$, $0 < \kappa_i^{01} < 1$. In case we have m levels of strength, we have to assess $\frac{m(m+1)}{2}$ coefficients. On the right-hand side is the corresponding table when the variables X_i are Boolean (then the middle column is dropped).

The global conditional possibility tables are thus obtained by applying Eq. 5, using the values of $\pi(Z_i \mid X_i)$, as given in the above table.

$$\pi(Y = j \mid \mathbf{x}) = \max_{i=1}^{n} \pi(Z_i = j \mid x_i) * (*_{\ell \neq i} \Pi(Z_\ell \leq j \mid x_\ell)).$$

For $n = 2$, $m = 2$, when the X_i's are three-valued and Boolean, respectively, the following conditional tables are obtained (in the Boolean case, only 4 lines remain):

$\mathbf{x}$	$\pi(2 \mid \mathbf{x})$	$\pi(1 \mid \mathbf{x})$	$\pi(0 \mid \mathbf{x})$
$(2,2)$	1	$\max(\kappa_1^{12}, \kappa_2^{12})$	$\kappa_1^{02} * \kappa_2^{02}$
$(2,1)$	1	1	$\kappa_1^{02} * \kappa_2^{01}$
$(2,0)$	1	κ_1^{12}	κ_1^{02}
$(1,2)$	1	1	$\kappa_1^{01} * \kappa_2^{02}$
$(1,1)$	0	1	$\kappa_1^{01} * \kappa_2^{01}$
$(1,0)$	0	1	κ_1^{01}
$(0,2)$	1	κ_2^{12}	κ_2^{02}
$(0,1)$	0	1	κ_2^{01}
$(0,0)$	0	0	1

$\mathbf{x}$	$\pi(2 \mid \mathbf{x})$	$\pi(1 \mid \mathbf{x})$	$\pi(0 \mid \mathbf{x})$
$(2,2)$	1	$\max(\kappa_1^{12}, \kappa_2^{12})$	$\kappa_1^{02} * \kappa_2^{02}$
$(2,0)$	1	κ_1^{12}	κ_1^{02}
$(0,2)$	1	κ_2^{12}	κ_2^{02}
$(0,0)$	0	0	1

More generally, If we have m levels of strength, and n causal variables, we need $\frac{nm(m+1)}{2}$ coefficients for defining the uncertain MAX. If we take into account the leak, we have to add $\frac{m(m+1)}{2}$ coefficients per variable, in order to replace the 0 by a leak coefficient in the conditional tables $\pi(Z_i \mid X_i)$ (assuming that an effect of strong severity may take place even if the causes present have a weak intensity).

As for the uncertain MAX wrt uncertain OR, the uncertain MIN is a multiple-valued extension of the uncertain AND, where variables are valued on a the intensity scale $L = \{0 < 1 < \ldots < m\}$. We assume that $Y = \min(Z_1, \ldots, Z_n)$. We can then compute the conditional tables, as

$$\pi(y \mid x_1, \ldots, x_n) = \max_{z_1, \ldots, z_n : y = \min(z_1, \ldots, z_n)} *_{i=1}^{n} \pi(z_i \mid x_i)$$

$$= \max_{i=1}^{n} \pi(Z_i = y \mid x_i) * (*_{j \neq i} \Pi(Z_j \geq y \mid x_j)).$$

The conditional possibility tables are thus obtained by applying Eq. 5, using the same values of $\pi(Z_i \mid X_i)$, as in the case of the uncertain MAX. For $n = 2$, $m = 2$, this gives the following conditional tables (for ternary and binary inputs, respectively):

x	$\pi(2\mid\mathbf{x})$	$\pi(1\mid\mathbf{x})$	$\pi(0\mid\mathbf{x})$
(2,2)	1	$\max(\kappa_1^{12}, \kappa_2^{12})$	$\max(\kappa_1^{02}, \kappa_2^{02})$
(2,1)	0	1	$\max(\kappa_1^{02}, \kappa_2^{02})$
(2,0)	0	κ_1^{12}	1
(1,2)	0	1	$\max(\kappa_1^{01}, \kappa_2^{02})$
(1,1)	0	1	$\max(\kappa_1^{01}, \kappa_2^{01})$
(1,0)	0	0	1
(0,2)	0	κ_2^{12}	1
(0,1)	0	0	1
(0,0)	0	0	1

x	$\pi(2\mid\mathbf{x})$	$\pi(1\mid\mathbf{x})$	$\pi(0\mid\mathbf{x})$
(2,2)	1	$\max(\kappa_1^{12}, \kappa_2^{12})$	$\max(\kappa_1^{02}, \kappa_2^{02})$
(2,0)	0	κ_1^{12}	1
(0,2)	0	κ_2^{12}	1
(0,0)	0	0	1

As observed in the previous section when comparing the uncertain OR to the noisy OR, the simultaneous presence of a number of causes, which, taken in isolation, do not normally produce an effect, may lead to a plausible effect under a noisy MAX, which can never be the case with an uncertain MAX. Yet situations of this kind do arise in applications and are fully compatible with possibility theory. In order to make the elicitation of possibility tables describing such situations easy, an appropriate uncertain gate has to be designed, by providing a suitable uncertain function f which can trigger an effect through the accumulation of enough weak causes. One idea we have tested in order to approximate such behavior is the proposal of the uncertain MAX with thresholds, described in Sect. 4, which, in addition to the usual parameters of an uncertain MAX, takes a threshold θ_j for each value y_j of the effect variable Y (threshold gates also exist in the probabilistic setting [6]). Such threshold is an integer expressing the minimum number of causes that have to concur in order for effect y_j to become possible.

4   Implementation

A prototype involving the uncertain connectives defined above, allowing to execute possibilistic models such as the one described in Sect. 5 has been implemented in R. Here, we give some details about the practical implementation

---

Algorithm 1. UNCERTAIN-MAX(Y, prm).

Generate a conditional possibility table for variable Y given its causes $X_1, \ldots, X_n$ using the uncertain MAX with the given parameters prm.

Input: Y: the effect variable; $prm = \{\langle cond_i, \mathbf{k}_i \rangle\}$: a set of normalized possibility distributions
  $\mathbf{k}_i = (\kappa_{i1}, \ldots, \kappa_{i\|Y\|})$, $\max_{j=1,\ldots,\|Y\|}\{\kappa_{ij}\} = 1$, which apply when condition $cond_i$ holds;
  $cond_i = (\langle X_{ij}, x_{ij} \rangle)$, a (possibly empty) array of pairs of a cause variable X_{ij} and one of its
  values x_{ij}; $cond_i$ holds if $X_{ij} = x_{ij}$ holds for all j; an empty condition always holds.
Output: $\pi(Y \mid X_1, \ldots, X_n)$: a conditional possibility distribution of Y given its causes $X_1, \ldots, X_n$.
  1: $\pi(Y|X_1, \ldots, X_n) \leftarrow \mathbf{0}$
  2: **for all** $\mathbf{x} \in X_1 \times \ldots \times X_n$ **do**
  3:    $K \leftarrow \{\mathbf{k} : \langle cond_i, \mathbf{k} \rangle \in prm, \mathbf{x} \models cond_i\}$ {Select the parameters that apply to $\mathbf{x}$}
  4:    **for all** $\mathbf{y} = (y_1, \ldots, y_{\|K\|}) \in Y^{\|K\|}$ **do**
  5:       $\beta \leftarrow \min_{i=1,\ldots,\|K\|}\{\kappa_{iy_i}\}$
  6:       $\bar{y} \leftarrow \max_{i=1,\ldots,\|K\|}\{y_i\}$
  7:       $\pi(\bar{y} \mid \mathbf{x}) \leftarrow \max\{\beta, \pi(\bar{y} \mid \mathbf{x})\}$
  8:    **end for**
  9: **end for**
 10: **return** $\pi(Y \mid X_1, \ldots, X_n)$

---

of the uncertain connectives defined in the paper. Due to space limitations, we focus in particular on the uncertain MAX (and its variant with thresholds), whose implementation is non-trivial.

The way the uncertain MAX is implemented is shown in Algorithm 1. The parameter prm taken as input by this algorithm may be thought of as representing a set of rules of the form

$$X_{i_1} = x_{i_1} \wedge \ldots \wedge X_{i_m} = x_{i_m} \Rightarrow Y \sim (\kappa(y_1), \ldots, \kappa(y_n)), \tag{7}$$

where the X_{i_j} on the left-hand side are parent variables of Y in the possibilistic graphical model, the x_{i_j} are one of their values, and $(\kappa_1(y_1), \ldots, \kappa_1(y_n))$ is a normalized possibility distribution over the values of variable Y, i.e., for all $y \in Y$, $\kappa(y) \in [0, 1]$, and $\max_{y \in Y} \kappa(y) = 1$. Notice that this generalizes the uncertain gates to the case of multivalued variables. The left-hand side of a rule may be empty (i.e., $m = 0$): in that case, the rule is interpreted as if it were

$$\top \Rightarrow Y \sim (\kappa_\ell(y_1), \ldots, \kappa_\ell(y_n)). \tag{8}$$

Such rules may be used to represent leak coefficients, which apply to all possible combinations of causes.

The antecedents of the rules fed into the uncertain MAX must cover all possible combinations $\mathbf{x} \in X_1 \times \ldots \times X_n$ of the values of the parent variables of Y in order to ensure that the resulting conditional possibility distribution $\pi(Y \mid X_1, \ldots, X_n)$ be normalized. We may notice that, if a leak rule of the form of Eq. 8 is given, that rule alone already covers all combinations of parent variable values and is thus a sufficient condition for the normalization of $\pi(Y \mid X_1, \ldots, X_n)$; in that case, the parameters of the uncertain MAX may be underspecified.

The uncertain MAX with thresholds, whose implementation is shown in Algorithm 2, has an additional parameter, which consists of an array of thresholds $(\theta_1, \ldots, \theta_{\|Y\|})$, with $\theta_i \in \{1, 2, \ldots, \|X_1 \times \ldots \times X_n\|\}$. Each threshold is

---

Algorithm 2. UNCERTAIN-MAX-THRESHOLD(Y, *prm*, *thr*).

Generate a conditional possibility table for variable Y given its causes $X_1, \ldots, X_n$ using the uncertain MAX with thresholds with parameters *prm* and thresholds *thr*.

Input: Y: the effect variable; $prm = \{\langle cond_i, \mathbf{k}_i \rangle\}$: a set of normalized possibility distributions $\mathbf{k}_i = (\kappa_{i1}, \ldots, \kappa_{i\|Y\|})$, $\max_{j=1,\ldots,\|Y\|}\{\kappa_{ij}\} = 1$, which apply when condition $cond_i$ holds; $cond_i = (\langle X_{ij}, x_{ij} \rangle)$, a (possibly empty) array of pairs of a cause variable X_{ij} and one of its values x_{ij}; $cond_i$ holds if $X_{ij} = x_{ij}$ holds for all j; an empty condition always holds; $thr = (\theta_1, \ldots, \theta_{\|Y\|})$: the minimal number of combinations of values of the causes for which each value of Y is more possible than the leak.

Output: $\pi(Y \mid X_1, \ldots, X_n)$: a conditional possibility distribution of Y given its causes $X_1, \ldots, X_n$.

```
 1: π(Y|X₁,...,Xₙ) ← 0
 2: κℓ ← 0
 3: for all ⟨condᵢ, k⟩ ∈ prm : condᵢ = ⊤ do
 4:     κℓ ← max{κℓ, k}
 5: end for
 6: for all x ∈ X₁ × ... × Xₙ do
 7:     cnt ← 0 {A vector of counters, one for each y ∈ Y}
 8:     K ← {k : ⟨condᵢ, k⟩ ∈ prm, x ⊨ condᵢ} {Select the parameters that apply to x}
 9:     for all y = (y₁,...,y∥K∥) ∈ Y∥K∥ do
10:         β ← min_{i=1,...,∥K∥}{κᵢyᵢ}
11:         ȳ ← max_{i=1,...,∥K∥}{yᵢ}
12:         if β > κℓ(ȳ) then
13:             cnt_ȳ ← cnt_ȳ + 1
14:         end if
15:         if cnt_ȳ ≥ θ_ȳ then
16:             β ← 1
17:         end if
18:         π(ȳ | x) ← max{β, π(ȳ | x)}
19:     end for
20: end for
21: return π(Y | X₁,...,Xₙ)
```

---

associated with one value y of Y and represents the minimal number of combinations of the causes for which y is more possible than the baseline possibility given by the leak coefficients ($\kappa(y) > \kappa_\ell(y)$), or zero if no leak is provided.

5 Application

The metropolitan area of Aix-Marseille in southern France has experienced ongoing social polarization since the 1980s. The geography of unemployment, on the one hand, and the concentration of high-skilled professionals, on the other, contribute considerably to the structuring of a contrasted metropolitan social morphology [5,9]. Knowledge of factors inducing social polarization of the municipalities in the metropolitan area is nevertheless uncertain. Several factors contribute to the valorization or to the devalorization of the municipal residential space. But these factors have "soft", uncertain impacts on the phenomena under investigation: the same causes can not always lead to the same effects. A probabilistic model of these socio-spatial mechanisms has already been proposed [15] (cf. Fig. 1) in the form of a Bayesian network (BN). The BN was built using expert knowledge elicited through noisy logical gates (OR, AND, and MAX). We thus developed a possibilistic network (PN) using uncertain logical gates (OR, AND, standard MAX, and MAX-threshold) in order to link the same 26

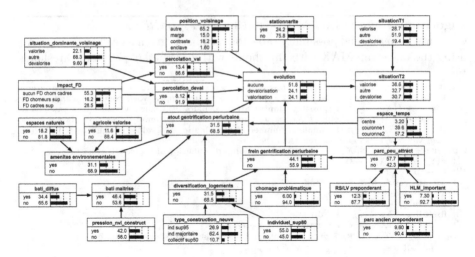

Fig. 1. The BN model for the valorization/devalorization of municipalities in the study area (from [15]).

variables of the BN (there is only one (ternary) MAX-threshold, with 7 parents). The parametrization of the PN was made compatible with the BN parametrization using the "most prudent" probability-to-possibility preference-preserving transformation, i.e., the T_1^{-1} converse transformation of [8], well suited to treating subjective probabilities, in order to transform probabilistic parameters into possibilistic ones.

In Fig. 2 we show how an uncertain OR logical gate can be used to generate a TPC. Only three parameters must be elicited: the possibilistic force of the two parent variables on the child variable (necessity of the consequence given that the parents are sufficient causes) and the leak parameter, which takes into account the activation of the consequence from secondary causes not included in the model. This table allows possibilistic inference from uncertain knowledge. If, for example, for a given municipality of the study area, we are relatively certain of having natural areas ($\Pi = 1$, $N = 0.5$) and if it is only partially possible that we have valorized agricultural areas ($\Pi = 0.5$), we can infer that it is relatively certain ($N = 0.5$) that the municipality in question has environmental amenities.

Another difference with the probabilistic model is the possibility of keeping track of the k_i parameters in the inference, in order to follow the sensitivity of results to the parameters of uncertain causation. The advantage of uncertain logical gates can be better appreciated in the whole model (Fig. 1). *Evolution* is, for example, a ternary variable (having three values: *no evolution*, *valorization*, and *devalorization*) depending on 5 binary variables and a 4-value variable. The TPC is thus made of $3 \times 2^5 \times 4 = 384$ parameters, whereas the uncertain MAX-threshold gate used in our PN model only requires 27 parameters.

Both the BN and the PN model were thus used to produce trend scenarios for social polarization in the 439 municipalities of the Aix-Marseille

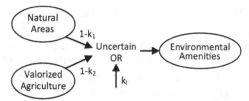

Where (1-k$_i$) = N (z$_i$|x$_i$) can be interpreted as « possibilistic force» of cause x$_i$
Π (*Environmental_Amenities* | *Natural_Areas, Valorized_Agriculture*) =
Uncertain OR (*leak* 0.1, *Natural_Areas* 0.8,*Valorized_Agriculture* 0.9)

Conditional Possibility Table for variable *Environmental_amenities*

Natural_Areas	Val_Agriculture	yes	no
yes	yes	1	min(k$_1$,k$_2$)=0.1
yes	no	1	K$_1$=0.1
no	yes	1	K$_2$=0.2
no	no	K$_l$=0.1	1

Fig. 2. Generation of a TPC through an uncertain OR logical gate.

metropolitan area. The future state variable that is inferred in these scenarios is the ternary variable *Situation T2*, having three possible values: *Valorized* (*V*), *Devalorized* (*D*) or *Other* (*O*).

Both scenarios are based on uncertain knowledge of relationships among variables and produce uncertain evaluation of the future state of the metropolitan area in terms of social polarization. Nevertheless, the probabilistic model infers a most probable value of *Situation T2* for each municipality. This often gives a fallacious impression of certainty: probability differences between inferred values can be relatively small. The possibilistic model, using a min-max logic, produces in many cases sets of completely possible values ($\Pi = 1$). We thus decided to test the significance of the probability differences in the BN model: only probability differences exceeding a given threshold were considered different. For a given threshold, we could thus infer even with the BN small sets of most probable values for some municipalities.

If no threshold is considered, the most probable values inferred by the BN and the completely possible values inferred by the PN coincide only in 54.7 % of cases. In the remaining cases, possibilistic results are more uncertain and always include probabilistic results (most probable values are always completely possible for the PN).

The best agreement between the two models is obtained with thresholds 0.20 and 0.25 (lower and higher values give worse results). 72.4 % and 77.2 % of the inferred values are then identical. Most probable values are almost always compatible with PN solutions: they are included in the completely possible values as, for example, when $\{V, O\}$ are the most probable values and $\{V, O, D\}$ is the set of completely possible values. The inverse is not always the case: depending on threshold value, 24 % and 18 % of possibilistic solutions are not included in the most probable values.

In conclusion, uncertain logical gates made the construction of the PN model possible. The use of most probable solutions of the BN model often gives a false impression of certainty. In order to compare results from the BN and the PN models, we need to enlarge the notion of most probable values: solutions whose probabilities differ less than 0.20/0.25 must be considered as equally probable. In this case, the solutions of the two models are identical for around three quarters of the municipalities of the study area. Despite this, the possibilistic model integrates a larger amount of uncertainty in the solutions inferred. Indeed, in the remaining quarter of municipalities, completely possible values inferred by the PN are normally larger sets than most probable values inferred by the BN. The BN model also tends to overestimate the valorization of municipalities in the study area: the PN model often infers complete uncertainty ($\{V, O, D\}$ all equally possible) whereas the most possible values are just V or $\{V, O\}$. A further analysis of the parametrization of the two models is nevertheless necessary in order to assess the origin of such a bias.

6   Conclusion

This is the first detailed study of the counterpart of the main probabilistic noisy gates for possibilistic networks, together with an illustrative implementation on a human geography application. Uncertain possibilistic gates are of primary interest for the practical use of possibilistic networks, when uncertainty has an epistemic flavor. The study has revealed some noticeable differences of behavior between noisy gates and uncertain possibilistic gates, in particular when the cumulation of causes having a rare effect may increase the plausibility of the effect. Generally speaking, possibilistic modeling appears to be more cautious. A detailed comparative study of the expressive power of Bayesian nets and possibilistic networks is a topic for further investigation, as well as the development of a complete panoply of uncertain possibilistic gates.

Acknowledgments. This work has been partially funded by CNRS PEPS Project Geo-Incertitude.

References

1. Antonucci, A.: The imprecise noisy-OR gate. In: Proceedings of 14th International Conference on Information Fusion (FUSION'11), Chicago, IL, 5–8, 1–7 July 2011
2. Amor, N., Benferhat, S., Mellouli, K.: Anytime propagation algorithm for min-based possibilistic graphs. Soft Comput. **8**, 150–161 (2003)
3. Benferhat, S., Dubois, D., Garcia, L., Prade, H.: On the transformation between possibilistic logic bases and possibilistic causal networks. Int. J. Approximate Reasoning **29**(2), 135–173 (2002)

4. Caglioni, M., Dubois, D., Fusco, G., Moreno, D., Prade, H., Scarella, F., Tettamanzi, A.: Mise en œuvre pratique de réseaux possibilistes pour modéliser la spécialisation sociale dans les espaces métropolisés. In: LFA 2014, pp. 267–274 (2014)
5. Centi, C.: Le Laboratoire Marseillais: Chemins d'Intégration Métropolitaine et de Segmentation Sociale. L'Harmattan, Paris (1996)
6. Díez, F., Drudzel, M.: Canonical probabilistic models for knowledge engineering. Technical report CISIAD-06-01 (2007)
7. Dubois, D., Prade, H.: Possibility Theory. Plenum Press, New York (1988)
8. Dubois, D., Prade, H., Sandri, S.: On possibility/probability transformations. In: Lowen, R., Roubens, M. (eds.) Fuzzy Logic, pp. 103–112. Kluwer Academic Publishers, Dordrecht (1993)
9. Fusco, G., Scarella, F.: Métropolisation et ségrégation sociospatiale. Les flux des migrations résidentielles en PACA. L'Espace Géographique 40(4), 319–336 (2011)
10. Henrion, M.: Some practical issues in constructing belief networks. In: Kanal, L., Levitt, T., Lemmer, J. (eds.) Uncertainty in Artificial Intelligence, vol. 3, pp. 161–173. Elsevier, New York (1989)
11. Jensen, F.: Bayesian Networks and Decision Graphs. Springer, New York (2001)
12. Piatti, A., Antonucci, A., Zaffalon, M.: Building knowledge-based expert systems by credal networks: a tutorial. In: Baswell, A.R. (ed.) Advances in Mathematics Research, vol. 11. Nova Science Publishers, New York (2010)
13. Parsons, S., Bigham, J.: Possibility theory and the generalized noisy OR model. In: Proceedings of 16th International Conference Information Processing and Management of Uncertainty (IPMU 1996), Granada, pp. 853–858 (1996)
14. Renooij, S., van der Gaag, L.C., Parsons, S.: Context-specific sign-propagation in qualitative probabilistic networks. Artif. Intell. 140(1/2), 207–230 (2002)
15. Scarella, F.: La ségrégation résidentielle dans l'espace-temps métropolitain: analyse spatiale et géo-prospective des dynamiques résidentielles de la métropole azuréenne. Ph.D. thesis, Université Nice Sophia Antipolis (2014)
16. Zadeh, L.A.: Fuzzy sets as a basis for a theory of possibility. Fuzzy Sets Syst. 1, 3–28 (1978)

Argumentation

Undercutting in Argumentation Systems

Leila Amgoud[1]([✉]) and Farid Nouioua[2]

[1] IRIT – CNRS, Toulouse, France
amgoud@irit.fr
[2] LSIS – Aix-Marseille University, Marseille, France
farid.nouioua@lsis.org

Abstract. Rule-based argumentation systems are developed for reasoning about defeasible information. They take as input a *theory* made of a set of *strict rules*, which encode strict information, and a set of *defeasible rules* which describe general behaviour with exceptional cases. They build *arguments* by chaining such rules, define *attacks* between them, use a *semantics* for evaluating the arguments, and finally identify the *plausible conclusions* that follow from the rules.

One of the main attack relations of such systems is the so-called *undercutting* which blocks the application of defeasible rules in some contexts. In this paper, we show that this relation is powerful enough to capture alone all the different conflicts in a theory. We present the first argumentation system that uses only undercutting and fully characterize both its extensions and its plausible conclusions under various acceptability semantics.

Keywords: Rule-based argumentation · Undercutting · Acceptability semantics

1 Introduction

Rule-based argumentation systems are developed for reasoning about defeasible information. As a major feature, they take as input a *theory* made of a set of *facts*, a set of *strict rules*, which encode strict information, and a set of *defeasible rules* which describe general behaviour with exceptional cases. They build *arguments* by chaining such rules, define *attacks* between them, use a *semantics* for evaluating the arguments, and finally identify the *plausible conclusions* that follow from the rules. Examples of such systems are ASPIC [2], its extended version ASPIC+ [14], Delp [8] and the system developed in [11]. Some of these systems satisfy the rationality postulates proposed in [3]. However, the plausible conclusions of any of these systems have never been characterized. Thus, despite the wide use of these systems, their outputs are still unknown.

Besides that, systems like Delp use *rebuttal* as attack relation between arguments. Rebuttal captures the fact that the conclusions of two arguments are inconsistent. Systems like ASPIC [2] and Pollock's system [13] use, in addition to rebuttal, *undercut* which blocks the application of defeasible rules in particular contexts. Let us illustrate this relation by an example borrowed from [13]. Consider the following argument (*a*):

© Springer International Publishing Switzerland 2015
C. Beierle and A. Dekhtyar (Eds.): SUM 2015, LNAI 9310, pp. 267–281, 2015.
DOI: 10.1007/978-3-319-23540-0_18

"The object is red (*or*) because it looks red (*lr*)".

This argument uses of the defeasible rule $lr \Rightarrow or$ (meaning that generally, if an object looks red, then it is red). Assume now another argument (*b*) which states the following:

"The rule $lr \Rightarrow or$ is inapplicable since the object is illuminated by a red light".

The argument *b* undercuts *a* and the conclusion (*or*) of *a* is not drawn from the theory. Undercut deals with the *exceptions* of defeasible rules. Indeed, every exception of a defeasible rule gives birth to an attack from any argument involving the exception toward any argument using the rule. In the example, being illuminated by a red light is a specific case where the rule $lr \Rightarrow or$ cannot be applied.

In this paper, we argue that undercut can do more than dealing with exceptions of defeasible rules. It can also perfectly play the role of rebuttal, and deal thus with inconsistency in a theory. The basic idea is the following: any defeasible rule $x \Rightarrow y$ should be blocked when $\neg y$ follows from the theory. We propose the first rule-based argumentation system that uses undercutting as its single attack relation. We show that it satisfies the rationality postulates discussed in [3] under naive, stable and preferred semantics. From a conceptual point of view, this system is much simpler than existing ones that combine rebuttal and undercut. For instance, in order to satisfy the postulates, ASPIC and ASPIC+ require a different variant of rebuttal for each semantics. Our system satisfies the postulates under all semantics. Moreover, restricted rebut, one of the variants of rebuttal, is based on an assumption which is not intuitive. Indeed, this relation compares only the rules whose heads are inconsistent, and neglects the remaining structure of the arguments. For instance, it considers that the argument $(x_1, x_1 \Rightarrow y_1, y_1 \rightarrow z)$ attacks the argument $(x_2, x_2 \rightarrow y_2, y_2 \Rightarrow \neg z)$ since z follows from a strict rule while $\neg z$ follows from a defeasible one. Note that the converse is not true even if the first rule of the first argument is defeasible while that of the second argument is strict. In our system, we do not make such assumptions. The second main contribution of the paper consists of providing the first and full characterizations of the extensions as well as the set of plausible conclusions of our system under naive, stable and preferred semantics proposed in [7].

The paper is organized as follows: Sect. 2 defines the rule-based system we are interested in, Sect. 3 analyses its properties, Sect. 4 characterizes its outputs (extensions and plausible conclusions), and Sect. 5 compares it with existing systems. The proofs can be downloaded from http://www.irit.fr/~Leila.Amgoud/sum15.pdf.

2   Rule-Based Systems

As in [1], three kinds of information are distinguished: *Facts* representing factual information like 'Tweety is a bird', *strict rules* representing strict information like 'Penguins do not fly' and *defeasible rules* describing general behavior with exceptional cases like 'Birds fly'. In what follows, $\mathcal{L}$ is a set of *literals*, i.e. atoms or negation of atoms, representing knowledge. The negation of an atom x from $\mathcal{L}$ is denoted $\neg x$. $\mathcal{L}'$ is a set of atoms used for naming rules. The two sets satisfy the constraint $\mathcal{L} \cap \mathcal{L}' = \emptyset$. Every rule has a single name and two rules cannot have the same name. Throughout the paper, rules are named $r, r_1, r_2, \ldots$. The function $\texttt{Rule}(r_i)$ returns the rule whose name is r_i.

- Facts are elements of $\mathcal{L}$.
- Defeasible rules are of the form $x_1, \ldots, x_n \Rightarrow x$ and $x, x_1 \ldots, x_n$ are literals in $\mathcal{L}$.
- Strict rules are of the form $x_1, \ldots, x_n \rightarrow x$ where $x_1, \ldots, x_n$ are literals of $\mathcal{L}$ and

$$\begin{cases} x \in \mathcal{L} & \text{or} \\ x \in \mathcal{L}' & \text{and } \texttt{Rule}(x) \text{ is defeasible.} \end{cases}$$

Note that the names of rules cannot appear in bodies of (strict or defeasible) rules. This means that it is not possible to represent information of the form "if rule r is applied (or is blocked), then y holds". Moreover, strict rules cannot be blocked. By default, any defeasible rule can be applied, unless explicitly mentioned in the language by strict rules $x_1, \ldots, x_n \rightarrow x$ with $x \in \mathcal{L}'$. Such a rule is read as follows: If $x_1, \ldots, x_n$ hold, then the defeasible rule x is always *not applicable*.

Definition 1 (Theory). *A theory is a triple $\mathcal{T} = (\mathcal{F}, \mathcal{S}, \mathcal{D})$ where $\mathcal{F} \subseteq \mathcal{L}$ is a set of facts and $\mathcal{S} \subseteq \mathcal{L}'$ (respectively $\mathcal{D} \subseteq \mathcal{L}'$) is a set of strict (respectively defeasible) rules.*

Notations: For each rule $x_1, \ldots, x_n \rightarrow x$ (as well as $x_1, \ldots, x_n \Rightarrow x$) whose name is r, the *head* of the rule is $\texttt{Head}(r) = x$ and the *body* of the rule is $\texttt{Body}(r) = \{x_1, \ldots, x_n\}$. Let $\mathcal{T} = (\mathcal{F}, \mathcal{S}, \mathcal{D})$ and $\mathcal{T}' = (\mathcal{F}', \mathcal{S}', \mathcal{D}')$ be two theories. We say that $\mathcal{T}$ is a *sub-theory* of $\mathcal{T}'$, written $\mathcal{T} \sqsubseteq \mathcal{T}'$, iff $\mathcal{F} \subseteq \mathcal{F}'$ and $\mathcal{S} \subseteq \mathcal{S}'$ and $\mathcal{D} \subseteq \mathcal{D}'$. The relation $\sqsubset$ is the strict version of $\sqsubseteq$ (i.e., it is the case that at least one of the three inclusions is strict). Finally, $\texttt{Defs}(\mathcal{T}) = \mathcal{D}$.

Let us now show how new information is produced from a given theory. This is generally the case when (strict and/or defeasible) rules are fired in a *derivation schema*.

Definition 2 (Derivation schema). *Let $\mathcal{T} = (\mathcal{F}, \mathcal{S}, \mathcal{D})$ be a theory and $x \in \mathcal{L} \cup \mathcal{L}'$. A derivation schema for x from $\mathcal{T}$ is a finite sequence $d = \langle (x_1, r_1), \ldots, (x_n, r_n) \rangle$ s.t.*

- $x_n = x$
- *for $i = 1 \ldots n$,*

- $x_i \in \mathcal{F}$ *and* $r_i = \emptyset$, *or*
- $r_i \in \mathcal{S} \cup \mathcal{D}$ *and* $\texttt{Head}(r_i) = x_i$ *and* $\texttt{Body}(r_i) \subseteq \{x_1, \ldots, x_{i-1}\}$

$\text{Seq}(d) = \{x_1, \ldots, x_n\}$.
$\text{Facts}(d) = \{x_i \mid i \in \{1, \ldots, n\}, r_i = \emptyset\}$.
$\text{Strict}(d) = \{r_i \mid i \in \{1, \ldots, n\}, r_i \in \mathcal{S}\}$.
$\text{Def}(d) = \{r_i \mid i \in \{1, \ldots, n\}, r_i \in \mathcal{D}\}$.
$\text{CN}(\mathcal{T})$ denotes the set of all literals that have a derivation schema from $\mathcal{T}$.

It is clear from the definition that CN is *monotonic*.

Example 1. Let $\mathcal{T}_1 = (\mathcal{F}_1, \mathcal{S}_1, \mathcal{D}_1)$ be a theory such that $\mathcal{F}_1 = \{p, b\}$, $\mathcal{S}_1 = \{(r_1)\ p \rightarrow \neg f\}$ and $\mathcal{D}_1 = \{(r_2)\ b \Rightarrow f\}$. From $\mathcal{T}_1$, we have the following minimal derivations:

- $d_1 = \langle (p, \emptyset) \rangle$
- $d_2 = \langle (b, \emptyset) \rangle$
- $d_3 = \langle (p, \emptyset), (\neg f, r_1) \rangle$
- $d_4 = \langle (b, \emptyset), (f, r_2) \rangle$

A notion of *consistency* and another of *coherence* are associated with this language.

Definition 3 (Consistency–Coherence). *A set $X \subseteq \mathcal{L}$ is consistent iff $\nexists x, y \in \mathcal{L}$ such that $x = \neg y$. It is inconsistent otherwise. A theory $\mathcal{T} = (\mathcal{F}, \mathcal{S}, \mathcal{D})$ is consistent iff $\text{CN}(\mathcal{T})$ is consistent. It is coherent iff $\text{CN}(\mathcal{T}) \cap \mathcal{D} = \emptyset$.*

The set of strict rules should be closed under transposition. This is required for ensuring the rationality postulates proposed in [3].

Definition 4 (Closure under transposition). *Let $\mathcal{S}$ be a set of strict rules. For any rule $r = x_1, \ldots, x_n \rightarrow x$ with $x \in \mathcal{L}$, r' is a transposition of r iff $r' = x_1, \ldots, x_{i-1}, \neg x, x_{i+1}, \ldots, x_n \rightarrow \neg x_i$ for some $1 \leq i \leq n$. We define $Cl_t(\mathcal{S})$ as the minimal set such that:*

- $\mathcal{S} \subseteq Cl_t(\mathcal{S})$, and
- *If $r \in Cl_t(\mathcal{S})$ and r' is a transposition of r then $r' \in Cl_t(\mathcal{S})$.*

We say that $\mathcal{S}$ is closed under transposition iff $Cl_t(\mathcal{S}) = \mathcal{S}$.

Throughout the paper, we will consider undercut for capturing *all* the possible conflicts between arguments. Thus, undercut will be used both for blocking general rules in presence of exceptions of such rules, and also for handling inconsistency. For that purpose, for each defeasible rule r, the theory should contain the strict rule $\neg\text{Head}(r) \rightarrow r$. This closure captures simply the fact that the two literals $\text{Head}(r)$ and $\neg\text{Head}(r)$ cannot hold at the same time.

Definition 5 (Closed theory). *A theory $\mathcal{T} = (\mathcal{F}, \mathcal{S}, \mathcal{D})$ is closed iff*

- $\mathcal{S}$ *is closed under transposition, and*
- *for every defeasible rule $r = x_1, \ldots, x_n \Rightarrow x \in \mathcal{D}$, $\neg x \rightarrow r \in \mathcal{S}$.*

Example 1 (Cont). The closed version of T_1 is $T_1' = (\mathcal{F}_1, \mathcal{S}_1', \mathcal{D}_1)$ such that $\mathcal{S}_1' = \{(r_1)\ p \rightarrow \neg f, (r_3)\ f \rightarrow \neg p, (r_4)\ \neg f \rightarrow r_2\}$.

The backbone of an argumentation system is naturally the notion of *arguments*. They are built from a closed theory using the notion of derivation schema as follows.

Definition 6 (Argument). *Let* $T = (\mathcal{F}, \mathcal{S}, \mathcal{D})$ *be a closed theory. An* argument *defined from* T *is a pair* (d, x) *s.t.*

- $x \in \mathcal{L} \cup \mathcal{L}'$
- d *is a derivation schema for* x *from* T
- $\sharp T' \sqsubset (\texttt{Facts}(d), \texttt{Strict}(d), \texttt{Def}(d))$ *s.t.* $x \in \texttt{CN}(T')$

An argument (d, x) *is* strict *iff* $\texttt{Def}(d) = \emptyset$.

Unlike ASPIC and ASPIC+ systems, arguments are minimal in our system. An argument may have several sub-parts, each of which is called *sub-argument*.

Definition 7 (Sub-argument). *An argument* (d, x) *is a* sub-argument *of* (d', x') *iff* $(\texttt{Facts}(d), \texttt{Strict}(d), \texttt{Def}(d)) \sqsubseteq (\texttt{Facts}(d'), \texttt{Strict}(d'), \texttt{Def}(d'))$.

Notations: $\texttt{Arg}(T)$ denotes the set of all arguments built from theory T in the sense of Definition 6. If $a = (d, x)$ is an argument, $\texttt{Conc}(a) = x$ and $\texttt{Sub}(a)$ is the set of all its sub-arguments. For a set $\mathcal{E}$ of arguments, $\texttt{Concs}(\mathcal{E}) = \{x \mid (d, x) \in \mathcal{E}\}$ and $\texttt{Th}(\mathcal{E})$ is a theory such that:

$$\texttt{Th}(\mathcal{E}) = (\bigcup_{(d,x) \in \mathcal{E}} \texttt{Facts}(d), \bigcup_{(d,x) \in \mathcal{E}} \texttt{Strict}(d), \bigcup_{(d,x) \in \mathcal{E}} \texttt{Def}(d)).$$

The undercutting relation is defined as follows:

Definition 8 (Undercutting). *Let* $T = (\mathcal{F}, \mathcal{S}, \mathcal{D})$ *be a closed theory and* $(d, x), (d', x') \in \texttt{Arg}(T)$. (d, x) undercuts (d', x'), *denoted by* $(d, x)\ \mathcal{R}_u\ (d', x')$, *iff* $x \in \texttt{Def}(d')$.

Example 1 (Cont). The set $\texttt{Arg}(T_1')$ contains:

- $a_1 : (\langle (b, \emptyset) \rangle, b)$
- $a_2 : (\langle (p, \emptyset) \rangle, p)$
- $a_3 : (\langle (p, \emptyset), (\neg f, r_1) \rangle, \neg f)$
- $a_4 : (\langle (p, \emptyset), (\neg f, r_1), (r_2, r_4) \rangle, r_2)$
- $a_5 : (\langle (b, \emptyset), (f, r_2) \rangle, f)$
- $a_6 : (\langle (b, \emptyset), (f, r_2), (\neg p, r_3) \rangle, \neg p)$

a_4 undercuts both a_5 and a_6 since $r_2 \in \texttt{Def}(d_5)$ and $r_2 \in \texttt{Def}(d_6)$.

Strict arguments cannot be attacked using this relation.

Proposition 1. *Let* $T = (\mathcal{F}, \mathcal{S}, \mathcal{D})$ *be a theory. For any argument* $a \in \texttt{Arg}((\mathcal{F}, \mathcal{S}, \emptyset))$, $\nexists b \in \texttt{Arg}(T)$ *such that* $b \mathcal{R}_u a$.

Note that self-attacking arguments may exist.

Example 2. *Consider the theory* $\mathcal{T}_2 = (\mathcal{F}_2, \mathcal{S}_2, \mathcal{D}_2)$ *s.t.* $\mathcal{F}_2 = \{x\}$, $\mathcal{S}_2 = \{(r_1)\ t \rightarrow r_2\}$, *and* $\mathcal{D}_2 = \{(r_2)\ x \Rightarrow t\}$. *The set* $\mathrm{Arg}(\mathcal{T}_2)$ *contains the three arguments:*

- $a_1 : (\langle\langle(x, \emptyset)\rangle, x\rangle)$
- $a_2 : (\langle\langle(x, \emptyset), (t, r_2)\rangle, t\rangle)$
- $a_3 : (\langle\langle(x, \emptyset), (t, r_2), (r_2, r_1)\rangle, r_2\rangle)$

The argument a_3 *undercuts itself and* a_2.

Throughout the paper, we study the following rule-based argumentation system.

Definition 9 (AS). *An argumentation system (AS) defined over a closed theory* $\mathcal{T} = (\mathcal{F}, \mathcal{S}, \mathcal{D})$ *is a pair* $\mathcal{H} = (\mathrm{Arg}(\mathcal{T}), \mathcal{R}_u)$ *where* $\mathcal{R}_u \subseteq \mathrm{Arg}(\mathcal{T}) \times \mathrm{Arg}(\mathcal{T})$.

Arguments are evaluated using extension-based semantics [7]. These semantics are based on two key notions:

- *Conflict-freeness*: A set $\mathcal{E}$ of arguments is conflict-free iff $\nexists a, b \in \mathcal{E}$ s.t. $a\mathcal{R}_u b$.
- *Defence*: A set $\mathcal{E}$ of arguments defends an argument a iff for all argument b s.t. $b\mathcal{R}_u a$, $\exists c \in \mathcal{E}$ s.t. $c\mathcal{R}_u b$.

Definition 10 (Semantics). *Let* $\mathcal{H} = (\mathrm{Arg}(\mathcal{T}), \mathcal{R}_u)$ *be an argumentation system defined over a closed theory* $\mathcal{T}$ *and* $\mathcal{E} \subseteq \mathrm{Arg}(\mathcal{T})$.

- $\mathcal{E}$ *is a* naive *extension iff it is a maximal (w.r.t. set* $\subseteq$) *conflict-free set.*
- $\mathcal{E}$ *is a* preferred *extension iff it is a maximal (w.r.t. set* $\subseteq$) *conflict-free set which defends all its elements.*
- $\mathcal{E}$ *is a* stable *extension iff* $\mathcal{E}$ *is conflict-free and* $\forall a \in \mathrm{Arg}(\mathcal{T}) \setminus \mathcal{E}$, $\exists b \in \mathcal{E}$ *such that* $b\mathcal{R}_u a$.

Notations: $\mathrm{Ext}_x(\mathcal{H})$ denotes the set of all extensions of system $\mathcal{H}$ under semantics x where $x \in \{n, p, s\}$, n (resp. p, s) stands for naive (resp. preferred, stable). When we do not need to refer to a particular semantics, we write $\mathrm{Ext}(\mathcal{H})$ for short.

The extensions of a system are used for defining the *plausible conclusions* to be drawn from the theory over which the system is built. A literal is a plausible conclusion iff it is a common conclusion to all the extensions.

Definition 11 (Plausible conclusions). *The set of* plausible conclusions *of an argumentation system* $\mathcal{H}$ *is*

$$\mathrm{Output}(\mathcal{H}) = \begin{cases} \emptyset & \text{if } \mathrm{Ext}(\mathcal{H}) = \emptyset \\ \bigcap_{\mathcal{E}_i \in \mathrm{Ext}(\mathcal{H})} \mathrm{Concs}(\mathcal{E}_i) & \text{else.} \end{cases}$$

Example 1 (Cont). The argumentation system $\mathcal{H}_1 = (\mathsf{Arg}(\mathcal{T}_1'), \mathcal{R}_u)$ has a single stable extension which is also preferred: $\mathcal{E} = \{a_1, a_2, a_3, a_4\}$. Thus, $\mathsf{Output}(\mathcal{H}_1) = \{p, b, \neg f, r_2\}$.

Example 2 (Cont). The argumentation system $\mathcal{H}_2 = (\mathsf{Arg}(\mathcal{T}_2), \mathcal{R}_u)$ has a single preferred extension: $\mathcal{E} = \{a_1\}$ and thus $\mathsf{Output}(\mathcal{H}_2) = \{x\}$. However, $\mathsf{Output}(\mathcal{H}_2) = \emptyset$ under stable semantics since $\mathsf{Ext}_s(\mathcal{H}) = \emptyset$.

3   Properties of the System

Let us now analyse the properties of the argumentation system defined in the previous section. We show that it satisfies all the rationality postulates proposed in [3]. Indeed, every extension (under any of the reviewed semantics) contains all the sub-arguments of its arguments. The system is also coherent, that is it is not possible for an extension to use a defeasible rule in one of its arguments, and at the same time to block that rule by another argument. In addition, for preferred and stable semantics, every extension returns a consistent set of conclusions (unless the strict part of the theory is inconsistent) and the set of conclusions of every extension is closed under strict rules (under stable and preferred semantics), that is it is not possible that an extension supports a conclusion x and forgets y if $x \to y \in \mathcal{S}$.

Theorem 1. *Let $\mathcal{H} = (\mathsf{Arg}(\mathcal{T}), \mathcal{R}_u)$ be an argumentation system built over a closed theory $\mathcal{T} = (\mathcal{F}, \mathcal{S}, \mathcal{D})$ s.t. $\mathsf{Ext}(\mathcal{H}) \neq \emptyset$. For all $\mathcal{E} \in \mathsf{Ext}(\mathcal{H})$, the following hold:*

- *The theory $\mathsf{Th}(\mathcal{E})$ is coherent,*
- *For each $a \in \mathcal{E}$, $\mathsf{Sub}(a) \subseteq \mathcal{E}$,*

Under stable and preferred semantics, consistency and closure under strict rules are also satisfied. However, both properties are violated under naive semantics. This is not surprising since naive semantics does not take into account the orientation of attacks, and thus the crucial distinction between strict rules and defeasible ones.

Theorem 2. *Let $\mathcal{H} = (\mathsf{Arg}(\mathcal{T}), \mathcal{R}_u)$ be an argumentation system built over a closed theory $\mathcal{T} = (\mathcal{F}, \mathcal{S}, \mathcal{D})$ s.t. $\mathsf{Ext}_x(\mathcal{H}) \neq \emptyset$ with $x \in \{s, p\}$. For each $\mathcal{E} \in \mathsf{Ext}_x(\mathcal{H})$, the following hold:*

- $\mathsf{Concs}(\mathcal{E})$ *is consistent iff* $\mathsf{CN}((\mathcal{F}, \mathcal{S}, \emptyset))$ *is consistent,*
- $\mathsf{Concs}(\mathcal{E}) = \mathsf{CN}((\mathsf{Concs}(\mathcal{E}), \mathcal{S}, \emptyset))$,

The following properties follow from the previous theorem.

Corollary 1. *Let $\mathcal{H} = (\mathsf{Arg}(\mathcal{T}), \mathcal{R}_u)$ be an argumentation system built over a closed theory $\mathcal{T} = (\mathcal{F}, \mathcal{S}, \mathcal{D})$ s.t. $\mathsf{Ext}_x(\mathcal{H}) \neq \emptyset$ with $x \in \{s, p\}$. The following hold:*

- Output($\mathcal{H}$) *is consistent iff* CN($(\mathcal{F}, \mathcal{S}, \emptyset)$) *is consistent,*
- Output($\mathcal{H}$) $=$ CN((Output($\mathcal{H}$), $\mathcal{S}, \emptyset$)).

The previous results show that the outcomes of the argumentation system (its extensions and set of plausible conclusions) satisfy nice properties under stable and preferred semantics. However, they do not say anything about the kind of conclusions the system draws from a theory. We answer this question in the next section.

4   The Outputs of the System

This section provides formal characterizations of the outputs of the system under the three reviewed semantics. For each semantics, we characterize the extensions in terms of sub-theories of the theory over which the system is built, delimit the number of extensions, and fully characterize the set of plausible conclusions.

4.1   Naive Semantics

A sub-theory that corresponds to a naive extension is called *option*.

Definition 12 (Option). *An* option *of a closed theory* $\mathcal{T} = (\mathcal{F}, \mathcal{S}, \mathcal{D})$ *is a sub-theory* $(\mathcal{F}', \mathcal{S}', \mathcal{D}')$ *such that*

- $\mathcal{F}' = \mathcal{F}$, $\mathcal{S}' \subseteq \mathcal{S}$ *and* $\mathcal{D}' \subseteq \mathcal{D}$
- $(\mathcal{F}', \mathcal{S}', \mathcal{D}')$ *is coherent*
- $\forall r \in \mathcal{S}' \cup \mathcal{D}'$, Body($r$) $\subseteq$ CN($(\mathcal{F}', \mathcal{S}', \mathcal{D}')$)
- $\nexists \mathcal{S}'', \mathcal{D}''$ *such that* $(\mathcal{F}', \mathcal{S}', \mathcal{D}') \sqsubset (\mathcal{F}', \mathcal{S}'', \mathcal{D}'')$ *and* $(\mathcal{F}', \mathcal{S}'', \mathcal{D}'')$ *satisfies the previous conditions.*

Opt($\mathcal{T}$) *denotes the set of options of the closed theory* $\mathcal{T}$.

Thus, an option is obtained by taking all the facts and a maximal (w.r.t set inclusion) subset of (strict and defeasible) rules so that the sub-theory remains coherent and all the added rules are applicable. Notice that no priority is given to strict rules over defeasible ones. This is explained by the fact that naive semantics does not distinguish between attackers and attacked arguments.

Example 3. *Consider the closed theory* $\mathcal{T}_3 = (\mathcal{F}_3, \mathcal{S}_3, \mathcal{D}_3)$:

$$\mathcal{F}_3 \begin{cases} x \\ y \end{cases} \qquad \mathcal{S}_3 \begin{cases} t \to r_2 & (r_4) \\ u \to r_1 & (r_5) \\ s \to r_3 & (r_6) \end{cases} \qquad \mathcal{D}_3 \begin{cases} x \Rightarrow t & (r_1) \\ y \Rightarrow u & (r_2) \\ t \Rightarrow s & (r_3) \end{cases}$$

The theory $\mathcal{T}_3$ *has three options:*

- $\mathcal{O}_1 = (\mathcal{F}_3, \emptyset, \{r_1, r_2, r_3\})$ $\qquad\qquad$ CN($\mathcal{O}_1$) $= \{x, y, t, u, s\}$
- $\mathcal{O}_2 = (\mathcal{F}_3, \{r_4\}, \{r_1, r_3\})$ $\qquad\qquad$ CN($\mathcal{O}_2$) $= \{x, y, t, s, r_2\}$
- $\mathcal{O}_3 = (\mathcal{F}_3, \{r_5\}, \{r_2\})$ $\qquad\qquad$ CN($\mathcal{O}_3$) $= \{x, y, u, r_1\}$

Let us now establish the relationship between naive extensions of an argumentation system and the options of the closed theory over which it is built. Each naive extension returns one option and two naive extensions cannot return the same option.

Theorem 3. *Let $\mathcal{H} = (\mathrm{Arg}(\mathcal{T}), \mathcal{R}_u)$ be an AS built over a closed theory $\mathcal{T}$.*

- *For all $\mathcal{E} \in \mathrm{Ext}_n(\mathcal{H})$, there exists a single option $\mathcal{O} \in \mathrm{Opt}(\mathcal{T})$ such that $\mathrm{Th}(\mathcal{E}) = \mathcal{O}$ and $\mathrm{Concs}(\mathcal{E}) = \mathrm{CN}(\mathcal{O})$. We put: $\mathrm{Option}(\mathcal{E}) \overset{\mathrm{def}}{=} \mathcal{O}$.*
- *For all $\mathcal{E}, \mathcal{E}' \in \mathrm{Ext}_n(\mathcal{H})$, if $\mathrm{Option}(\mathcal{E}) = \mathrm{Option}(\mathcal{E}')$ then $\mathcal{E} = \mathcal{E}'$.*
- *For all $\mathcal{E} \in \mathrm{Ext}_n(\mathcal{H})$, $\mathcal{E} = \mathrm{Arg}(\mathrm{Option}(\mathcal{E}))$.*

The following theorem shows that inversely, each option leads to one naive extension and two different options do not return the same naive extension.

Theorem 4. *Let $\mathcal{H} = (\mathrm{Arg}(\mathcal{T}), \mathcal{R}_u)$ be an AS built over a closed theory $\mathcal{T}$.*

- *For all $\mathcal{O} \in \mathrm{Opt}(\mathcal{T})$, $\mathrm{Arg}(\mathcal{O}) \in \mathrm{Ext}_n(\mathcal{H})$.*
- *For all $\mathcal{O} \in \mathrm{Opt}(\mathcal{T})$, $\mathcal{O} = \mathrm{Option}(\mathrm{Arg}(\mathcal{O}))$.*
- *For all $\mathcal{O}_1, \mathcal{O}_2 \in \mathrm{Opt}(\mathcal{T})$, if $\mathrm{Arg}(\mathcal{O}_1) = \mathrm{Arg}(\mathcal{O}_2)$, $\mathcal{O}_1 = \mathcal{O}_2$.*

Example 3 (Cont). The arguments built from $\mathcal{T}_3$ are summarized below.

- $a_1 : (\langle (x, \emptyset) \rangle, x)$
- $a_2 : (\langle (y, \emptyset) \rangle, y)$
- $a_3 : (\langle (x, \emptyset), (t, r_1) \rangle, t)$
- $a_4 : (\langle (x, \emptyset), (t, r_1), (r_2, r_4) \rangle, r_2)$
- $a_5 : (\langle (y, \emptyset), (u, r_2) \rangle, u)$
- $a_6 : (\langle (y, \emptyset), (u, r_2), (r_1, r_5) \rangle, r_1)$
- $a_7 : (\langle (x, \emptyset), (t, r_1), (s, r_3) \rangle, s)$
- $a_8 : (\langle (x, \emptyset), (t, r_1), (s, r_3), (r_3, r_6) \rangle, r_3)$

The graph of attacks is depicted in the Fig. 1 below:

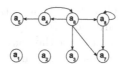

Fig. 1. Graph of attacks built from the theory $\mathcal{T}_3$

The AS $\mathcal{H}_3 = (\mathrm{Arg}(\mathcal{T}_3), \mathcal{R}_u)$ has three naive extensions $\mathcal{E}_1 = \{a_1, a_2, a_3, a_5, a_7\}$, $\mathcal{E}_2 = \{a_1, a_2, a_3, a_4, a_7\}$ and $\mathcal{E}_3 = \{a_1, a_2, a_5, a_6\}$ which capture the options $\mathcal{O}_1$, $\mathcal{O}_2$ and $\mathcal{O}_3$ respectively. Indeed, $\mathrm{Th}(\mathcal{E}_1) = \mathcal{O}_1$ (resp. $\mathrm{Th}(\mathcal{E}_2) = \mathcal{O}_2$, $\mathrm{Th}(\mathcal{E}_3) = \mathcal{O}_3$) and $\mathrm{Concs}(\mathcal{E}_1) = \mathrm{CN}(\mathcal{O}_1)$ (resp. $\mathrm{Concs}(\mathcal{E}_2) = \mathrm{CN}(\mathcal{O}_2)$, $\mathrm{Concs}(\mathcal{E}_3) = \mathrm{CN}(\mathcal{O}_3)$).

From the previous correspondence, the number of naive extensions is delimited.

Corollary 2. *Let* $\mathcal{H} = (\mathrm{Arg}(\mathcal{T}), \mathcal{R}_u)$ *be an AS. It holds that* $|\mathrm{Ext}_n(\mathcal{H})| = |\mathrm{Opt}(\mathcal{T})|$.

The plausible conclusions of an argumentation system under naive semantics are the literals that follow from all the options of the theory over which the system is built.

Corollary 3. *Let* $\mathcal{H} = (\mathrm{Arg}(\mathcal{T}), \mathcal{R}_u)$ *be an AS.* $\mathrm{Output}(\mathcal{H}) = \bigcap_{\mathcal{O} \in \mathrm{Opt}(\mathcal{T})} \mathrm{CN}(\mathcal{O})$.

Example 3 (Cont). Under naive semantics, $\mathrm{Output}(\mathcal{H}) = \mathrm{CN}(\mathcal{O}_1) \cap \mathrm{CN}(\mathcal{O}_2) \cap \mathrm{CN}(\mathcal{O}_3) = \{x, y\}$.

4.2   Stable Semantics

The sub-theories of a closed theory that capture stable extensions are called *strong options* and are defined as follows:

Definition 13 (Strong Option). *A strong option of a closed theory* $\mathcal{T} = (\mathcal{F}, \mathcal{S}, \mathcal{D})$ *is a sub-theory* $(\mathcal{F}', \mathcal{S}', \mathcal{D}')$ *such that*

- $\mathcal{F}' = \mathcal{F}$, $\mathcal{S}' = \mathcal{S}$ *and* $\mathcal{D}' \subseteq \mathcal{D}$
- $(\mathcal{F}', \mathcal{S}', \mathcal{D}')$ *is coherent*
- $\forall r \in \mathcal{D}'$, $\mathrm{Body}(r) \subseteq \mathrm{CN}((\mathcal{F}', \mathcal{S}', \mathcal{D}'))$
- $\forall r \notin \mathcal{D}'$ *we have: either* $r \in \mathrm{CN}(\mathcal{F}', \mathcal{S}', \mathcal{D}')$ *or* $\exists x \in \mathrm{Body}(r)$ *such that* $x \notin \mathrm{CN}(\mathcal{F}', \mathcal{S}', \mathcal{D}')$

$\mathrm{SOpt}(\mathcal{T})$ *denotes the set of strong options of theory* $\mathcal{T}$.

In a strong option $\mathcal{O} = (\mathcal{F}, \mathcal{S}, \mathcal{D}')$, it is not necessary that all the strict rules of $\mathcal{S}$ are applicable. Let S'' be the subset of strict rules that are applicable in $\mathcal{O}$, i.e., $S'' = \{r \in \mathcal{S} \mid \mathrm{Body}(r) \subseteq \mathrm{CN}(\mathcal{O})\}$. Then, the sub-theory $\mathcal{O}' = (\mathcal{F}, S'', \mathcal{D}')$ is an option of $\mathcal{T}$ which clearly has the same conclusions as $\mathcal{O}$ (i.e., $\mathrm{CN}(\mathcal{O}) = \mathrm{CN}(\mathcal{O}')$). In addition, every strict (resp. defeasible) rule r which is kept outside $\mathcal{O}'$ is not applicable (resp. is not applicable or is such that $r \in \mathrm{CN}(\mathcal{O}')$). This latter constraint does not hold necessarily for every option. Accordingly, every strong option corresponds to a single option but the converse is not true.

Thus, in addition to an "internal condition" (coherence) satisfied by both options and strong options, the latter require an additional "external condition" which consists of *justifying* each rule kept outside. Notice, that this idea is not new in non-monotonic reasoning. We find it namely in the distinction between Reiter's extensions [15] and Lukaszewicz's extensions [12] in default logic as well as between answer sets [10] and ι-answer sets [9] in logic programming. Let us illustrate strong options and their relationship with options in our running example.

Example 3 (Cont). The theory $\mathcal{T}_3$ has one strong option $\mathcal{O} = (\mathcal{F}_3, \mathcal{S}_3, \{r_2\})$. Note that the only strict rule in $\mathcal{S}_3$ which is applicable for $\mathcal{O}$ is r_5. If we discard from $\mathcal{O}$ the remaining non-applicable strict rules, we get exactly the option $\mathcal{O}_3$

($\text{CN}(\mathcal{O}) = \text{CN}(\mathcal{O}_3)$). Note also that each rule which is not included in $\mathcal{O}_3$ is justified. Namely, the strict rules r_4 and r_6 are note applicable ($t \in \text{Body}(r_4)$, $t \notin \text{CN}(\mathcal{O}_3)$, $s \in \text{Body}(r_6)$, and $s \notin \text{CN}(\mathcal{O}_3)$); the defeasible rule r_1 is such that $r_1 \in \text{CN}(\mathcal{O}_3)$ and the defeasible rule r_3 is not applicable ($t \in \text{Body}(r_3)$ and $t \notin \text{CN}(\mathcal{O}_3)$). So $\mathcal{O}_3$ gives rise to a strong option by adding all the non-applicable strict rules. This is not the case for $\mathcal{O}_1$ and $\mathcal{O}_2$. Indeed, adding the missing strict rules to them leads to incoherent sub-theories.

It is worthy to say that a closed theory may not have strong options. This is not surprising since as we will show, there is a bijection between the set of stable extensions and the set of strong options. Indeed, every stable extension gives birth to a strong option and two stable extensions cannot return the same strong option.

Theorem 5. *Let* $\mathcal{H} = (\text{Arg}(\mathcal{T}), \mathcal{R}_u)$ *be an argumentation system built over a closed theory* $\mathcal{T}$ *s.t.* $\text{Ext}_s(\mathcal{H}) \neq \emptyset$.

- *For all* $\mathcal{E} \in \text{Ext}_s(\mathcal{H})$, *there exists a single strong option* $\mathcal{O} \in \text{SOpt}(\mathcal{T})$ *s.t.* $\text{Th}(\mathcal{E}) \sqsubseteq \mathcal{O}$ *and* $\text{Concs}(\mathcal{E}) = \text{CN}(\mathcal{O})$. *We put* $\text{SOption}(\mathcal{E}) \stackrel{\text{def}}{=} \mathcal{O}$.
- *For all* $\mathcal{E}, \mathcal{E}' \in \text{Ext}_s(\mathcal{H})$, *if* $\text{SOption}(\mathcal{E}) = \text{SOption}(\mathcal{E}')$ *then* $\mathcal{E} = \mathcal{E}'$.
- *For all* $\mathcal{E} \in \text{Ext}_s(\mathcal{H})$, $\mathcal{E} = \text{Arg}(\text{SOption}(\mathcal{E}))$.

Inversely, every strong option leads to one stable extension and two strong options cannot lead the same stable extension.

Theorem 6. *Let* $\mathcal{H} = (\text{Arg}(\mathcal{T}), \mathcal{R}_u)$ *be an argumentation system built over a closed theory* $\mathcal{T}$ *s.t.* $\text{Ext}_s(\mathcal{H}) \neq \emptyset$.

- *For all* $\mathcal{O} \in \text{SOpt}(\mathcal{T})$, $\text{Arg}(\mathcal{O}) \in \text{Ext}_s(\mathcal{H})$.
- *For all* $\mathcal{O} \in \text{SOpt}(\mathcal{T})$, $\mathcal{O} = \text{SOption}(\text{Arg}(\mathcal{O}))$.
- *For all* $\mathcal{O}_1, \mathcal{O}_2 \in \text{SOpt}(\mathcal{T})$, *if* $\text{Arg}(\mathcal{O}_1) = \text{Arg}(\mathcal{O}_2)$ *then* $\mathcal{O}_1 = \mathcal{O}_2$.

Example 3 (Cont). Among the three naive extensions of the argumentation system $\mathcal{H}_3$ built from $\mathcal{T}_3$, the only stable extension is $\mathcal{E}_3$ which captures the strong options $\mathcal{O}$. Indeed, $\text{Th}(\mathcal{E}_3) \sqsubseteq \mathcal{O}$ and $\text{Concs}(\mathcal{E}_3) = \text{CN}(\mathcal{O})$.

We have seen so far that there is a one to one correspondence between naive (resp. stable) extensions and options (resp. strong options). We have also shown that every strong option is a sub-theory of one option. Thus, the number of stable extensions of a rule-based system is delimited as follows.

Corollary 4. *Let* $\mathcal{H} = (\text{Arg}(\mathcal{T}), \mathcal{R}_u)$ *be an argumentation system built over a closed theory* $\mathcal{T}$. *The following holds:* $0 \leq |\text{Ext}_s(\mathcal{H})| = |\text{SOpt}(\mathcal{T})| \leq |\text{Opt}(\mathcal{T})|$.

Under stable semantics, the plausible conclusions of an AS are the literals that follow from all the strong options of the theory over which the system is built.

Corollary 5. *Let* $\mathcal{H} = (\text{Arg}(\mathcal{T}), \mathcal{R}_u)$ *be an argumentation system built over a closed theory* $\mathcal{T}$ *s.t.* $\text{Ext}_s(\mathcal{H}) \neq \emptyset$. $\text{Output}(\mathcal{H}) = \bigcap_{\mathcal{O} \in \text{SOpt}(\mathcal{T})} \text{CN}(\mathcal{O})$.

Example 3 (Cont). $\mathcal{O}$ is the only strong option of $\mathcal{T}_3$. Thus, $\texttt{Output}(\mathcal{H}) = \texttt{CN}(\mathcal{O}) = \{x, y, u, r_1\}$.

Let us summarize: rule-based argumentation systems may not have stable extensions in which case they miss intuitive conclusions like facts. Systems that do have stable extensions return exactly the literals that follow from all the strong options of the closed theory at hand.

4.3   Preferred Semantics

We show next that the sub-theories that capture preferred extensions are the so-called *preferred options*.

Definition 14. (Preferred Option). *A* preferred option *of a closed theory* $\mathcal{T} = (\mathcal{F}, \mathcal{S}, \mathcal{D})$ *is a sub-theory* $(\mathcal{F}', \mathcal{S}', \mathcal{D}')$ *s.t.*

- $\mathcal{F}' = \mathcal{F}$, $\mathcal{S}' = \mathcal{S}$ *and* $\mathcal{D}' \subseteq \mathcal{D}$
- $(\mathcal{F}', \mathcal{S}', \mathcal{D}')$ *is coherent*
- $\forall r \in \mathcal{D}'$, $\texttt{Body}(r) \subseteq \texttt{CN}((\mathcal{F}', \mathcal{S}', \mathcal{D}'))$
- $\forall \mathcal{D}'' \subseteq \mathcal{D}$, *if* $\exists r' \in \mathcal{D}'$ *such that* $r' \in \texttt{CN}(\mathcal{F}, \mathcal{S}, \mathcal{D}'')$ *then* $\exists r'' \in \mathcal{D}''$ *such that* $r'' \in \texttt{CN}(\mathcal{F}, \mathcal{S}, \mathcal{D}')$
- $\nexists \mathcal{D}''$ *such that* $\mathcal{D}' \subset \mathcal{D}''$ *and* $(\mathcal{F}', \mathcal{S}', \mathcal{D}'')$ *satisfies the previous conditions.*

$\texttt{POpt}(\mathcal{T})$ *denotes the set of preferred options of theory* $\mathcal{T}$.

Preferred options are between options and strong options of a theory $\mathcal{T}$.

- Every strong option of $\mathcal{T}$ is a preferred option of $\mathcal{T}$. The converse is not true.
- Every preferred option is a sub-part of an option. More precisely, for every preferred option $\mathcal{O} = (\mathcal{F}, \mathcal{S}, \mathcal{D}')$, if $\mathcal{S}''$ is the subset of strict rules that are applicable in $\mathcal{O}$, i.e., $\mathcal{S}'' = \{r \in \mathcal{S} \mid \texttt{Body}(r) \subseteq \texttt{CN}(\mathcal{O})\}$, then there is a unique option $\mathcal{O}'$ such that $\mathcal{O}'' = (\mathcal{F}, \mathcal{S}'', \mathcal{D}') \sqsubseteq \mathcal{O}'$ and $\texttt{CN}(\mathcal{O}) = \texttt{CN}(\mathcal{O}'') \subseteq \texttt{CN}(O')$.

Example 3 (Cont). There are three sub-theories of $\mathcal{T}_3$ that satisfy the four first conditions of Definition 14: $\mathcal{O}p_0 = (\mathcal{F}_3, \mathcal{S}_3, \emptyset)$, $\mathcal{O}p_1 = (\mathcal{F}_3, \mathcal{S}_3, \{r_2\})$ and $\mathcal{O}p_2 = (\mathcal{F}_3, \mathcal{S}_3, \{r_1\})$. The maximal ones (that satisfy also the last condition of Definition 14) are $\mathcal{O}p_1$ and $\mathcal{O}p_2$. Notice that $\mathcal{O}p_1$ is exactly the unique strong option of $\mathcal{T}_3$. The other preferred option $\mathcal{O}p_2$ captures a sub-part of the option $\mathcal{O}_2 = (\mathcal{F}_3, \{r_4\}, \{r_1, r_3\})$. Indeed, by keeping in $\mathcal{O}p_2$ only the strict rues that are applicable we obtain: $\mathcal{O}p'_2 = (\mathcal{F}_3, \{r_4\}, \{r_1\})$. We have : $\mathcal{O}p'_2 \sqsubseteq \mathcal{O}_2$ and $\texttt{CN}(\mathcal{O}p_2) = \texttt{CN}(\mathcal{O}p'_2) \subseteq \texttt{CN}(\mathcal{O}_2)$.

Now, we show that every preferred extension leads to a preferred option and two preferred extensions cannot return the same preferred option.

Theorem 7. *Let* $\mathcal{H} = (\texttt{Arg}(\mathcal{T}), \mathcal{R}_u)$ *be an AS built over a closed theory* $\mathcal{T}$.

- *For all* $\mathcal{E} \in \texttt{Ext}_p(\mathcal{H})$, *there exists a single preferred option* $\mathcal{O} \in \texttt{POpt}(\mathcal{T})$ *s.t.* $\texttt{Th}(\mathcal{E}) \sqsubseteq \mathcal{O}$ *and* $\texttt{Concs}(\mathcal{E}) = \texttt{CN}(\mathcal{O})$. *We put:* $\texttt{POption}(\mathcal{E}) \overset{\text{def}}{=} \mathcal{O}$.
- *For all* $\mathcal{E}, \mathcal{E}' \in \texttt{Ext}_p(\mathcal{H})$, *if* $\texttt{POption}(\mathcal{E}) = \texttt{POption}(\mathcal{E}')$ *then* $\mathcal{E} = \mathcal{E}'$.

- *For all $\mathcal{E} \in \mathtt{Ext}_p(\mathcal{H})$, $\mathcal{E} = \mathtt{Arg}(\mathtt{POption}(\mathcal{E}))$.*

Inversely, every preferred option corresponds to a unique preferred extension and two preferred options cannot return the same preferred extension.

Theorem 8. *Let $\mathcal{H} = (\mathtt{Arg}(\mathcal{T}), \mathcal{R}_u)$ be an AS built over a closed theory $\mathcal{T}$.*

- *For all $\mathcal{O} \in \mathtt{POpt}(\mathcal{T})$, $\mathtt{Arg}(\mathcal{O}) \in \mathtt{Ext}_p(\mathcal{H})$.*
- *For all $\mathcal{O} \in \mathtt{POpt}(\mathcal{T})$, $\mathcal{O} = \mathtt{POption}(\mathtt{Arg}(\mathcal{O}))$.*
- *For all $\mathcal{O}_1, \mathcal{O}_2 \in \mathtt{POpt}(\mathcal{T})$, if $\mathtt{Arg}(\mathcal{O}_1) = \mathtt{Arg}(\mathcal{O}_2)$ then $\mathcal{O}_1 = \mathcal{O}_2$.*

Example 3 (Cont). The system $\mathcal{H}_3$ constructed from $\mathcal{T}_3$ has two preferred extensions: $\mathcal{E}p_1 = \{a_1, a_2, a_5, a_6\}$ and $\mathcal{E}p_2 = \{a_1, a_2, a_3, a_4\}$. They capture the preferred options $\mathcal{O}p_1$ and $\mathcal{O}p_2$ respectively. Indeed, $\mathtt{Th}(\mathcal{E}p_1) \sqsubseteq \mathcal{O}p_1$ (resp. $\mathtt{Th}(\mathcal{E}p_2) \sqsubseteq \mathcal{O}p_2$) and $\mathtt{Concs}(\mathcal{E}p_1) = \mathtt{CN}(\mathcal{O}p_1)$ (resp. $\mathtt{Concs}(\mathcal{E}p_2) = \mathtt{CN}(\mathcal{O}p_2)$).

The number of preferred extensions of an argumentation system $\mathcal{H}$ is exactly the number of preferred options of the theory over which the system is built.

Corollary 6. *Let $\mathcal{H} = (\mathtt{Arg}(\mathcal{T}), \mathcal{R}_u)$ be an argumentation system built over a closed theory $\mathcal{T}$. It holds that $|\mathtt{Ext}_p(\mathcal{H})| = |\mathtt{POpt}(\mathcal{T})|$.*

The plausible conclusions of an argumentation system, under preferred semantics, are the literals that follow from all the preferred options of the theory at hand.

Corollary 7. *Let $\mathcal{H} = (\mathtt{Arg}(\mathcal{T}), \mathcal{R}_u)$ be an argumentation system built over a closed theory $\mathcal{T}$. $\mathtt{Output}(\mathcal{H}) = \bigcap_{\mathcal{O} \in \mathtt{POpt}(\mathcal{T})} \mathtt{CN}(\mathcal{O})$.*

Example 3 (Cont). $\mathtt{Output}(\mathcal{H}_3) = \mathtt{CN}(\mathcal{O}p_1) \cap \mathtt{CN}(\mathcal{O}p_2) = \{x, y\}$.

5   Related Work

There are a couple of rule-based argumentation systems in the literature. Some of them like ASPIC and its extended version ASPIC+ are shown to satisfy the rationality postulates defined in [3], namely the consistency and closure under strict rules of their sets of plausible conclusions. While this is testimony to some strength of these formalisms, it does not say anything about the *kind* of plausible conclusions they draw from a theory. Surprisingly, the outputs of these systems (their extensions and their plausible conclusions) have never been characterized. The authors of those systems provide only examples to show that the outputs are meaningful. This is certainly not sufficient. Our paper is the first that attempts a systematic study of the outcomes of rule-based systems under naive, stable and preferred semantics. There are two notable exceptions. The first work, done in [1], considered a *fragment* of our logical language and rebuttal as attack relation. Blocking rules was not allowed. Extensions were characterized in terms of sub-theories. However, some sub-theories may not have corresponding extensions. Thus, there is no bijection between the two. Our formalism is thus more general and our characterisations of its outcomes are more accurate since they are one-to-one correspondences. The second work, done in [4], investigated the link between the logic programming semantics and argumentation ones. The theory over which an argumentation system is built is a logic program, that is, only one type of rules is used. Thus, the logical language is very different from ours.

In addition to the characterizations of the system's outcomes, the other main novelty of our paper is the exclusive use of undercut for encoding conflicts between arguments. This relation is always coupled with rebuttal which handles inconsistency in other systems. In our paper, we have shown that undercut is powerful enough to perfectly fulfil the role of rebuttal. Moreover, the system satisfies all the rationality postulates under any semantics while in ASPIC and ASPIC+, for each semantics, one should use a different definition of rebuttal in order to satisfy the postulates.

Regarding the definition of undercut, there are three proposals in the literature which are all equivalent. The first definition is the one followed in our paper and in [14]. The idea is to assign a name to every defeasible rule and to allow these names to be in heads of other rules. Unlike in [14], in our paper, names of rules may only be in heads of strict rules. The reason is that undercut shows exceptions of defeasible rules, and exceptions are certain information. For instance, in case of penguin, the rule "birds fly" is not applicable. The second proposal, given in [13] and followed in [3], uses an objectivation operator which transforms any defeasible rule into a literal. The latter plays the role of the name of the rule in our system. The last definition, proposed in [5,6], extends the logical language by a new form of rules with which one can block defeasible rules. Whatever the definition is, none of these systems characterized its outcomes.

Acknowledgments. This work benefited from the support of AMANDE ANR-13-BS02-0004 and ASPIQ ANR-12-BS02-0003 projects of the French National Research Agency.

References

1. Amgoud, L., Besnard, P.: A formal characterization of the outcomes of rule-based argumentation systems. In: Liu, W., Subrahmanian, V.S., Wijsen, J. (eds.) SUM 2013. LNCS, vol. 8078, pp. 78–91. Springer, Heidelberg (2013)
2. Amgoud, L., Caminada, M., Cayrol, C., Lagasquie, M.C., Prakken, H.: Towards a Consensual Formal Model: inference part. Deliverable of ASPIC project (2004)
3. Caminada, M., Amgoud, L.: On the evaluation of argumentation formalisms. Artif. Intell. J. **171**(5–6), 286–310 (2007)
4. Caminada, M., Sá, S., Alcântara, J.: On the equivalence between logic programming semantics and argumentation semantics. In: van der Gaag, L.C. (ed.) ECSQARU 2013. LNCS, vol. 7958, pp. 97–108. Springer, Heidelberg (2013)
5. Cohen, A., García, A.J., Simari, G.R.: Backing and undercutting in defeasible logic programming. In: Liu, W. (ed.) ECSQARU 2011. LNCS, vol. 6717, pp. 50–61. Springer, Heidelberg (2011)

6. Cohen, A., García, A.J., Simari, G.R.: Backing and undercutting in abstract argumentation frameworks. In: Lukasiewicz, T., Sali, A. (eds.) FoIKS 2012. LNCS, vol. 7153, pp. 107–123. Springer, Heidelberg (2012)

7. Dung, P.M.: On the acceptability of arguments and its fundamental role in nonmonotonic reasoning, logic programming and n-person games. Artif. Intell. J. **77**(2), 321–357 (1995)

8. García, A.J., Simari, G.R.: Defeasible logic programming: an argumentative approach. Theor. Pract. Logic Program. **4**(1–2), 95–138 (2004)

9. Gebser, M., Gharib, M., Mercer, R., Schaub, T.: Monotonic answer set programming. J. Logic Comput. **19**(4), 539–564 (2009)

10. Gelfond, M., Lifschitz, V.: Classical negation in logic programs and disjunctive databases. New Gener. Comput. **9**, 365–385 (1991)

11. Governatori, G., Maher, M.J., Antoniou, G., Billington, D.: Argumentation semantics for defeasible logic. J. Logic Comput. **14**(5), 675–702 (2004)

12. Lukaszewicz, W.: Considerations on default logic: an alternative approach. Comput. Intell. **4**, 1–16 (1988)

13. Pollock, J.L.: How to reason defeasibly. Artif. Intell. J. **57**(1), 1–42 (1992)

14. Prakken, H.: An abstract framework for argumentation with structured arguments. J. Argum. Comput. **1**(2), 93–124 (2010)

15. Reiter, R.: A logic for default reasoning. Artif. Intell. J. **13**(1–2), 81–132 (1980)

Formalizing Explanatory Dialogues

Abdallah Arioua[1,2(✉)] and Madalina Croitoru[2]

[1] INRA, UMR IATE, Montpellier, France
[2] University of Montpellier, Montpellier, France
abdallaharioua@gmail.com

Abstract. Many works have proposed architectures and models to incorporate explanation within agent's design for various reasons (i.e. *human-agent teamwork improvement, training in virtual environment* [10], *belief revision* [8], etc.), with this novel architectures a problematic is emerged: how to communicate these explanations in a goal-directed and rule-governed dialogue system? In this paper we formalize Walton's CE dialectical system of explanatory dialogues in the framework of Prakken. We extend this formalization within the Extended CE system by generalizing the protocol and incorporating a general account of dialectical shifts. More precisely, we show how a shift to any dialogue type can take place, as an example we describe a shift to argumentative dialogue with the goal of giving the explainee the possibility to challenge explainer's explanations. In addition, we propose the use of commitment and understanding stores to avoid circular and inconsistent explanations and to judge the success of explanation. We show that the dialogue terminates, under specific conditions, in finite steps and the space complexity of the stores evolves polynomially in the size of the explanatory model.

1 Introduction

The design of explanation facilities for intelligent systems is an active research area and a widely recognized problem [11,14] in Artificial Intelligence. In multi-agent systems (MAS), following the influential Walton and Krabbe typology of dialogues [21], different dialogue types have been proposed. *Negotiation dialogues* deal with resource limitation. *Deliberation dialogues* deal with planning collaborative actions. *Persuasion dialogues* deal with resolution of conflicts of opinion.

When it comes to explanation between autonomous agents, the concept of *dialectical explanatory dialogue* has been addressed by [19,20] as a way to formalize explanatory dialogues within a dialectical system called CE. The dialogue takes place between an explainer and an explainee, the goal is to get the explainee to understand something *whose truthfulness is agreed upon*. As stated by Walton "CE *represents a basic or minimal system of explanation dialogue that provides a beginning framework that is very simple, but can be extended by constructing more complex systems*" [19].

Building upon the state of the art, the objective of the paper is to provide a *formal framework* of explanatory dialogue called ECE system (**E**xtended

© Springer International Publishing Switzerland 2015
C. Beierle and A. Dekhtyar (Eds.): SUM 2015, LNAI 9310, pp. 282–297, 2015.
DOI: 10.1007/978-3-319-23540-0_19

CE system) that extends and generalizes the CE system. The guidelines of the contribution lay in the following points:

- **Generalization:** we generalize the sequential protocol of [19,20] and introduce a more flexible protocol (liberal protocol) where the explainee and the explainer can backtrack to early stages in the dialogue. We give a **general** account of dialectical shifts in ECE and *as an example* we describe a shift to argumentative dialogue to facilitate arguing over explanations (as argued for in [14]).
- **Extension:** we introduce commitment and understanding stores to avoid circular and inconsistent explanations and to judge the success of explanation. We allow for nested explanation requests and feedback when the explainee cannot understand something in the explanation.

We formalize the ECE dialogue system in the general framework of [16] and modify it to suit the formal specification of an explanatory dialogue. We choose Prakken's framework for its flexibility and implementability in Prolog [4]. The ECE dialogue is assumed to take place between two autonomous agents (i.e. humans or intelligent agents) without adhering to a specific internal model.

This work complements the efforts [8–10] of equipping agents with explanation facilities by facilitating explanation exchange in a goal-directed and rule-governed dialogue system. Furthermore, this work contributes to the enrichment of communication in multi-agent systems by promoting a new type of dialogues that intends to capture the concept of explanation. In knowledge-based systems, the state of the art covers extensively explanatory dialogues [5,6,13] but none of the existing approaches has formally studied these dialogues by abstracting away from any domain-specific knowledge. Our work can serve as a theoretical background under which these systems can be evaluated and compared.

The paper is organized as follows. In Sect. 2 we recall Prakken's system for argumentative dialogues [16] and the CE system of explanatory dialogues [19,20]. Then, in Sect. 3 we present the formalization of the Extended CE (ECE) system and we study its properties. Next, in Sect. 4 we present the second component of ECE system, i.e. dialectical shift. In Sect. 5 we apply our system on a detailed example. Section 6 concludes the paper.

2 Background

2.1 Argumentative Dialogue (ARG System)

The system of argumentative dialogues (denoted as ARG) is a many-player turn-taking game between *proponents* and *opponents* arguing in favor or against a statement. We consider here the formal dialogue system for argumentation defined in [16] (denoted as ARG). ARG has a *topic language* L_t, a logic $\mathscr{L}$, a context $K \subseteq L_t$ (assumed consistent and remains the same throughout a dialogue) and a topic $T \in L_t$. L_t is a logical language whose well-formed formulae are denoted by Greek letters, ψ, φ, ϕ, etc.[1]. The logic $\mathscr{L}$ is assumed to be an argumentation logic with compliance to Dung [7] where arguments can be attacked and defended. For an argument A we denote by $prem(A)$, $conc(A)$ its premises and conclusion respectively.

The system ARG has also a *communication language* L_c along with a *protocol P*. The communication language specifies the utterances used throughout the dialogue and P organizes their use (which utterance succeeds the other). According to P, for each utterance the replying utterances can be seen either as an *attack* or a *surrender* (see Table 1). The dialogue incorporates participants. In our case we consider only two participants $Pr = \{p, o\}$, the **p**roponent and the **o**pponent, each of which has a commitment store $C_i \subseteq L_c$ such that $i \in Pr$ (similar to commitment function of [16]). The stores *publicly* indicate statements within the topic language L_t a participant is committed to (i.e. committed to their truthfulness). ARG has effect rules that specify for an arbitrary utterance $l \in L_c$ its effect on the commitments of the participants. At the beginning of the dialogue the stores of the participants are empty. Then, they get updated within the dialogue. Formally, the commitment store $C_i \in \{C_p, C_o\}$ stays intact if and only if the participant $i \in Pr$ utters WHY(φ). Otherwise, it is changed as follows:

- $C_i = C_i \cup \{\varphi\}$ iff the participant i has put forward CLAIM(φ) or CONCEDE(φ).
- $C_i = C_i \setminus \{\varphi\}$ iff the participant i has put forward RETRACT(φ) (i is no longer committed to φ).
- $C_i = C_i \cup prem(A) \cup conc(A)$ iff the participant i has put forward ARGUE(A) (i is committed to the premises and the conclusion of the argument A).

The dialogue has a *turntaking rule* that specifies who is allowed to talk next. The dialogue has also *termination rules* that indicate when the dialogue terminates. The *outcome rules* are activated after the termination of the dialogue, they determine the winner of the dispute in the dialogue.

Table 1. Reply structure. The arguments A and B and the attack relation are defined according to $\mathscr{L}$. ∎

Utterances	Attacks	Surrenders
CLAIM(φ)	WHY(φ)	CONCEDE(φ)
WHY(φ)	ARGUE(A) $(conc(A) = \varphi)$	RETRACT(φ)
ARGUE(A)	WHY(φ) $(\varphi \in prem(A))$, ARGUE(B) (B attacks A)	CONCEDE(φ) $(\varphi \in perm(A)$ or $\varphi = conc(A))$
CONCEDE(φ)	no attack	no surrender
RETRACT(φ)	no attack	no surrender

---

[1] Throughout the paper we always use Greek letters ψ, φ, ϕ, etc. as metavariables for syntactically different well-formed formula (wff), and $\Gamma, \Gamma_0, \ldots$ for sets of wffs.

When it comes to the dialectical shift, we shall deal with the ARG dialogue system as described above with: (1) a *liberal protocol* P and a reply structure as mentioned in Table 1[2]; (2) CLAIM(φ) or ARGUE(A) as opening moves where φ and $conc(A)$ are the topic of the dialogue; and (3) the non-deterministic turntaking rule that dictates that the proponent starts with a single move, then the dialogue switches to the opponent and then it becomes everyone's turn.

2.2 Explanatory Dialogue (CE System)

The system of explanatory dialogues (denoted as CE) is a two-player turn-taking formal dialogue system of **explanation** [19,20]. It takes place between an *explainer* and an *explainee*. The speech acts of requesting and providing an explanation are represented as dialogue moves in the system.

The moves allowed within CE are two distinct sets of moves: one for the explainer and another set for the explainee. The dialogue always starts with an assertion of a statement by the explainer, i.e. ASSERT(φ) and then the explainee requests an explanation for φ, i.e. EXPLAIN(φ) (φ is accessible by the two parties and believed to be true). Next, the explainer can offer an explanation attempt or declares her/his inability to explain. In first case the explainee can ask for further explanations or acknowledge her/his understanding. In [20] a shift to examination dialogue is introduced allowing to test explainee's understanding and to judge the success of the explanation.

In this paper we build upon the CE system described in [19,20] and extend and generalize it as mentioned in the introduction.

3 The Extended CE System

Relying on Prakken's framework for formalizing dialogue systems [16], in this section we formalize the ECE system (**Extended CE**) of explanatory dialogues.

3.1 The Formal Framework

Topic Language, Participants and the Logic. The ECE system of explanatory dialogues takes place between two participants $Pr = \{E, X\}$, the explainer E and the explainee X. ECE has a *topic language* L_t and a logic $\mathscr{L}$ and a context $K \subseteq L_t$ which is assumed to be consistent throughout the dialogue and it is *shared* between E and X. The **purpose** of the dialogue is to facilitate understanding transference by means of explanation about a statement $T \in K$ (closed wff if L_t is a first-order or higher language), this statement is *assumed to be true* by both participants.

The Explanatory Model. Each participant $i \in \{E, X\}$ in ECE has an explanatory model $\mathcal{E}_i = \langle L_t, \Vdash_x, E \rangle$ which consists of the topic language L_t and a finite *explanatory relation* denoted as $\Vdash_x$ and defined over $2^{L_t} \times L_t{}'$ such that $L_t{}' \subseteq L_t$

---

[2] See [16] for a full description of the protocol.

is the set of closed wffs of L_t. The parameter x varies over a common and non-empty set E of explanation types. $\Vdash_x$ intends to identify those wffs in L_t that can be considered as an *explanation* for another *closed* wff in L_t. An explanation contains an explanandum which is the thing to be explained and explanans which are the facts and rules that together *bear* explanatory relevance to the explanandum. The parameter x defines $|E|$ explanatory relations, e.g. *mechanistic, terminological*, etc. (see [11] for explanation types).

Due to the controversy around explanatory models [15], in this formalization we just consider an abstract setting where the model $\mathcal{E}_i$ can provide an explanation for an arbitrary explanandum. Formally, given a set Γ of wffs and a closed wff φ we read $\Gamma \Vdash_x \varphi$ as "Γ is an x-explanation of φ" such that $x \in E$.

The Communication Language. The dialogue is endowed with a *communication language* L_c where $l \in L_c$ is of the form as described in Table 2: "Utterances". In fact, $L_c = L_c^E \cup L_c^X$ where L_c^E (resp. L_c^X) is the performative utterances of the explainer (resp. the explainee). For a given communication language a *reply relation* $\mathscr{R}$ specifies for each $l \in L_c$ its appropriate replies. $\mathscr{R}$ allocates replies according to the syntax and the content of the utterance (Table 2: "Reply"). Please notice that in the reply relation the explainee cannot ask for an explanation if she/he possesses or has already acquired one, this will prevent redundant requests. It is formally defined as follows, the explainee X asks EXPLAIN(ρ) iff $\nexists \Gamma'$ such that $\Gamma' \Vdash_x \rho$ in $\mathcal{E}_X$.

The Protocol. The dialogue is governed by a *protocol P* that organizes the use of L_c. To define P we need to define the notion of a dialogue, which in turn is based on the notion of moves.

A move [16] is a tuple $m = \langle ID, p, l, t \rangle$ such that: (1) $ID \in \mathbb{N}^*$, the identifier of the move, (2) $p \in \{E, X\}$, the participant p who played the move, (3) $l \in L_c$, the utterance l put forward by the participant p and (4) $t \in \mathbb{N}$, the target move t. For a given move m we denote $id(m) = ID$, $pr(m) = p$, $sp(m) = l$ and $tr(m) = t$. We denote by M the set of all moves.

An explanatory dialogue in ECE is a dialogue in the sense of [16], that is, a sequence of moves where the explainer/explainee can reply to each other in a non-sequential way. This generalizes CE by rendering the dialogue *liberal* in the sense that it gives the liberty to the two participants to backtrack to early stages in the dialogue.

Definition 1 (Explanatory Dialogue). *An explanatory dialogue is a sequence of moves $d = \langle m_1, \ldots, m_n \rangle$. The sequence $d_i = \langle m_1, \ldots, m_i \rangle$ such that $i < n$ is denoted by d_i, where d_0 is the empty dialogue. The set of all explanatory dialogues, denoted by $M^{<\infty}$, is the set of all sequences d_i such that $i \in \mathbb{N}^*$ and for each jth element in d_i where $0 < j \leqslant i$, it is the case that (1) $id(m_j) = j$; (2) $tr(m_1) = 0$; and:*

(3) $tr(m_j) = k$ for some m_k preceding m_j in the sequence.

If $(sp(m_j), sp(m_k)) \in \mathscr{R}$ (in the reply relation) we say that m_j replies to m_k in d.

Table 2. The communication language L_c of ECE. ∎

Utterances	Description	Reply
ASSERT(φ)	E reports a statement φ that is accepted as factual by both parties	EXPLAIN(φ) iff $\nexists \Gamma$ s.t $\Gamma \Vdash_x \varphi$ in $\mathcal{E}_x$
EXPLAIN(φ)	X requests an explanation for φ	ATTEMPT(Γ, φ) iff $\Gamma \Vdash_x \varphi$ in $\mathcal{E}_E$ otherwise INABILITY(φ)
ATTEMPT(Γ, φ)	E explains φ by Γ	POSITIVE(φ), NEGATIVE(ρ, φ) s.t $\rho \in \Gamma$, NEGATIVE(Γ, φ)
INABILITY(φ)	E has no explanation	no reply and the dialogue terminates if φ is the topic
POSITIVE(φ)	X understands the explanation of φ	no reply and the dialogue terminates if φ is the topic
NEGATIVE(ρ, φ)	X doesn't understand ρ in the explanation of φ	EXPLAIN(ρ) iff $\nexists \Gamma'$ s.t $\Gamma' \Vdash_x \rho$ in $\mathcal{E}_x$
NEGATIVE(Γ, φ)	X doesn't understand the whole explanation	no reply

Unlike the turntaking function defined in [19,20] which allows *one move at a turn* policy, we define a non-deterministic turn taking policy.

Definition 2 (Turntaking Function). *A turntaking function T is defined as follows $T : M^{<\infty} \longrightarrow 2^{\{E,X\}}$. T assigns to every dialogue the next legal turn as follows:*

– $T(d_0) = \{E\}$, $T(d_1) = \{X\}$, *else* $T(d_i) = \{E, X\}$.

Let us recall the concept of protocol from [16] and then define ECE's protocol. We denote by $dom(X)$ the domain of the function X. A *protocol P* for a dialogue system is a function P from a nonempty subset $D \subseteq M^{<\infty}$ to 2^M where for every dialogues $d = \langle m_1, \ldots, m_n \rangle$ and moves m' we have $d \in dom(P)$ and $m \in P(d)$ iff $d = \langle m_1, \ldots, m_n, m' \rangle \in dom(P)$. The elements of $dom(P)$ are the *legal dialogues* while those of $P(d)$ are the moves allowed after d. If d is a legal dialogue and $P(d) = \emptyset$, then d is a terminated dialogue.

Definition 3 (ECE's Protocol). *A protocol P for the ECE system is defined as follows: for all moves m and all legal dialogues d. $m \in P(d)$ iff:*

$R_1 : pr(m) \in T(d)$ *(it is the turn of $pr(m)$);*
$R_2 :$ *If $d = d_0$ then $sp(m)$ is of the form* ASSERT (φ);
$R_3 :$ *If $d \neq d_0$ and $m \neq m_1$, then m replies to $tr(m)$;*

R_4 : If m replies to m', then $pr(m) \neq pr(m')$ (one cannot respond to one's own moves);

R_5 : If there is m' in d such that $tr(m) = tr(m')$ then $sp(m) \neq sp(m')$ (two replies to a move should be different).

R_6 : For any $m' \in d$ such that $tr(m') = tr(m)$ and $sp(m') = \text{POSITIVE}(\varphi)$, $sp(m) \neq \text{NEGATIVE}(\Gamma, \varphi)$ (understanding cannot be revoked).

A comment about R_6 is in order here. The underlying assumption of this rule is that the agent is prudent in the sense that he/she declares his/her understanding iff she/he is sure about it. This rule may seem restrictive in certain cases where one can have the illusion of understanding and he/she should be provided with a second chance by revoking understanding, despite the fact that this could be an interesting phenomenon to study we limit the scope of the paper to the aforementioned assumption for the sake of simplicity.

The Stores, Effect Rules and Outcome Rules. In CE system [19, 20] stores have not been proposed as part of the system. In the ECE system we extend CE by adding commitment and understanding stores to:

- Keep a clear view of explainee's state of understanding so he/she can backtrack and request more explanations.
- Judge the success of the explanatory dialogue.
- Track the consistency of the explanation. For example, imagine that the explainer is explaining φ by an explanation $\Gamma = \{\psi, \beta\}$ where he/she is committed to the truthfulness of $\neg\psi$, this would be contradictory.
- Avoid circular explanations. This means that it is forbidden to explain ψ by $\{\varphi\}$ such that φ is asked to be explained (this could provoke the infinite chain $\text{EXPLAIN}(\varphi)$, $\text{ATTEMPT}(\{\psi\}, \varphi)$, $\text{EXPLAIN}(\psi)$, $\text{ATTEMPT}(\{\varphi\}, \psi)$, ..., etc.).

Let us formally introduce the notion of stores.

Definition 4 (Stores). The sets NUS_X, $CS_E \subseteq L_t$ denote respectively the understanding and commitment stores where the subscribes refer to the participants.

A store $st \in \{NUS_X, CS_E\}$ is inconsistent iff $st \vdash \psi$ and $st \vdash \neg\psi$ for some $\psi \in L_t$ ($\vdash$ is the inference relation of $\mathscr{L}$).

For the explainee, an understanding store NUS_X serves as an understanding indicator of his/her current understanding state. Note that NUS_X represents what is **not yet understood** instead of what has been understood. For the explainer, a commitment store CS_E represents explainer's commitments to the truthfulness of certain statements. The explainee (resp. explainer) does not have a commitment (resp. understanding) store. Let us specify the rules to update the stores.

Definition 5 (Effect Rules). Let d be a legal dialogue, NUS_X and CS_E be explainee's and explainer's current stores and m is the next legal move after d.

- If $sp(m) = \text{EXPLAIN}(\varphi)$ then $NUS_X = NUS_X \cup \{\varphi\}$,
- If $sp(m) = \text{POSITIVE}(\varphi)$ then (1) $NUS_X = NUS_X \setminus \{\varphi\}$.
- If $sp(m) = \text{ASSERT}(\varphi)$ then $CS_E = CS_E \cup \{\varphi\}$,
- If $sp(m) = \text{ATTEMPT}(\Gamma, \varphi)$ then $CS_E = CS_E \cup \Gamma \cup \{\varphi\}$.

The first set of effect rules on NUS_X indicate that when the explainee requests an explanation about φ we presume that he/she could not understand φ, thus we add it to NUS_X and we revoke it when he/she acknowledge understanding. The second set of effect rules on CS_E state that the explainer is committed to the truthfulness of the explanans (elements of the explanation) and the explanandum.

In what follows we extend Definition 3 with the following rule that considers the stores to avoid circular explanation.

Definition 6 (ECE's Protocol Extended Rules). *Let P be the protocol of ECE, d be a legal dialogue and m be a move. Then m is a legal move after d iff $m \in P(d)$ and:*

R_7 : *If $sp(m) = \text{ATTEMPT}(\Gamma, \varphi)$ then there is no $\psi \in \Gamma$ such that $\psi \in NUS_X$.*

From now on we say that a move m is legal after a dialogue d if and only if it satisfies protocol rules R_1-R_7.

A successful explanatory dialogue is a dialogue where the explainee's *understanding store* is empty. Certainly, we cannot be sure whether the understanding has really taken place but it is one way to quantify the success and failure of an explanatory dialogue. Another alternative would be the use of examination dialogue as proposed in [20]. In our system, instead of limiting shifts to examination dialogues we provide a *general account* of dialectical shifts which can be instantiated to capture any shift (including the one of examination dialogue).

3.2 Properties

In what follows we present interesting results of the ECE system. We investigate termination, number of steps before termination and space complexity of the stores.

As one may notice the protocol of ECE induces a tree structure on any legal explanatory dialogue (see the example in Sect. 5), this is due to the possibility of backtracking and multiple replies to certain moves, e.g. the move ATTEMPT can be answered by at least two moves NEGATIVE and POSITIVE. Therefore, in this section we deal with this induced tree structure in which the nodes correspond to moves and an edge from a move m to m' means m' replies to m.

One of the interesting properties of ECE is termination, that means whenever two participants start an explanatory dialogue and certain conditions are respected we can guarantee termination in finite steps.

Lemma 1. *Let φ be an explanandum and let X be the set of all explanations of φ in $\mathcal{E}_E$. If X is finite then EXPLAIN(φ) has a finite number of child nodes.*

Lemma 2. *Let the explanandum φ be the topic of a legal explanatory dialogue d and let Γ be an explanation of φ in $\mathcal{E}_E$. If Γ is finite then every branch in the dialogue that starts with* ATTEMPT(Γ, φ) *terminates.*

To study termination we define the explanans relation between L_t's elements.

Definition 7 (Explanans and Explanans Path). *Let $\mathcal{E}_E = \langle L_t, \Vdash_x, E \rangle$ be the explanatory model of the explainer. We define the binary relation $\mathcal{N} \subseteq L_t \times L_t$ such that $(\varphi', \varphi) \in \mathcal{N}$ iff there exists an explanation Γ such that $\varphi' \in \Gamma$ and Γ explains φ, and we read it "φ' is an explanan of φ". We denote by $\mathbb{D}(\varphi)$ the explanatory depth of φ which corresponds to the length of the longest explanans path in $\mathcal{N}$ that starts with φ.*

Corollary 1. *Let $\mathcal{E}_E = \langle L_t, \Vdash_x, E \rangle$ be the explanatory model of the explainer. If $\Vdash_x$ is finite then so is $\mathcal{N}$. Consequently, for every explanandum φ in $\mathcal{E}_E$, $\mathbb{D}(\varphi)$ is finite.*

The previous lemmas guides as towards the termination property. The intuition is that if the width of the corresponding tree of the dialogue is finite then the dialogue terminates. Note that the depth of the tree is also finite because (a) no repetition is allowed, (b) understanding cannot be revoked and (c) the explanatory model of the explainer is finite (Sect. 3) hence the depth of the tree is finite.

Proposition 1 (Termination). *If the conditions in Lemmas 1 and 2 hold for every explanandum φ then any legal explanatory dialogue d will terminate in finite steps.*

Note that a step here corresponds to a move at a given turn. We consider in what follows the maximum number of steps (in worst-case) the dialogue will undertake until the termination. The worst-case scenario is when the dialogue is of the shape of a somewhat saturated tree, this corresponds to the case where for every explanation request EXPLAIN there is an explanation attempt ATTEMPT and for every explanation attempt there are two negative acknowledgments NEGATIVE each of which are followed by an explanation request. In fact this happens when the explainee has requested an explanation about every statement made by the explainer and in return he/she obtained explanations about every request he/she made but unfortunately understood nothing. Considering an arbitrary explanandum φ as an input the following holds.

Proposition 2 (Termination Steps). *Let $\mathcal{E}_E$ be the explanatory model of the explainer, $\mathbb{D}(\varphi)$ be the explanatory depth of an arbitrary φ and X be the set of all its explanations. Assume that $\forall \Gamma \in X, |\Gamma| = |X| = k$. Then every legal explanatory dialogue d with topic φ will terminate at most in $O(k^{\mathbb{D}(\varphi)})$ steps.*

We consider the space complexity of the stores CS_E and NUS_x. In the worst-case scenario (the same as the previous) the size of CS_E and NUS_x will converge to the size of the content of the explanatory model of the explainer, this is explained as follows (1) the size of NUS_X increases due to the nested explanation requests made by the explainee, (2) the size of CS_E increases also because the explainer will provide explanations for every request, this results in an update of CS_E. In what follows we consider as inputs the explanatory model and an arbitrary explanandum φ, but since the size of the explanatory model is much bigger than the size of the memory allocated to φ, then φ will not be considered. We show in what follows that the stores polynomially evolve in the size of the explanatory model.

Proposition 3 (Evolution of Stores). *Let $\mathcal{E}_E = \langle L_t, \Vdash_x, E \rangle$ be the explanatory model of the explainer and $\Sigma = \{\Gamma, \psi \mid \exists x((\Gamma, \psi) \in \Vdash_x)\}$ be the content of the explanatory model $\mathcal{E}_E$. In the worst-case scenario $|CS_E| = |NUS_X| = |\Sigma|$. Consequently, any legal explanatory dialogue d has an $O(|\Sigma|)$ worst-case space complexity.*

This happens in the worst-case when the dialogue charges the whole content of the explanatory model twice, one corresponds to the CS_E and the other for NUS_X.

4   Dialectical Shifts in ECE System

In this section we present the second extension of CE [19,20] by introducing and formalizing the concept of a dialectical shift within ECE. We start by a formal account of dialectical shift then we show how a simple shift from ECE to ARG can be instantiated in such formalism.

4.1   Dialectical Shifts in ECE

Generally, a shift between two distinct systems SYS and SYS' should consider the following questions: (1) what is the direction of the shift? (2) when the shift is *licit* [21]? (3) what happens to the stores when we shift? (4) what are the effects of the outcome of one system on the other? To answer these questions we need to introduce the notion of state, licit states and receiving states.

Definition 8 (State). *A state of a dialogue system SYS is a tuple $\langle T, C, M \rangle$ such that $T \in L_t$ is the topic, C is the set of current stores, M is the current move (the most recent move in the dialogue).*

For instance, if SYS is the ECE system then $C = \{CS_E, C_X, NUS_X\}$ such that C_X is the commitment store of the explainee in the last argumentative dialogue. If SYS is the ARG system then $C = \{C_o, C_p\}$ (opponent's and proponent's stores). The set of all possible states of a given dialogue system SYS is denoted as $\mathscr{C}_{SYS}$. The sets $S_{SYS}, R_{SYS} \subseteq \mathscr{C}_{SYS}$ are called the set of licit states and receiving states of SYS respectively.

Licit states are states from which one can shift to another dialogue. R_{SYS} represents the set of states a given dialogue system can begin with when a shift occurs.

For instance, the state $s = \langle T, C, M \rangle$ where $T = \{\varphi\}$, $C = \{C_o = \emptyset, C_p = \emptyset\}$, $M = \{m = \langle 1, p, \text{CLAIM}(\varphi), 0 \rangle\}$ is a receiving state of the argumentative dialogue which happens to be also an initial state as defined in Subsect. 2.1. For any dialogue system SYS that anticipates a shift to another dialogue system SYS', the sets S_{SYS} and $R_{\text{SYS}'}$ should be nonempty. At least, $R_{\text{SYS}'}$ is set to $I_{\text{SYS}'}$ such that $I_{\text{SYS}'}$ is the set of all initial states of SYS'. Nevertheless, providing $R_{\text{SYS}'}$ with more states stays a matter of choice.

After defining the licit and receiving states we present the *general* definition of a *shift*. A shift is a transition from one system to another under a specific condition. the first system should be in a state where the shift is allowed (licit states).

Definition 9 (Shift Function). *Let* SYS *and* SYS' *be two distinct dialogue systems and let* S_{SYS} *and* $R_{\text{SYS}'}$ *be the sets of licit states (resp. receiving states) of the dialogue system* SYS *(resp.* SYS'*). A shift is a function* $\mathbb{S} : S_{\text{SYS}} \rightarrow 2^{R_{\text{SYS}'}}$.

From Definitions 8 and 9, on can see that the content of S_{SYS}, R_{SYS}, $S_{\text{SYS}'}$ and $R_{\text{SYS}'}$ for two *distinct* dialogue systems defines the type of the shift (one-way or two-way) and the direction (from which to which system) and *nested* or *not nested*. If $S_{\text{SYS}} \neq \emptyset$ and $R_{\text{SYS}'} \neq \emptyset$ and the other sets are empty, then this is a *one-way* shift from SYS to SYS'. If $S_{\text{SYS}'} \neq \emptyset$ and $R_{\text{SYS}} \neq \emptyset$ and the other sets are empty then this is a *one-way* shift from SYS' to SYS. If all of these sets are not empty, then this is a *two-way* shift in both directions, and it is a *nested shift* where one can shift from SYS to SYS' then shift back to SYS' and so on. Otherwise the shift does not occur.

4.2   Dialectical Shift from ECE to ARG

Consistency, plausibility and sense-making are among the important conditions for an explanatory dialogue as mentioned in [19]. Our hypothesis is that a dialectical shift from ECE to ARG could help in satisfying such conditions by giving the explainer (resp. explainee) the possibility to provide support (resp. questions) for (resp. the) explanation by means of arguments.

The shift is one-way from ECE to ARG where we cannot shift back until the argumentative dialogue within ARG comes to an end. This means that the argumentative dialogue is embodied in ECE and we cannot call an instance of ECE from within an instance of an argumentative dialogue. The commitment store CS_E of the explainer in ECE dialogue persists in the argumentative dialogue and will be used and updated. In other words the explainer will not change his commitments if a shift occurs. Finally, at the end of the argumentative dialogue two things will happen. Firstly, the explainee will have a commitment store C_X that will be shared between all argumentative dialogues (in case of multiple shifts). Secondly, explainee's *understanding store* will be updated at the end with respect to the outcome of the argumentative dialogue. For instance if the explainee had doubts about a statement ψ in the explanation and the explainer wins the argumentative dialogue then ψ will be deleted from the explainee's understanding store NUS_X. Otherwise NUS_X will still have ψ and if the ECE dialogue ends, the explanation will be judged unsuccessful.

Since we are dealing in our case with one-way shift from ECE to the argumentative dialogue we only need to set $R_{ECE} = \emptyset$ and $S_{ARG} = \emptyset$ and define the rest, i.e. $S_{ECE} \neq \emptyset$ and $R_{ARG} \neq \emptyset$.

Definition 10 (ECE's Licit and Receiving States). *Let $\mathscr{C}_{ECE}$, S_{ECE} and R_{ECE} be respectively the set of all states, licit and receiving states of the ECE system, let $\mathscr{C}_{ARG}$, S_{ARG} and R_{ARG} be respectively the set of all states, licit and receiving states of the ARG system and let $s = \langle T, C, M \rangle$ be a state. Then:*

- $S_{ECE} = \{s | s \in \mathscr{C}_{ECE}, sp(M) = \text{ATTEMPT}(\Gamma, \varphi)\}$.
- $R_{ARG} = \{s | s \in \mathscr{C}_{ARG}, sp(M) \in \{\text{CLAIM}(\varphi), \text{ARGUE}(A)\}\}$.

Such that Γ is an x-explanation of φ and A is an argument.

As one may notice, S_{ECE} contains those states where the move is $\text{ATTEMPT}(\varphi)$ (φ is an arbitrary wff) and R_{ARG} contains states which represent the initial states of the ARG dialogue (states where M is either $\text{CLAIM}(\varphi)$ or $\text{ARGUE}(A)$).

Under the specifications of Definition 10, in what follows we instantiate the shift function in our context (from ECE to ARG).

Definition 11 (ECE's Shift Function). *Let S_{ECE} and R_{ARG} be the sets of licit states (resp. receiving states) of the dialogue system ECE (resp. the argumentative dialogue system ARG). Let $s = \langle T, C, M \rangle$ be a state of ECE such that $sp(M)$ is $\text{ATTEMPT}(\Gamma, \varphi)$. Then, the shift function $\mathbb{S}$ is specified as follows: $\mathbb{S}(s) = R'$ such that for each $s' = \langle T', C', M' \rangle \in R'$:*

- $T' = \psi$ *such that* $\psi \in \Gamma$,
- $C' = \{C_E', C_X'\}$ *such that* $C_E' = CS_E$ *and* $C_X' = C_X$ *where* $\{CS_E, C_X\} \subset C$,
- $M' = m$ *such that* $m = \langle 1, p, X, 0 \rangle$, $X \in \{\text{CLAIM}(T'), \text{ARGUE}(A)\}$, $conc(A) = T'$.

The function dictates that if the utterance of the current move is $\text{ATTEMPT}(\Gamma, \varphi)$ then we can shift to an argumentative dialogue where the participants are the explainer (as the proponent) and the explainee (as the opponent) and the topic is arguing over one of the explanans (say ψ) of the explanation Γ such that the proponent starts either by $\text{CLAIM}(\psi)$ or $\text{ARGUE}(A)$ $(conc(A) = \psi)$. The shift function also specifies the migration of stores from one dialogue to another. In our case the commitment store C_E' of the explainer in ARG is set to his commitment store CS_E of ECE, similarly the commitment store C_X' of the explainee in ARG is set to the commitment store of the previous shift.

When the argumentative dialogue ARG comes to an end, the stores of the ECE dialogue are updated as follows:

- If the explainer wins then we update NUS_X according to Definition 5, else NUS_X persists as it was. In all cases, $CS_E = C_E'$ and $C_X = C_X'$.

The commitment store of the explainee within the argumentative dialogue will be kept within ECE for further shifts. Both understanding NUS_X and commitment CS_E stores of ECE will be updated according to the outcome of ARG dialogue as indicated above. When we shift back to ECE, the dialogue continues from where is left off according to the protocol P.

It is noteworthy that explainer's commitment store CS_E is shared between all instances of the argumentative dialogue because before any shift CS_E is migrated to C_E' and updated within the shift and then C_E' is migrated back to CS_E at the end of the shift (the same applies to C_X).

5 Example Dialogue

In this section we apply the ECE dialogue system to an example about explaining why coal is black (inspired from [20]).

Figure 1(a) is a tree representation of a segment of an ECE dialogue where the subscript in participants name refers to dialogue stages (i.e. E_1 means E at stage 1), an edge between two nodes means that the lower one replies to the higher one. The gray dashed box represents the ARG dialogue after a shift. Figure 1(b) explains the meaning of the logical symbols. Figure 1(c) shows the evolution of stores within ECE and ARG (in ARG, stages 4–9 are replaced by $4'$–$9'$), column S refers to the stage, the 2nd–3rd columns represent the stores of ECE and the rest represent the stores of ARG, the brace ARG focuses on the content of the stores of ARG within the shift. "n/a" means that the content is unavailable (because the shift has not taken place yet), Ans at stage n refers to the content of the store at stage $n-1$, we may not use Ans when it's clear.

In Fig. 1(a), the explainer E states a fact which the explainee doesn't understand, hence the explainee X requests an explanation at stage 2. Next, at stage 3, E offers an explanation Γ. Since we are in a licit state, we have two scenarios: either (1) continue within ECE dialogue or (2) shift to ARG dialogue. Let us start with (1), X at stage 4 says he doesn't understand ρ and he requests an explanation for it. Next at stage 6, the explanation Γ' is presented. After that, X acknowledges the understanding of ρ, although it seems that the whole explanation doesn't make sense to him, thus at stage 8 he declares that he doesn't understand the whole explanation of φ. At stage 9, E gives another explanation Γ'' which X could understand, the dialogue can terminate and the explanation is judged successful.

Let us see scenario (2): at stage 3, E might have doubts about ψ, maybe it seems implausible that the *earth is aged more than million years*. Thus the shift takes place where X asks "why" in which he demands a justification (not explanation). Next, E (proponent) presents two arguments at stage $6', 7'$ after which X (opponent) concedes. Now the ARG dialogue ends and the commitment store C_X' will persist in ECE and will be used in future shifts. Note that nothing prevents us from continuing the ECE dialogue. The evolution of the stores is presented in Fig. 1(c) (4th–5th columns) where the stores at stage 1–2 haven't been set since the shift hasn't started. At stage 3 (where the shift starts) the commitment store of E in ECE is migrated to C_E' and updated (stages $6', 7'$) by adding the premises of arguments A, B. When X concedes (at stages $8',9'$) ψ is added to C_X' and at the end of the shift (stage 10) C_E' is migrated back to CS_E.

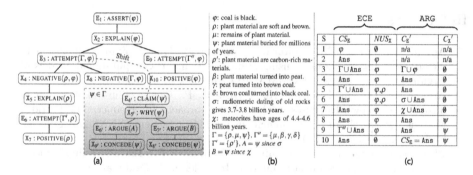

Fig. 1. An example of an ECE dialogue.

6   Conclusion and Future Work

In this paper we have proposed a dialectical system for explanatory dialogue called ECE. This system captures and generalizes the dialectical system CE [19,20] by incorporating a more general protocol, a new component (dialectical shift), an additional structure (stores). We have proposed the use of commitment and understanding stores to avoid circular and inconsistent explanations. We introduced and formalized dialectical shifts and we applied it to capture the argumentative aspects of explanatory dialogues. We have shown that the dialogue terminates and the space complexity is polynomial.

We left, for future work, the study of the previous properties in the presence of a dialectical shift and multi-shifts wihtin ECE. The paper provides no semantic for the dialogue, a good starting point would be [8,12] where a change in beliefs can occur if an explanation is provided. This could give raise to an operational semantics for ECE system.

In previous work [1,2] we have proposed explanation facilities based on a custom-tailored dialogue for inconsistent-knowledge bases, we focused in this work on the bigger picture where a more general setting is considered, i.e. a dialogue between an explainer and an explainee within a formal framework which is independent from any domain-related specifications. This framework can be enriched by investigating explainee mental models that account for reasoning fallacies (such as the work described in [3]). We plan to test such explanation dialogue primarily in the DUR-DUR project which aims at providing decision-support systems in Agronomy. Although the specificity of this application, the generic approach presented here is promising for other Agronomy related real world cases such as [17,18].

Acknowledgment. Financial support from the French National Research Agency (ANR) for the project DUR-DUR (ANR-13-ALID-0002) is gratefully acknowledged. We are also grateful to Nouredine Tamani for his valuable comments on the paper.

References

1. Arioua, A., Tamani, N., Croitoru, M., Buche, P.: Query failure explanation in inconsistent knowledge bases using argumentation. In: Computational Models of Argument: Proceedings of COMMA 2014, vol. 266, p. 101 (2014)
2. Arioua, A., Tamani, N., Croitoru, M., Buche, P.: Query failure explanation in inconsistent knowledge bases: a dialogical approach. In: Bramer, M., Petridis, M. (eds.) Research and Development in Intelligent Systems XXXI, pp. 119–133. Springer, Heidelberg (2014)
3. Bisquert, P., Croitoru, M., de Saint Cyr-Bannay, F.D.: Towards a dual process cognitive model for argument evaluation. In: Beierle, C., Dekhtyar, A. (eds.) SUM 2015. LNAI, vol. 9310, pp. XX–YY. Springer, Heidelberg (2015)
4. Bodenstaff, L., Prakken, H., Vreeswijk, G.: On formalising dialogue systems for argumentation in the event calculus. In: Proceedings of the 11th International Workshop on Nonmonotonic Reasoning, pp. 374–382 (2006)
5. Cawsey, A.: Explanation and Interaction: The Computer Generation of Explanatory Dialogues. MIT Press, Cambridge (1992)
6. de Vries, E., Lund, K., Baker, M.: Computer-mediated epistemic dialogue: explanation and argumentation as vehicles for understanding scientific notions. J. Learn. Sci. 11(1), 63–103 (2002)
7. Dung, P.M.: On the acceptability of arguments and its fundamental role in nonmonotonic reasoning, logic programming and n-person games. Artif. Intell. 77(2), 321–357 (1995)
8. Falappa, M.A., Kern-Isberner, G., Simari, G.R.: Explanations, belief revision and defeasible reasoning. Artif. Intell. 141, 1–28 (2002)
9. Harbers, M., Bradshaw, J.M., Johnson, M., Feltovich, P., van den Bosch, K., Meyer, J.-J.: Explanation in human-agent teamwork. In: Cranefield, S., van Riemsdijk, M.B., Vázquez-Salceda, J., Noriega, P. (eds.) COIN 2011. LNCS, vol. 7254, pp. 21–37. Springer, Heidelberg (2012)
10. Harbers, M., van den Bosch, K., Meyer, J.-J.C.: A study into preferred explanations of virtual agent behavior. In: Ruttkay, Z., Kipp, M., Nijholt, A., Vilhjálmsson, H.H. (eds.) IVA 2009. LNCS, vol. 5773, pp. 132–145. Springer, Heidelberg (2009)
11. Haynes, S.R., Cohen, M.A., Ritter, F.E.: Designs for explaining intelligent agents. Int. J. Hum.-Comput. Stud. 67(1), 90–110 (2009)
12. Khemlani, S., Johnson-Laird, P.N.: Cognitive changes from explanations. J. Cogn. Psychol. 25(2), 139–146 (2013)
13. Moore, J.D.: Participating in Explanatory Dialogues: Interpreting and Responding to Questions in Context. MIT Press, Cambridge (1995)
14. Moulin, B., Irandoust, H., Bélanger, M., Desbordes, G.: Explanation and argumentation capabilities: towards the creation of more persuasive agents. Artif. Int. Rev. 17(3), 169–222 (2002)
15. Pitt, J.C.: Theories of Explanation. Oxford University Press, New York (1988)
16. Prakken, H.: Coherence and flexibility in dialogue games for argumentation. J. Log. Comput. 15(6), 1009–1040 (2005)
17. Tamani, N., Mosse, P., Croitoru, M., Buche, P., Guillard, V., Guillaume, C., Gontard, N.: An argumentation system for eco-efficient packaging material selection. Comput. Electron. Agric. 113, 174–192 (2015)
18. Thomopoulos, R., Croitoru, M., Tamani, N.: Decision support for agri-food chains: a reverse engineering argumentation-based approach. Ecol. Inform. 26, 182–191 (2015)

19. Walton, D.: Dialogical models of explanation. In: Proceedings of the AAAI Workshop on Explanation-Aware Computing (ExaCt 2007), vol. 2007, pp. 1–9 (2007)
20. Walton, D.: A dialogue system specification for explanation. Synthese **182**(3), 349–374 (2011)
21. Walton, D., Krabbe, E.: Commitment in Dialogue: Basic Concepts of Interpersonal Reasoning. SUNY Press, Albany (1995)

Towards a Dual Process Cognitive Model for Argument Evaluation

Pierre Bisquert[1]([✉]), Madalina Croitoru[2],
and Florence Dupin de Saint-Cyr[3]

[1] INRA, Montpellier, France
`pierre.bisquert@supagro.inra.fr`
[2] University Montpellier, Montpellier, France
`croitoru@lirmm.fr`
[3] IRIT, Toulouse, France
`florence.bannay@irit.fr`

Abstract. In this paper we are interested in the computational and formal analysis of the persuasive impact that an argument can produce on a human agent. We propose a dual process cognitive computational model based on the highly influential work of Kahneman and investigate its reasoning mechanisms in the context of argument evaluation. This formal model is a first attempt to take a greater account of human reasoning and is a first step to a better understanding of persuasion processes as well as human argumentative strategies, which is crucial in collective decision making domain.

Keywords: Cognitive computational models · Dual process reasoning · Persuasion · Argument

1 Introduction

Gaining more and more attention, persuasion is a crucial aspect of human interaction and is closely linked to social groups creation and dynamics [30,33]. With the recent rise of computer science technology, the study of persuasion began to transcend its original fields (including psychology, rhetoric and political sciences) and to take lasting root in the artificial intelligence (AI) domain.

In the AI domain, two predominant trends may be identified: *interactive technologies for human behavior* and *dialogue protocols for persuasion*. The former trend aims at producing systems able to persuade humans to change their behavior for another one considered better [21]. It has often been used in the context of health-care [19], environment [5] or education [12]. Such an approach, by definition, is human-machine oriented. The latter trend, derived from logic and philosophy authors such as Hamblin [13], Perelman [22] or

This work has been supported by the Agence Nationale de la Recherche (grant ANR-12-CORD-0012) and has benefited from useful discussion in Dagstuhl Seminar 15221 "Multi-disciplinary approaches to reasoning".

C. Beierle and A. Dekhtyar (Eds.): SUM 2015, LNAI 9310, pp. 298–313, 2015.
DOI: 10.1007/978-3-319-23540-0_20

Walton [31], aims at creating normative dialogue protocols ensuring rational interactions between agents [1,20,24]. The proposed protocols regulate the persuasion processes engaged between agents such that conflicts are resolved in a fair manner. These approaches are often machine-machine oriented and prescriptive.

In this paper we are interested in the computational and formal analysis of the persuasive interactions that occur between humans. Since humans are known to be subject to reasoning biases, we are interested in the link between persuasion and cognitive biases. The importance of this subject has, in particular, been highlighted in the field of law in the context of a court [8] or psychology [15]. This formalisation is a first step towards a better understanding of human persuasion strategies and may help to detect and notify cognitive biases, e.g. in protocols handling collective decision making.

Several works in psychology analyze cognitive biases with the help of *dual process theory* [2,9–11,26,29], where reasoning may be achieved thanks to two different processes, one being heuristic, superficial and fast, and the other being scrupulous, thorough and slow. Indeed, according to Kahneman [29], the first system (called S1) deals with quick and instinctive thoughts and is based on associations such as cause-effect, resemblance, valence, etc. The second system (called S2) is used as little as possible and is a slow and conscious process that deals with what we commonly call reason. Cognitive biases arise mostly when the superficial reasoning is used. In their seminal article [29], Tversky and Kahneman explain how supposedly "rational" judgments are based on data with limited validity and processed according to heuristic rules. They illustrate their thesis with a number of biases empirically demonstrated (such as the illusion of validity, retrievability of instances, anchoring, framing, etc.). This diptych has been popularized in many domains including persuasion [6,23]. In the Elaboration Likelihood Model [23], two *routes* might be used to persuade someone: the *central route*, which calls for a careful examination of the received message, and the *peripheral route*, using simple cues to evaluate the position advocated by an orator. While works such as [23,29] coincide in spirit, our aim is to unify them into a formal framework with four cognitive profiles for evaluating an argument such that a more engaged agent will use a deeper reasoning (S2) while a quiescent agent will only use associations (S1).

After defining a new cognitive model and two reasoning processes based on [29] as well as [23] in Sect. 2, we present how an argument might be evaluated and its effect on the agent's mind in Sect. 3. Finally, some properties are shown in Sect. 4.

2 Towards a Computational Model of Cognitive Evaluation

2.1 Cognitive Model

In this paper, our aim is to define a computational cognitive model of the evaluation of an argument. Based on Kahneman's theory, we propose to define an agent cognitive model as two components: AT (an association table linking a formula to an ordered set of formulae and to a flag encoding an appreciation) and KB (a logical knowledge base) in order to encode S1 and S2 respectively.[1] Formally, we consider a propositional language and we denote by $\mathcal{L}$ the set of well formed formulae of this language given the usual connectives $\wedge$, $\vee$, $\rightarrow$, $\neg$ and the constants $\bot$ and $\top$. The set of symbols in the language is denoted by $\mathcal{V}$. $\vdash$ denotes classical inference. The fact that a symbol s appears in a formula φ is denoted by $s \in \varphi$. We also consider a propositional language, denoted $\mathcal{L}_G$, based on a set of symbols $\mathcal{V}_G$ distinct from $\mathcal{V}$ ($\mathcal{V}_G \cap \mathcal{V} = \varnothing$). Formulae of $\mathcal{L}_G$ are called *generic formulae*.

Definition 1 (Association Table). *An agent's association table AT is a set of triples of the form $(\varphi, (S, \succ_S), f)$ where:*

- *$\varphi \in \mathcal{L}$ is a well formed formula representing a piece of knowledge,*
- *$S \subseteq \mathcal{L}$ is a set of well formed formulae associated to φ endowed with a total strict order $\succ_S \subseteq S \times S$, the pair $(S, \succ_S)$ is called a stack (when there is no ambiguity, the total order will be omitted),*
- *$f \in \{\oplus, \ominus, \odot\}$ is a flag stating that φ is respectively accepted, rejected or not specified (also called empty flag) in the association table.*

The set of all well formed formulae in the association table is denoted by L_{AT}, i.e., $L_{AT} = \bigcup_{(\varphi, S, f) \in AT} \{\varphi\}$. Given a formula $\varphi \in L_{AT}$, the stack S associated with φ in AT will be denoted by $AT(\varphi)$, the ith element of S is denoted $AT(\varphi, i)$, and the top element of this stack is denoted $\text{Top}(\varphi)$ ($\text{Top}(\varphi) = AT(\varphi, 1)$). Formally, $\text{Top}(\varphi) = \varphi_0$ s.t. $\forall \varphi' \neq \varphi_0 \in AT(\varphi), \varphi_0 \succ_S \varphi'$. The flag f associated to φ is denoted by $\text{flag}(\varphi)$. If f is a flag then $-f$ is a flag such that $-\oplus = \ominus$, $-\ominus = \oplus$ and $-\odot = \odot$. Note that AT is implicit in the definitions of Top and flag.

A knowledge base contains Strict and Defeasible Beliefs, Appreciations (*i.e.* associations of formulae to flags) and a set of Appreciation Rules[2] called *a-rules* as described below.

Definition 2 (Knowledge Base). *A knowledge base KB built on $\mathcal{L}$ and $\mathcal{L}_G$ is a quadruplet $KB = (F, \Delta, A, R)$ s.t. $F \subseteq \mathcal{L}$ is a set of formulae, Δ is a set of default rules, A is a set of appreciations and R is a set of a-rules, where*

- *A default rule is denoted $a \leadsto b$ with $(a, b) \in \mathcal{L} \times \mathcal{L}$ with the intended meaning "if a is true then generally b holds".*
- *An appreciation is a pair $(\varphi, f) \in \mathcal{L} \times \{\oplus, \ominus, \odot\}$ meaning that φ is associated to the flag f.*
- *An a-rule has the form $(E_K, E_A) \Rightarrow (\psi, f)$ where $E_K \subseteq \mathcal{L}_G \times \mathcal{L}_G$ is a set of pairs of generic formulae (called generic default rules), $E_A \subseteq \mathcal{L}_G \times \{\oplus, \ominus, \odot\}$ is a set of generic appreciations, $\psi \in \mathcal{L}_G$ is a generic formula and $f \in \{\oplus, \ominus\}$ is a flag. This kind of rule has the intended meaning "if all the default rules E_K apply in a given context and if all the appreciations E_A hold then generally the new appreciation (ψ, f) is valid".*

---

[1] Note that S1 and S2 are linked as we will see in **(1)** of Definition 3.

[2] Inspired from the Desire-Generation rules (of Rahwan and Amgoud [25]).

The use of default rules has two main interests. First, it simplifies the writing: it allows us to express a rule without mentioning every exception to it. Second, it allows us to reason with incomplete descriptions of the world: if nothing is known about the exceptional character of the situation, it is assumed to be normal, and reasoning can be completed.

Definition 3 (Cognitive Model). *A cognitive model is a tuple* $\kappa = (KB, AT, \lambda, i)$:

- $KB = (F, \Delta, A, R)$ *is a knowledge base,*
- AT *is an association table such that:* $\forall \varphi, \varphi' \in \mathcal{L}, \forall f \in \{\oplus, \ominus, \odot\}$,

$$\left.\begin{array}{l} \bullet \ \textit{if } \varphi \in F \textit{ then } \forall s, s' \in \varphi, \ s \in AT(s'), \\ \bullet \ \textit{if } \varphi \rightsquigarrow \varphi' \in \Delta \textit{ then } \varphi' \in AT(\varphi), \\ \bullet \ \textit{if } (\varphi, f) \in A \textit{ then } \texttt{flag}(\varphi) = f, \end{array}\right\} \quad (1)$$

- $\lambda \in \mathbb{N}$ *is an integer value representing the threshold above which the agent feels to be enough aware about the topic of a formula to be able to reason rationally,*
- $i : \mathcal{L} \rightarrow \{0, 1, 2\}$ *is a three value marker that gives the interest level of the agent relatively to a formula.*

In other words, (1) expresses the link between KB and AT, more precisely, every pair of symbols belonging to a given formula in F, and every pair of formulae in Δ linked by a default rule, are associated in AT and the flags in AT comply with A. In case of ambiguity about the current cognitive model, the symbols AT, $\texttt{Top}$, $\texttt{flag}$ will be indexed by the cognitive model κ they refer to.

Example 1. *We illustrate here the question of performing the separation of durum wheat cereal (or other plants in the field such as peas) after the harvest that was done within an ANR DUR-DUR[3] meeting. As our keen internship student was performing his literature review, he quickly learned that post harvest separation (phs) is efficient (eff), which implies a process that is not expensive ($\neg exp$). His KB contains formulae such as phs $\rightsquigarrow$ eff and eff$\rightsquigarrow \neg exp$. However, during a coffee break, he heard a colleague working on post harvest separation with optical harvest devices (opt) and learned that these instruments are generally very long to produce (ltp): phs $\wedge$ opt $\rightsquigarrow$ ltp. He is certain that long production is not efficient: ltp $\rightarrow \neg eff$. While he still does not know whether to accept or reject the post harvest separation, the first thing he now associates post harvest with is the long time to produce, something he disapproves of. This is represented by the flag $\ominus$ in AT (see Fig. 1) and by the appreciation (ltp, $\ominus$) in KB.*

2.2 System 1 and System 2 Reasoning

Let us see how to use this representation framework in order to reason. In this paper, we call *reasoning* the process of evaluating the acceptability of a formula

[3] French funded project aiming at improving durum wheat sustainability (http://www.agence-nationale-recherche.fr/?Project=ANR-13-ALID-0002).

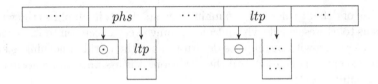

Fig. 1. Partial representation of the associative table.

$\varphi \in \mathcal{L}$, *i.e.*, mapping φ to a flag in $\{\oplus, \ominus, \odot\}$. The reasoning is not the same in S1 and S2. In S1, reasoning is based on the association table AT while in S2 it is based on an inference principle. We propose to encode S1-reasoning as follows: if the current formula has a non-empty flag, then this flag is returned; else, if the current concept has an empty flag, the concepts of the stack associated to the current concept are evaluated recursively, in an order relative to their position in the stack and the number of iterations.

We first define a *reflection path* R_φ associated to a concept φ thanks to a sequence D_φ of iterations from the initial formula φ. This sequence contains the successive depths d_i in the stacks corresponding to formulae with an emptyflag that are necessary to follow in order to find a formula with a non-empty flag. The reflection path jumps recursively from a formula φ_i to a formula φ_{i+1} if φ_{i+1} appears in the stack of φ_i at the depth d_i (each depth d_i in the sequence should not exceed the total depth of each stack $AT(\varphi_i)$). Note that many reflection paths can be built from a formula φ; this is why we will select the cheapest one in terms of cognitive effort.

Definition 4 (Reflection Path). *A reflection path* $R_\varphi = (\varphi_1, \ldots, \varphi_n)$ *from* φ *is a sequence of* $n \geq 1$ *formulae corresponding to a sequence* $(d_1, \ldots, d_{n-1})$ *of* $n - 1$ *integers such that* $\varphi_1 = \varphi$ *and recursively*

$$\forall 1 \leq i < n, \quad \varphi_{i+1} = AT(\varphi_i, d_i), \quad \text{with } d_i \leq |AT(\varphi_i)| \text{ and } \mathtt{flag}(\varphi_i) = \odot.$$

We denote $\mathtt{flag}(R_\varphi)$ *the flag associated to the last concept reached by the sequence* R_φ, *hence,* $\mathtt{flag}(R_\varphi) = \begin{cases} \mathtt{flag}(\varphi_n) \neq \odot & \text{if } n \text{ is finite,} \\ \odot & \text{otherwise.} \end{cases}$

The cognitive weight associated *to a reflection path* $R_\varphi = (\varphi_1, \ldots \varphi_n)$ associated *to the integers* $(d_i)_{1 \leq i < n}$ *is* $\mathtt{weight}(D) = \sum_{i=1}^{n-1} d_i + n$.

The *cognitive weight* associated to a sequence allows to take into account both the depth in the stack and the number of iterations. The more deep and long is the sequence, the more it requires an effort to the agent. S1-reasoning will amount to find and follow reflection paths of *minimal cognitive weight* until a non-empty flag is reached. Hence, S1-reasoning consists in finding a non-empty flag to associate to a concept while minimizing[4] the cognitive effort.

---

[4] Note that we could also have given more weight to the depth in the stack than to iteration or conversely, hence transform the equation into $\mathtt{weight}(D) = \alpha . \sum_{i=1}^{n} d_i + \beta . n$ with a "smart" tuning of the ratio between α and β (this tuning should be based on psychological experiments).

Definition 5 (S1-reasoning). *Given a cognitive model* $\kappa = (KB, AT, \lambda, i)$, *We call S1-entailment, denoted by* $\vdash_1$, *the inference obtained by following a reflection path:* $\varphi \vdash_1 \psi$ *iff* $\psi \in R_\varphi$ *and* R_φ *is finite.*

We define S1-reasoning[5], about a formula φ, *denoted* $eval_1(\varphi, \kappa)$, *as* $eval_1(\varphi) = \mathtt{flag}(R_\varphi)$ *where* R_φ *is a reflection path from* φ *s.t. there is no reflection path* R'_φ *from* φ *with* $\mathtt{weight}(R'_\varphi) < \mathtt{weight}(R_\varphi)$.

Example 2. *Given the association table shown in Fig. 1, the result of* $eval_1(phs)$ *is* $\ominus$. *Indeed, since the formula phs has the flag* $\odot$, *the S1-reasoning gets the top formula of the stack associated to phs, which is ltp; the reflection path is* $R_{phs} = (phs, ltp)$ *and its associated sequence is* (1). *The flag of ltp being different than* $\odot$, *it is the result of the evaluation.*

Concerning S2, the study of the best rational model among all the proposals done in the AI literature is out of the scope of the paper. We propose to use, for the sake of illustration, the idea of defeasible approach of [3], called "contextual entailment" which is an extension of the "preferential entailment" [17]. Preferential entailment is an inference relation satisfying "desirable" postulates (listed in Sect. 4).

The set of conclusions that one can obtain by using a "preferential entailment" is usually regarded as the minimal set of conclusions that any reasonable non-monotonic consequence relation for default reasoning should generate. Moreover, it correctly addresses the specificity problem: results issued from subclasses override those obtained from super-classes [28]. Unfortunately, in spite of these two advantages, "preferential entailment" is too cautious and suffers from the so-called irrelevance problem: from a rule "generally, if a then b", it is not possible to deduce that b follows from $a \wedge d$ even if d is irrelevant to a and b. A typical example of irrelevance problem is that from "generally, birds fly" it is not possible to deduce that "red birds fly".

The approach proposed in [3] has shown to be an extension of "preferential entailment" which corrects this problem. This is why we choose to build S2 on the same idea. This is based on the identification of default rules having exceptions in a given context:

Definition 6 ([3]). *Let c be a consistent formula considered as the current context, let* Δ *be a set of default rules. A default rule* $a \rightsquigarrow b \in \Delta$ *has an exception with c if and only if one of the two conditions holds:*

---

[5] In practice, a constructive method to obtain R_φ could be an adaptation of Dijkstra algorithm on a graph where the vertices are partial reflection paths. An arc would link a vertex to another vertex if it corresponds to an extension of the path of one iteration (hence there would be as many arcs starting from a given vertex as the stack corresponding to this vertex is deep), namely there would be an arc between (φ_1, φ_2) and $(\varphi_1, \varphi_2, \varphi_3)$. The algorithm should start from the vertex corresponding to the empty path (*i.e.* it corresponds to the initial concept φ) and find a shortest path to a vertex with a non-empty flag. The length of a path would be the $\mathtt{weight}$ of the reflection path R_φ contained in the last vertex of the path.

1. $a \wedge c \wedge b$ is inconsistent,
2. $\exists \varphi \in \mathcal{L}, c \vdash \varphi$ and $a \wedge \varphi \hspace{0.5mm}|\hspace{-1mm}\sim_{\Delta} \neg b$,

where $|\hspace{-1mm}\sim_{\Delta}$ is the inference relation defined by the closure of the preference entailment relation $|\hspace{-1mm}\sim$ over the set obtained by interpreting each default $a \rightsquigarrow b \in \Delta$ as $a \hspace{0.5mm}|\hspace{-1mm}\sim b$.

Definition 7 (S2-entailment). *Given a knowledge base $KB = (F, \Delta, A, R)$, S2-entailment, denoted $|\hspace{-1mm}\sim_2$, is defined by $\forall \varphi, \varphi' \in \mathcal{L}, \varphi \hspace{0.5mm}|\hspace{-1mm}\sim_2 \varphi'$ iff $F_\varphi \cup \{\varphi\} \nvdash \bot$ and $F_\varphi \cup \{\varphi\} \vdash \varphi'$, where $F_\varphi = F \cup \{a \rightarrow b | a \rightsquigarrow b \in \Delta$ has no exception with $\varphi\}$.*

Example 3. *The student's KB is s.t. $\Delta = \{phs \rightsquigarrow eff, eff \rightsquigarrow \neg exp, phs \wedge opt \rightsquigarrow ltp\}$ and $F = \{ltp \rightarrow \neg eff\}$. It holds that $phs \hspace{0.5mm}|\hspace{-1mm}\sim_2 \neg exp$ (using Cautious monotony on $phs \hspace{0.5mm}|\hspace{-1mm}\sim eff$ and $eff \hspace{0.5mm}|\hspace{-1mm}\sim \neg exp$ and Cut on $phs \wedge eff \hspace{0.5mm}|\hspace{-1mm}\sim \neg exp$ and $phs \hspace{0.5mm}|\hspace{-1mm}\sim eff$ and due to the fact that Contextual entailment generalizes Preferential entailment, see Proposition 1).*

Note that we are not yet in position to define *S2-reasoning*, which could evaluate the flag of a formula φ given a cognitive model κ. In order to do so we should define an aggregation function that combines all the possible flags that could be obtained for φ given the available beliefs, appreciations and a-rules. However, we have enough material to define the evaluation of one argument as shown in the next section.

3   Argument Evaluation

3.1   Argument and Profiles

We first give a (restrictive) definition of an argument, since we only consider arguments in favor of appreciations and not in favor of beliefs as it is the case in, for instance, [1].

Definition 8. (Argument). *An argument is a tuple $(s, h, w, (c, f))$ where s is a formula (the speaker enunciating the argument), h is a pair (K_h, A_h) with a set of default rules K_h and a set of appreciations $A_h \subseteq \mathcal{L} \times \{\oplus, \ominus, \odot\}$ (the premise of the argument), w is an a-rule (the warrant), c is a formula (the conclusion) and $f \in \{\oplus, \ominus\}$ is a flag stating that the argument conclusion should be accepted or rejected.*

This definition is syntactic. Hence, quadruplets containing premises not linked with the conclusion may comply with our definition. It is up to the listener to declare if the argument is valid semantically. This is the aim of this section. In the ELM model [23], the determination of the "route" for persuasion is made thanks to two main factors: the *interest* in processing the message and the *ability* (wrt knowledge and cognitive availability) to process it. In our model, the **interest** is given by the function i (see Definition 3). An agent may be not interested by a formula φ ($i(\varphi) = 0$), interested ($i(\varphi) = 1$) or "fanatic"

$(i(\varphi) = 2)$. The **knowledge** is represented by the size of the stack related to φ in AT. This size is compared to the agent's threshold λ (see Definition 3) in order to link the quantity of information the agent has to his feeling about being sufficiently aware on φ.

We use these factors for distinguishing several profiles of agents (note that we leave the cognitive availability for future work). In order to make a clear-cut categorisation of the possible engagements and to comply with the notions used in the ELM model, we define four levels of engagement: unconcerned, enthusiastic, quiescent or engaged with increasing involved level of cognition (see Definitions 11–14). Such profiles represent typical (and extreme) dispositions wrt the evaluation of an argument which goes beyond the classical idea to propose credulous and sceptical attitudes (see e.g. [1]).

Definition 9 (Profile). *The* profile *of an agent is a function that maps a formula* $\varphi \in \mathcal{L}$ *and a cognitive model* $\kappa = (KB, AT, \lambda, i)$ *to an element of* $\{unc, ent, qui, eng\}$:

$$\texttt{profile}(\varphi, \kappa) = \begin{cases} unc & if \ i(\varphi) = 0 \\ qui & if \ i(\varphi) = 1 \ and \ |AT(\varphi)| < \lambda \\ eng & if \ i(\varphi) = 1 \ and \ |AT(\varphi)| \geq \lambda \\ ent & if \ i(\varphi) = 2 \end{cases}$$

The following postulate expresses that if an agent is **enthusiastic** about a formula φ, then she has an opinion about φ.

Postulate 1. $\texttt{profile}(\varphi, \kappa) = ent$ *implies* $\texttt{flag}_\kappa(\varphi) \neq \odot$.

The next section details the value of the function $\texttt{evalarg}$ defined below.

Definition 10 (Evaluation of an Argument). *Given a cognitive model* $\kappa = (KB, AT, \lambda, i)$, *an argument* $a = (s, h, w, (c, f))$ *and a profile* $p = \texttt{profile}(c, \kappa)$, *let* $\texttt{evalarg}$ *be a function that maps* a *and* p *to an evaluation of the argument in* $\{\oplus, \ominus, \odot\}$, *denoted as* $\texttt{evalarg}(a, p)$.

3.2 Argument Evaluation According to Profiles

In this section, we define formally how the evaluation is done with respect to the four profiles.

Unconcerned. As its name implies, the unconcerned profile represents the fact that no interest is given by the agent in the received argument. Hence, an unconcerned agent will not bother trying to evaluate this argument and will just discard it.

Definition 11 (Unconcerned Evaluation). *Given an argument* $a = (s, h, w, (c, f))$, *the evaluation of* a *by an unconcerned agent unc is never done.*

Enthusiastic. The enthusiastic profile represents the fact that an agent is already convinced. As such, she does not feel the need to evaluate rationally the argument and will just check if the flag of the argument's conclusion correspond to the flag in her AT.

Definition 12 (Enthusiastic Evaluation). *Given an argument* $a = (s, h, w, (c, f))$*, the evaluation of* a *by an enthusiastic agent* $\texttt{evalarg}(a, ent) = \oplus$ *iff* $eval_1(c) = f$ *else* $\texttt{evalarg}(a, ent) = \ominus$.

Quiescent. A quiescent profile represents an "ideally instinctive" agent evaluating an argument thanks to her S1. More precisely, when receiving an argument, the agent evaluates the argument's conclusion and the speaker. She will accept the argument if she agrees with the conclusion and does not reject the speaker, or vice-versa.

Definition 13 (Quiescent Evaluation). *Given an argument* $a = (s, h, w, (c, f))$*, the evaluation of* a *by a quiescent agent with a cognitive model* κ *is defined as follows:*

$$
\texttt{evalarg}(a, qui) = \begin{cases} \oplus & \text{if } (eval_1(c, \kappa) = f \text{ and } eval_1(s, \kappa) \neq \ominus) \text{ or} \\ & (eval_1(c, \kappa) \neq -f \text{ and } eval_1(s, \kappa) = \oplus), \\ \ominus & \text{if } (eval_1(c, \kappa) = -f \text{ and } eval_1(s, \kappa) \neq \oplus) \text{ or} \\ & (eval_1(c, \kappa) \neq f \text{ and } eval_1(s, \kappa) = \ominus), \\ \odot & \text{otherwise} \end{cases}
$$

In future work, we plan to take into account the extra sources of persuasion such as the context created by the source of information including trustworthiness and charisma of the source, the contextual mood of the agent, etc.

Example 4. *During a long and very technical meeting, when a partner said that "since post harvest separation is highly expensive, which is undesirable, post harvest has to be rejected", our internship student did not have the cognitive ability to rationally consider this argument. While he would not have agreed with a deeper analysis, he instead relied on his S1, where post harvest separation is associated with something he rejects (see Fig. 1), and therefore accepted the argument.*

Engaged. An engaged profile represents an "ideally rational" agent evaluating an argument exclusively thanks to her knowledge base. In this work, we propose to define an engaged agent as someone who evaluates an argument wrt its set of warrants that are encoded in a way to capture critical questions (see [4, 32]). An engaged agent has to pass three steps before validating an argument: validity of the warrant ("Am I able to recognize this scheme of thought as a valid one?" translated into "Does it already exists in my personal base of a-rules")[6]; a syntactic validity of the use of the warrant in the argument ("Is the warrant conform with the premises and conclusions of the argument?" translated in terms of existence of a unification function σ); rational validation of applicability ("Are the premises correct and necessary ?" translated into the use of contextual inference in order to prove them).

---

[6] Note that we propose to be neutral wrt an argument that uses an unknown warrant.

Definition 14 (Engaged Evaluation). *Given an argument* $a = (s, h, w, (c, f))$, *with* $h = (K_h, A_h)$, *the evaluation of* a *by an engaged agent with a cognitive model* $\kappa = (KB, AT, \lambda, i)$ *with* $KB = (F, \Delta, A, R)$ *is defined as follows:*

$$
\mathrm{evalarg}(a, eng) = \begin{cases} \oplus \;\; if \begin{cases} w \in R \text{ and} \\ \exists \sigma : \mathcal{V}_G \to \mathcal{V} \text{ s.t. } \sigma(w) = (h \Rightarrow (c, f)) \text{ and} \\ \forall (x \rightsquigarrow y) \in K_h, x \mid\sim_2 y \text{ and } \neg x \mid\not\sim_2 y \text{ and } A_h \subseteq A \end{cases} \\ \ominus \;\; if \begin{cases} w \in R \text{ and} \\ \nexists \sigma : \mathcal{V}_G \to \mathcal{V} \text{ s.t. } \sigma(w) = (h \Rightarrow (c, f)) \text{ or} \\ \exists (x \rightsquigarrow y) \in K_h, x \mid\not\sim_2 y \text{ or } \neg x \mid\sim_2 y \text{ or } A_h \nsubseteq A \end{cases} \\ \odot \;\; otherwise \end{cases}
$$

Example 5. *Several days after the meeting, our internship student thought of the partner's argument again. Now that he is able to analyze the argument more rationally, he can recognize its type* ($w \in R$)*: his set of warrants* R *contains two a-rules,* $w_1 = (\{a \rightsquigarrow b\}, \{(b, \ominus)\}) \Rightarrow (a, \ominus)$ *and* $w_2 = (\{a \rightsquigarrow b\}, \{(b, \oplus)\}) \Rightarrow (a, \oplus)$ *which encode the schemes associated to* arguments from positive or negative consequences *(see [32] for a definition of these argumentation schemes). Since* $h = (\{phs \rightsquigarrow exp\}, \{(exp, \ominus)\})$ *and the conclusion is* $(phs, \ominus)$*, the argument is well formed wrt* w_1*; however,* w_2 *is not applicable. Then, he checks if the premise holds: as seen in Ex. 3,* $phs \mid\sim_2 \neg exp$*, and thus* $phs \mid\not\sim_2 exp$*. Hence, he rejects the argument.*

3.3 Argument Influence on the Agent's Mind

Once the argument has been evaluated by an agent, her cognitive model may have to be modified to account for the persuasive impact of the argument. Such modifications can either be the change of a flag value, the addition of a new association or the addition of a new rule. Definition 15 gives the functions representing these modifications.

Definition 15 (Update Operations). *Given two cognitive states* $\kappa = (KB, AT, \lambda, i)$ *with* $KB = (F, \Delta, A, R)$ *and* κ'*, two formulae* $x, y \in \mathcal{L}$*, a set of default rules* $D \subseteq \mathcal{L} \times \mathcal{L}$ *and a flag* $f \in \{\oplus, \ominus, \odot\}$*, we define:*

- $\mathrm{noop}(\kappa) = \kappa$
- $\mathrm{setflag}(\kappa, x, f) = \kappa'$ *where* $\kappa' = ((F, \Delta, A', R), AT', \lambda, i)$ *with*
  - $L_{AT'} = L_{AT} \cup \{x\}$,
  - $\forall \varphi \in L_{AT}$ *s.t.* $\varphi \neq x$, $\mathrm{flag}_{\kappa'}(\varphi) = \mathrm{flag}_{\kappa}(\varphi)$ *and* $AT'(\varphi) = AT(\varphi)$,
  - $\mathrm{flag}_{\kappa'}(x) = f$ *and* $A' = A \setminus \{(x, \mathrm{flag}_{\kappa}(x))\} \cup \{(x, f)\}$ *and* $AT'(x) = AT(x)$.
- $\mathrm{push}(\kappa, (x, y)) = \kappa'$ *where* $\kappa' = (KB', AT', \lambda, i)$ *with*
  - *if* $x \notin L_{AT}$ *then* $AT' = AT \cup \{(x, S_x, \odot)\}$ *with* $S_x = \{y\}$,
  - *else*
    - $\forall \varphi \in L_{AT}$ *s.t.* $\varphi \neq x$, $\mathrm{flag}_{\kappa'}(\varphi) = \mathrm{flag}_{\kappa}(\varphi)$ *and* $AT'(\varphi) = AT(\varphi)$,
    - $\mathrm{flag}_{\kappa'}(x) = \mathrm{flag}_{\kappa}(x)$ *and* $AT'(x) = AT(x) \cup \{y\}$ *with* $\mathrm{Top}(x) = y$,
- $\mathrm{addrule}(\kappa, D) = \kappa'$ *s.t.* $\kappa' = (F, \Delta \cup D, A, R), AT, \lambda, i)$.

Depending on the profile, the cognitive model will be modified in different ways. These differences aim at representing the fact that the persuasion may be deeper depending on the cognitive involvement of the agent. Table 1 gives the functions to apply to κ in order to update it, according to the possible evaluations of an argument by an agent and her profile. The "$\times$" in the *ent* and *unc* lines corresponds to impossible cases due to, respectively, Postulate 1 and Definition 11.

Table 1. Update of a cognitive state κ.

profile(c, κ)	evalarg$((s, h, w, (c, f)))$		
	$\odot$	$\ominus$	$\oplus$
unc	push$(\kappa, (c, h))^a$	$\times$	$\times$
ent	$\times$	push$(\kappa, (c, h))$	push$(\kappa, (c, h))$
		setflag$(\kappa, s, \ominus)$	push$(\kappa, (h, c))$
			setflag$(\kappa, s, \oplus)$
qui	$\times$	push$(\kappa, (c, h))$	push$(\kappa, (c, h))$
		setflag$(\kappa, c, -f)$	push$(\kappa, (h, c))$
		setflag$(\kappa, s, \ominus)$	setflag(κ, c, f)
			setflag$(\kappa, s, \oplus)$
eng	noop	noop	addrule(κ, K_h)
			setflag(κ, c, f)

[a] An argument is never evaluated by an unconcerned agent. However, we represent the fact that, like enthusiastic and quiescent agents, she is unconsciously influenced by what she hears.

4   Properties and Postulates

We have not yet been able to experiment in presence of human subjects in order to validate our model, but we have started to explore its rational properties.

4.1   Entailment Properties

Let us examine the properties of S1 and S2-entailment. Due to the construction of $\mid\sim_2$ on the basis of contextual entailment, it follows that $\mid\sim_2$ is obeying the same properties.

Proposition 1. $\mid\sim_2$ *obeys the axiom and the five inference postulates of [17]:*

- Reflexivity: $a \mid\sim_2 a$,
- Left logical equivalence: *if* $\vdash a \leftrightarrow b$ *and* $a \mid\sim_2 c$ *then* $b \mid\sim_2 c$,
- Right weakening: *if* $a \vdash b$ *and* $c \mid\sim_2 a$ *then* $c \mid\sim_2 b$,
- Cut: *if* $a \wedge b \mid\sim_2 c$ *and* $a \mid\sim_2 b$ *then* $a \mid\sim_2 c$,

- Cautious monotony: *if* $a \mathrel{|\!\sim_2} b$ *and* $a \mathrel{|\!\sim_2} c$ *then* $a \wedge b \mathrel{|\!\sim_2} c$,
- Or: *if* $a \mathrel{|\!\sim_2} c$ *and* $b \mathrel{|\!\sim_2} c$ *then* $a \vee b \mathrel{|\!\sim_2} c$.

It is not the same for $\mathrel{|\!\sim_1}$, since it may be sensitive to the syntax, *i.e.*, nothing prevents to have a different stack for two equivalent formulae.

Proposition 2. – $\mathrel{|\!\sim_1}$ *obeys* Reflexivity *only for the formulae that admit finite reflection paths*
- $\mathrel{|\!\sim_1}$ *obeys* Left logical *equivalence only if AT is syntax dependent* i.e. $\varphi \leftrightarrow \psi$ *iff* $AT(\varphi) = AT(\psi)$,
- $\mathrel{|\!\sim_1}$ *does not obey* Right weakening
- Transitivity *holds, namely,* $a \mathrel{|\!\sim_1} b$ *and* $b \mathrel{|\!\sim_1} c$ *implies* $a \mathrel{|\!\sim_1} c$
- Cut, Cautious monotony *and* Or*: do not necessarily hold.*

Proof. *Reflexivity*: if $\exists R_a$ s.t. $\mathtt{flag}(R_a) \neq \odot$ then $a \in R_a$ hence $a \mathrel{|\!\sim_1} a$ otherwise it is not the case. *Right weakening*: since b can be deducible logically from a but not in $AT(a)$. ı*Transitivity*: it means that $b \in R_a$ and $c \in R_b$, hence if $\mathtt{flag}(b) = \odot$ then $c \in R_a$ else $c = b$ hence $c \in R_a$ as well. *Cut, Cautious Monotony* and *Or*: it is due to the independence of associations wrt logic (hence "logical and" is not necessarily compatible with associations),   □

4.2 Incorporation Property

Let us notice that after receiving an argument, the knowledge of an agent can only increase: more precisely, among the formulae that were already present, the number of flags that are not empty decreases (however some new formula may be added with an empty flag) and the number of associations grows. Moreover some rules can also be added in the case of an engaged profile.

Proposition 3. *Let* $\kappa = ((F, \Delta, A, R), AT, \lambda, i)$, $\kappa' = ((F', \Delta', A', R'), AT',$ $\lambda', i')$ *such that* κ' *is the cognitive model obtained from* κ *after the utterance of an argument. It holds that* $L_{AT} \subseteq L_{AT'}$, $\forall \varphi \in L_{AT}, AT(\varphi) \subseteq AT'(\varphi)$, *and* $F = F'$, $\Delta \subseteq \Delta'$, $R = R'$, $\lambda = \lambda'$ *and* $i = i'$.

Note that the flag values are non-monotonic since a formula can obtain either an accepted, rejected or empty flag depending on the engagement profile.

4.3 Public Opinion Axioms

According to [34], the model of how information is transformed in public opinion follows four axioms mentioned below. Our proposal satisfies these axioms:

Reception Axiom: *The greater the level a person's level of cognitive engagement with an issue the more likely he/she will be exposed to and comprehend political messages concerning that issue.* It holds since an unconcerned agent does not evaluate the argument, an enthusiastic agent takes it into account if she agrees with the conclusion, a quiescent agent evaluates it with S1-reasoning and an engaged agent evaluates it with S2-entailment. Hence, the more engaged an agent is, the more information she takes into account (in the following order: unconcerned, enthusiastic, quiescent, engaged).

Resistance Axiom: *People tend to resist arguments that are inconsistent with their political predispositions but they do so only to the extent that they posses the contextual information necessary to perceive a relationship between the message and their predispositions.* Unconcerned, enthusiastic and engaged agents may resist an argument since they are not influenced by its flag. A quiescent agent resists arguments that are against her opinion or uttered by a source she rejects (see Definition 13).

Accessibility Axiom: *The more recently a consideration has been called to mind, or thought about, the less time it takes to retrieve that consideration or related considerations from memory and bring them to the top of the head for use.* This axiom is satisfied concerning the association table AT since every kind of profile add the new piece of information at the top of the stack (see Table 1).

Response Axiom: *Individuals answer survey questions by averaging across the considerations that are immediately salient or accessible to them.* It holds for quiescent and enthusiastic: a quiescent agent evaluates a formula by considering the most immediately accessible information and an enthusiastic agent evaluates only the immediate value of a formula. However, it does not hold for unconcerned and engaged agents: one does not evaluate the formula, and the other evaluates the formula with her knowledge base.

5   Conclusion

This paper is a first proposal of a formalization of dual process theory and its link with human persuasion. Based on the ELM model of persuasion, we define four profiles evaluating an argument in different ways. One of the profiles aims at reasoning thanks to an association table, and another is based on a logical inference mechanism named contextual entailment. This mechanism is a possible implementation of S2 and can be changed without jeopardizing the cognitive model. Moreover, each profile integrates the contents of the received argument differently. Accordingly to public opinion axioms, the more cognition was involved in its evaluation, the more persuasive content will take root in the mind of the agent.

Related Work. Dual process theories have already been implemented for problem solving. Namely, [14] with an extension of the CLARION architecture that relies on two modules: a bottom-level (resp. top-level) module handling implicit knowledge (resp. explicit knowledge), which recall the S1 and S2 systems but is not based on formal logic. [27] proposes a general intelligence cognitive architecture composed of a long-term memory independent of specific tasks and a capacity-limited working memory. The S1 and S2 systems allow them to distinguish between perception and imagination and are represented thanks to two binary relations on the element of the long-term memory and two propagation processes. Some works, similarly to ours but not in a logical framework, aim at explaining purely human processes. For instance, [18] studies the emergence of emotions thanks to a three-levels cognitive architecture: S1 (the *reactive* level) and S2, subdivided into the *algorithmic* level and the *reflective* level. The first one is responsible for fast and instinctive behaviours, the second one is used for cognitive control and the last one handles rational behavior.

In [16] is a different approach for persuasion, since the NAG program is able to analyze and generate arguments with the aim of persuading a human user. In order to do so, NAG comprises two different models, a normative model that is able to judge the correctness of an argument (in terms of links between the premises and the conclusion), and a user model, that is able to evaluate the persuasion capability of an argument on the user. Hence, NAG is interestingly able to analyze an argument given by the user and to try to generate a counterargument which is at the same time correct and specifically designed to be effective on the user. Since NAG has to persuade a human user, it requires a representation of her cognitive profile, in particular her reasoning errors such as cognitive biases. Major differences exist between our approach and NAG. First, NAG is intended to interact with users, and as such it is human-machine oriented. Then, the model does not rely on a logical dual process but is based on a Bayesian network; cognitive biases are thus taken into account by the modification of probability degrees while, in our framework, biases are due to faulty appreciations, warrants or beliefs. Finally, the authors do not use argumentation schemes (encoded in our warrants base R) and thus do not have a clear definition of argument and ways to evaluate them.

Perspectives. Since this work is a first attempt to formalize a two-process cognitive model and its links with argument evaluation, numerous perspectives can be envisaged. Namely, a refined definition of the weights associated to the reflection paths could help to account for the various heuristics Kahneman and Tversky described. Moreover, we would like to investigate a way to compute the cognitive availability of an agent in order to determine her ability to engage in the argument evaluation. Such study would benefit from the definition of a profile corresponding to a continuum between the quiescent and the engaged profiles. Such a profile might be more adapted to represent human reasoning with its bounded rationality, and may help to better capture the difference of persuasion strategies that can be used according to the proximity to S1 or S2-reasoning. Moreover, a more realistic model should take into account social influence [7] between agents when exchanging arguments. In the same way, the study of rhetorical mechanisms could improve the evaluation of argument with another dimension. Finally, the public opinion axioms of [34] show that results from psychology studies can be used to guide our research, and as such it would be useful to validate our proposal by an empirical study with human beings.

References

1. Amgoud, L., Maudet, N., Parsons, S.: An argumentation-based semantics for agent communication languages. In: ECAI 2002, pp. 38–42. IOS Press (2002)
2. Beevers, C.G.: Cognitive vulnerability to depression: a dual process model. Clin. Psychol. Rev. **25**(7), 975–1002 (2005)
3. Benferhat, S., Dupin de Saint Cyr - Bannay, F.: Contextual handling of conditional knowledge. In: Proceedings of IPMU 1996, Granada, Spain, July 1996
4. Blair, J.A., Johnson, R.H.: Informal logic: an overview. Informal Logic **20**(2), 93–108 (2000)
5. Burrows, R., Johnson, H., Johnson, P.: Developing an online social media system to influence pro-environmental behaviour based on user values. In: ICPT (2014)
6. Chaiken, S.: The heuristic model of persuasion. In: Social influence: The Ontario Symposium, vol. 5, pp. 3–37 (1987)
7. Cialdini, R.: Influence: Science and Practice. Allyn and Bacon, Boston (2001)
8. Clements, C.S.: Perception and persuasion in legal argumentation: using informal fallacies and cognitive biases to win the war of words. BYU Law Rev. **2013**(2), 319 (2013)
9. Croskerry, P., Singhal, G., Mamede, S.: Cognitive debiasing 1: origins of bias and theory of debiasing. BMJ Qual. Saf. **22**(Suppl 2), 58–64 (2013)
10. Epstein, S.: Integration of the cognitive and the psychodynamic unconscious. Am. Psychol. **49**(8), 709–724 (1994)
11. Evans, J.S.B.T., Curtis-Holmes, J.: Rapid responding increases belief bias: evidence for the dual-process theory of reasoning. Think. Reasoning **11**(4), 382–389 (2005)
12. Forget, A., Chiasson, S., van Oorschot, P.C., Biddle, R.: Persuasion for stronger passwords: motivation and pilot study. In: Oinas-Kukkonen, H., Hasle, P., Harjumaa, M., Segerståhl, K., Øhrstrøm, P. (eds.) PERSUASIVE 2008. LNCS, vol. 5033, pp. 140–150. Springer, Heidelberg (2008)
13. Hamblin, C.: Fallacies. University paperback, Methuen (1970)
14. Hélie, S., Sun, R.: Incubation, insight, and creative problem solving: a unified theory and a connectionist model. Psychol. Rev. **117**(3), 994–1024 (2010)
15. Hornikx, J., Hahn, U.: Reasoning and argumentation: towards an integrated psychology of argumentation. Think. Reasoning **18**(3), 225–243 (2012)
16. Korb, K.B., Mcconachy, R., Zukerman, I.: A cognitive model of argumentation. In: Proceedings of the Nineteenth Annual Conference of the Cognitive Science Society, pp. 400–405 (1997)
17. Kraus, S., Lehmann, D., Magidor, M.: Nonmonotonic reasoning, preferential models and cumulative logics. Artif. Intell. **44**, 167–207 (1990)
18. Larue, O., Poirier, P., Nkambou, R.: Emotional emergence in a symbolic dynamical architecture. In: Chella, A., Pirrone, R., Sorbello, R., Jóhannsdóttir, K.R. (eds.) Biologically Inspired Cognitive Architectures 2012. AISC, vol. 196, pp. 199–204. Springer, Heidelberg (2013)
19. Lehto, T., Oinas-Kukkonen, H.: Explaining and predicting perceived effectiveness and use continuance intention of a behaviour change support system for weight loss. Behav. Inf. Technol. **34**(2), 176–189 (2015)
20. Mackenzie, J.: Four dialogue systems. Stud. Logica **49**(4), 567–583 (1990)
21. Oinas-Kukkonen, H.: A foundation for the study of behavior change support systems. Pers. Ubiquit. Comput. **17**(6), 1223–1235 (2013)
22. Perelman, C., Olbrechts-Tyteca, L.: The New Rhetoric: A Treatise on Argumentation. University of Notre Dame Press, Notre Dame (1969)

23. Petty, R., Cacioppo, J.: The elaboration likelihood model of persuasion. Adv. Exp. Soc. Psychol. **19**(C), 123–205 (1986)
24. Prakken, H.: Formal systems for persuasion dialogue. Knowl. Eng. Rev. **21**(2), 163–188 (2006)
25. Rahwan, I., Amgoud, L.: An argumentation based approach for practical reasoning. In: Proceedings of the Fifth International Joint Conference on Autonomous Agents and MultiAgent Systems, pp. 347–354 (2006)
26. Sloman, S.A.: The empirical case for two systems of reasoning. Psychol. Bull. **119**(1), 3–22 (1996)
27. Strannegård, C., von Haugwitz, R., Wessberg, J., Balkenius, C.: A cognitive architecture based on dual process theory. In: Kühnberger, K.-U., Rudolph, S., Wang, P. (eds.) AGI 2013. LNCS, vol. 7999, pp. 140–149. Springer, Heidelberg (2013)
28. Touretzky, D.: Implicit ordering of defaults in inheritance systems. In: Proceedings of AAAI 1984. University of Texas at Austin (1984)
29. Tversky, A., Kahneman, D.: Judgment under uncertainty: heuristics and biases. Science **185**(4157), 1124–1131 (1974)
30. van Knippenberg, D.: Social identity and persuasion: reconsidering the role of group membership. In: Social Identity and Social Cognition, vol. XVII, pp. 315–331 (1999)
31. Walton, D.: Logical Dialogue: Games and Fallacies. University Press of America, Lanham (1984)
32. Walton, D., Reed, C., Macagno, F.: Argumentation Schemes. Cambridge University Press, Cambridge (2008)
33. Wood, W.: Attitude change: persuasion and social influence. Annu. Rev. Psychol. **51**(1), 539–570 (2000)
34. Zaller, J.: The Nature and Origins of Mass Opinion. Cambridge Studies in Political Psychology Series. Cambridge University Press, Cambridge (1992)

Change in Abstract Bipolar Argumentation Systems

Claudette Cayrol and Marie-Christine Lagasquie-Schiex$^{(\boxtimes)}$

IRIT-UPS, Toulouse, France
{ccayrol,lagasq}@irit.fr

Abstract. An argumentation system can undergo changes (addition or removal of arguments/interactions), particularly in multiagent systems. In this paper, we are interested in dynamics of abstract bipolar argumentation systems, *i.e.* argumentation systems using two kinds of interaction: attacks and supports. We propose change characterizations that use and extend previous results defined in the case of Dung abstract argumentation systems.

Keywords: Dynamics of bipolar Argumentation · Deductive support

1 Introduction

The main feature of argumentation is the ability to deal with incomplete and/or contradictory information, especially for reasoning [1,19]. Moreover, argumentation can be used to formalize dialogues between several agents by modeling the exchange of arguments in, *e.g.*, negotiation between agents [3,4]. An argumentation system (AS for short) consists of a collection of arguments interacting with each other through a relation reflecting conflicts between them, called *attack*. An issue of argumentation is then to determine "acceptable" sets of arguments (*i.e.*, sets able to defend themselves collectively while avoiding internal attacks), called *"extensions"*, and thus to reach a coherent conclusion. Formal frameworks have greatly eased the modeling and study of AS. In particular, the framework of [19] allows for abstracting from the "concrete" meaning of the arguments and relies only on binary interactions that may exist between them. This approach enables the user to focus on other aspects of argumentation, including its dynamic side. Indeed, in the course of a discussion or due to the acquisition of new pieces of information, an AS can undergo changes such as the addition of a new argument or the removal of an argument considered as illegal. This is of particular interest for dialogs in a multiagent system since it is unrealistic to consider that the argumentation system reflecting the dialog can be statically defined. Moreover, it is important to reuse as far as possible computations carried out in the original system. That's why it is interesting to characterize these changes by giving properties describing a change operation and to provide conditions under which these properties hold. This has been done in several papers[1], especially [9], for Dung AS with only attacks.

---

[1] See for instance [7,8,11,17,18].

© Springer International Publishing Switzerland 2015
C. Beierle and A. Dekhtyar (Eds.): SUM 2015, LNAI 9310, pp. 314–329, 2015.
DOI: 10.1007/978-3-319-23540-0_21

In this paper, we are interested in the extension of this work to bipolar AS (BAS for short), *i.e.* AS augmented with a second kind of interaction, the support relation. This relation represents a positive interaction between arguments and has been first introduced by [21,29]. In [12], the support relation is left general so that the resulting bipolar framework keeps a high level of abstraction. However there is no single interpretation of the support, and a number of researchers proposed specialized variants of the support relation: deductive support [10], necessary support [23,24], evidential support [25,26]. Each specialization can be associated with an appropriate modelling using appropriate complex attacks. These proposals have been developed quite independently, based on different intuitions and with different formalizations. [14] presents a comparative study in order to restate these proposals in a common setting, the *bipolar argumentation framework*. The idea is to keep the original arguments, to add complex attacks defined by the combination of the original attacks and the supports, and to modify the classical notions of acceptability. An important contribution of [14] is to highlight a kind of duality between the deductive and the necessary interpretations of support, which results in a duality in the modelling by complex attacks. Handling support is a growing concern: [27] gives a translation between necessary supports and evidential supports; [28] proposes a justification of the necessary support using the notion of subarguments; [22] studies an extension of the necessary support; [20] gives a logical study of bipolar systems; [16] proposes a general framework for taking into account recursive attacks and supports. However, there is no work concerning the study of the dynamics of a bipolar AS while it is an essential issue for modelling the actions of the participants to a multiagent system:

Example 1. *Journalists during an editorial board discuss about the publication of an information I:*

Journalist J_1 (Argument a): *I is important, we must publish it;*
Journalist J_2 (Argument b): *I is about a person X, it is forbidden to publish without the agreement of the concerned person and X disagrees with the publication;*
Journalist J_1 (Argument c): *X is a public person (she is the Prime Minister); in this case, her agreement is not mandatory;*
Journalist J_2 (Argument d): *However, I have heard about X's resignation;*
Journalist J_3 (Argument e): *I now understand why CNN has announced yesterday the postponement of the Council of Ministers;*
Journalist J_4 (Argument f): *However, yesterday was April Fools' Day; so CNN news announced yesterday are not reliable.*

This example illustrates a typical situation between agents that exchange arguments in order to take a decision (here, publish or not publish information I). In this dialog, one can see arguments (here, informal arguments corresponding to pieces of dialog), attacks (for instance Argument b attacks Argument a), supports (between Argument d and Argument e); and the dynamics of argumentation is illustrated by the dynamics of the dialog: at each step of the dialog, the *global argumentation system evolves (here, by the addition of an argument and an interaction).*

In this paper, we define the update of BAS and characterize it in a special case: a BAS reduced to an AS that is changed by the introduction of a new argument that interacts with another argument using supports. Such an update is realized using a combination of the works of both domains (bipolar argumentation and dynamics of argumentation).

Background is given in Sect. 2 for AS and BAS, and in Sect. 3 for change operations. Section 4 proposes a change operation concerning a BAS. Characterizations of this new change operation are presented in Sect. 5. Finally, Sect. 6 concludes and suggests perspectives. The proofs are given in [15].

2   Abstract Bipolar Argumentation System

2.1   Abstract Argumentation System

Dung's abstract framework consists of a set of arguments and only one type of interaction between these arguments, these interactions representing attacks.

Definition 1 (Dung AS). *A Dung argumentation system (AS, for short) is a pair* $\langle \mathbf{A}, \mathbf{R} \rangle$ *where* $\mathbf{A}$ *is a finite and non-empty set of arguments and* $\mathbf{R}$ *is a binary relation over* $\mathbf{A}$ *(a subset of* $\mathbf{A} \times \mathbf{A}$*), called the* attack *relation.*

An AS can be represented by a directed graph denoted by $\mathcal{G}$, in which nodes represent arguments and edges are defined by the attack relation: $\forall a, b \in \mathbf{A}$, $a\mathbf{R}b$ is represented by $a \nrightarrow b$. Semantics introduced by Dung enable to characterize admissible sets of arguments that satisfy a form of optimality. Here we only use (see [6] for a survey of semantics in abstract AS):

Definition 2 (Admissibility, Extensions). *Given* AS $= \langle \mathbf{A}, \mathbf{R} \rangle$ *and* $S \subseteq \mathbf{A}$,

- *S is* conflict-free *in* AS *if and only if (iff for short) there are no arguments* $a, b \in S$, *such that (s.t. for short)* $a\mathbf{R}b$.
- $a \in \mathbf{A}$ *is* acceptable *in* AS *with respect to (wrt for short)* S *iff* $\forall b \in \mathbf{A}$ *s.t.* $b\mathbf{R}a$, $\exists c \in S$ *s.t.* $c\mathbf{R}b$. $\mathcal{F}$ *denotes the characteristic function of* AS *defined by* $\forall S \subseteq \mathbf{A}$, $\mathcal{F}(S) = \{x \text{ s.t. } x \text{ is acceptable in } \mathsf{AS} \text{ wrt } S\}$.
- *S is* admissible *in* AS *iff* S *is conflict-free and each argument in* S *is acceptable in* AS *wrt* S.
- *S is a* preferred extension *of* AS *iff it is a maximal (wrt* $\subseteq$*) admissible set in* AS.
- *S is a* stable extension *of* AS *iff it is conflict-free and for each* $a \notin S$, *there is* $b \in S$ *s.t.* $b\mathbf{R}a$.
- *S is the* grounded extension *of* AS *iff it is the least fixpoint of* $\mathcal{F}$.

Example 2. *Let* AS *be represented by the following graph.* $\{a\}$ *and* $\{b, d\}$ *are the two preferred extensions,* $\{b, d\}$ *is also stable and* $\varnothing$ *is the grounded extension.*

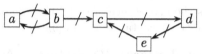

The status of an argument is determined by its membership to the extensions of the selected semantics: *e.g.*, an argument is "skeptically accepted" (resp. "credulously") if it belongs to all the extensions (resp. at least to one extension) and "rejected" if it does not belong to any extension.

2.2   Abstract Bipolar Argumentation System

The abstract bipolar argumentation framework presented in [13] extends Dung's framework in order to take into account both negative interactions expressed by the attack relation and positive interactions expressed by a support relation (see [2] for a more general survey about bipolarity in argumentation).

Definition 3 (BAS). *A bipolar argumentation system (BAS, for short) is a tuple* $\langle \mathbf{A}, \mathbf{R}_{att}, \mathbf{R}_{sup} \rangle$ *where* $\mathbf{A}$ *is a finite and non-empty set of arguments,* $\mathbf{R}_{att}$ *is a binary relation over* $\mathbf{A}$ *called the* attack *relation and* $\mathbf{R}_{sup}$ *is a binary relation over* $\mathbf{A}$ *called the* support *relation.*

A BAS can still be represented by a directed graph $\mathcal{G}_b$, with two kinds of edges: let a and $b \in \mathbf{A}$, $a\mathbf{R}_{att}b$ (resp. $a\mathbf{R}_{sup}b$) means that a attacks b (resp. a supports b) and it is represented by $a \not\rightarrow b$ (resp. by $a \rightarrow b$).

Among the different variants defined for interpreting a support between arguments, [10] proposed the notion of deductive support. This notion is intended to enforce the following constraint: If $b\mathbf{R}_{sup}c$ then the acceptance of b implies the acceptance of c, and as a consequence the non-acceptance of c implies the non-acceptance of b. The support used in Example 1 can be considered as a deductive one (If X has resigned then the Council of Ministers must be postponed):

Example 1 (cont'd). *The bipolar argumentation system corresponding to the editorial board can be represented by:* $\boxed{f} \not\rightarrow \boxed{e} \leftarrow \boxed{d} \not\rightarrow \boxed{c} \rightarrow \boxed{b} \not\rightarrow \boxed{a}$

In order to compute semantics of a BAS, one of the main proposals is to translate the BAS into an AS expressing the new attacks due to the presence of supports (this kind of "flattening" is studied for instance in [20]). For deductive support, two kinds of attack can be added. The first one, called mediated attack in [10], corresponds to the case when $b\mathbf{R}_{sup}c$ and $a\mathbf{R}_{att}c$: the acceptance of a implies the non-acceptance of c and so the non-acceptance of b.

Definition 4 (Mediated attack). *[10] Let* BAS $= \langle \mathbf{A}, \mathbf{R}_{att}, \mathbf{R}_{sup} \rangle$. *There is a mediated attack from a to b iff there is a sequence* $a_1 \mathbf{R}_{sup} \dots \mathbf{R}_{sup} a_{n-1}$, *and* $a_n \mathbf{R}_{att} a_{n-1}$, $n \geq 3$, *with* $a_1 = b$, $a_n = a$. $\mathbf{M}^{\mathbf{R}_{sup}}_{\mathbf{R}_{att}}$ *denotes the set of mediated attacks generated by* $\mathbf{R}_{sup}$ *on* $\mathbf{R}_{att}$.

Moreover, the deductive interpretation of support justifies the introduction of another attack (called supported attack in [13]): if $a\mathbf{R}_{sup}c$ and $c\mathbf{R}_{att}b$, the acceptance of a implies the acceptance of c and the acceptance of c implies the non-acceptance of b; so, the acceptance of a implies the non-acceptance of b.

Definition 5 (Supported attack). *[13] Let* $\mathsf{BAS} = \langle \mathbf{A}, \mathbf{R}_{\mathrm{att}}, \mathbf{R}_{\mathrm{sup}} \rangle$. *There is a supported attack from a to b iff there is a sequence* $a_1 \mathbf{R}_{\mathrm{sup}} \dots \mathbf{R}_{\mathrm{sup}} a_{n-1} \mathbf{R}_{\mathrm{att}} a_n$, $n \geq 3$, *with* $a_1 = a$, $a_n = b$. $\mathbf{S}\mathbf{R}_{\mathbf{R}_{\mathrm{att}}}^{\mathbf{R}_{\mathrm{sup}}}$ *denotes the set of supported attacks generated by* $\mathbf{R}_{\mathrm{sup}}$ *on* $\mathbf{R}_{\mathrm{att}}$.

So, the deductive interpretation of support produces new kinds of attack, from a to b, in the following cases:

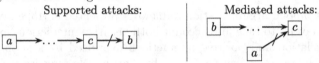

Supported attacks: Mediated attacks:

By iterating the construction, d-attacks can be defined:[2]

Definition 6 (d-attacks). *[14] Let* $\mathsf{BAS} = \langle \mathbf{A}, \mathbf{R}_{\mathrm{att}}, \mathbf{R}_{\mathrm{sup}} \rangle$ *with* $\mathbf{R}_{\mathrm{sup}}$ *being a set of deductive supports. There exists a d-attack from a to b iff*

- *either* $a\mathbf{R}_{\mathrm{att}}b$, *or* $a\mathbf{S}\mathbf{R}_{\mathbf{R}_{\mathrm{att}}}^{\mathbf{R}_{\mathrm{sup}}}b$, *or* $a\mathbf{M}\mathbf{R}_{\mathbf{R}_{\mathrm{att}}}^{\mathbf{R}_{\mathrm{sup}}}b$ **(Basic case)**,
- *or there exists an argument c s.t. there is a sequence of supports from a to c and c d-attacks b* **(Case 1)**,
- *or there exists an argument c s.t. a d-attacks c and there is a sequence of supports from b to c* **(Case 2)**.

$\mathbf{D}\mathbf{R}_{\mathbf{R}_{\mathrm{att}}}^{\mathbf{R}_{\mathrm{sup}}}$ *denoted the set of d-attacks generated by* $\mathbf{R}_{\mathrm{sup}}$ *on* $\mathbf{R}_{\mathrm{att}}$. $\langle \mathbf{A}, \mathbf{D}\mathbf{R}_{\mathbf{R}_{\mathrm{att}}}^{\mathbf{R}_{\mathrm{sup}}} \rangle$ *is called the deductive associated Dung AS of* BAS *and denoted by* $\mathsf{AS}^{\mathsf{BAS}}$.

Example 1 (Cont'd). *The deductive associated Dung AS can be represented by (a mediated attack appears from f to d):*

Then, in this system, using for instance the preferred semantics, one can conclude to the acceptability of a (so the information I will be published).

Turning BAS into $\mathsf{AS}^{\mathsf{BAS}}$ enables to consider the semantics defined by Dung. Moreover, the first step leading to add new attacks, it falls within works about dynamics of AS.

3 Dynamics in Argumentation Systems

When studying argumentation dynamics, an important issue is to save computation, that is to reuse as far as possible previous computations carried out in the original argumentation system. This issue has been extensively discussed in [9] with the following methodology: A typology of change operations has been proposed and the impact of each change operation on the computation of the

---

[2] It generalizes mediated, supported and also the "super-mediated attack" defined in [14].

extensions has been studied. So, the work of [9] is particularly suitable for our purpose and easily adaptable.[3] In this paper, following Example 1, we use the change operations corresponding to either the addition of an argument and the interactions (only attacks) involving it, or the addition of some interactions:

Definition 7 (Addition in an AS). *Let* $\mathsf{AS} = \langle \mathbf{A}, \mathbf{R} \rangle$.

1. *Let* z *be an argument and* $\mathcal{I}_z$ *be a set of interactions s.t.* $\mathcal{I}_z \subseteq (\mathbf{A} \times \{z\}) \cup (\{z\} \times \mathbf{A})$. *Adding* z *and* $\mathcal{I}_z$ *is a change operation, denoted by* $\oplus_{\mathcal{I}_z}^z$, *providing a new system s.t.:* $\oplus_{\mathcal{I}_z}^z \langle \mathbf{A}, \mathbf{R} \rangle = \langle \mathbf{A} \cup \{z\}, \mathbf{R} \cup \mathcal{I}_z \rangle$.
2. *Let* $\mathcal{I}$ *be a set of interactions s.t.* $\mathcal{I} \subseteq (\mathbf{A} \times \mathbf{A})$ *and* $\mathcal{I} \cap \mathbf{R} = \varnothing$. *Adding* $\mathcal{I}$ *is a change operation, denoted by* $\oplus_{\mathcal{I}}$, *providing a new system s.t.:* $\oplus_{\mathcal{I}} \langle \mathbf{A}, \mathbf{R} \rangle = \langle \mathbf{A}, \mathbf{R} \cup \mathcal{I} \rangle$.

The AS resulting of a change, denoted by $\mathsf{AS}' = \langle \mathbf{A}', \mathbf{R}' \rangle$, *is represented by* $\mathcal{G}'$.

In each case, given a semantics, the set of extensions of AS (resp. AS') is denoted by $\mathbf{E}$ (resp. $\mathbf{E}'$), with $\mathcal{E}_1, \ldots, \mathcal{E}_n$ (resp. $\mathcal{E}_1', \ldots, \mathcal{E}_n'$) standing for the extensions. We consider the same semantics before and after the change.

The impact of a change operation has been studied in [9] through the notion of *change property* that can be seen as a set of pairs $(\mathcal{G}, \mathcal{G}')$, where $\mathcal{G}$ and $\mathcal{G}'$ are argumentation graphs. Here we just recall some of these properties.

Properties about the set of extensions. Change properties express structural modifications of an AS that are caused by a change operation. For that purpose, a partition based on three possible cases of evolution of the set of extensions has been defined in [9]: the *extensive* (resp. *restrictive, constant*) case, in which the number of extensions increases (resp. decreases, remains the same).

For each case, numerous sub-cases are proposed and denoted by a letter (*e* for the *extensive* case, *r* for the *restrictive* case and *c* for the *constant* case) subscripted by the expression $\gamma - \gamma'$, where γ (resp. γ') describes the set of extensions before (resp. after) the change. Thus γ and γ' can be:

- $\varnothing$: the set of extensions is empty,
- $1e$: the set of extensions is reduced to one empty extension,
- $1ne$: the set of extensions is reduced to one non-empty extension,
- k (resp. j): the set of extensions contains k (resp. j) extensions s.t. $1 < k$ (resp. $1 < j < k$: note that the symbol j is used only if the symbol k belongs also to the expression $\gamma - \gamma'$).

For instance, the notation $e_{\varnothing - 1ne}$ means that the change increases the number of extensions (so it is an *extensive* case), with no initial extension ($\varnothing$) and

---

[3] Other works could be considered for addressing the issue of incremental computation in a dynamic context. [5] for instance presents a more general approach dealing with modularity in abstract argumentation, based on the partition of an argumentation framework in interacting subframeworks. However, the application to our purpose is not straightforward and requires further investigation.

one non-empty final extension ($1ne$). Nevertheless, some special sub-cases of the *constant* case are denoted by another method since they are based on notions distinct from the emptiness or the number of the extensions; for these sub-cases, the subscript is replaced by a qualifier. For instance, the *c-conservative* case describes the case where the extensions remain unchanged after the change.

Here is the formal definition of these changes:

Definition 8 (Extensive, Restrictive and Constant changes). *The change from $\mathcal{G}$ to $\mathcal{G}'$ is extensive (resp. restrictive, constant) iff $|\mathbf{E}| < |\mathbf{E}'|$ (resp. $|\mathbf{E}| > |\mathbf{E}'|$, $|\mathbf{E}| = |\mathbf{E}'|$).*[4]

1. *The sub-cases of* extensive *changes from $\mathcal{G}$ to $\mathcal{G}'$ are:*
   (a) $e_{\varnothing-1ne}$ *iff $|\mathbf{E}| = 0$ and $|\mathbf{E}'| = 1$, with $\mathcal{E}' \neq \varnothing$.*
   (b) $e_{\varnothing-k}$ *iff $|\mathbf{E}| < |\mathbf{E}'|$, $|\mathbf{E}| = 0$ and $|\mathbf{E}'| > 1$.*
   (c) e_{1e-k} *iff $|\mathbf{E}| < |\mathbf{E}'|$ and $|\mathbf{E}| = 1$, with $\mathcal{E} = \varnothing$.*
   (d) e_{1ne-k} *iff $|\mathbf{E}| < |\mathbf{E}'|$ and $|\mathbf{E}| = 1$, with $\mathcal{E} \neq \varnothing$.*
   (e) e_{j-k} *iff $1 < |\mathbf{E}| < |\mathbf{E}'|$.*
2. *The sub-cases of* restrictive *changes from $\mathcal{G}$ to $\mathcal{G}'$ are:*
   (a) $r_{1ne-\varnothing}$ *iff $|\mathbf{E}| = 1$, with $\mathcal{E} \neq \varnothing$, and $|\mathbf{E}'| = 0$.*
   (b) $r_{k-\varnothing}$ *iff $|\mathbf{E}| > |\mathbf{E}'|$, $|\mathbf{E}| > 1$ and $|\mathbf{E}'| = 0$.*
   (c) r_{k-1e} *iff $|\mathbf{E}| > |\mathbf{E}'|$ and $|\mathbf{E}'| = 1$, with $\mathcal{E}' = \varnothing$.*
   (d) r_{k-1ne} *iff $|\mathbf{E}| > |\mathbf{E}'|$ and $|\mathbf{E}'| = 1$, with $\mathcal{E}' \neq \varnothing$.*
   (e) r_{k-j} *iff $1 < |\mathbf{E}'| < |\mathbf{E}|$.*
3. *The sub-cases of* constant *changes from $\mathcal{G}$ to $\mathcal{G}'$ are:*
   (a) *c-conservative iff $\mathbf{E} = \mathbf{E}'$.*
   (b) c_{1e-1ne} *iff $\mathbf{E} = \{\{\}\}$ and $\mathbf{E}' = \{\mathcal{E}'\}$, with $\mathcal{E}' \neq \varnothing$.*
   (c) c_{1ne-1e} *iff $\mathbf{E} = \{\mathcal{E}\}$, with $\mathcal{E} \neq \varnothing$ and $\mathbf{E}' = \{\{\}\}$.*
   (d) *c-expansive iff $\mathbf{E} \neq \varnothing$ and $|\mathbf{E}| = |\mathbf{E}'|$ and $\forall \mathcal{E}_i \in \mathbf{E}, \exists \mathcal{E}'_j \in \mathbf{E}', \varnothing \neq \mathcal{E}_i \subset \mathcal{E}'_j$ and $\forall \mathcal{E}'_j \in \mathbf{E}', \exists \mathcal{E}_i \in \mathbf{E}, \varnothing \neq \mathcal{E}_i \subset \mathcal{E}'_j$.*
   (e) *c-narrowing iff $\mathbf{E} \neq \varnothing$ and $|\mathbf{E}| = |\mathbf{E}'|$ and $\forall \mathcal{E}_i \in \mathbf{E}, \exists \mathcal{E}'_j \in \mathbf{E}', \varnothing \neq \mathcal{E}'_j \subset \mathcal{E}_i$ and $\forall \mathcal{E}'_j \in \mathbf{E}', \exists \mathcal{E}_i \in \mathbf{E}, \varnothing \neq \mathcal{E}'_j \subset \mathcal{E}_i$.*
   (f) *c-altering iff $|\mathbf{E}| = |\mathbf{E}'|$ and it is neither c-conservative, nor c_{1e-1ne}, nor c_{1ne-1e}, nor c-expansive, nor c-narrowing.*

Definition 8.3a–c and 3f are fairly straightforward. Definition 8.3d states that a *c-expansive* change is a change where all the extensions of $\mathcal{G}$, which are initially not empty, are increased by some arguments. A *c-narrowing* change, according to Definition 8.3e, is a change where all the extensions of $\mathcal{G}$ are reduced by some arguments without becoming empty.

Example 1 (Cont'd). *All agents always propose constant changes, since they want to take a decision without ambiguity. For instance, consider the second turn of the dialog: using the preferred semantics, the current extension is $\{c, a\}$, and J_2 chooses a c-altering change because she totally disagrees with this extension.*

---

[4] Let S be a set, $|S|$ denotes the cardinality of S.

Properties about the acceptability of a set of arguments. A change can also have an impact on the acceptability of sets of arguments. For instance, in a dialog, it would be interesting to know whether the addition (or the removal) of an argument modifies the acceptability of the arguments previously accepted. We say "monotony from $\mathcal{G}$ to $\mathcal{G}'$" when every argument accepted *before* the change is still accepted *after* the change, *i.e.*, no accepted argument is lost and there is a (not necessarily strict) *expansion* of acceptability.[5]

Definition 9 (Simple expansive monotony). *The change from $\mathcal{G}$ to $\mathcal{G}'$ satisfies the property of* simple expansive monotony *iff* $\forall \mathcal{E}_i \in \mathbf{E}, \exists \mathcal{E}'_j \in \mathbf{E}', \mathcal{E}_i \subseteq \mathcal{E}'_j$.

Note that [9] describes many other properties such as, for instance, a property of "enforcement"[6] that would be interesting for J_1 in Example 1 in order to obtain the acceptability of Argument a.

4 A Change Operation Taking into Account Support

First of all, it should be noted that turning $\mathsf{BAS} = \langle \mathbf{A}, \mathbf{R}_{\mathrm{att}}, \mathbf{R}_{\mathrm{sup}} \rangle$ into its deductive associated Dung system $\mathsf{AS}^{\mathsf{BAS}}$ corresponds to the *update of a specific system*, $\mathsf{AS} = \langle \mathbf{A}, \mathbf{R}_{\mathrm{att}} \rangle$, the reduction of BAS to its direct attacks (see Fig. 1). The next step is to allow for updating a BAS. So Definition 7 is generalized:

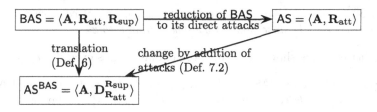

Fig. 1. The translation of BAS into $\mathsf{AS}^{\mathsf{BAS}}$ is an update

Definition 10 (Addition in a BAS). *Let* $\mathsf{BAS} = \langle \mathbf{A}, \mathbf{R}_{\mathrm{att}}, \mathbf{R}_{\mathrm{sup}} \rangle$.

1. *Let z be an argument, $\mathcal{I}a_z$ (resp. $\mathcal{I}s_z$) be a set of attacks (resp. supports) concerning z. $\mathcal{I}s_z \cup \mathcal{I}a_z$ is denoted by $\mathcal{I}_z$. We assume that $\mathcal{I}_z \subseteq (\mathbf{A} \times \{z\}) \cup (\{z\} \times \mathbf{A})$.*
   Adding z and $\mathcal{I}_z$ is a change operation, denoted by $\oplus^z_{(\mathcal{I}a, \mathcal{I}s)}$, providing a new BAS s.t.: $\oplus^z_{(\mathcal{I}a, \mathcal{I}s)} \langle \mathbf{A}, \mathbf{R}_{\mathrm{att}}, \mathbf{R}_{\mathrm{sup}} \rangle = \langle \mathbf{A} \cup \{z\}, \mathbf{R}_{\mathrm{att}} \cup \mathcal{I}a_z, \mathbf{R}_{\mathrm{sup}} \cup \mathcal{I}s_z \rangle$.

---

[5] A second case, referred as "monotony from $\mathcal{G}'$ to $\mathcal{G}$", has been described in [9]. It is not used in this paper.

[6] This property is described in [8] and only considers the status of an argument after the change without taking into account the evolution of extensions. Of course, many other possibilities could be defined (*e.g.* combining extensiveness and monotony).

2. Let $\mathcal{I}a$ (resp. $\mathcal{I}s$) be a set of attacks (resp. supports). $\mathcal{I}s \cup \mathcal{I}a$ is denoted by $\mathcal{I}$. We assume that $\mathcal{I} \subseteq (\mathbf{A} \times \mathbf{A})$ and $\mathcal{I} \cap (\mathbf{R}_{att} \cup \mathbf{R}_{sup}) = \varnothing$.
   Adding $\mathcal{I}$ is a change operation, denoted by $\oplus_{(\mathcal{I}a, \mathcal{I}s)}$, providing a new BAS s.t.: $\oplus_{(\mathcal{I}a, \mathcal{I}s)} \langle \mathbf{A}, \mathbf{R}_{att}, \mathbf{R}_{sup} \rangle = \langle \mathbf{A}, \mathbf{R}_{att} \cup \mathcal{I}a, \mathbf{R}_{sup} \cup \mathcal{I}s \rangle$.

The system resulting of a change is denoted by $\mathsf{BAS}' = \langle \mathbf{A}', \mathbf{R}_{att}', \mathbf{R}_{sup}' \rangle$ and its deductive associated Dung AS is denoted by $\mathsf{AS}^{\mathsf{BAS}'}$.

   Due to lack of place, **in this paper, we only study the case corresponding to Definition** 10.1. As we consider deductive support and from Definitions 10 and 6, the following consequence obviously holds:

Consequence 1. Let $\mathsf{BAS} = \langle \mathbf{A}, \mathbf{R}_{att}, \mathbf{R}_{sup} \rangle$. Let $\oplus_{(\mathcal{I}a, \mathcal{I}s)}^z$ be a change operation on BAS producing BAS'. $\mathsf{AS}^{\mathsf{BAS}'} = \langle \mathbf{A} \cup \{z\}, \mathbf{D}_{\mathbf{R}_{att} \cup \mathcal{I}a_z}^{\mathbf{R}_{sup} \cup \mathcal{I}s_z} \rangle$.

   Due to the above result, it seems natural to study the update of BAS by comparing $\mathsf{AS}^{\mathsf{BAS}}$ and $\mathsf{AS}^{\mathsf{BAS}'}$. However, it is not always possible to identify a *unique change* on $\mathsf{AS}^{\mathsf{BAS}}$, *as defined in Definition* 7, that produces $\mathsf{AS}^{\mathsf{BAS}'}$. Indeed, the addition of an argument with interactions in BAS can induce the addition in $\mathbf{D}_{\mathbf{R}_{att}}^{\mathbf{R}_{sup}}$ of new attacks between arguments of $\mathbf{A}$ (see Example 3).

Example 3. Let $\mathsf{BAS} = \langle \{a, b\}, \varnothing, \varnothing \rangle$, let us apply on BAS the change $\oplus_{(\mathcal{I}a, \mathcal{I}s)}^z$ with $\mathcal{I}a_z = \{(a, z)\}$ and $\mathcal{I}s_z = \{(b, z)\}$; in this case, following Definitions 10.1 and 6, $\mathsf{AS}^{\mathsf{BAS}'}$ contains the new attack (a, b) that does not concern z (and this attack appears only because there is a support from b to z).

   Another example shows that this problem also exists even if $\mathcal{I}a_z = \varnothing$:

Example 4. Consider $\mathsf{BAS} = \langle \{a, b, c\}, \{(c, a)\}, \varnothing \rangle$, and apply on BAS the change $\oplus_{(\mathcal{I}a, \mathcal{I}s)}^z$ with $\mathcal{I}a_z = \varnothing$ and $\mathcal{I}s_z = \{(b, z), (z, c)\}$; in this case, following Definitions 10.1 and 6, $\mathsf{AS}^{\mathsf{BAS}'}$ contains the new attack (b, a) that does not concern z.

So, if we add an argument z with at least one support in BAS, the change of $\mathsf{AS}^{\mathsf{BAS}}$ into $\mathsf{AS}^{\mathsf{BAS}'}$ cannot always be expressed using either Definition 7.1 (since attacks are added that do not concern z), or Definition 7.2 (since the argument z is added). The links between the different systems are illustrated by Fig. 2.
   This suggests to consider elementary changes (addition of one attack or one support). In this paper, we consider two particular cases. The first one concerns a BAS with only one support *from z to a*, z being unattacked. In this case, Definition 6 obviously implies that z has in $\mathsf{AS}^{\mathsf{BAS}}$ exactly the same role as a in AS:

Proposition 1. Let $\mathsf{BAS} = \langle \mathbf{A}, \mathbf{R}_{att}, \mathbf{R}_{sup} \rangle$ with $\mathbf{R}_{sup} = \{(z, a)\}$ and z is not attacked in BAS. The following properties hold:

- if a is unattacked in BAS then z is unattacked in $\mathsf{AS}^{\mathsf{BAS}}$ (no direct attack, no direct or inductive supported or mediated attack on z);

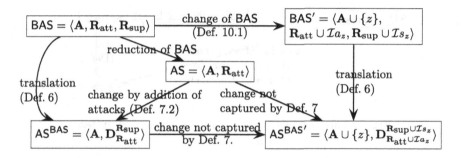

Fig. 2. Links between the different systems

- *if a is attacked by b in* BAS *then z is attacked by b in* $\mathsf{AS}^{\mathsf{BAS}}$ *(this is a mediated attack on z);*
- *if a attacks b in* BAS *then z attacks b in* $\mathsf{AS}^{\mathsf{BAS}}$ *(this is a supported attack).*
- *if a is defended by c against b in* BAS *then z is defended by c against b in* $\mathsf{AS}^{\mathsf{BAS}}$ *(the defence of a direct attack on a can be used for the defence of the mediated attack on z).*
- *if c is defended by b against a in* BAS *then c is defended by b against z in* $\mathsf{AS}^{\mathsf{BAS}}$ *(a mediated attack can be used as a defence against a supported attack).*

A second particular case concerns a BAS with only one support *on* an unattacked argument. In this case, Definition 6 obviously implies that the set of attacks remains unchanged:

Proposition 2. *Let* $\mathsf{BAS} = \langle \mathbf{A}, \mathbf{R}_{\mathrm{att}}, \mathbf{R}_{\mathrm{sup}} \rangle$ *with* $\mathbf{R}_{\mathrm{sup}} = \{(a, z)\}$ *and z unattacked by* BAS. *Then* $\mathbf{D}_{\mathbf{R}_{\mathrm{att}}}^{\mathbf{R}_{\mathrm{sup}}} = \mathbf{R}_{\mathrm{att}}$.

Moreover, in these particular cases, following Definition 10.1, Propositions 1 and 2, the addition of one argument involved in only one support in BAS cannot add attacks between arguments of $\mathbf{A}$ and preserves acceptability:

Proposition 3. *Let* $\mathsf{BAS} = \langle \mathbf{A}, \mathbf{R}_{\mathrm{att}}, \mathbf{R}_{\mathrm{sup}} \rangle$ *s.t.* $\mathbf{R}_{\mathrm{sup}} = \varnothing$.[7] *Let* $\oplus_{(\mathcal{I}a, \mathcal{I}s)}^{z}$ *be a change operation defined on* BAS *with* $\mathcal{I}a_z = \varnothing$, $|\mathcal{I}s_z| = 1$ *and producing* BAS′.

- $\forall x, y \in \mathbf{A}$, *s.t. y does not attack x in* BAS *then there is no attack from y to x in* $\mathsf{AS}^{\mathsf{BAS}'}$.
- $\forall y \in \mathbf{A}$, *if y is unattacked in* BAS *then it remains unattacked in* $\mathsf{AS}^{\mathsf{BAS}'}$.
- *Consider* $\mathcal{F}$ *(resp. $\mathcal{F}'$) the characteristic function of* AS *(resp. $\mathsf{AS}^{\mathsf{BAS}'}$). $\forall S \subseteq \mathbf{A}$, $\mathcal{F}(S) \subseteq \mathcal{F}'(S)$.*

Thus, considering a BAS reduced to an AS (*i.e. without any support*), if we add only one argument with one support, the links between the different systems are given by Fig. 3.

So we are able to characterize the addition of a support by an addition of attacks. In the next section, we study this simplified change operation.

---

[7] In this case, BAS is reduced to an AS. So BAS, its reduction AS and $\mathsf{AS}^{\mathsf{BAS}}$ collapse.

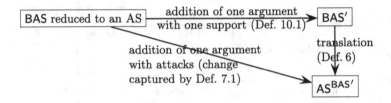

Fig. 3. Links between systems if there is no support in BAS

5   Characterizing the Addition of an Argument and a Support

In Sect. 5.1 (resp. Sect. 5.2), we give some results about the characterization of the addition of a supported (resp. supporting) argument in a BAS.

5.1   Case of an Added Supported Argument

In this case, as a direct application of Proposition 2, we prove that the update of a BAS without supports has a deductive associated Dung AS that corresponds to the addition of an argument without interaction into the initial BAS.

Proposition 4. *Let* $\mathsf{BAS} = \langle \mathbf{A}, \mathbf{R}_{\mathrm{att}}, \mathbf{R}_{\mathrm{sup}} \rangle$ *s.t.* $\mathbf{R}_{\mathrm{sup}} = \varnothing$. *Let* $\oplus^z_{(\mathcal{I}a, \mathcal{I}s)}$ *be a change operation defined on* BAS *with* $\mathcal{I}a_z = \varnothing$ *and* $\mathcal{I}s_z = \{(a, z)\}$ *and producing* BAS'. $\mathsf{AS}^{\mathsf{BAS}'} = \oplus^z_{\varnothing} \langle \mathbf{A}, \mathbf{R}_{\mathrm{att}} \rangle$.

Due to Proposition 4, Definitions 7.1 and 10.1, we have:

Proposition 5. *Let* $\mathsf{BAS} = \langle \mathbf{A}, \mathbf{R}_{\mathrm{att}}, \mathbf{R}_{\mathrm{sup}} \rangle$ *s.t.* $\mathbf{R}_{\mathrm{sup}} = \varnothing$. *Let* $\oplus^z_{(\mathcal{I}a, \mathcal{I}s)}$ *be a change operation defined on* BAS *with* $\mathcal{I}a_z = \varnothing$ *and* $\mathcal{I}s_z = \{(a, z)\}$ *and producing* BAS'. *Let* s *be a semantics* $\in \{grounded, preferred, stable\}$. $\mathcal{E}$ *is an extension of* AS *under* s *iff* $\mathcal{E}' = \mathcal{E} \cup \{z\}$ *is an extension of* $\mathsf{AS}^{\mathsf{BAS}'}$ *under* s. *Moreover, there is no stable extension in* AS *iff there is no stable extension in* $\mathsf{AS}^{\mathsf{BAS}'}$.

And an obvious consequence of Proposition 5 is:

Consequence 2. *The change* $\oplus^z_{(\varnothing, \{(a,z)\})}$ *is only either c-expansive, or* c_{1e-1ne}, *or c-conservative. In the last case, the only possibility is* $\mathbf{E} = \mathbf{E}' = \varnothing$.

Some examples of this change are given in Table 1.

5.2   Case of an Added Supporting Argument

In this case, the existence of cycles is preserved as shown by:

Proposition 6. *Let* $\mathsf{BAS} = \langle \mathbf{A}, \mathbf{R}_{\mathrm{att}}, \mathbf{R}_{\mathrm{sup}} \rangle$ *s.t.* $\mathbf{R}_{\mathrm{sup}} = \varnothing$. *Let* $\oplus^z_{(\mathcal{I}a, \mathcal{I}s)}$ *be a change operation defined on* BAS *with* $\mathcal{I}a_z = \varnothing$, $\mathcal{I}s_z = \{(z, a)\}$ *and producing* BAS'.

- If a belongs to a cycle of attacks in BAS then z belongs to a new cycle of attacks in $\mathsf{AS}^{\mathsf{BAS}'}$ and the length of both cycles is the same.
- If a does not belong to a cycle of attacks in BAS then there is no cycle of attacks in $\mathsf{AS}^{\mathsf{BAS}'}$ involving z.

This result is proven using Definitions 4–6 and by *reductio ad absurdum* for the second item. Moreover, following Definition 6 and Proposition 1, we can characterize the impact of this change for stable semantics:

Table 1. Addition of a supported argument in an AS

BAS (reduced to an AS) updated with z and the support (a, z)	$\mathsf{AS}^{\mathsf{BAS}'}$	Extensions	
		before change	after change
		$\{a, c\}$ is the grounded, preferred and stable extension	$\{a, c, z\}$ is the grounded, preferred and stable extension
		The change is *c-expansive*	
		$\varnothing$ is the grounded extension; $\{a\}$ and $\{c\}$ are the preferred and stable extensions	$\{z\}$ is the grounded extension; $\{a, z\}$ and $\{c, z\}$ are the preferred and stable extensions
		The change is *c-expansive*(preferred, stable) or c_{1e-1ne}(grounded)	
		$\varnothing$ is the grounded and preferred extensions; there is no stable extension	$\{z\}$ is the grounded and preferred extensions; there is no stable extension
		The change is *c-expansive*(preferred), or c_{1e-1ne}(grounded), or *c-conservative*(stable)	

Proposition 7. Let $\mathsf{BAS} = \langle \mathbf{A}, \mathbf{R}_{\mathrm{att}}, \mathbf{R}_{\mathrm{sup}} \rangle$ s.t. $\mathbf{R}_{\mathrm{sup}} = \varnothing$. Let $\oplus^z_{(\mathcal{I}a, \mathcal{I}s)}$ be a change operation defined on BAS with $\mathcal{I}a_z = \varnothing$ and $\mathcal{I}s_z = \{(z, a)\}$ and producing BAS'. Let $\mathcal{E}$ be a stable extension of AS:

- if $a \notin \mathcal{E}$ then $\mathcal{E}$ is a stable extension of $\mathsf{AS}^{\mathsf{BAS}'}$;
- if $a \in \mathcal{E}$ then $\mathcal{E} \cup \{z\}$ is a stable extension of $\mathsf{AS}^{\mathsf{BAS}'}$.

And more generally, the *simple expansive monotony* of the change operation can be proven:

Table 2. Addition of a supporting argument in an AS

BAS (reduced to an AS) updated with z and the support (z, a)	$AS^{BAS'}$	Extensions	
		before change	after change
		$\{a\}$ is the grounded, preferred and stable extension	$\{a, z\}$ is the grounded, preferred and stable extension
		The change is c-*expansive*	
		$\varnothing$ is the grounded and preferred extension; there is no stable extension	$\{z\}$ is the grounded, preferred and stable extension
		The change is c_{1e-1ne} (grounded, preferred) or $e_{\varnothing-1ne}$ (stable)	
		$\{b\}$ is the grounded, preferred and stable extension	$\{b\}$ is the grounded, preferred and stable extension
		The change is c-*conservative*	
		$\varnothing$ is the grounded and preferred extension; there is no stable extension	$\varnothing$ is the grounded extension; $\{z, c\}$ and $\{z, d\}$ are the preferred and stable extensions
		The change is c-*conservative* (grounded) or e_{1e-k} (preferred), or $e_{\varnothing-k}$ (stable)	
		$\varnothing$ is the grounded extension; $\{b\}$ is the preferred and stable extension	$\varnothing$ is the grounded extension; $\{b\}$ and $\{z\}$ are the preferred and stable extensions
		The change is c-*conservative* (grounded) or e_{1ne-k} (preferred, stable)	
		$\varnothing$ is the grounded extension; $\{b\}$ and $\{c\}$ are the preferred and stable extensions	$\varnothing$ is the grounded extension; $\{b\}$, $\{c\}$ and $\{z\}$ are the preferred and stable extensions
		The change is c-*conservative* (grounded) or e_{j-k} (preferred, stable)	

Proposition 8. *Let* $\mathsf{BAS} = \langle \mathbf{A}, \mathbf{R}_{att}, \mathbf{R}_{sup} \rangle$ *s.t.* $\mathbf{R}_{sup} = \varnothing$. *Let* s *be a semantics belonging to* {*grounded, preferred, stable*}. *Let* $\oplus^z_{(\mathcal{I}a, \mathcal{I}s)}$ *be a change operation defined on* BAS *with* $\mathcal{I}a_z = \varnothing$ *and* $\mathcal{I}s_z = \{(z, a)\}$ *and producing* BAS'.

$\forall \mathcal{E}$ *extension of* AS *under* s, $\exists \mathcal{E}'$ *an extension of* $\mathsf{AS}^{\mathsf{BAS}'}$ *under* s *s.t.* $\mathcal{E} \subseteq \mathcal{E}'$.

This result is proven using Definition 2, Propositions 1 and 3, by induction on the characteristic function for the grounded semantics, showing that $\mathcal{E}$ is admissible in $\mathsf{AS}^{\mathsf{BAS}'}$ for the preferred semantics and following Proposition 7 for the stable semantics. An obvious consequence of the two previous results is:

Consequence 3. *The change* $\oplus^z_{(\varnothing, \{(z,a)\})}$ *cannot be restrictive, nor c-narrowing, nor c-altering, nor* c_{1ne-1e}.

Some examples of this change are given in Table 2.

6   Conclusion and Future Works

This paper presents preliminary work about change for abstract bipolar argumentation systems, *i.e.* where there exist two kinds of interaction, attacks and supports. The central idea is to take advantage of two kinds of previous works, works about dynamics in argumentation systems (AS) and works about bipolar argumentation systems (BAS). Indeed, it has been shown that a BAS can be turned into a standard Dung's AS by adding appropriate attacks. Our main contribution is to show how the addition of one argument together with one support involving it (and without any attack) impacts the extensions of the resulting system. In this particular case, we have clearly identified the attacks that must be added and we have obtained specific properties which enable to characterize this change. These characterizations refine and complete the results presented in [9] that cannot be used directly for characterizing the impact of these new attacks (the conditions used in [9] are too strong with regard to our case and thus they cannot be satisfied here). Our work is of particular interest in a multiagent context if we do not want to recompute the extensions when a agent gives a new argument that supports (or is supported by) an already existing argument.

Although our results are given for elementary changes (addition of one argument and one support), they can be generalized considering that the addition of a set of arguments with interactions can be viewed as a sequence of elementary additions. Nevertheless, in order to achieve this generalization, there are two issues to be solved: (1) characterize the addition of an argument with attacks (as was done for AS; results given in [9] will be useful) and (2) study the addition of interactions (this operation has been defined in [9] for AS and in our paper for BAS but not completely studied). This future study could also give a way for computing directly the $\mathsf{AS}^{\mathsf{BAS}}$ of a BAS.

Moreover, our work concerns only a special variant of support, the deductive one. Using the duality between necessary and deductive supports, our results can be easily translated for necessary support. However, it remains to adapt them to the case of a generalized support (a support from a set of arguments to an argument as proposed by [22]).

And finally, it would be interesting to extend this study to the case of non abstract BAS.

References

1. Amgoud, L., Cayrol, C.: A reasoning model based on the production of acceptable arguments. Ann. Math. Artif. Intell. **34**, 197–216 (2002)
2. Amgoud, L., Cayrol, C., Lagasquie-Schiex, M.C., Livet, P.: On bipolarity in argumentation frameworks. Intl. J. Intell. Syst. **23**, 1062–1093 (2008)
3. Amgoud, L., Maudet, N., Parsons, S.: Modelling dialogues using argumentation. In: Proceedings of ICMAS, pp. 31–38 (2000)
4. Amgoud, L., Vesic, S.: A formal analysis of the role of argumentation in negotiation dialogues. J. Logic Comput. **22**, 957–978 (2012)
5. Baroni, P., Boella, G., Cerutti, F., Giacomin, M., van der Torre, L., Villata, S.: On the input/output behavior of argumentation frameworks. Artif. Intell. **217**, 144–197 (2014)
6. Baroni, P., Caminada, M., Giacomin, M.: An introduction to argumentation semantics. Knowl. Eng. Rev. **26**(4), 365–410 (2011)
7. Baroni, P., Giacomin, M., Liao, B.: On topology-related properties of abstract argumentation semantics. A correction and extension to dynamics of argumentation systems: a division-based method. Artif. Intell. **212**, 104–115 (2014)
8. Baumann, R.: What does it take to enforce an argument? Minimal change in abstract argumentation. In: Proceedings of ECAI, pp. 127–132. IOS Press (2012)
9. Bisquert, P., Cayrol, C., Dupin de Saint Cyr Bannay, F., Lagasquie-Schiex, M.C.: Characterizing change in abstract argumentation systems. In: Ferm, E., Gabbay, D., Simari, G. (eds.) Trends in Belief Revision and Argumentation Dynamics. Studies in Logic, vol. 48, pp. 75–102. College Publications (2013)
10. Boella, G., Gabbay, D.M., van der Torre, L., Villata, S.: Modelling defeasible and prioritized support in bipolar argumentation. Ann. Math. AI **66**, 163–197 (2012)
11. Booth, R., Kaci, S., Rienstra, T., van der Torre, L.: A logical theory about dynamics in abstract argumentation. In: Liu, W., Subrahmanian, V.S., Wijsen, J. (eds.) SUM 2013. LNCS, vol. 8078, pp. 148–161. Springer, Heidelberg (2013)
12. Cayrol, C., Lagasquie-Schiex, M.C.: On the acceptability of arguments in bipolar argumentation frameworks. In: Godo, L. (ed.) ECSQARU 2005. LNCS (LNAI), vol. 3571, pp. 378–389. Springer, Heidelberg (2005)
13. Cayrol, C., Lagasquie-Schiex, M.C.: Coalitions of arguments: a tool for handling bipolar argumentation frameworks. Intl. J. Intell. Syst. **25**, 83–109 (2010)
14. Cayrol, C., Lagasquie-Schiex, M.C.: Bipolarity in argumentation graphs: towards a better understanding. IJAR **54**(7), 876–899 (2013)
15. Cayrol, C., Lagasquie-Schiex, M.C.: Change in abstract bipolar argumentation systems. Technical report RR-2015-02-FR, IRIT (2015). http://www.irit.fr/publis/ADRIA/PapersMCL/Rapport-IRIT-2015-02.pdf
16. Cohen, A., Gottifredi, S., García, A.J., Simari, G.R.: An approach to abstract argumentation with recursive attack and support. J. Appl. Logic (2014)
17. Coste-Marquis, S., Konieczny, S., Mailly, J.-G., Marquis, P.: A translation-based approach for revision of argumentation frameworks. In: Fermé, E., Leite, J. (eds.) JELIA 2014. LNCS, vol. 8761, pp. 397–411. Springer, Heidelberg (2014)

18. Doutre, S., Herzig, A., Perrussel, L.: A dynamic logic framework for abstractargumentation. In: Proceedings of KR, pp. 62–71. AAAI Press (2014)
19. Dung, P.M.: On the acceptability of arguments and its fundamental role in nonmonotonic reasoning, logic programming and n-person games. Artif. Intell. **77**, 321–357 (1995)
20. Gabbay, D.M.: Logical foundations for bipolar and tripolar argumentation networks: preliminary results. J. Logic Comput. (2013)
21. Karacapilidis, N., Papadias, D.: Computer supported argumentation and collaborative decision making: the hermes system. Inf. Syst. **26**(4), 259–277 (2001)
22. Nouioua, F.: AFs with necessities: further semantics and labelling characterization. In: Liu, W., Subrahmanian, V.S., Wijsen, J. (eds.) SUM 2013. LNCS, vol. 8078, pp. 120–133. Springer, Heidelberg (2013)
23. Nouioua, F., Risch, V.: Bipolar argumentation frameworks with specialized supports. In: Proceedings of ICTAI, pp. 215–218. IEEE Computer Society (2010)
24. Nouioua, F., Risch, V.: Argumentation frameworks with necessities. In: Benferhat, S., Grant, J. (eds.) SUM 2011. LNCS, vol. 6929, pp. 163–176. Springer, Heidelberg (2011)
25. Oren, N., Norman, T.J.: Semantics for evidence-based argumentation. In: Proceedings of COMMA, pp. 276–284 (2008)
26. Oren, N., Reed, C., Luck, M.: Moving between argumentation frameworks. In: Proceedings of COMMA, pp. 379–390. IOS Press (2010)
27. Polberg, S., Oren, N.: Revisiting support in abstract argumentation systems. In: Proceedings of COMMA, pp. 369–376. IOS Press (2014)
28. Prakken, H.: On support relations in abstract argumentation as abstraction of inferential relations. In: Proceedings of ECAI, pp. 735–740 (2014)
29. Verheij, B.: Deflog: on the logical interpretation of prima facie justified assumptions. J. Logic Comput. **13**, 319–346 (2003)

On Argumentation with Purely Defeasible Rules

Zimi Li[1]([⊠]) and Simon Parsons[2]

[1] Department of Computer Science, Graduate Center,
City University of New York, New York, USA
zli2@gc.cuny.edu
[2] Department of Computer Science, University of Liverpool,
Liverpool, UK
s.d.parsons@liverpool.ac.uk

Abstract. ASPIC$^+$ is one of the most widely used systems for structured arguments and includes the use of both strict and defeasible rules. Here we consider using just the defeasible part of ASPIC$^+$. We show that using the resulting system, it is possible, in a well defined sense, to capture the same information as using ASPIC$^+$ with strict rules.

1 Introduction

Argumentation theory is concerned with the way that intelligent agents discuss whether some statement holds. In the past few years, formal argumentation frameworks have been heavily studied and applications have been proposed in fields such as natural language processing, the semantic web and multi-agent systems. Studying argumentation provides results which help in developing tools and applications in these areas. Dung's seminal work [8] tells us how to handle the conflicts between arguments. However, it says nothing about the structure of arguments, or how to construct arguments and attack relationships from a knowledge base. Providing the logical basis for argumentation has been the subject of several authors, including [4,9,10]. This paper is concerned with we the work that started with the ASPIC [2] framework, and we briefly summarise this work below.

Following [2,6] pointed out that ASPIC may lead to some non-intuitive results, suggested that all argumentation frameworks must satisfy three rationality postulates in order to avoid these anomalies, and showed how ASPIC could be modified to satisfy them. [12] presented an extension of ASPIC, called ASPIC$^+$, which also satisfies the postulates under certain restrictions. [1,13] provide further discussion of the approach. [11] modified the ASPIC$^+$ framework, to develop a more general structured framework for argumentation with preferences. [5,14] presented some examples where ASPIC-like systems could lead to non-intuitive results and gave solutions. Finally, [7] looked at a new variation of ASPIC$^+$ which still satisfies the rationality postulates while loosening the restriction on rebut that ASPIC$^+$ requires to satisfy the rationality postulates.

Here we continue this line of work, considering another variation of ASPIC$^+$ which only contains defeasible elements. We find that, like the system in [7], our

C. Beierle and A. Dekhtyar (Eds.): SUM 2015, LNAI 9310, pp. 330–343, 2015.
DOI: 10.1007/978-3-319-23540-0_22

system can both loosen the restrictions on ASPIC$^+$ and still satisfy the rationality postulates, while being able to establish exactly the same set of conclusions as ASPIC$^+$ from the corresponding knowledge-base.

2   Background

2.1   Abstract Argumentation

An abstract argumentation framework [8] is a pair $AF = \langle \mathcal{A}, Defeats \rangle$, where $\mathcal{A}$ is a set of arguments, and $Defeats$ is a binary relation collecting all pairs of arguments A and B such that A defeats B, i.e. $Defeats \subseteq \mathcal{A} \times \mathcal{A}$. An argument is called *acceptable* iff it can defend itself, that is, all of its defeaters have been defeated. A subset S of $\mathcal{A}$ is said to be *conflict-free* if there are no arguments in S that defeat an argument in S. Given an abstract argumentation framework, one is typically interested in which of the arguments are acceptable. This is done through argument-based *semantics*, which define different ways to determine acceptability. [8] defines several semantics — complete, grounded, preferred and stable. A given semantics will specify some (possibly empty) sets of acceptable arguments for a given argumentation framework. These sets are also called argument-based extensions, or simply *extensions*. The conclusions of the arguments in an extension are called the *justified conclusions*.

The state-of-the-art way to establish the extensions is through the labeling approach, which is nicely summarized by [3]. This approach can be described in terms of a *labeling function LF* which maps from arguments to a set of labels $\{\texttt{IN}, \texttt{OUT}, \texttt{UNDEC}\}$. Not all labelings are helpful in determining acceptability, and we determine the helpful labelings through the idea of *legality*. For a legal labeling LF, an argumentation framework, $\langle \mathcal{A}, Defeats \rangle$, and an argument $x \in \mathcal{A}$:

1. x is legally $\texttt{IN}$ iff x is labeled $\texttt{IN}$ and every $y \in \mathcal{A}$ that defeats x is labeled $\texttt{OUT}$.
2. x is legally $\texttt{OUT}$ iff x is labeled $\texttt{OUT}$ and there is at least one $y \in \mathcal{A}$ that defeats x and is labeled $\texttt{IN}$.
3. x is legally $\texttt{UNDEC}$ iff there is no $y \in \mathcal{A}$ that defeats x such that y is labeled $\texttt{IN}$, and there is at least one $y \in \mathcal{A}$ that defeats x such that y is labeled $\texttt{UNDEC}$.

Note that the $\texttt{UNDEC}$ state occurs when x cannot be labeled $\texttt{IN}$ (because it has at least one defeater that is not $\texttt{OUT}$), and cannot be labeled $\texttt{OUT}$ (because it has no $\texttt{IN}$ defeater). If an argument is not legally labeled, it is said to be *illegally* labeled. More precisely, an argument is illegally labeled l, where $l \in \{\texttt{IN}, \texttt{OUT}, \texttt{UNDEC}\}$, if it is not legally labeled l.

With the notion of legality tying labelings to $Defeats$ relations, we can identify acceptable sets of arguments through the notions of *admissibility* and *completeness*. An *admissible* labeling has no arguments that are illegally $\texttt{IN}$, and no arguments that are illegally $\texttt{OUT}$. A *complete* labeling is an admissible labeling that, in addition, has no arguments that are illegally $\texttt{UNDEC}$. Then, given a complete labeling LF, we have: (1) LF is a *grounded* labeling iff there is no complete labeling with a smaller set of IN arguments; (2) *LF* is a *preferred* labeling iff there is no complete labeling with a larger set of IN arguments; and (3) *LF* is a *stable* labeling if it contains no UNDEC arguments.

The labeling approach exactly matches Dung's semantics [8]. If *LF* is a complete labeling, then every *x* labeled IN by *LF* is in the complete extension, and so on for grounded, preferred and stable labelings. If an argument is in a given extension, we say that it is justified in the corresponding semantics.

2.2   ASPIC$^+$ Argumentation Framework

Next, we review the ASPIC$^+$ argumentation framework in [11]. This defines two kinds of inference rules: strict rules (denoted $\rightarrow$), meaning the conclusion is always accepted without any exception, and defeasible rules (denoted $\Rightarrow$), meaning the conclusion is accepted unless there is an exception.

Definition 1 (ASPIC$^+$ Argumentation System). *An* argumentation system *is a triple* $AS = \langle \mathcal{L}, \mathcal{R}, n \rangle$ *where:*

- $\mathcal{L}$ *is a logical language closed under negation* $^{-}$.
- $\mathcal{R} = \mathcal{R}_s \cup \mathcal{R}_d$ *is a set of strict (*$\mathcal{R}_s$*) and defeasible (*$\mathcal{R}_d$*) inference rules of the form* $\phi_1, \ldots, \phi_n \rightarrow \phi$ *and* $\phi_1, \ldots, \phi_n \Rightarrow \phi$ *respectively (where* ϕ_i, ϕ *are meta-variables ranging over wff in* $\mathcal{L}$*), and* $\mathcal{R}_s \cap \mathcal{R}_d = \emptyset$.
- $n : \mathcal{R}_d \mapsto \mathcal{L}$ *is a naming convention for defeasible rules.*

We say that a set of propositions in $\mathcal{L}$ is *consistent* iff there do not exist two propositions a and a' such that $a = \overline{a'}$, and it is helpful to think of completing the set of strict rules by considering all the negative connections between propositions mentioned in a strict rule:

Definition 2 (Closure under Transposition). *If* $\mathcal{R}_s$ *is a set of strict rules, we say* $\mathcal{R}_s$ *is* closed under transposition *iff if* $\phi_1, \ldots, \phi_n \rightarrow \phi \in \mathcal{R}_s$, *then* $\phi_1, \ldots, \phi_{i-1}, \overline{\phi}, \phi_{i+1}, \ldots, \phi_n \rightarrow \overline{\phi_i} \in \mathcal{R}_s$ $(i = 1 \ldots n)$

Definition 3 (ASPIC$^+$ Knowledge Base). *A* knowledge base *in an argumentation system* $\langle \mathcal{L}, \mathcal{R}, n \rangle$ *is a set* $\mathcal{K} \subseteq \mathcal{L}$ *consisting of two disjoint subsets* $\mathcal{K}_n$ *(the axioms) and* $\mathcal{K}_p$ *(the ordinary premises).*

The above definitions distinguish the premises and the inference rules into two sets, the set of strict elements ($\mathcal{R}_s$ and $\mathcal{K}_n$) and the set of defeasible elements ($\mathcal{R}_d$ and $\mathcal{K}_p$).

Definition 4 (ASPIC$^+$ Argumentation Theory). *An* argumentation theory *AT is a pair* $\langle AS, \mathcal{K} \rangle$ *of an argumentation system AS and a knowledge base* $\mathcal{K}$.

Before defining precisely what an argument is, we need to introduce some notions which can be defined just understanding that an argument is made up of some subset of the knowledge $\mathcal{K}$, along with a sequence of rules, that lead to a conclusion. Given this, $\texttt{Prem}(\cdot)$ returns all the premises, $\texttt{Conc}(\cdot)$ returns the conclusion and $\texttt{TopRule}(\cdot)$ returns the last rule in the argument. $\texttt{Sub}(\cdot)$ returns all the sub-arguments of a given argument, that is all the arguments that are subset of the given argument.

Definition 5 (Argument). *An argument A from of an argumentation theory $AT = \langle\langle\mathcal{L}, \mathcal{R}, n\rangle, \mathcal{K}\rangle$ is:*

1. *ϕ if $\phi \in \mathcal{K}$ with: $\texttt{Prem}(A) = \{\phi\}$; $\texttt{Conc}(A) = \{\phi\}$; $\texttt{Sub}(A) = \{A\}$; $\texttt{TopRule}(A)$ = undefined.*
2. *$A_1, \ldots, A_n \to$ (or $\Rightarrow$) ϕ if A_i are arguments such that there exists a strict (or defeasible) rule $\texttt{Conc}(A_1), \ldots, \texttt{Conc}(A_n) \to$ (or $\Rightarrow$) ϕ in $\mathcal{R}_s$ (or $\mathcal{R}_d$). $\texttt{Prem}(A) = \texttt{Prem}(A_1) \cup \ldots \cup \texttt{Prem}(A_n)$; $\texttt{Conc}(A) = \phi$; $\texttt{Sub}(A) = \texttt{Sub}(A_1) \cup \ldots \cup \texttt{Sub}(A_n) \cup \{A\}$; $\texttt{TopRule}(A) = \texttt{Conc}(A_1), \ldots, \texttt{Conc}(A_n) \to$ (or $\Rightarrow$) ϕ.*

We write $\mathcal{A}(AT)$ to denote the set of arguments from the theory AT.

We say that an argument A is *consistent* iff $\{\texttt{Conc}(A') | A' \in \texttt{Sub}(A)\}$ is consistent. We further say that an argument A is *strict* if A only contains strict rules, that is $\mathcal{R}_s \neq \emptyset$ and $\mathcal{R}_d = \emptyset$;[1]. Similarly, we say: A is *defeasible* if A contains at least one defeasible rule, $\mathcal{R}_d \neq \emptyset$; A is *firm* if A only contains axioms, $\mathcal{K}_n \neq \emptyset$, $\mathcal{K}_p = \emptyset$; A is *plausible* if A contains ordinary premises.

An argument can be attacked in three ways: on its ordinary premises, on its conclusion, or on its inference rules:

Definition 6 (ASPIC$^+$ Attack). *An argument A attacks an argument B iff A undermines, rebuts or undercuts B, where:*

- *A undermines B (on B') iff $\texttt{Conc}(A) = \overline{\phi}$ for some $B' = \phi \in \texttt{Prem}(B)$ and $\phi \in \mathcal{K}_p$.*
- *A rebuts B (on B') iff $\texttt{Conc}(A) = \overline{\phi}$ for some $B' \in \texttt{Sub}(B)$ of the form $B''_1, \ldots, B''_2 \Rightarrow \phi$.*
- *A undercuts B (on B') iff $\texttt{Conc}(A) = \overline{n(r)}$ for some $B' \in \texttt{Sub}(B)$ such that $\texttt{TopRule}(B)$ is a defeasible rule r of the form $\phi_1, \ldots, \phi_n \Rightarrow \phi$.*

We denote "A attacks B" by (A, B).

Note that, in the ASPIC$^+$ attack relation, rebutting is *restricted*. That is an argument with a strict $\texttt{TopRule}$ can rebut an argument with a defeasible $\texttt{TopRule}$, but not vice versa.

Attacks can be distinguished as to whether they are preference-dependent (rebutting and undermining) or preference-independent (undercutting). The former succeed only when the attacker is preferred. The latter succeed whether or not the attacker is preferred.

---

[1] This is not same as definition of "strict" as in [11] where the only condition was that $\mathcal{R}_d = \emptyset$. Here we insist that a strict argument includes at least one strict rule. As a consequence, the notions of "strict" and "defeasible" are not duals, and an argument can be neither strict or defeasible — but only if it contains only premises and/or axioms.

Definition 7 (Preference Ordering). *A preference ordering $\preceq$ is a binary relation over arguments, i.e., $\preceq\, \subseteq \mathcal{A} \times \mathcal{A}$, where $\mathcal{A}$ is the set of all arguments constructed from the knowledge base in an argumentation system. We say A's preference level is less than or equal to that of B iff $A \preceq B$.*

In general, neither ASPIC$^+$ nor our defeasible system make any assumptions on the properties of the preference ordering, but in establishing a relationship between the two systems, we make use of the *weakest link* principle from [11]. This assumes two pre-orderings $\leq, \leq'$ over $\mathcal{R}_d$ and $\mathcal{K}_p$ respectively, and combines them into $A \prec B$ if:

- the defeasible rules in A include a rule which is weaker than (strictly less than according to $\leq$) all the defeasible rules in B, and
- the ordinary premises in A include an ordinary premise which is weaker (strictly less than according to $\leq'$) all the ordinary premises in B.

$A \prec B$ is then defined as usual as $A \preceq B$ and $B \npreceq A$. By combining the definition of arguments, attack relations and preference ordering, we have the following definitions:

Definition 8 (Structured Argumentation Framework). *A structured argumentation framework is a triple $\langle \mathcal{A}, att, \preceq \rangle$, where $\mathcal{A}$ is the set of all arguments constructed from the knowledge in the argumentation system, att is the attack relation, $\preceq$ is an preference ordering on $\mathcal{A}$.*

Definition 9 (ASPIC$^+$ Defeat). *A defeats B iff A undercuts B, or if A rebuts/undermines B on B' and B''s preference level is less than or equal to that of A ($B' \preceq A$).*

Then the idea of an argumentation framework follows from Definitions 5 and 9.

Definition 10 (Argumentation Framework). *An (abstract) argumentation framework AF corresponding to a structured argumentation framework $SAF = \langle \mathcal{A}, att, \preceq \rangle$ is a pair $\langle \mathcal{A}, Defeats \rangle$ such that Defeats is the defeat relation on $\mathcal{A}$ determined by SAF.*

Example 1 (adapted from [11]). Consider that we have the argumentation system $AS = \langle \mathcal{L}, \mathcal{R}, n \rangle$ where: $\mathcal{L} = \{a, b, c, d, e, f, \bar{a}, \bar{b}, \bar{c}, \bar{d}, \bar{e}, \bar{f}, \}$, $\mathcal{R}$ is $\mathcal{R}_s = \{d, f \to \bar{b}\}$ and $\mathcal{R}_d = \{a \Rightarrow b; \bar{c} \Rightarrow d; e \Rightarrow f; a \Rightarrow \overline{nd}\}$ and $n(\bar{c} \Rightarrow d) = nd$. We then add the knowledge-based $\mathcal{K}$ such that $\mathcal{K}_n = \emptyset$ and $\mathcal{K}_p = \{a; \bar{c}; e; \bar{e}\}$ to get the argumentation theory $AT = \langle AS, \mathcal{K} \rangle$. From this we can construct the arguments:

$$A_1 = [a]; A_2 = [A_1 \Rightarrow b]; A_3 = [A_1 \Rightarrow \overline{nd}];$$
$$B_1 = [\bar{c}]; B_2 = [B_1 \Rightarrow d]; B_1' = [e]; B_2' = [B_1' \Rightarrow f]; B = [B_2, B_2' \to \bar{b}];$$
$$C = [\bar{e}];$$

Let's call this set of arguments **A**, so that: $\mathbf{A} = \{A_1, A_2, A_3, B_1, B_2, B_1', B_2', B, C\}$. Note that $\mathtt{Prem}(B) = \{\bar{c}; e\}$, $\mathtt{Sub}(B) = \{B_1; B_2; B_1'; B_2'; B\}$, $\mathtt{Conc}(B) = \bar{b}$, and $\mathtt{TopRule}(B) = d, f \to \bar{b}$. The attacks between these arguments are shown in

Fig. 1(a). These make up the set $att = \{(C, B'_1), (B'_1, C), (C, B'_2), (C, B), (B, A_2), (A_3, B_2), (A_3, B)\}$ With a preference order $\preceq$ defined by : $A_2 \prec B; C \prec B; C \prec B'_1; C \prec B'_2$, we have the structured argumentation framework $\langle \mathbf{A}, att, \preceq \rangle$. This structured argumentation framework establishes a defeat relation $Defeats = \{(B'_1, C), (B, A_2), (A_3, B), (A_3, B_2)\}$ which is shown in Fig. 1(b). With this, we can finally write down the argumentation framework $\langle \mathbf{A}, Defeats \rangle$. Note that this is not a rational ASPIC$^+$ framework, since the strict rules are not closed under transposition, but serves to explain the concepts introduced above.

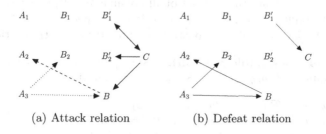

(a) Attack relation   (b) Defeat relation

Fig. 1. The attack and defeat relations from Example 1. A dotted arrow shows undercutting, a dashed arrow shows rebutting, and a solid arrow shows undermining.

3   ASPIC$_D^+$: A Purely Defeasible System

3.1   Definition

The full definition of ASPIC$_D^+$ starts from a variation on the ASPIC$^+$ notion of an argumentation system where there are only defeasible elements:

Definition 11 (ASPIC$_D^+$ Argumentation System). *An argumentation system is a triple* $AS_D = \langle \mathcal{L}, \mathcal{R}_d, n \rangle$ *where:*

- *$\mathcal{L}$ is a logical language closed under negation $\bar{\cdot}$.*
- *$\mathcal{R}_d$ is a set of defeasible inference rules of the form $\phi_1, \ldots, \phi_n \Rightarrow \phi$ (where ϕ_i, ϕ are meta-variables ranging over wff in $\mathcal{L}$).*
- *$n : \mathcal{R}_d \mapsto \mathcal{L}$ is a naming convention for defeasible rules.*

Definition 12 (ASPIC$_D^+$ Knowledge Base). *A knowledge base in an argumentation system* $\langle \mathcal{L}, \mathcal{R}_d, n \rangle$ *is a set $\mathcal{K}_p$ of ordinary premises.*

Definition 13 (ASPIC$_D^+$ Argumentation Theory). *An argumentation theory AT_D is a pair $\langle AS, \mathcal{K}_p \rangle$ of an argumentation system AS and a set of ordinary premises $\mathcal{K}_p$.*

Arguments in ASPIC$_D^+$ are then defined as in Definition 5, but there are no strict rules or axioms so there are no strict or firm arguments.

Since any ASPIC$_D^+$ argumentation theory is an ASPIC$^+$ argumentation theory with an empty set of strict rules and an empty set of axioms, we have:

Proposition 1. *For a given language $\mathcal{L}$, $\mathbf{AT_D}$, the set of all possible ASPIC$_D^+$ argumentation theories, is a subset of $\mathbf{AT}$, the set of all possible ASPIC$^+$ argumentation theories.*

Proof. Pick any ASPIC$_D^+$ theory $AT_D \in \mathbf{AT_D}$. By definition this is a pair $\langle AS_D, \mathcal{K}_p \rangle$ where $AS_D = \langle \mathcal{L}, \mathcal{R}_d, n \rangle$. It is also an ASPIC$^+$ theory $AT \in \mathbf{AT}$ where $AT = \langle AS, \mathcal{K}_p \rangle$ (an ASPIC$^+$ theory with no axioms) and $AS = \langle \mathcal{L}, \mathcal{R}_d, n \rangle$ (an ASPIC$^+$ theory with no strict rules). Having made no specific assumptions about the composition of AT_D, the result holds for all possible ASPIC$_D^+$ theories.

However, despite the fact that the set of all possible ASPIC$_D^+$ theories is a subset of all possible ASPIC$^+$ theories, we can translate any specific ASPIC$^+$ theory into a specific ASPIC$_D^+$ theory. We demonstrate this by defining a translation:

Definition 14 (Defeasible version). ASPIC$_D^+$ *theory AT_D is the* defeasible version *of* ASPIC$^+$ *theory $AT = \langle AS, \mathcal{K}_n \cup \mathcal{K}_p \rangle$ where $AS = \langle \mathcal{L}, \mathcal{R}_s \cup \mathcal{R}_p, n \rangle$ iff:*

- $AS_D = \langle \mathcal{L}, \mathcal{R}_d \cup \mathcal{R}_{d'}, n' \rangle$, *where* $\mathcal{R}_{d'} = \{\phi_1, \ldots, \phi_n \Rightarrow \phi \mid \phi_1, \ldots, \phi_n \rightarrow \phi \in \mathcal{R}_s\}$ *and n' is n extended to name all the rules in $\mathcal{R}_{d'}$.*
- $AT_D = \langle AS_D, \mathcal{K}_p \cup \mathcal{K}_{p'} \rangle$, *where* $\mathcal{K}_{p'} = \{\phi \mid \phi \in \mathcal{K}_n\}$.

If AT_D is the defeasible version of AT, we call AS_D the defeasible version *of AS and write $AT_D = def(AT)$ and $AS_D = def(AS)$. We call the set of rules $\mathcal{R}_{d'}$ that were strict in AT the set of* converted rules, *and the set of premises $\mathcal{K}_{p'}$ that were axioms in AT are the set of* converted premises. *The defeasible version of an argument $A \in \mathcal{A}(AT)$ is an argument $A_D \in \mathcal{A}(AT_D)$ such that every axiom in A is replaced by the corresponding converted premise, and every struct rule in A is replaced by the corresponding converted rule.*

In other words, AT_D is the defeasible version of AT, if every axiom of AT becomes an ordinary premise of AT_D, and every strict rule in AT becomes a defeasible rule of AT, while all other components of AT are unchanged.

Given a preference order $\preceq$ over the elements of an ASPIC$^+$ theory AT, we will need to specify the preference order $\preceq_D$ over the defeasible version of the theory. One way to specify $\preceq_D$ is as follows in terms of the pre-orderings over the rules and premises of AT_D.

Definition 15 (Strict-first preference ordering). *Given an* ASPIC$^+$ *theory $AT = \langle \langle \mathcal{L}, \mathcal{R}_s \cup \mathcal{R}_d, n \rangle, \mathcal{K}_n \cup \mathcal{K}_p \rangle$ and preference orders $\leq$ and $\leq'$ over the defeasible rules and premises of that theory, the* strict-first preference order-ings $\leq_{sf}$ *and $\leq'_{sf}$ over the rules and premises of the defeasible version of AT, $AT_D = \langle \langle \mathcal{L}, \mathcal{R}_d \cup \mathcal{R}_{d'}, n' \rangle, \mathcal{K}_p \cup \mathcal{K}_{p'} \rangle$ are such that:*

- *For every $r, r' \in \mathcal{R}_d$, $r \leq_{sf} r'$ iff $r \leq r'$, and for every $k, k' \in \mathcal{K}_p$, $k \leq'_{sf} k'$ iff $k \leq' k'$.*
- *For any $r \in \mathcal{R}_d$ and any $r' \in \mathcal{R}_{d'}$, $r <_{sf} r'$, and for every $r', r'' \in \mathcal{R}_{d'}$, $r' =_{sf} r''$.*
- *For any $k \in \mathcal{K}_p$ and any $k' \in \mathcal{K}_{p'}$, $k <'_{sf} k'$, and for every $k', k'' \in \mathcal{K}'_p$, $k' ='_{sf} k''$.*

where $r =_{sf} r'$ *if* $r \leq_{sf} r'$ *and* $r' \leq_{sf} r$, $r <_{sf} r'$ *if* $r \leq_{sf} r'$ *and* $r' \nleq_{sf} r$, $k ='_{sf} k'$ *if* $k \leq'_{sf} k'$ *and* $k' \leq'_{sf} k$, *and* $k <'_{sf} k'$ *if* $k \leq'_{sf} k'$ *and* $k' \nleq'_{sf} k$.

In other words, all the elements of AT_D that were defeasible in AT have the same preference order as in AT, and all elements that were strict in AT are strictly higher in the preference order than any element that was defeasible in AT. The notion of attack in ASPIC$_D^+$ differs from that in ASPIC$^+$ in that there is no restriction on rebut, and any rule can be undercut:

Definition 16 (ASPIC$_D^+$ Attack). *An argument A attacks an argument B iff A undermines, rebuts or undercuts B, where:*

- *A undermines B (on B') iff $\mathrm{Conc}\,(A) = \overline{\phi}$ for some $B' = \phi \in \mathrm{Prem}(B)$.*
- *A rebuts B (on B') iff $\mathrm{Conc}\,(A) = \overline{\phi}$ for some $B' \in \mathrm{Sub}(B)$.*
- *A undercuts B (on B') iff $\mathrm{Conc}\,(A) = \overline{n(r)}$ for some $B' \in \mathrm{Sub}(B)$.*

With these definitions, we can once again combine the definition of arguments, attack relations and the preference ordering from Definition 7 to get notions of a structured argumentation framework and defeat that are the same as for ASPIC$^+$. To begin to understand the relationship between ASPIC$^+$ and ASPIC$_D^+$, consider this version of Example 1:

Example 2. Consider the ASPIC$_D^+$ argumentation system AS_D which is the defeasible version of the system AS in Example 1. We have $\mathcal{R}_d = \{a \Rightarrow b; \overline{c} \Rightarrow d; e \Rightarrow f; a \Rightarrow \overline{nd}; d, f \Rightarrow \overline{b}\}$, $\mathcal{K}_p = \{a; \overline{c}; e; \overline{e}\}$, and $n(\overline{c} \Rightarrow d) = nd$. We can construct the arguments:

$$A_1 = [a]; A_2 = [A_1 \Rightarrow b]; A_3 = [A_1 \Rightarrow \overline{nd}];$$
$$B_1 = [\overline{c}]; B_2 = [B_1 \Rightarrow d]; B'_1 = [e]; B'_2 = [B'_1 \Rightarrow f]; B = [B_2, B'_2 \Rightarrow \overline{b}];$$
$$C = [\overline{e}];$$

Compared with the attacks in Example 1, there is an additional attack here: A_2 rebuts B. With the same preference ordering $\preceq$ over arguments as in Example 1, the defeat relation remains same.

3.2   Properties of ASPIC$_D^+$

We begin by showing that ASPIC$_D^+$ satisfies the three rationality postulates that were introduced in [6] and since then have been considered the basic requirement of a sensible argumentation system. Without strict rules, two of these postulates follow immediately.

Proposition 2 (Closure under Strict Rules). *The conclusions of any extension an ASPIC$_D^+$ theory are closed under strict rules.*

Proof. With no strict rules, the conclusion follows immediately.

Proposition 3 (Direct Consistency). *The conclusions of any extension of an* ASPIC$_D^+$ *theory are consistent.*

Proof. Suppose the conclusions of one of the extensions E are inconsistent, i.e., there exist two arguments $A, A' \in E$ such that $\text{Conc}(A) = \overline{\text{Conc}(A')}$. If $\text{Conc}(A) \in \mathcal{K}$, by Definition 6, then A' undermines A. On the other hand, if $\text{Conc}(A) \notin \mathcal{K}$, by Definition 6, then A' rebuts A. Either way, A' attacks A in any case. Similarly, A attacks A'.

According to Definition 9, at least one of the attack relations is a defeat relation. Therefore, E is not conflict-free and thus E is not an extension under Dung's semantics. The contradicition defeats the assumption of inconsistency and the result holds.

Proposition 4 (Indirect Consistency). *The closure under strict rules of the conclusions of any extension of an* ASPIC$_D^+$ *theory is consistent.*
Proof. With no strict rules, this follows immediately from Proposition 3.

Despite the triviality of two of the results, it is worth noting that there are no restrictions on the semantics for which these results hold — they hold for all the standard Dung semantics. Thus ASPIC$_D^+$ goes further than the ASPIC- of [7] in extending the scope of reasoning possible with unrestricted rebut since ASPIC-only satisfied the rationality postulates for the grounded semantics. Of course, this extension is achieved by giving up strict rules, and it is natural to ask what the consequence is for what can be represented in an ASPIC$_D^+$ theory. Would using ASPIC$_D^+$ mean any restriction on what can be represented? Our main result is to show that there is no restriction on what can be represented in ASPIC$_D^+$ compared with what can be represented in ASPIC$^+$ in the sense that for any ASPIC$^+$ theory we can build an ASPIC$_D^+$ theory with the same justified conclusions. We start with the observation that:

Proposition 5. *For a given language $\mathcal{L}$, there is a defeasible version AT_D of any* ASPIC$^+$ *argumentation theory AT.*

Proof. Consider the clauses of Definition 14 as a series of rewrite rules. Any AT can be converted into its defeasible version by turning every axiom into an ordinary premise and every strict rule into a defeasible rule.

This means that whatever information we have in an ASPIC$^+$ theory, we can capture it in an ASPIC$_D^+$ theory — we don't lose the ability to represent information about the world by using ASPIC$_D^+$ rather than ASPIC$^+$. However, it is not just representing information that is important. The set of arguments that can be constructed from a theory, and, in particular, the justified conclusions of a theory are also important.

Proposition 6. *Given an* ASPIC$^+$ *theory AT and its defeasible version AT_D, $|\mathcal{A}(AT)| = |\mathcal{A}(AT_D)|$ and for every $A \in AT$ there is exactly one $A_D \in \mathcal{A}(AT_D)$ such that A_D is the defeasible version of A.*

Proof. We show there is a 1-to-1 map between $\mathcal{A}(AT)$ and $\mathcal{A}(AT_D)$. For each argument that is just a premise or an axiom $A = [\phi]$, we have $A_D = [\phi]$ that *is just a premise; for each argument* $A = [A_1, \ldots, A_n \Rightarrow \phi]$, *we have* $A_D = [A_{1_D}, \ldots, A_{n_D} \Rightarrow \phi]$; *for each argument* $A = [A_1, \ldots, A_n \rightarrow \phi]$, *we have* $A_D = [A_{1_D}, \ldots, A_{n_D} \Rightarrow \phi]$.

Thus any ASPIC$^+$ theory can be turned into an ASPIC$_D^+$ theory, and we can generate the same number of arguments, but arguments that had strict components will now only have defeasible components. Furthermore, there are preference orderings such that the same preferences exist between ASPIC$_D^+$ arguments as between the corresponding ASPIC$^+$ arguments:

Proposition 7. *Consider the set of arguments* $\mathcal{A}$ *of an* ASPIC$^+$ *theory* AT *and the set of arguments* $\mathcal{A}_D$ *constructed from the defeasible version of the theory* AT_D. *If the preference order over* AT_D *is the strict-first version of that over* AT, *then using the weakest link principle, for any* $A, B \in \mathcal{A}$, *and* $A_D, B_D \in \mathcal{A}_D$ *where* A_D, B_D *are the defeasible versions of* A *and* B, $A_D \preceq_D B_D$ *iff* $A \preceq B$.

Proof. Let $AT = \langle AS, \mathcal{K}_n \cup \mathcal{K}_p \rangle$ *and* $AT_D = \langle AS_D, \mathcal{K}_{p'} \cup \mathcal{K}_p \rangle$. *Consider the preference order* $\leq$ *over rules in* AT, *and the preference order* $\leq'$ *over premises. Let* $\langle \leq_D, \leq'_D \rangle$ *contain all the relations in* $\langle \leq, \leq' \rangle$. *Since* AF_D *has more defeasible elements than* AF, *we need to determine where these elements fit in the ordering. With a strict-first ordering, the translated strict rules/axioms have the highest preference ordering, and so the weakest links in* $\mathcal{A}(AT_D)$ *are not the translated strict rules/axioms. Furthermore, all the remaining rules/premises in* AT_D *have the same preference ordering as in* AT. *Therefore,* AT_D *and* AT *have the same preference ordering over arguments.*

In other words, using the weakest link principle, we can take a set of ASPIC$^+$ arguments create the defeasible versions of those arguments and still have the same preference ordering as over the original set of arguments. This allows us to show our main result, that we can construct a defeasible version of a given ASPIC$^+$ framework such that the justified conclusions of both theories are the same.

Proposition 8. *Consider a rational* ASPIC$^+$ *theory* AT *and its defeasible version* AT_D *where the preference ordering over* AT_D *is the strict-first version of the ordering over* AT. *Under the weakest link principle, the justified conclusions of* AT *and* AT_D *are the same.*

Proof. From Proposition 6 we know that for every argument in $\mathcal{A}(AT_D)$, there is an argument in $\mathcal{A}(AT)$, with the same conclusion, and vice versa. From Proposition 7, we know that under the weakest link principle, the preference order $\preceq$ over $\mathcal{A}(AT_D)$ is the same as the preference order $\preceq_D$ over $\mathcal{A}(AT_D)$. Now, consider the attack relations att and att$_D$ over $\mathcal{A}(AT)$ and $\mathcal{A}(AT_D)$. If $(A, A') \in$ att, then $(A_D, A'_D) \in$ att$_D$ and there is an attack between the defeasible versions of the arguments A_D and A'_D. However, att$_D$ can contains more attacks. (A_D, A'_D) can be in att$_D$ when $(A, A') \notin$ att iff (1) A' is (just) an axiom in AT or (2) A' has a strict `TopRule` and the attack is not permitted by restricted rebut. We now show, in turn, that these additional attacks do not affect the justified conclusions.*

First, if A' is an axiom, then A'_D, which as a lone premise that is the defeasible version of an axiom, has the highest possible preference. Thus it can only be defeated by an A_D that has the highest level of preference. Such an argument is the defeasible version of a strict argument. However, if A was strict, AT would not be rational (it would have two strict elements in conflict). Therefore, we have the same defeat relations over $\mathcal{A}(AT)$ and $\mathcal{A}(AT_D)$ and hence the same justified conclusions for AT and AT_D.

Second, if $\texttt{TopRule}(A')$ is strict, there are two sub-cases that concern us. (a) If $A_D \prec A'_D$, the attack does not become a defeat. Thus AF_D and AF have the same defeat relation, therefore they have the same justified conclusions. (b) If $A'_D \preceq A_D$, then there is one more defeat relation over $\mathcal{A}(AT_D)$ than over $\mathcal{A}(AT)$. We will show that this additional defeat relation has no effect. Consider applying all the defeat relations except this additional one — there are three possibilities for the status of A'_D which will be mirrored by the status of A' which does not have to contend with this additional defeat, and for each of these, we have to consider all three possibilities for the status of A_D.

(1) A'_D is labeled IN. If A_D is labeled IN, then AT has two IN arguments, A and A', and the conclusions of these arguments are in the set of justified conclusions. However, since A_D and A'_D rebut one another, the conclusions of A and A' are contradictory, violating direct consistency. Thus A and A' cannot both be IN, and so neither can A'_D and A_D before the application of the new defeat. If A_D is labeled OUT then adding the defeat relation (A_D, A'_D) has no effect. If A_D is labelled IN, the situation is more complicated. We start by noting that A will also be UNDEC, and then consider how this can be the case. A has a strict top rule, so $A' = [A'_1, \ldots, A'_n \rightarrow a]$ where the top rule is $p_1, \ldots, p_n \rightarrow a$. Similarly, $A = [A_1, \ldots, A_n \Rightarrow \overline{a}]$ with a top rule $q_1, \ldots, q_n \Rightarrow \overline{a}$. By closure under transposition, there exists a strict rule $p_1, \ldots, p_{i-1}, \overline{a}, p_{i+1}, \ldots, p_n \rightarrow \overline{p_i}$ in AT. Since $A'_D \preceq A_D$, it is not possible for A' to be strict, so A' has at least one defeasible sub-argument, and hence a sub-argument with a defeasible top-rule. Lets assume that this is one of the $A'_1, \ldots, A'_n$ that combine with the strict top rule, and call it A'_i. Using the strict rule from the transposition of the top rule, we get an argument $B = [A'_1, \ldots, A'_{i-1}, A, A'_{i+1}, \ldots, A'_n \rightarrow \overline{p_i}]$ which rebuts A'. B is A plus the $A'_j, j \neq i$, and the transposed strict rule. If $A'_1, \ldots, A'_n$ do not have defeasible top rules, then we chain the corresponding transposed strict top rule(s) to B to build an argument that attacks A' further down the argument tree until we get an argument, call it B', which rebuts A' on its defeasible sub-argument. Now, $A' \preceq B'$ since $A' \preceq A$ and B' is A plus some sub-arguments of A' and a sequence of strict (transposed) rules. Therefore B' defeats A'. Moreover, any defeater of B' must be a defeater of A or A'. Next we consider the labeling. Since A' is labeled IN, all the defeaters of A' are labeled OUT. Since A is labeled UNDEC, all the defeaters of A are labeled OUT or UNDEC. Therefore, the defeaters of B, which are the defeaters of A or A', are labelled OUT or UNDEC. Thus B is labeled IN or UNDEC. Since B defeats A', A' can not be labeled IN, contradicting what we started with.

(2) A'_D is labeled OUT. Adding one more defeat relation (A_D, A'_D) has no effect.

(3) A'_D is labeled UNDEC. If A_D is labeled OUT or UNDEC, then adding the defeat relation (A_D, A'_D) has no effect. However, if A_D is labeled IN, then applying the last defeat relation means that A'_D will now be labeled OUT while A', which does not have to contend with (A, A'), will be UNDEC.

So A'_D cannot be initially labeled IN. If it is labelled OUT, the status of A'_D cannot change as a result of the additional defeat. If A'_D is initially labeled UNDEC, the status of A'_D can change. However, by showing that A'_D does not defeat any other arguments we can show that this change does not affect the justified conclusions. Consider an argument $B \in \mathcal{A}(AT_D)$ that is attacked by A'_D. A'_D cannot undercut B since the conclusion of A'_D is not a "rule" (if it were a rule, there would be no rebut between A_D and A'_D and there would be no new defeat relation to consider). A'_D can not undermine B since the conclusion of A'_D is not a premise because we know that A' and hence A'_D has a TopRule. So we can only be dealing with a rebut, and since we already know that A_D rebuts A_D, B has to be an argument of which A_D is a sub-argument. Since $A'_D \prec A_D$, A'_D does not defeat B.

Thus, in all of these three sub-cases of (b), the additional defeat (A_D, A'_D) has no effect on the status of the arguments in $\mathcal{A}(AT_D)$, again there is no difference between the justified conclusions of AT and AT_D, and the result holds.

This result justifies our claim that ASPIC$_D^+$ makes it possible to represent the same information as ASPIC$^+$. Given an ASPIC$^+$ theory, we can encode the information in purely defeasible form in an ASPIC$_D^+$ theory that gives us exactly the same set of (justified) conclusions. The following example helps to show how this is possible.

Example 3. Consider that we start with the following theory AT_1 (given the same language as before), closed under transposition:

$$\mathcal{R}_d = \{a \Rightarrow b; b' \Rightarrow \bar{c}\} \qquad \mathcal{R}_s = \{a' \to b'; b \to c; \bar{b'} \to \bar{a'}; \bar{c} \to \bar{b}\}$$
$$\mathcal{K}_n = \{a; a'\} \qquad \mathcal{K}_p = \emptyset$$

Then we can construct the following arguments:

$$A_1 = [a] \quad A_2 = [A_1 \Rightarrow b] \quad A_3 = [A_2 \to c]$$
$$B_1 = [a'] \quad B_2 = [B_1 \to b'] \quad B_3 = [B_2 \Rightarrow \bar{c}] \quad B_4 = [B_3 \to \bar{b}]$$

The attack relations are shown in Fig. 2(a). Now we translate this framework to the ASPIC$_D^+$ theory AT_2:

$$\mathcal{R}_d = \{a \Rightarrow b; b' \Rightarrow \bar{c}\} \qquad \mathcal{R}'_d = \{a' \Rightarrow b'; b \Rightarrow c; \bar{b'} \Rightarrow \bar{a'}; \bar{c} \Rightarrow \bar{b}\}$$
$$\mathcal{K}'_n = \{a; a'\} \qquad \mathcal{K}_p = \emptyset$$

Then we can construct the following arguments:

$$A_1 = [a] \quad A_2 = [A_1 \Rightarrow b] \quad A_3 = [A_2 \Rightarrow c]$$
$$B_1 = [a'] \quad B_2 = [B_1 \Rightarrow b'] \quad B_3 = [B_2 \Rightarrow \bar{c}] \quad B_4 = [B_3 \Rightarrow \bar{b}]$$

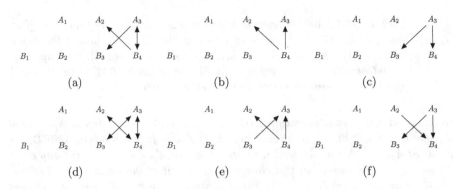

Fig. 2. Attack and defeat relations for AT_1 and AT_2. (a) Attack relations for AT_1 and Defeat relations for AT_1 in case (a), (b) Defeat relations for AT_1 case (b), (c) Defeat relations for AT_1 in case (c), (d) Attack relations for AT_2 and Defeat relations for AT_2 in case (a), (e) Defeat relations for AT_2 case (b), (f) Defeat relations for AT_2 case c)

The attack relations are shown in Fig. 2(d). Now let's consider the different possible preference orderings over rules:

(a) $a \Rightarrow b = b' \Rightarrow \bar{c}^2$. By the weakest link principle, all the attack relations are defeat relations, see Fig. 2(a) and (d). Here AT_2 has additional defeat relations, but they are directed at arguments that are already defeated. Under the grounded semantics, the set of arguments in the extension is $\{A_1, B_1, B_2\}$ and $\{A_1, B_1, B_2\}$, and the justified conclusions are $\{a, a', b'\}$ and $\{a, a', b'\}$.

(b) $a \Rightarrow b < b' \Rightarrow \bar{c}$. By the weakest link principle, the defeat relations are shown in Fig. 2(b) and (e). Again AT_2 has an additional defeat relation, but again it has no effect on the justified conclusions. Under the grounded semantics, the set of arguments in the extension is $\{A_1, B_1, B_2, B_3, B_4\}$, $\{A_1, B_1, B_2, B_3, B_4\}$ and $\{A_1, A_2, B_1, B_2, B_3\}$, and the justified conclusions are $\{a, \bar{b}, a', b', \bar{c}\}$ and $\{a, \bar{b}, a', b', \bar{c}\}$.

(c) $a \Rightarrow b > b' \Rightarrow \bar{c}$. By the weakest link principle, the defeat relations are shown in Fig. 2(c) and (f). As before AT_2 has additional defeats, but they have no effect. Under the grounded semantics, the set of arguments in the extension is $\{A_1, A_2, A_3, B_1, B_2\}$, and $\{A_1, A_2, A_3, B_1, B_2\}$, and the justified conclusions are $\{a, b, a', b', c\}$ and $\{a, b, a', b', c\}$.

For all the cases, the justified conclusions of AT_1 and AT_2 are exactly same.

4   Conclusion

We have shown that ASPIC$_D^+$, the defeasible subset of ASPIC$^+$, has the same functionality in terms of knowledge representation as ASPIC$^+$. Both formalisms

---

[2] $A = B$ is defined as $A \leq B$ and $B \leq A$.

draw the same justified conclusions from corresponding argumentation systems. In addition, ASPIC_D^+ goes further than ASPIC^+ and ASPIC- in the sense of satisfying the rationality postulates with unrestricted rebut for all of Dung's semantics. In our view, this justifies the choice of ASPIC_D^+ in any application that might use ASPIC^+. Proposition 5 tells us that using ASPIC_D^+ means we can represent exactly the same information that we could in ASPIC^+, and Proposition 8 tells us that provided that we encode strict rules as defeasible rules with the highest level of preference and use the weakest link principle, the same set of justified conclusions will be obtained as if we had used ASPIC^+. In that sense, we do not lose anything over what is possible in ASPIC^+ by using ASPIC_D^+. In addition, and in contrast to ASPIC^+, we do not have to impose any restrictions on ASPIC_D^+ in order for it to accord to the rationality principles, which, as [7] points out, is rather counter-intuitive and hard to explain to users.

References

1. Amgoud, L.: Five weaknesses of ASPIC$^+$. In: Greco, S., Bouchon-Meunier, B., Coletti, G., Fedrizzi, M., Matarazzo, B., Yager, R.R. (eds.) IPMU 2012, Part III. CCIS, vol. 299, pp. 122–131. Springer, Heidelberg (2012)
2. Amgoud, L., Bodenstaff, L., Caminada, M., McBurney, P., Parsons, S., Prakken, H., van Veenen, J., Vreeswijk, G.A.W.: Final review and report on formal argumentation system. Deliverable D2.6. Technical report, ASPIC IST-FP6-002307 (2006)
3. Baroni, P., Caminada, M., Giacomin, M.: An introduction to argumentation semantics. Knowl. Eng. Rev. **26**(4), 365–410 (2011)
4. Besnard, P., Hunter, A.: A logic-based theory of deductive arguments. Artif. Intell. **128**, 203–235 (2001)
5. Caminada, M.: Contamination in formal argumentation systems. In: Proceedings of the 17th Belgium-Netherlands Conference on Artificial Intelligence (2005)
6. Caminada, M., Amgoud, L.: On the evaluation of argumentation formalisms. Artif. Intell. **171**(5), 286–310 (2007)
7. Caminada, M., Modgil, S., Oren, N.: Preferences and unrestricted rebut. In: Computational Models of Argument: Proceedings of COMMA 2014 (2014)
8. Dung, P.M.: On the acceptability of arguments and its fundamental role in non-monotonic reasoning, logic programming and n-persons games. Artif. Intell. **77**(2), 321–358 (1995)
9. Dung, P.M., Kowalski, R.A., Toni, F.: Assumption-based argumentation. In: Rahwan, I., Simari, G.R. (eds.) Argumentation in Artificial Intelligence, pp. 199–218. Springer, New York (2009)
10. García, A.J., Simari, G.R.: Defeasible logic programming: An argumentative approach. Theory Pract. Logic Programm. **4**(1+2), 95–138 (2004)
11. Modgil, S., Prakken, H.: A general account of argumentation with preferences. Artif. Intell. **195**, 361–397 (2012)
12. Prakken, H.: An abstract framework for argumentation with structured arguments. Argument Comput. **1**(2), 93–124 (2010)
13. Prakken, H., Modgil, S.: Clarifying some misconceptions on the ASPIC+ framework. In: Computational Models of Argument: Proceedings of Comma 2012 (2012)
14. Wu, Y., Podlaszewski, M.: Implementing crash-resistance and non-interference in logic-based argumentation. J. Logic Comput. 1–31 (2014)

Dealing with Inconsistency

A Possibilistic Analysis of Inconsistency

Didier Dubois and Henri Prade[✉]

IRIT–CNRS, 118, Route de Narbonne, Toulouse, France
{dubois,prade}@irit.fr

Abstract. Central in standard possibilistic logic (where propositional logic formulas are associated with lower bounds of their necessity measures), is the notion of inconsistency level of a possibilistic logic base. Formulas whose level is strictly above this inconsistency level constitute a sub-base free of any inconsistency. Some extensions, based on the notions of paraconsistent completion of a possibilistic logic base, and of safely supported formulas, have been proposed for handling formulas below the level of inconsistency. In this paper we further explore these ideas, and show the interest of considering the minimal inconsistent subsets in this setting. Lines for further research are also outlined.

1 Introduction

Reasoning under inconsistency [6,13], or evaluating the inconsistency of a knowledge base [10,11] have raised a lot of interest in artificial intelligence for a long time. However, the different approaches which have been proposed do not usually take into account the fact that all the formulas in a knowledge base are not necessarily equally certain. Possibilistic logic [8] provides a simple way for a partial handling of inconsistency by taking advantage of a stratification of the knowledge base according to the certainty level associated to the logical formulas. Then we can compute an inconsistency level for a propositional knowledge base, and all the formulas whose certainty is strictly above this inconsistency level form a consistent sub-base. The formulas whose certainty is equal to or smaller than the inconsistency level remain drown in inconsistency, including formulas that are not involved in any minimal inconsistent subsets. This state of fact can be somewhat remedied by defining a paraconsistent completion of the knowledge base, and by using a so-called safely supported entailment relation [3,5]. Strangely enough, this entailment is more productive than the possibilistic logic entailment, but it nevertheless preserves the consistency of the set of consequences. Yet it has remained largely ignored. This short paper revisits the approach and shows its relation with minimal inconsistent subsets.

The paper is structured as follows. Section 2 deals with the flat case [4] where formulas are not associated with certainty levels. We present the idea of paraconsistent completion as a basis for analyzing the conflicts, and then identify the safely supported consequences. In Sect. 3, we deal with possibilistic logic formulas, and extend the previous definitions. Then a new characterization of safely supported entailment is proposed. Lines for further research are also discussed.

© Springer International Publishing Switzerland 2015
C. Beierle and A. Dekhtyar (Eds.): SUM 2015, LNAI 9310, pp. 347–353, 2015.
DOI: 10.1007/978-3-319-23540-0_23

2   Flat Propositional Knowledge Bases

Let $\Sigma = \{p_i \mid i = 1, \ldots, n\}$ denote a propositional logic knowledge base. Σ may be inconsistent. Let us first recall two basic notions, needed in the forthcoming discussion: the notions of support for a proposition and of minimal inconsistent subset. A Σ-based *support* (or *reason*, or argument) for a proposition p is a subset S_p of propositions in Σ such that (i) S_p is consistent; (ii) $S_p \vdash p$ (where $\vdash$ is the classical logic consequence relation); (iii) $\nexists S' \subset S_p$ such that $S' \vdash p$. In other words, S_p is a minimal consistent subset of propositions in Σ that together entail p. Likewise, a minimal inconsistent subset of Σ is a minimal subset of propositions that entail $\bot$: a non empty subset $S^\bot$ of Σ such that (i) $S^\bot$ is inconsistent ($S^\bot \vdash \bot$); (ii) $\nexists S' \subset S^\bot$ such that $S' \vdash \bot$.

For a complete analysis of the inconsistency situation of formulas in Σ, we need to define the "paraconsistent completion" Σ^{comp} of Σ.

2.1   Paraconsistent Completion

For analyzing the potential conflicts in Σ, it is convenient to proceed with the following construction. The paraconsistent completion Σ^{comp} of Σ is obtained by applying the following procedure: to each formula p_i in Σ, one associates i) the set of reasons for p_i, and ii) the set of reasons for $\neg p_i$. More formally, $\Sigma^{comp} = \{(p_i, \{P_1, \cdots, P_r\}, \{C_1, \cdots, C_s\}) \mid p_i \in \Sigma, P_i \text{ is a reason for } p_i, C_j \text{ is a reason for } \neg p_i\}$.

Clearly, if $p_i \in \Sigma$, then $(p_i, \{P_1, \cdots, P_r\}, \{C_1, \cdots, C_s\}) \in \Sigma^{comp}$, and if $\exists j$ s.t. $p_j \equiv \neg p_i$ then $(\neg p_i, \{P'_1, \cdots, P'_s\}, \{C'_1, \cdots, C'_r\}) \in \Sigma^{comp}$ with $\forall i \ P'_i = C_i, \forall j \ C'_j = P_j$. Note that as soon as $p_i \in \Sigma$, the set of reasons for p_i is not empty: it contains at least $\{p_i\}$.

The reasons for and against p_i can be summarized by triples of the form (p_i, π_i, γ_i) for $i = 1, \ldots, n$ where $\pi_i \in \{0, 1\}, \gamma_i \in \{0, 1\}$, and: (i) $p_i \in \Sigma$; (ii) $\pi_i = 1$ for acknowledging the fact that $\exists P_k$, a reason for p_i; (iii) $\gamma_i = 1$ if $\exists C_l$ that is a reason for $\neg p_i$, and $\gamma_i = 0$ if $\nexists C_l$ (no reason for $\neg p_i$). Let $\Sigma^{para} = \{(p_i, \pi_i, \gamma_i) : i = 1, \ldots, n\}$. Note that $\pi_i \neq 0$ (hence $= 1$), since each $p_i \in \Sigma$ supports itself.

If $\min(\pi_i, \gamma_i) = 1$, then p_i is said to be *paraconsistent* (in the sense of "conflicting"). Thus in Σ, there are two kinds of propositions, the formulas p_i such as $\gamma_i = 0$ which should be considered as true, and the formulas that are paraconsistent. Note that strictly speaking there is no formula of the form $(p_j, 0, 1)$ in Σ^{para} since the information that p_j is false appears there only under the form $(\neg p_j, 1, 0)$, i.e. $\neg p_j$ is true. However, note also that one may have $(p_i, \{\{p_j\}\}, \{C_1, \cdots, C_s\}) \in \Sigma^{comp}$, where no C_k contains $\neg p_j$, which might be understood as suggesting that p_j, being only supported by itself, is questionable. Besides, there is no formula of the form $(p_k, 0, 0)$ in Σ^{para} (it would express that there is no reason for p_k, nor for $\neg p_k$).

2.2 Safely Supported Propositions

Once Σ^{comp} and Σ^{para} are built from Σ, one can evaluate reasons S in favor of a proposition p by means of the two evaluations, $Def(S)$ and $Uns(F)$, respectively revealing the potential weakness of its support and its lack of safety:

- $Def(S) = \min_i \{\pi_i | (p_i, \pi_i, \gamma_i) \in \Sigma^{para}$ and $p_i \in S\}$.

In fact, one always have $Def(S) = 1$ since $\forall i, p_i \in S$, we have $\pi_i = 1$, and the case $Def(S) = 0$ is impossible here since p_i is in Σ ($p_i \in \Sigma$ is understood as $(p_i, 1)$). We shall see that when propositional formulas become weighted, we still always have $Def(S) > 0$, but $Def(S)$ may be "close to 0".

- $Uns(S) = \max_i \{\gamma_i | (p_i, \pi_i, \gamma_i) \in \Sigma^{para}$ and $p_i \in S\}$.

Clearly, $Uns(S) = 0$ if $\forall i | p_i \in S$ $\gamma_i = 0$, i.e. if S does not contain any paraconsistent formula, while $Uns(S) = 1$ if $\exists i | p_i \in S$ $\gamma_i = 1$, i.e. there is at least one paraconsistent formula in S. Thus, $Uns(S)$ reflects if there are a reason pro and a reason against an element of S that can be both built from formulas in Σ.

A reason S in favor of proposition p is *free* iff $Def(S) > Uns(S)$, i.e. iff all the formulas in S are believed to be true and none is inconsistent with other formulas in Σ. By extension, in this case, we shall say all the formulas in S are *free* as well. Moreover, any formula in a minimal inconsistent subset $S^\perp = \{r_1, r_2, \cdots, r_k\}$ of Σ is not free, since $S^\perp \setminus \{r_j\}$ is consistent and $\forall j, S^\perp \setminus \{r_j\} \vdash \neg r_j$. Thus, if $\exists S^\perp, r_j \in S^\perp \subseteq \Sigma$ then $(r_j, 1, 1) \in \Sigma^{para}$, i.e., r_j is a paraconsistent formula in Σ. If a formula is involved in several minimal inconsistent subsets, one might think that this formula could be considered as more "paraconsistent" since there exists several distinct reasons against it. However, this looks debatable since a "basic" piece of information often used in inferences may have some chance to be, on the contrary, strongly established.

In the classical case, $Def(S) > Uns(S) \Leftrightarrow Uns(S) = 0$, since then $Def(S) = 1$. Thus, a proposition p is *safely supported* if it exists a reason S for it which is free. The safely supported propositions are just the consequences of the set of free ones. It follows that the set of safely supported formulas in Σ is always *consistent*. So in particular, p and $\neg p$ cannot be both safely supported.

This departs from the so-called *argumentative inference* [2], which is more adventurous than the safely supported inference, since it may lead to an inconsistent set of conclusions, but not to direct contradictions such as p and $\neg p$. The argumentative inference amounts to conclude p if there is a reason for p and no reason for $\neg p$ in Σ.

For instance, consider the base $\Sigma = \{r, \neg r \vee p, \neg r, r \vee q\}$. Then, we can infer both p and q argumentatively from Σ. In contrast, the reader can check that $\Sigma^{para} = \{(r, 1, 1), (\neg r, 1, 1), (\neg r \vee p, 1, 0), (r \vee q, 1, 0)\}$, from which one can infer neither that p nor q is safely supported.

Still, as recalled in the discussion section, one can also infer $(p, 1, 1)$ and $(q, 1, 1)$ from Σ^{para}, thus acknowledging that p and q are indeed paraconsistent conclusions.

We now examine how the notions of reason, of paraconsistent completion, and of safely supported proposition can become graded.

3   Possibilistic Logic Bases

We now assume that the propositions that are elements of a reason supporting a proposition may be pervaded with uncertainty. More precisely, the propositions p_i are now replaced by possibilistic logic formulas [8] of the form (p_i, a_i), i.e., p_i is believed with certainty at least a_i, a_i's belonging to a linearly ordered, bounded scale $S = \{s_1 = 1 > s_2 > \cdots > s_{n+1} = 0\}$, with top and bottom elements denoted by 1 and 0 respectively.

Let $\Sigma = \{(p_i, a_i) \mid i = 1, \cdots, m\}$, where a_i is the strength with which p_i is believed to be true in Σ. The higher a_i, the higher the strength. Thus, (p, a) is subsumed by (p, b) as soon as $b > a$. So, it is assumed that Σ does not contain both (p_i, a_i) and (p_j, a_j) with $p_i \equiv p_j$ and $a_i \neq a_j$. Let $\Sigma^* = \{p_i \mid (p_i, a_i) \in \Sigma\}$. Similarly, if $S \subseteq \Sigma$, S^* denotes the set of propositions appearing in the possibilistic formulas in S without their weight. The set of propositions Σ^* is *not* assumed to be consistent. In possibilistic logic, this amounts to say that the inconsistency level of Σ is strictly positive [8].

A subset S of Σ is said to be a *reason* for p iff (i) S^* is consistent; (ii) $\exists a > 0, S \vdash_\pi (p, a)$ where $\vdash_\pi$ is the possibilistic logic entailment[1]; (iii) $\nexists S' \subset S$ such that $S' \vdash_\pi (p, b)$ with $b > 0$.

In other words, S is such that S^* is a minimal consistent subset of propositions that entail p and a is the minimum of the weights of the formulas in S. a is the weight of the reason. Clearly there may exist distinct reasons S and S' (with $S^* \neq S'^*$) for p in Σ. Thus the pair $(S, (p, a))$ is a (possibilistic) argument for p with strength a, with $a = \min\{a_i \mid (p_i, a_i) \in S\}$.

3.1   Graded Paraconsistent Completions

On this basis, one can extend the completions Σ^{comp} and Σ^{para} to a possibilistic logic base Σ. Namely to each formula p_i in Σ^*, one may associate i) the set of reasons for p_i, and the set of reasons for $\neg p_i$, or ii) or only the weights of the best reason for p_i and of the best reason for $\neg p_i$.

More formally, the first one is defined by

$$\Sigma^{comp} = \{(p_i, \{P_1, \cdots, P_r\}, \{C_1, \cdots, C_s\}) \mid (p_i, a_i) \in \Sigma, P_i \text{ is a (graded)}$$
reason for p_i, C_j is a (graded) reason for $\neg p_i\}$.

The second completion is defined by

$$\Sigma^{para} = \{(p_i, \pi_i, \gamma_i) \mid (p_i, a_i) \in \Sigma, \pi_i \text{ is the greatest weight of a reason for } p_i$$
in Σ, γ_i is the greatest weight of a reason for $\neg p_i$ in $\Sigma\}$.

Note that $\pi_i \geq a_i$.

---

[1] Possibilistic inference is governed by the resolution rule $(\neg p \lor q, a), (p \lor r, b) \vdash_\pi (q \lor r, \min(a, b))$ [8].

Example. $\Sigma = \{(p, s_1), (\neg p \vee q, s_2), (\neg p, s_3), (\neg r, s_4), (r, s_5), (\neg r \vee q, s_6)\}$ (with $s_6 > 0$)

Then $\Sigma^{para} =$
$\{(p, s_1, s_3), (\neg p \vee q, s_2, 0), (\neg p, s_3, s_1), (\neg r, s_4, s_5), (r, s_5, s_4), (\neg r \vee q, s_2, 0)\}$.
$\Sigma^{comp} = \{(p, \{(p, s_1)\}, \{(\neg p, s_3)\}),$
$(\neg p \vee q, \{(\neg p \vee q, s_2)\}, \emptyset), (\neg p, \{(\neg p, s_3), \{(p, s_1)\}\}), (\neg r, \{(\neg r, s_4)\}, \{(r, s_5)\}),$
$(r, \{(r, s_5)\}, \{(\neg r, s_4)\}), (\neg r \vee p, \{\{(p, s_1), (\neg p \vee q, s_2)\}, \{(\neg r, s_4)\}\}, \emptyset)\}$.

3.2   Graded Safely Supported Propositions

The notion of safely supported proposition then extends to possibilistic propositional formulas with weights. Once Σ^{para} is built from Σ, one can evaluate reasons S in favor of p_i in the following way, by means of the two measures [3,5]:

- $Def(S) = \min\{\pi_i \mid ((p_i, \pi_i, \gamma_i) \in \Sigma^{para}$ and $p_i \in S^*\}$.
- $Uns(S) = \max\{\gamma_i \mid ((p_i, \pi_i, \gamma_i) \in \Sigma^{para}$ and $p_i \in S^*\}$

$Def(S)$ reflects the less certain belief in S, $Uns(S)$ the most strongly attacked belief in S. Note that we always have $Def(S) > 0$, but $Def(S)$ may be equal to s_n, and thus now "close to 0".

A reason is *free* iff $Def(S) > Uns(S)$, i.e. iff its certainty is above the strength of the strongest attack. Then a proposition p is *safely supported* if it exists a reason S that is free for it. It can be shown [5] that the set of safely supported consequences of a base Σ is always *consistent*. So in particular, p and $\neg p$ cannot be both safely supported.

It clearly generalizes the case of a binary scale, i.e. a scale S with only two levels 1 and 0 , (where the condition $Def(S) > Uns(S)$ can only hold under the form $Uns(S) = 0$), which means that all the formulas in S are fully believed and none is attacked. In the graded case, the formulas involved in S are only more believed than they are attacked.

Let us come back to minimal inconsistent subsets. Let S be a minimal inconsistent subset in Σ^*, and let $inc(S) = \min\{a_j \mid (p_j, a_j) \in \Sigma, p_j \in S\}$ be the level of inconsistency of S. Then, $inc(\Sigma) = \max\{inc(S) : S$ minimal inconsistent subset of $\Sigma\}$, where $inc(\Sigma) = \max\{a \mid \Sigma \vDash_\pi (\bot, a)\}$ and $\vDash_\pi$ is the standard possibilistic entailment defined by possibilistic resolution [8]. Moreover, it appears that if $(p_i, \pi_i, \gamma_i) \in \Sigma^{para}$, we have

$\gamma_i = \max\{inc(C_k) : (p_i, a_i) \in \Sigma, p_i \in C_k, C_k$ minimal inconsistent subset of $\Sigma\}$

with $inc(C_k) = \min\{a_j \mid (p_j, a_j) \in \Sigma, p_j \in C_k\}$. In fact we have the following result: *the safely supported entailment from Σ coincides with the possibilistic entailment from the consistent possibilistic logic base Σ^{cons} obtained from Σ by deleting, in all minimal inconsistent subsets S of Σ, the formulas with a certainty level equal to inc(S).* Namely $\Sigma^{cons} = \Sigma \setminus \{(p_i, a_i) \mid (p_i, a_i) \in S, S$ minimal inconsistent subset of $\Sigma, a_i = inc(S)\}$.

3.3   Lines for Further Research

The construction of Σ^{comp} and of Σ^{para} is reminiscent of the motivations of Belnap for introducing his well-known four-valued logic [1]. Belnap was considering several sources of information for which an atomic formula p may be known to be true, known to be false, or unknown. This may be naturally encoded by one of the four triples $(p, 1, 0)$ (p is held for true according to sources), $(p, 0, 1)$ (p is held for false according to sources), $(p, 1, 1)$ (this is the paraconsistent case p is true according to some sources and false according to others), and $(p, 0, 0)$ stands for the case where the truth status of p is unknown for sources. In Belnap's calculus $(p, 1, 1)$ and $(q, 0, 0)$ yields $(p \wedge q, 0, 1)$, which may appear strange at first glance. As pointed out in [7], this may be understood in the following way. On the one hand, we have both an argument in favor of p true and an argument in favor of p false. On the other hand we have no argument either in favor of q true or in favor of q false. This is enough to build an argument in favor of $p \wedge q$ false (from the argument in favor of p false) and we cannot build any argument in favor of $p \wedge q$ true (since one has no argument in favor of q true).

Yet, there already exists an extension of possibilistic logic inference that can be defined from Σ^{para} (and then extended to Σ^{comp}). It is based on the following generalized resolution rule [9] where the paraconsistency of formulas can be propagated

$$(\neg p \vee q, \pi_1, \gamma_1), (p \vee r, \pi_2, \gamma_2) \vdash (q \vee r, \min(\pi_1, \pi_2), \max(\gamma_1, \gamma_2)).$$

There is also another inference rule that holds in the logic of supporters [12], a logic closely related to possibilistic logic, which corresponds to the case where there are no reasons against in Σ^{comp}, and where the scale S is binary:

$$(\neg p \vee q, P_1), (\neg p \vee r, P_2) \vdash (\neg p \vee r, P_1 \cup P_2).$$

This rule was proposed moreover in an ATMS-like perspective, where two kinds of literals are distinguished, as in the following example:
Example. Given $Assumptions = \{A, B, C\}$, and the knowledge base $\Sigma = \{(p, A), (q, B), (\neg q \vee p, C)\}$, p in Σ^{comp} is then supported by two reasons, i.e., we have $(p, \{\{A\}, \{B, C, \}\}, \emptyset)$.

Such inference rules may provide the starting point for reasoning directly in terms of arguments, and not only about arguments.

4   Concluding Remarks

This short paper is intended to show that the benefit of taking into account the certainty levels of formulas when reasoning under inconsistency may be still much higher than the one already obtained by applying standard possibilistic logic where only formulas strictly above the inconsistency level of the knowledge base are salvaged. Indeed when inconsistency takes place, it is often due to the presence of formulas in which we are not fully confident. Considering minimal inconsistent subsets provides a local view of where the conflicts take place, and then the deletion of the less certain formulas inside these subsets enables us to restore consistency while keeping more information than with the standard possibilistic logic view.

References

1. Belnap, N.D.: A useful four-valued logic. In: Dunn, J.M., Epstein, G. (eds.) Modern Uses of Multiple-Valued Logic, pp. 7–37. D. Reidel, Dordrecht (1977)
2. Benferhat, S., Dubois, D., Prade, H.: Argumentative inference in uncertain and inconsistent knowledge base. In: Proceeding of the 9th Conference on Uncertainty in Artificial Intelligence, Washington, DC, 9–11 July, pp. 411–419. Morgan Kaufmann, San Mateo (1993)
3. Benferhat, S., Dubois, D., Prade, H.: Reasoning in inconsistent stratified knowledge bases. In: Proceeding of the 26 IEEE International Symposium on Multiple-Valued Logic (ISMVL 1996), Santiago de Compostela, Spain, pp. 184–189, 29–31 May 1996
4. Benferhat, S., Dubois, D., Prade, H.: Some syntactic approaches to the handling of inconsistent knowledge bases: a comparative study. Flat Case Stud. Logica **58**, 17–45 (1997)
5. Benferhat, S., Dubois, D., Prade, H.: An overview of inconsistency-tolerant inferences in prioritized knowledge bases. In: Dubois, D., Prade, H., Klement, E.P. (eds.) Fuzzy Sets, Logic and Reasoning about Knowledge, pp. 395–417. Kluwer Academic Publisher, Dordrecht (1999)
6. Besnard, P., Hunter, A. (eds.): Reasoning with Actual and Potential Contradictions:Handbook of Defeasible Reasoning and Uncertainty Management Systems., vol. 2. Kluwer, Dordrecht (1998)
7. Dubois, D.: On ignorance and contradiction considered as truth-values. Logic J. IGPL **16**(2), 195–216 (2008)
8. Dubois, D., Lang, J., Prade, H.: Possibilistic logic. In: Gabbay, D.M., Hogger, C.J., Robinson, J.A., Nute, D. (eds.) Handbook of Logic in Artificial Intelligence and Logic Programming, vol. 3, pp. 439–513. Oxford Sci. Publ, Oxford Univ. Press, New York (1994)
9. Dubois, D., Lang, J., Prade, H.: Handling uncertainty, context, vague predicates, and partial inconsistency in possibilistic logic. In: Driankov, D., L. Ralescu, A., Eklund, P. (eds.) IJCAI-WS 1991. LNCS, vol. 833, pp. 45–55. Springer, Heidelberg (1994)
10. Hunter, A., Konieczny, S.: On the measure of conflicts: Shapley inconsistency values. Artif. Intell. **174**(14), 1007–1026 (2010)
11. Jabbour, S., Ma, Y., Raddaoui, B., Sais, L.: Prime implicates based inconsistency characterization. In: Proceeding of the 21st European Conference on Artificial Intelligence (ECAI 2014), pp. 1037–1038. IOS Press, Prague (2014)
12. Lafage, C., Lang, J., Sabbadin, R.: A logic of supporters. In: Bouchon-Meunier, B., Yager, R.R., Zadeh, L.A. (eds.) Information, Uncertainty and Fusion, pp. 381–392. Kluwer, Dordrecht (1999)
13. Rescher, N., Manor, R.: On inference from inconsistent premises. Theor. Decis. **1**, 179–219 (1970)

First-Order Under-Approximations
of Consistent Query Answers

Floris Geerts[1], Fabian Pijcke[2], and Jef Wijsen[2(✉)]

[1] Universiteit Antwerpen, Antwerpen, Belgium
[2] Université de Mons, Mons, Belgium
jef.wijsen@umons.ac.be

Abstract. Consistent Query Answering (CQA) has by now been widely adopted as a principled approach for answering queries on inconsistent databases. The consistent answer to a query q on an inconsistent database **db** is the intersection of the answers to q on all repairs, where a repair is any consistent database that is maximally close to **db**. Unfortunately, computing consistent answers under primary key constraints has already exponential data complexity for very simple conjunctive queries, which is completely impracticable.

In this paper, we propose a new framework for divulging an inconsistent database to end users, which adopts two postulates. The first postulate complies with CQA and states that inconsistencies should never be divulged to end users. Therefore, end users should only get consistent query answers. The second postulate states that the data complexity of user queries must remain tractable (i.e., in **P** or even in **FO**). User queries with exponential data complexity will be rejected. We investigate which consistent query answers can still be obtained under such access postulates.

1 Introduction

Inconsistent, incomplete and uncertain data is widespread in the internet and social media era. This has given rise to a new paradigm for query answering, called *Consistent Query Answering* (CQA). This paradigm starts with the notion of *repair*, which is a new consistent database that minimally differs from the original inconsistent database. In general, an inconsistent database can have many repairs. In this respect, database repairing is different from data cleaning which aims at a unique cleaned database.

In this paper, we assume that the only constraints are primary keys, one per relation. A repair of an inconsistent database **db** is a maximal subset of **db** that satisfies all primary key constraints. Primary keys will be underlined. For example, the database of Fig. 1 stores ages and cities of residence of male and female persons. For simplicity, assume that persons have unique names (attribute N). Every person has exactly one age (attribute A) and city (attribute C). However, distinct tuples may agree on the primary key N, because there can be uncertainty about ages and cities. In the database of Fig. 1, there is uncertainty about

© Springer International Publishing Switzerland 2015
C. Beierle and A. Dekhtyar (Eds.): SUM 2015, LNAI 9310, pp. 354–367, 2015.
DOI: 10.1007/978-3-319-23540-0_24

M	N	A	C
	Ed	48	Mons
	Ed	48	Paris
	Dirk	29	Mons

F	N	A	C
	An	37	Mons
	Iris	37	Paris

Fig. 1. Example database with primary key violations.

the city of Ed (it can be Mons or Paris). The database can be repaired in two ways: delete either $M(\underline{Ed}, 48, Mons)$ or $M(\underline{Ed}, 48, Paris)$.

When database repairing results in multiple repairs, CQA shifts from standard semantics to certainty semantics. Given a query, the *certain answer* (also called *consistent answer*) is defined as the intersection of the answers on all repairs. That is, for a query q on an inconsistent database **db**, CQA replaces the standard query answer $q(\mathbf{db})$ with the certain answer, defined by the following intersection:

$$\bigcap \{q(\mathbf{r}) \mid \mathbf{r} \text{ is a repair of } \mathbf{db}\}. \tag{1}$$

Thus, the certainty semantics exclusively returns answers that hold true in every repair. Given a query q, we will denote by $\lfloor q \rfloor$ the query that maps a database to the answer defined by (1).

A practical obstacle to CQA is that the shift to certainty semantics involves a significant increase of complexity. When we refer to complexity in this paper, we mean data complexity, i.e., the complexity in terms of the size of the database (for a fixed query) [1, p. 422]. It is known for long [7] that there exist conjunctive queries q that join two relations such that the data complexity of $\lfloor q \rfloor$ is already **coNP**-hard. If this happens, CQA is completely impracticable.

This paper investigates ways to circumvent the high data complexity of CQA in a realistic setting, which is based on the following assumptions:

- If a query returns an answer to a user, then every tuple in that answer should belong to the certain answer. In Libkin's terminology [16], query answers must not contain *false positives*, i.e., tuples that are not certain.
- The only queries that can be executed in practice are those with data complexity in **P** or, even better, in **FO**. **FO** is the descriptive complexity class that captures all queries expressible in relational calculus.

Therefore, if the data complexity of a query $\lfloor q \rfloor$ is not in **P**, then the best we can go for is an approximation without false positives (also called under-approximation), computable in polynomial time. The term *strategy* will be used for queries that compute such approximations. Intuitively, a strategy can be regarded as a two-step process in which one starts by issuing a number of well-behaved queries $\lfloor q_i \rfloor$, for $i \in \{1, \ldots, \ell\}$, which can then be subject to a post-processing step. In this paper, well-behaved queries are those that are accepted by a query interface, e.g., self-join-free conjunctive queries q_i such that $\lfloor q_i \rfloor$ is in **FO**, and post-processing is formalized as queries built-up from the $\lfloor q_i \rfloor$'s.

We next illustrate our setting by an example. Consider the following scenario with two persons, called *Bob* and *Alice*. The person called *Bob* owns a database that is publicly accessible only via a query interface which restricts the syntax of the queries that can be asked. Our main results concern the case where the interface is restricted to self-join-free conjunctive queries. The database schema including all primary key constraints is publicly available. However, *Bob* is aware that his database contains many mistakes which should not be divulged. Therefore, whenever some end user asks a query q, *Bob* will actually execute the query $\lfloor q \rfloor$. That is, end users will get exclusively consistent answers. But, for feasibility reasons, *Bob* will reject any query q for which the data complexity of $\lfloor q \rfloor$ is too high. In this paper, we assume that *Bob* considers that data complexity is too high when it is beyond **FO**. The person called *Alice* interrogates *Bob*'s database, and she will be happy to get exclusively consistent answers. Unfortunately, her query q will be rejected by *Bob* if the data complexity of $\lfloor q \rfloor$ is too high (i.e., not in **FO**). If this happens, *Alice* has to change strategy. Instead of asking q, she can ask a finite number of queries $q_1, q_2, \ldots, q_\ell$ such that for every $i \in \{1, \ldots, \ell\}$, the data complexity of $\lfloor q_i \rfloor$ is in **FO**, and hence the query q_i will be accepted by *Bob*. No restriction is imposed on the number ℓ of queries that can be asked. The best *Alice* can hope for is that she can compute herself the answer to $\lfloor q \rfloor$ (or even to q) from *Bob*'s answers to $\lfloor q_1 \rfloor, \ldots, \lfloor q_\ell \rfloor$ by means of some post-processing. The question addressed in this paper is: Given that *Alice* wants to answer q, what queries should she ask to *Bob*?

Here is a concrete example. Assume *Bob* owns the database of Fig. 1. Interested in stable couples[1], *Alice* submits the query q_1 which asks "Get pairs of ages of men and women living in the same city":

$$q_1 = \{y, w \mid \exists x \exists u \exists z \, (M(\underline{x}, y, z) \wedge F(\underline{u}, w, z))\}.$$

The consistent answer is $\{(48, 37), (29, 37)\}$. However, the query $\lfloor q_1 \rfloor$ that returns the certain answer is known to have **coNP**-hard data complexity [13,14]. Therefore, *Bob* will reject q_1. *Alice* changes strategy and asks the query q_2 which asks "Get pairs of ages and city of men and women living in the same city":

$$q_2 = \{y, w, z \mid \exists x \exists u \, (M(\underline{x}, y, z) \wedge F(\underline{u}, w, z))\}. \tag{2}$$

Since the data complexity of $\lfloor q_2 \rfloor$ is known to be in **FO** [13,14], *Bob* will execute $\lfloor q_2 \rfloor$. The query q_2 returns $\{(29, 37, \text{Mons}), (48, 37, \text{Mons})\}$ on one repair, and $\{(29, 37, \text{Mons}), (48, 37, \text{Paris})\}$ on the other repair, so the certain answer is $\{(29, 37, \text{Mons})\}$. This in turn allows *Alice* to derive a certain answer to the original query: since $(29, 37, \text{Mons})$ belongs to the answer to $\lfloor q_2 \rfloor$, it is correct to conclude that $(29, 37)$ belongs to the answer to $\lfloor q_1 \rfloor$. An interesting question is whether *Alice* has a better strategy that divulges even more answers to $\lfloor q_1 \rfloor$.

The technical contributions of this paper are as follows. We first show that the following problem is undecidable: Given a relational calculus query q, is $\lfloor q \rfloor$ in **FO**? In view of this undecidability result, we then limit our attention to strategies that are first-order combinations (using disjunction and existential

---

[1] According to [6], marital stability is higher when the wife is 5+ years younger than her husband.

quantification) of queries $\lfloor q \rfloor$ that are known to be in **FO**. We show how to build optimal strategies under such syntax restrictions.

This paper is organized as follows. Section 2 discusses related work. Section 3 provides some mathematical definitions. Section 4 introduces our new framework for studying consistent query answering under primary key constraints, and introduces the problem OPTSTRATEGY. Intuitively, OPTSTRATEGY asks, given a query q, to find a new query q' that gets the largest subset of consistent answers while still obeying the restrictions imposed by our framework. Section 5 provides ways to solve OPTSTRATEGY in restricted settings. Finally, Sect. 6 concludes the paper.

2 Related Work

Consistent query answering (CQA) was proposed in [2] as a principled approach to handle data quality problems that arise from violations of integrity constraints. See the textbooks [3,10] for comprehensive overviews of these domains.

Fuxman and Miller [11] were the first ones to focus on CQA under the restrictions that consistency is only with respect to primary keys and that queries are self-join-free conjunctive. See [21] for a survey on consistent query answering to conjunctive queries under primary key constraints. Some recent results not covered by this survey can be found in [13,14].

Instead of returning the query answers true in every repair, one could return the query answers true in, e.g., a majority of repairs. This leads to the counting variant of CQA, which has been studied in [17,18]. As observed in [20], the counting variant of CQA under primary key constraints is closely related to query answering in block-independent-disjoint (BID) probabilistic databases [8,9]. Alternatively, one can obtain approximations by restricting the set of repairs. This approach has been considered in [5] in the setting of ontology-based data access.

Our work can also be regarded as querying "consistent views," in the sense that *Bob* returns exclusively consistent answers. It has been observed long ago [19] that consistent views are not closed under relational calculus. In other words, the position of the $\lfloor \cdot \rfloor$ construct in a query does matter. For example, for the database of Fig. 1, the query $\{x \mid \exists y \exists z \lfloor M(\underline{x}, y, z) \rfloor\}$ returns only Dirk, while $\lfloor \{x \mid \exists y \exists z M(\underline{x}, y, z)\} \rfloor$ returns both Ed and Dirk. Bertossi and Li [4] have used views to protect the secrecy of data in a database. In our setting, the query answers that are to be hidden from end users are those that are not true in every repair.

3 Preliminaries

We assume disjoint sets of *variables* and *constants*. If $\boldsymbol{x}$ is a sequence containing variables and constants, then $\mathsf{vars}(\boldsymbol{x})$ denotes the set of variables that occur in $\boldsymbol{x}$. A *valuation* over a set U of variables is a total mapping θ from U to the set of constants.

Atoms and Key-equal Facts. Each *relation name* R of arity n, $n \geq 1$, has a unique *primary key* which is a set $\{1, 2, \ldots, k\}$ where $1 \leq k \leq n$. We say that R has *signature* $[n, k]$ if R has arity n and primary key $\{1, 2, \ldots, k\}$. We say that R is *all-key* if $n = k$. For all positive integers n, k such that $1 \leq k \leq n$, we assume denumerably many relation names with signature $[n, k]$.

If R is a relation name with signature $[n, k]$, then $R(s_1, \ldots, s_n)$ is called an *R-atom* (or simply atom), where each s_i is either a constant or a variable $(1 \leq i \leq n)$. Such an atom is commonly written as $R(\underline{x}, y)$ where the primary-key value $x = s_1, \ldots, s_k$ is underlined and $y = s_{k+1}, \ldots, s_n$. An *R-fact* (or simply fact) is an R-atom in which no variable occurs. Two facts $R_1(\underline{a_1}, b_1), R_2(\underline{a_2}, b_2)$ are *key-equal* if $R_1 = R_2$ and $a_1 = a_2$.

We will use letters F, G, H for atoms. For an atom $F = R(\underline{x}, y)$, we denote by $\mathsf{key}(F)$ the set of variables that occur in x, and by $\mathsf{vars}(F)$ the set of variables that occur in F, that is, $\mathsf{key}(F) = \mathsf{vars}(x)$ and $\mathsf{vars}(F) = \mathsf{vars}(x) \cup \mathsf{vars}(y)$.

Uncertain Databases, Blocks, and Repairs. A *database schema* is a finite set of relation names. All constructs that follow are defined relative to a fixed database schema.

A *database* is a finite set **db** of facts using only the relation names of the schema. We often refer to databases as "uncertain databases" to stress that such databases can violate primary key constraints.

A *block* of **db** is a maximal set of key-equal facts of **db**. The term R-block refers to a block of R-facts, i.e., facts with relation name R. An uncertain database **db** is *consistent* if no two distinct facts are key-equal (i.e., if every block of **db** is a singleton). A *repair* of **db** is a maximal (with respect to set containment) consistent subset of **db**. We write $\mathsf{rset}(\mathbf{db})$ for the set of repairs of **db**.

Queries and Consistent Query Answering. We assume that the reader is familiar with *relational calculus* [1, Chapter 5] and with the notion of *queries* [15, Definition 2.7]. By **FO**, we denote the descriptive complexity class that contains the queries expressible in relational calculus.

For every m-ary $(m \geq 0)$ relational calculus query q, we define $\lfloor q \rfloor$ as the m-ary query that maps every database **db** to $\bigcap \{q(\mathbf{r}) \mid \mathbf{r} \in \mathsf{rset}(\mathbf{db})\}$. Clearly, if **db** is a consistent database, then $\lfloor q \rfloor(\mathbf{db}) = q(\mathbf{db})$.

Given two m-ary queries q_1 and q_2, we say that q_1 is *contained in* q_2, denoted by $q_1 \sqsubseteq q_2$, if for every database **db**, $q_1(\mathbf{db}) \subseteq q_2(\mathbf{db})$. We write $q_1 \sqsubset q_2$ if $q_1 \sqsubseteq q_2$ and $q_2 \not\sqsubseteq q_1$. We say that q_1 and q_2 are *equivalent*, denoted by $q_1 \equiv q_2$, if $q_1 \sqsubseteq q_2$ and $q_2 \sqsubseteq q_1$.

A 0-ary query is called *Boolean*. If q is a Boolean query, then q maps any database to either $\{\langle\rangle\}$ or $\{\}$, corresponding to **true** and **false** respectively.

A *conjunctive query* is a relational calculus query of the form $\{z \mid \exists y B\}$ where B is a conjunction of atoms. The conjunction B and the query are said to be *self-join-free* if no relation name occurs more than once in B. We write $\mathsf{vars}(B)$ for the set of variables that occur in B. By a slight abuse of notation, we denote by B also the set of conjuncts that occur in B. For example, if $B_1 = R(x) \wedge R(x) \wedge R(y)$ and $B_2 = R(x) \wedge R(y) \wedge R(z)$, then we may write $B_1 \subseteq B_2$.

Significantly, the following example shows that $\lfloor q \rfloor$ may not be expressible in relational calculus, even if q is self-join-free conjunctive.

Example 1. Let $q_1 = \{\langle\rangle \mid \exists x \exists y \exists z \, (R(\underline{x}, z) \land S(\underline{y}, z))\}$. The query q_1 is self-join-free conjunctive. It follows from [13] that $\lfloor q_1 \rfloor$ is not in **FO** (i.e., not expressible in relational calculus).

Let $q_2 = \{\langle\rangle \mid \exists x \exists y \, (R(\underline{x}, y) \land S(\underline{y}, b))\}$, where b is a constant. Then, $\lfloor q_2 \rfloor$ is equivalent to the following relational calculus query:

$$\exists x \exists y (R(\underline{x}, y) \land \\ \forall y \, (R(\underline{x}, y) \to (S(\underline{y}, b) \land \forall z \, (S(\underline{y}, z) \to z = b)))) .$$ □

4   A Framework for Divulging Inconsistent Databases

In this section, we formalize the setting that was described and illustrated in Sect. 1. The setting is captured by the language called **CQAFO**, which consists of first-order quantification and Boolean combinations of atomic formulas of the form $\lfloor q \rfloor$, where q is any relational calculus query. The atomic formulas $\lfloor q \rfloor$ capture that the database owner *Bob* only returns certain answers. Subsequently, the end user *Alice*, who interrogates *Bob*'s database, can do some post-processing on *Bob*'s outputs. In our setting, we assume that *Alice* uses first-order quantification and Boolean combinations of *Bob*'s answers.

Example 2. The scenario in Sect. 1 is captured by the **CQAFO** query

$$\{y, w \mid \exists Z \lfloor \exists x \exists u \, (M(\underline{x}, y, Z) \land F(\underline{u}, w, Z)) \rfloor\}.$$

The formula within $\lfloor \cdot \rfloor$ is the query (2). The quantification $\exists Z$ corresponds to *Alice* projecting away the cities column returned by *Bob*. For readability, we will often use upper case letters for variables that are quantified outside the range of $\lfloor \cdot \rfloor$. □

Example 3. The following query allows *Alice* to find the names of men with more than two cities in the database:

$$\{x \mid \lfloor \exists y \exists z M(\underline{x}, y, z) \rfloor \land \neg \exists Z \lfloor \exists y M(\underline{x}, y, Z) \rfloor\}.$$

To understand this query, it may be helpful to notice that $\{x, Z \mid \lfloor \exists y M(\underline{x}, y, Z) \rfloor\}$ returns tuple (n, c) whenever c is the only city of residence encoded for the person named n.   □

4.1   The Language **CQAFO**

Syntax of **CQAFO**

- If q is a relational calculus query, then $\lfloor q \rfloor$ is a **CQAFO** formula.
- If φ_1 and φ_2 are **CQAFO** formulas, then $\varphi_1 \land \varphi_2$, $\varphi_1 \lor \varphi_2$, and $\neg \varphi_1$ are **CQAFO** formulas.

– If φ is a CQAFO formula, then $\exists x\varphi$ and $\forall x\varphi$ are CQAFO formulas.

If φ is a CQAFO formula, then free(φ) denotes the set of free variables of φ (i.e., the variables not bound by a quantifier). If x is a tuple containing the free variables of φ, we write $\varphi(x)$.

A CQAFO query is an expression of the form $\{x \mid \varphi\}$, where x is a sequence of variables and constants containing each variable of free(φ). If x contains no constants and no double occurrences of the same variable, then such query is also denoted $\varphi(x)$.

Semantics. Let **db** be an uncertain database. Let $\varphi(x)$ be a CQAFO formula, and a be a sequence of constants (of same length as x). We inductively define $\mathbf{db} \models \varphi(a)$.

– If $\varphi(x) = \lfloor q(x) \rfloor$ for some relational calculus query $q(x)$, then $\mathbf{db} \models \varphi(a)$ if for every repair $\mathbf{r}$ of **db**, $\mathbf{r} \models q(a);^2$
– $\mathbf{db} \models \neg\varphi(a)$ if $\mathbf{db} \not\models \varphi(a)$;
– $\mathbf{db} \models \varphi_1 \wedge \varphi_2$ if $\mathbf{db} \models \varphi_1$ and $\mathbf{db} \models \varphi_2$;
– $\mathbf{db} \models \varphi_1 \vee \varphi_2$ if $\mathbf{db} \models \varphi_1$ or $\mathbf{db} \models \varphi_2$;
– if $\psi(x) = \exists y\varphi(y, x)$, then $\mathbf{db} \models \psi(a)$ if $\mathbf{db} \models \varphi(a', a)$ for some a';
– if $\psi(x) = \forall y\varphi(y, x)$, then $\mathbf{db} \models \psi(a)$ if $\mathbf{db} \models \varphi(a', a)$ for all a'.

Let $Q = \{x' \mid \varphi(x)\}$ be a CQAFO query. The answer $Q(\mathbf{db})$ is the smallest set containing $\theta(x')$ for every valuation θ over vars(x) such that for some a, $\theta(x) = a$ and $\mathbf{db} \models \varphi(a)$. Notice that vars($x'$) = vars($x$), but x', unlike x, can contain constants and multiple occurrences of the same variable. If x' contains no variables, then Q is Boolean.

4.2   Restrictions on Data Complexity

The language CQAFO of Sect. 4.1 captures our first postulate which states that the database owner *Bob* returns exclusively certain answers. But we do not prohibit that end user *Alice* does some post-processing on *Bob*'s answers. In this section, we will add our second postulate which states that *Bob* rejects queries q if the data complexity of $\lfloor q \rfloor$ is not in **FO**. Unfortunately, *Bob* has to face the following undecidability result.

Theorem 1. *The following problem is undecidable. Given a relational calculus query q, is $\lfloor q \rfloor$ in **FO**?*

Proof. Let $q_1 = \{\langle\rangle \mid \exists x \exists y \exists z \, (R(\underline{x}, z) \wedge S(\underline{y}, z) \wedge \varphi)\}$ where φ is a closed relational calculus formula such that all relation names in φ are all-key. We show hereinafter that $\lfloor q \rfloor$ is in **FO** if and only if φ is unsatisfiable. The desired result then follows by [1, Theorem 6.3.1], which states that (finite) satisfiability of relational calculus queries is undecidable.

Obviously, if φ is unsatisfiable, then $\lfloor q_1 \rfloor \equiv \mathbf{false}$, and hence $\lfloor q_1 \rfloor$ is in **FO**.

---

^{2} $\mathbf{r} \models q(a)$ is defined in the standard way.

We show next that if φ is satisfiable, then $\lfloor q_1 \rfloor$ is not in **FO**. Assume that φ is satisfiable. Let $q_0 = \exists x \exists y \exists z \left(R(\underline{x}, z) \wedge S(\underline{y}, z) \right)$. Let CERTAIN0 and CERTAIN1 be the problems defined next.

– CERTAIN0: Given a database **db**, determine whether every repair of **db** satisfies q_0.
– CERTAIN1: Given a database **db**, determine whether every repair of **db** satisfies q_1.

Let $\mathbf{db}_0$ be a database that is input to CERTAIN0. We show a polynomial-time many-one reduction from CERTAIN0 to CERTAIN1. Let **S** be the database schema that contains the relation names occurring in φ. An algorithm can consider systematically every finite database $\mathbf{db}'$ over **S** and test $\mathbf{db}' \models \varphi$, until a database $\mathbf{db}'$ is found such that $\mathbf{db}' \models \varphi$. The algorithm terminates because φ is satisfiable. Since the computation of $\mathbf{db}'$ does not depend on $\mathbf{db}_0$, it takes $\mathcal{O}(1)$ time. Since all relation names in $\mathbf{db}'$ are all-key, we have that $\mathbf{db}'$ is consistent. Clearly, q_0 is true in every repair of $\mathbf{db}_0$ if and only if q_1 is true in every repair of $\mathbf{db}_0 \cup \mathbf{db}'$. So we have established a polynomial-time many-one reduction from CERTAIN0 to CERTAIN1. Since CERTAIN0 is **coNP**-hard [13], it follows that CERTAIN1 is **coNP**-hard. Since **FO** $\subsetneq$ **coNP** [12], it follows that CERTAIN1 is not in **FO**. □

By Theorem 1, there exists no algorithm that allows *Bob* to decide whether he has to accept or reject a relational calculus query. In general, little is known about the complexity of $\lfloor q \rfloor$ for relational calculus queries q. One of the stronger known results is the following.

Theorem 2. ([13]). *The following problem is decidable in polynomial time. Given a self-join-free conjunctive query q, is $\lfloor q \rfloor$ in* **FO***? Moreover, if $\lfloor q \rfloor$ is in* **FO***, then a relational calculus query equivalent to $\lfloor q \rfloor$ can be effectively constructed.*

In view of Theorems 1 and 2, the following scenario is the best we can go for with the current state of art.

1. The database owner *Bob* only accepts self-join-free conjunctive queries q such that $\lfloor q \rfloor$ is in **FO**. Thus, *Bob* rejects every query that is not self-join-free conjunctive, and rejects a self-join-free conjunctive query q if $\lfloor q \rfloor$ is not in **FO**.
2. As before, *Alice* can do some first-order post-processing on the answers obtained from *Bob*.

Under these restrictions, we focus on the following research task: given that *Alice* wants to answer a self-join-free conjunctive query q on a database owned by *Bob*, develop a *strategy* for *Alice* to get a subset (the greater, the better) of certain answers. Our framework applies to Boolean queries by representing **true** and **false** by $\{\langle\rangle\}$ and $\{\}$ respectively. A formal definition follows.

4.3   Strategies

Strategies for a query q are defined next as relational calculus queries that can be expressed in CQAFO and that are contained in $\lfloor q \rfloor$.

Definition 1. *Let q be a self-join-free conjunctive query. A strategy for q is a CQAFO query φ such that $\varphi \sqsubseteq \lfloor q \rfloor$ and for every atomic formula $\lfloor q' \rfloor$ in φ, we have that q' is a self-join-free conjunctive query such that $\lfloor q' \rfloor$ is in* **FO**.

A strategy φ for q is optimal *if for every strategy ψ for q, we have $\psi \sqsubseteq \varphi$. The problem* OPTSTRATEGY *takes in a self-join-free conjunctive query q and asks to determine an optimal strategy for q.*

Some observations are in place.

- If the input to OPTSTRATEGY is a self-join-free conjunctive q such that $\lfloor q \rfloor$ is in **FO**, then the CQAFO query $\lfloor q \rfloor$ is itself an optimal strategy.
- Every strategy φ is in **FO**, because all atomic formulas $\lfloor q' \rfloor$ are required to be in **FO**. Therefore, if *Alice* wants to answer a query q such that $\lfloor q \rfloor$ is not in **FO**, then there is no strategy φ such that $\varphi \equiv \lfloor q \rfloor$.
- There is no fundamental reason why the input query to OPTSTRATEGY is required to be self-join-free conjunctive query. However, developing strategies for more expressive queries is left as an open question.

5   How to Construct Good Strategies?

Let q be a self-join-free conjunctive query. In this section, we investigate ways for constructing good (if not optimal) strategies for q of a particular syntax. In Sect. 5.1, we take the most simple approach: take the union of queries $\lfloor q_i \rfloor$ contained in $\lfloor q \rfloor$, where q_i is self-join-free conjunctive and $\lfloor q_i \rfloor$ is in **FO**. We then show that the strategies obtained in this way cannot be optimal. Therefore, an enhanced approach is developed in Sect. 5.2.

5.1   Post-processing by Unions only

Assume that the input to OPTSTRATEGY is a self-join-free conjunctive query $q(z)$. In this section, we look at strategies of the form

$$\bigcup_{i=1}^{\ell} \lfloor q_i \rfloor, \tag{3}$$

where each q_i is of the form $\{z_i \mid \exists y_i B_i\}$ in which z_i has same length as z and B_i is a self-join-free conjunction of atoms.[3]

We use union (with its standard semantics) instead of disjunction to avoid notational difficulties. For example, the union

$$\{x, a \mid \lfloor R(\underline{x}, a) \rfloor\} \cup \{x, y \mid \lfloor S(\underline{x}, y) \rfloor\},$$

---

[3] Notice that is can be easily verified that $\lfloor \{z_i \mid \exists y_i B_i\} \rfloor \equiv \{z_i \mid \lfloor \exists y_i B_i \rfloor\}$.

where a is a constant, is semantically clear, and is equivalent to

$$\{x, y \mid \lfloor R(\underline{x}, y) \wedge y = a \rfloor \vee \lfloor S(\underline{x}, y) \rfloor\},$$

in which equality is needed. It would be wrong to write $\{x, y \mid \lfloor R(\underline{x}, a) \rfloor \vee \lfloor S(\underline{x}, y) \rfloor\}$, an expression that is even not domain independent [1, p. 79].

Clearly, a formula of the form (3) is a strategy if for every $i \in \{1, \ldots, \ell\}$, $\lfloor q_i \rfloor$ is in **FO** and $\lfloor q_i \rfloor \sqsubseteq \lfloor q \rfloor$. The latter condition is equivalent to $q_i \sqsubseteq q$ as shown next.

Lemma 1. *Let q and q' be self-join-free m-ary conjunctive queries. Then, $q \sqsubseteq q'$ if and only if $\lfloor q \rfloor \sqsubseteq \lfloor q' \rfloor$.*

Proof. Let $q = \{z \mid \exists y B\}$ and $q' = \{z' \mid \exists y' B'\}$, where z and z' both have the same length m.
$\boxed{\Longrightarrow}$ Straightforward. $\boxed{\Longleftarrow}$ Assume $\lfloor q \rfloor \sqsubseteq \lfloor q' \rfloor$. Let μ be an injective mapping with domain $\mathsf{vars}(B)$ that maps each variable to a fresh constant not occurring elsewhere. Since μ is injective, its inverse μ^{-1} is well defined. Let $\mathbf{db} = \mu(B)$. Clearly, $\mathbf{db}$ is consistent and $q(\mathbf{db}) = \{\mu(z)\} = \lfloor q \rfloor(\mathbf{db})$. From $\lfloor q \rfloor \sqsubseteq \lfloor q' \rfloor$, it follows $\mu(z) \in q'(\mathbf{db}) = \lfloor q' \rfloor(\mathbf{db})$. Then, there exists a valuation θ over $\mathsf{vars}(B')$ such that $\theta(B') \subseteq \mathbf{db}$ and $\theta(z') = \mu(z)$. Then $\mu^{-1} \circ \theta(B') \subseteq B$ and $\mu^{-1} \circ \theta(z') = z$. Since $\mu^{-1} \circ \theta$ is a homomorphism from q' to q, it follows $q \sqsubseteq q'$ by the Homomorphism Theorem [1, Theorem 6.2.3]. $\square$

Lemma 1 does not hold for conjunctive queries with self-joins, as shown next.

Example 4. Let $q = \{\langle\rangle \mid R(\underline{a}, b) \wedge R(\underline{a}, c)\}$. For every uncertain database $\mathbf{db}$, $\lfloor q \rfloor(\mathbf{db}) = \{\}$. Let q' be a query such that $q \not\sqsubseteq q'$ (such query obviously exists). Then, $\lfloor q \rfloor \sqsubseteq \lfloor q' \rfloor$ and $q \not\sqsubseteq q'$. $\square$

Lemma 1 allows us to construct strategies of the form (3), as follows. Assume that the input to **OPTSTRATEGY** is a self-join-free conjunctive query $q(z)$. For some positive integer ℓ, generate self-join-free conjunctive queries $q_1, \ldots, q_\ell$ such that for each $i \in \{1, \ldots, \ell\}$, $q_i \sqsubseteq q$ and $\lfloor q_i \rfloor$ is in **FO**. The condition $q_i \sqsubseteq q$ is decidable by [1, Theorem 6.2.3]; the condition that $\lfloor q_i \rfloor$ is in **FO** is decidable by Theorem 2. Then by Lemma 1, $\bigcup_{i=1}^{\ell} \lfloor q_i \rfloor$ is a strategy for q.

Unfortunately, Theorem 3 given hereinafter states that there are cases where no strategy of the form (3) is optimal. We first generalize Lemma 1 to unions.

Lemma 2. *Let $q_0, q_1, \ldots q_\ell$ be self-join-free m-ary conjunctive queries. Then, $\lfloor q_0 \rfloor \sqsubseteq \bigcup_{i=1}^{\ell} \lfloor q_i \rfloor$ if and only if for some $i \in \{1, \ldots, \ell\}$, $q_0 \sqsubseteq q_i$.*

Proof. $\boxed{\Longleftarrow}$ Straightforward. $\boxed{\Longrightarrow}$ Assume $\lfloor q_0 \rfloor \sqsubseteq \bigcup_{i=1}^{\ell} \lfloor q_i \rfloor$. Let $q_0 = \{z_0 \mid \exists y_0 B_0\}$, where B_0 is self-join-free. Let μ be an injective mapping with domain $\mathsf{vars}(B_0)$ that maps each variable to a fresh constant not occurring elsewhere. Since μ is injective, its inverse μ^{-1} is well defined. Let $\mathbf{db} = \mu(B_0)$. Clearly, $\mathbf{db}$ is consistent and $q_0(\mathbf{db}) = \{\mu(z_0)\} = \lfloor q_0 \rfloor(\mathbf{db})$. From $\lfloor q_0 \rfloor \sqsubseteq \bigcup_{i=1}^{\ell} \lfloor q_i \rfloor$, it follows that we can assume $i \in \{1, \ldots, \ell\}$ such that $\mu(z_0) \in q_i(\mathbf{db}) = \lfloor q_i \rfloor(\mathbf{db})$.

Let $q_i = \{z_i \mid \exists y_i B_i\}$. Then, there exists a valuation θ over $\mathsf{vars}(B_i)$ such that $\theta(B_i) \subseteq \mathbf{db}$ and $\theta(z_i) = \mu(z_0)$. Then $\mu^{-1} \circ \theta(B_i) \subseteq B_0$ and $\mu^{-1} \circ \theta(z_i) = z_0$. Since $\mu^{-1} \circ \theta$ is a homomorphism from q_i to q_0, it follows $q_0 \sqsubseteq q_i$.   □

Theorem 3. *There exists a self-join-free conjunctive query q such that for every strategy φ of the form (3) for q, there exists another strategy ψ of the form (3) for q such that $\varphi \sqsubset \psi$.*

Proof. Let $q = \{\langle\rangle \mid \exists x \exists y \exists z \left(R(\underline{x}, z) \wedge S(\underline{y}, z)\right)\}$. Then $\lfloor q \rfloor$ is not in **FO** [14]. For every constant c, let q_c be the query defined by $\{\langle\rangle \mid \exists y \exists z \left(R(\underline{c}, z) \wedge S(\underline{y}, z)\right)\}$. For every constant c, we have that $\lfloor q_c \rfloor \sqsubseteq \lfloor q \rfloor$ and $\lfloor q_c \rfloor$ is in **FO**.

Let φ be a strategy for q of the form (3). Let A be the greatest set of constants such that for all $c \in A$, there exists some $i \in \{1, \dots, \ell\}$ such that $q_i \equiv q_c$. Let b be a constant such that $b \notin A$. Clearly $\varphi \sqsubseteq \varphi \cup \lfloor q_b \rfloor \sqsubseteq \lfloor q \rfloor$. It suffices to show that $\varphi \sqsubset \varphi \cup \lfloor q_b \rfloor$, meaning that φ is not optimal.

Assume towards a contradiction that $\lfloor q_b \rfloor \sqsubseteq \varphi$. By Lemma 2, there exists $i \in \{1, \dots, \ell\}$ such that $q_b \sqsubseteq q_i \sqsubseteq q$. Let q_i be the existential closure of $(R(\underline{s}, t) \wedge S(\underline{u}, v))$. From $q_i \sqsubseteq q$, it follows that $t = v$. From $q_b \sqsubseteq q_i$ and $b \notin A$, it follows that s, t, u are pairwise distinct variables. But then $q_i \equiv q$, contradicting that $\lfloor q_i \rfloor$ is in **FO**. We conclude by contradiction that $\varphi \sqsubset \varphi \cup \lfloor q_b \rfloor$.   □

5.2   Post-processing by Unions and Quantification

The proof of Theorem 3 indicates that strategies of the form (3) lack expressiveness because the number of constants in such strategies is bounded. An obvious extension is to look for strategies that replace constants with existentially quantified variables. The following example shows how such extension solves the lack of expressiveness that underlies the proof of Theorem 3.

Example 5. Let $q = \exists x \exists y \exists z \left(R(\underline{x}, z) \wedge S(\underline{y}, z)\right)$. Let φ be the **CQAFO** formula defined by $\varphi := \exists X \lfloor \exists y \exists z \left(R(\underline{X}, z) \wedge S(\underline{y}, z)\right) \rfloor$. It can be shown that φ is a strategy for q, i.e., $\varphi \sqsubseteq \lfloor q \rfloor$ and $\lfloor \exists y \exists z \left(R(\underline{X}, z) \wedge S(\underline{y}, z)\right) \rfloor$ is in **FO**. Recall from Example 2 that the use of upper case X is for readability.   □

Assume that the input to **OPTSTRATEGY** is a self-join-free conjunctive query $q(z)$. In this section, we investigate strategies of the form

$$\bigcup_{i=1}^{\ell} Q_i, \tag{4}$$

where for each $i \in \{1 \dots, \ell\}$, Q_i is a **CQAFO** query of the form

$$\{z_i \mid \exists X_i \lfloor \exists y_i B_i \rfloor\}, \tag{5}$$

in which z_i has the same length as z, and B_i is a self-join-free conjunction of atoms. It is understood that z_i, X_i, and y_i have, pairwise, no variables in common, and that $\mathsf{vars}(z_i X_i y_i) = \mathsf{vars}(B_i)$. For readability, we will use upper case Q to refer to **CQAFO** queries of the form (5). The main tools for constructing strategies of the form (4) are provided by Theorems 4 and 5.

Theorem 4. *The following problem is decidable in polynomial time. Given a* CQAFO *query Q of the form (5), is Q in* **FO**? *Moreover, if Q is in* **FO**, *then a relational calculus query equivalent to Q can be effectively constructed.*

Proof. A CQAFO query Q of the form (5) is in **FO** if and only if $\lfloor \exists \boldsymbol{y}_i B_i \rfloor$ is in **FO**. The latter condition is decidable by Theorem 2.

Theorem 5. *Given a self-join-free conjunctive query q_1 and a* CQAFO *query Q_2 of the form (5), it can be decided whether $Q_2 \sqsubseteq \lfloor q_1 \rfloor$.*

Proof. (Crux.) Let $q_1 = \{\boldsymbol{z}_1 \mid \exists \boldsymbol{y}_1 B_1\}$ and $Q_2 = \{\boldsymbol{z}_2 \mid \exists \boldsymbol{X}_2 \lfloor \exists \boldsymbol{y}_2 B_2 \rfloor\}$. It can be shown that $Q_2 \sqsubseteq \lfloor q_1 \rfloor$ if and only if there exists a valuation θ over $\mathsf{vars}(B_1)$ such that $\theta(\boldsymbol{z}_1) = \boldsymbol{z}_2$ and $\theta(B_1) \subseteq B_2$. □

We point out that Theorem 5 is interesting in its own right. It is well known [1, Corollary 6.3.2] that containment of relational calculus queries is undecidable. A large fragment for which containment is decidable is the class of unions of conjunctive queries. Notice, however, that the queries in the statement of Theorem 5 need not be monotone (and even not first-order), and that decidability of query containment for such queries is not obvious.

Example 6. Let $Q = \{x \mid \exists Y \lfloor R(\underline{x}, Y) \rfloor\}$. Let $\mathbf{db} = \{R(\underline{a}, 1)\}$ and $\mathbf{db}' = \{R(\underline{a}, 1), R(\underline{a}, 2)\}$. Then $\mathbf{db} \subseteq \mathbf{db}'$, but $Q(\mathbf{db}) = \{a\}$ is not contained in $Q(\mathbf{db}') = \{\}$. Hence Q is not monotone. We have that Q is equivalent to the following relational calculus query:

$$\{x \mid \exists y \, (R(\underline{x}, y) \wedge \forall y' \, (R(\underline{x}, y') \to y = y'))\}. \qquad \square$$

Assume that the input to OPTSTRATEGY is a self-join-free conjunctive query $q(\boldsymbol{z})$. Theorem 5 allows us to build a strategy of the form (4) for q as follows. Let A be the set of constants that occur in q. Let φ be the disjunction of all (up to variable renaming) CQAFO formulas Q_i of the form (5) that use exclusively constants from A such that $Q_i \sqsubseteq \lfloor q \rfloor$ and Q_i is in **FO**. Clearly, there are at most finitely many such formulas (up to variable renaming). Containment of Q_i in $\lfloor q \rfloor$ is decidable by Theorem 5. Finally, the condition that Q_i is in **FO** is decidable by Theorem 4. The following theorem remedies the negative result of Theorem 3.

Theorem 6. *For every self-join-free conjunctive query q, there exists a computable strategy φ of the form (4) for q, such that for every strategy ψ of the form (4) for q, $\psi \sqsubseteq \varphi$.*

Proof. Assume that the input to OPTSTRATEGY is a self-join-free conjunctive query $q(\boldsymbol{z})$. Let φ be the strategy defined in the paragraph preceding this theorem. Let $Q = \{\boldsymbol{z}_0 \mid \exists \boldsymbol{X} \lfloor \exists \boldsymbol{y} B \rfloor\}$ be a query of the form (5) where B is a self-join-free conjunction of atoms such that Q is in **FO** and $Q \sqsubseteq \lfloor q \rfloor$. If all constants that occur in B also occur in q, then Q is already contained in some disjunct of φ (by construction of φ). Assume next that B contains some constants that do not occur in q, and let these constants be $a_1, \ldots, a_m$. For $i \in \{1, \ldots, m\}$, let X_i

be a new fresh variable. Let B' be the conjunction obtained from B by replacing each occurrence of each a_i with X_i. Let $Q' = \{z_0 \mid \exists \boldsymbol{X} \exists X_1 \cdots \exists X_m \lfloor \exists \boldsymbol{y} B' \rfloor\}$. From the proof of Theorem 2, it follows $Q' \sqsubseteq \lfloor q \rfloor$. It can be easily seen that $Q \sqsubseteq Q'$. Furthermore, from [13], it follows that Q' is in **FO**. Since all constants that occur in B' also occur in q, we have that Q' is already contained in some disjunct of φ (by construction of φ).

To conclude, whenever $Q = \{z_0 \mid \exists \boldsymbol{X} \lfloor \exists \boldsymbol{y} B \rfloor\}$ is a query of the form (5) where B is a self-join-free conjunction of atoms such that Q is in **FO** and $Q \sqsubseteq \lfloor q \rfloor$, we have that $\varphi \cup Q \sqsubseteq \varphi$. $\qquad\square$

So far, we have imposed no restrictions on the size of the computable strategy φ in the statement of Theorem 6. From a practical point of view, it is interesting to construct, among all optimal strategies φ of the form (4), the one with the smallest number ℓ of disjuncts. It is an open question, however, how to minimize strategies of the form (4).

6   Conclusion

We have studied a realistic setting for divulging an inconsistent database to end users. In this setting, users access the database exclusively via syntactically restricted queries, and get exclusively consistent answers computable in **FO** data complexity. If the data complexity is higher, then the query will be rejected, in which case users have to fall back on strategies that obtain a large (the larger, the better) subset of the consistent answer. Such strategies combine answers obtained from several "easier" queries.

Although our setting applies to arbitrary queries and constraints, we searched for strategies when constraints are primary keys, and the database is accessible only via self-join-free conjunctive queries for which consistent query answering is in **FO**. Under these access restrictions, we showed how to construct strategies that combine answers by means of union and quantification. It is an open question whether our strategies can still be improved, e.g., by using negation.

References

1. Abiteboul, S., Hull, R., Vianu, V.: Foundations of Databases. Addison-Wesley, Boston (1995)
2. Arenas, M., Bertossi, L.E., Chomicki, J.: Consistent query answers in inconsistent databases. In: PODS, pp. 68–79. ACM Press (1999)
3. Bertossi, L.E.: Database Repairing and Consistent Query Answering. Synthesis Lectures on Data Management. Morgan & Claypool Publishers, San Rafael (2011)
4. Bertossi, L.E., Li, L.: Achieving data privacy through secrecy views and null-based virtual updates. IEEE Trans. Knowl. Data Eng. **25**(5), 987–1000 (2013)
5. Bienvenu, M., Rosati, R.: Tractable approximations of consistent query answering for robust ontology-based data access. In: IJCAI. IJCAI/AAAI (2013)
6. Cao, N.V., Fragnire, E., Gauthier, J.-A., Sapin, M., Widmer, E.D.: Optimizing the marriage market: an application of the linear assignment model. Eur. J. Oper. Res. **202**(2), 547–553 (2010)

7. Chomicki, J., Marcinkowski, J.: Minimal-change integrity maintenance using tuple deletions. Inf. Comput. **197**(1–2), 90–121 (2005)
8. Dalvi, N.N., Ré, C., Suciu, D.: Probabilistic databases: diamonds in the dirt. Commun. ACM **52**(7), 86–94 (2009)
9. Dalvi, N.N., Re, C., Suciu, D.: Queries and materialized views on probabilistic databases. J. Comput. Syst. Sci. **77**(3), 473–490 (2011)
10. Fan, W., Geerts, F.: Foundations of Data Quality Management. Synthesis Lectures on Data Management. Morgan & Claypool Publishers, San Rafael (2012)
11. Fuxman, A.D., Miller, R.J.: First-order query rewriting for inconsistent databases. In: Eiter, T., Libkin, L. (eds.) ICDT 2005. LNCS, vol. 3363, pp. 337–351. Springer, Heidelberg (2005)
12. Immerman, N.: Descriptive Complexity. Graduate Texts in Computer Science. Springer, New York (1999)
13. Koutris, P., Wijsen, J.: The data complexity of consistent query answering for self-join-free conjunctive queries under primary key constraints. In: PODS, pp. 17–29. ACM (2015)
14. Koutris, P., Wijsen, J.: A trichotomy in the data complexity of certain query answering for conjunctive queries. CoRR, abs/1501.07864 (2015)
15. Libkin, L.: Elements of Finite Model Theory. Springer, New York (2004)
16. Libkin, L.: SQL's three-valued logic and certain answers. In: ICDT. LIPIcs, vol. 31, pp. 94–109. Schloss Dagstuhl - Leibniz-Zentrum fuer Informatik (2015)
17. Maslowski, D., Wijsen, J.: A dichotomy in the complexity of counting database repairs. J. Comput. Syst. Sci. **79**(6), 958–983 (2013)
18. Maslowski, D., Wijsen, J.: Counting database repairs that satisfy conjunctive queries with self-joins. In: ICDT, pp. 155–164. OpenProceedings.org (2014)
19. Wijsen, J.: Making more out of an inconsistent database. In: Benczúr, A.A., Demetrovics, J., Gottlob, G. (eds.) ADBIS 2004. LNCS, vol. 3255, pp. 291–305. Springer, Heidelberg (2004)
20. Wijsen, J.: Charting the tractability frontier of certain conjunctive query answering. In: PODS, pp. 189–200. ACM (2013)
21. Wijsen, J.: A survey of the data complexity of consistent query answering under key constraints. In: Beierle, C., Meghini, C. (eds.) FoIKS 2014. LNCS, vol. 8367, pp. 62–78. Springer, Heidelberg (2014)

Using Rules of Thumb for Repairing Inconsistent Answer Set Programs

Elie Merhej[1]([✉]), Steven Schockaert[2], and Martine De Cock[1,3]

[1] Ghent University, Ghent, Belgium
elie.merhej@ugent.be
[2] Cardiff University, Cardiff, UK
schockaerts1@cardiff.ac.uk
[3] University of Washington Tacoma, Tacoma, USA
martine.decock@ugent.be, mdecock@u.washington.edu

Abstract. Answer set programming is a form of declarative programming that can be used to elegantly model various systems. When the available knowledge about these systems is imperfect, however, the resulting programs can be inconsistent. In such cases, it is of interest to find plausible repairs, i.e. plausible modifications to the original program that ensure the existence of at least one answer set. Although several approaches to this end have already been proposed, most of them merely find a repair which is in some sense minimal. In many applications, however, expert knowledge is available which could allow us to identify better repairs. In this paper, we analyze the potential of using expert knowledge in this way, by focusing on a specific case study: gene regulatory networks. We show how we can identify the repairs that best agree with insights about such networks that have been reported in the literature, and experimentally compare this strategy against the baseline strategy of identifying minimal repairs.

1 Introduction

Answer Set Programming (ASP) is a form of declarative programming mainly oriented towards NP-hard problems [19]. It enables a form of non-monotonic reasoning by virtue of a negation-as-failure operator with a purely declarative semantics [12]. An ASP program is a set of rules that describes a problem [19]. This set of rules is fed to answer set solvers that find stable models (i.e. answer sets) of the program at hand. These answer sets then directly correspond to the solutions of the considered problem. Alternatively, answer set programs are sometimes also used to simulate systems (e.g. for solving planning problems [17,18]), in which case answer sets typically correspond to sequences of states.

We are interested in the case where ASP programs have no answer sets. We call these programs inconsistent, and we look for ways to restore their consistency. For example, in a search problem, having no answer sets could mean that the problem is over-constrained, and we may want to look at ways to relax the problem. In applications where ASP programs simulate a system, inconsistencies could mean that the rules describing the system being simulated are not in

© Springer International Publishing Switzerland 2015
C. Beierle and A. Dekhtyar (Eds.): SUM 2015, LNAI 9310, pp. 368–381, 2015.
DOI: 10.1007/978-3-319-23540-0_25

agreement with available observations. We may want to find a way to adapt the description of the system, which amounts to a form of belief revision for answer set programs [7]. In this paper, we will focus on the latter type of ASP programs.

While different methods exist for repairing ASP programs, most of them are based on finding some sort of minimal repair, e.g. adding or removing the smallest number of facts to ensure that the program has at least one answer set [1,2,10]. While this is a reasonable principle in the absence of any further information, in real-world applications, we often have access to some kind of expert knowledge about the system being modelled that can be exploited to identify the most plausible repair (which may not necessarily be minimal).

To demonstrate this idea, let us consider the following biological setup: in Fig. 1, we have a table containing time-series observed data about which of three genes were active at different time points, and a draft of a Gene Regulatory Network (GRN) which might not be correct. A GRN is a directed graph that represents the way a group of genes affect one another. GRNs can be modeled in different ways [4,6,13], with one popular model being boolean networks [24]. Treating a GRN as a boolean network implies that an edge from gene A to gene B can either represent a positive regulation, which means that A activates B, or a negative regulation, which means that A inhibits B. If A is active at a specific time step, and A activates B, B becomes active in the next time step. Similarly, if A is active and A inhibits B, B becomes inactive in the next time step. More details about activation rules are provided in Sect. 3. The GRN graph might have missing edges and/or erroneous edges due to the complexity of network generation methods [8,20], and as a result it might be inconsistent with the observed experimental data in the table. The task at hand is to repair the network to make it consistent with the table. A common method of repair would be to find the smallest number of modifications to the graph that makes it consistent with the table. Based on the GRN and table in Fig. 1, since gene 2 stays active from $t=1$ to $t=2$, a minimal repair would be to remove the edge $2 \dashv 2$. The repaired network is shown in Fig. 2(a).

However, there is a known property about GRNs that states that the diameter (i.e. the length of the longest of the shortest paths between two nodes in the graph) of a GRN tends to be very small. Considering this information leads to another repair, which is shown in Fig. 2(b). Notice that this repair is not min-

T	gene1	gene2	gene3
t=0	+	-	-
t=1	-	+	-
t=2	-	+	+

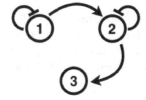

Fig. 1. A time-series table which is inconsistent with a given GRN. Edges with pointed endpoints denote activations. Edges with flat endpoints denote inhibitions. The edge $2 \dashv 2$ causes the inconsistency.

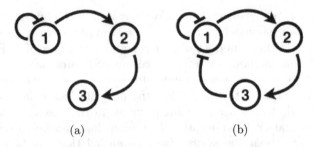

(a)   (b)

Fig. 2. Two possible repairs for the GRN from Fig. 1. (a) A minimal repair (diameter $= 2$). (b) A repair that minimizes the diameter of the graph (diameter $= 1$).

imal (we removed the edge $2 \dashv 2$ and added the edge $3 \dashv 1$), but the diameter of this new graph (diameter $= 1$) is smaller than the one in the previous repair (diameter $= 2$). We have thus generated a repair, which could be more plausible than the minimal repair.

The aim of this paper is to assess the viability of using informal, expert-provided rules of thumb for repairing inconsistent ASP programs, focusing on the specific use case of GRNs. In particular, we show how expert knowledge about GRNs found in the biological literature can be formalized in ASP and we experimentally compare the quality of the resulting repairs against baseline methods. Note that while we only consider GRNs in our experiments, the proposed method is entirely generic, being applicable to any setting where expert knowledge can be formalized in ASP.

This paper is structured as follows. First, in Sect. 2 we provide some background on answer set programming. In Sect. 3, we describe the considered use case of gene regulatory networks, summarizing in particular the available expert knowledge from the biological literature. Section 4 then shows how this expert knowledge can be encoded in ASP and how these encodings can be used to identify plausible repairs. Subsequently, in Sect. 5, we discuss our experimental results where we apply our approach on five known gene regulatory networks. Finally, we conclude in Sect. 6.

2   Answer Set Programming

Answer Set Programming (ASP) is a declarative problem solving language [12,19], which requires users to describe a problem as a set of rules. ASP solvers can then find the answer sets (see below), which correspond to the solutions of the encoded problem. An ASP rule has the form

$$h \leftarrow a_1, \ldots, a_j, \text{not } b_{j+1}, \ldots, \text{not } b_k. \tag{1}$$

where h, $a_1, \ldots, a_j$ and $b_{j+1}, \ldots, b_k$ are called *atoms*. Let r be an ASP rule of form (1). $head(r) = h$ is the *head* of r and $body(r) = \{a_1, \ldots, a_j, \text{not } b_{j+1}, \ldots, \text{not } b_k\}$ is the *body* of r. Let $body^+(r) = \{a_1, \ldots, a_j\}$ and $body^-(r) =$

$\{b_{j+1}, \ldots, b_k\}$. The "," in $body(r)$ represents a conjunction. If $body(r) = \emptyset$, then r is called a *fact*. For convenience, often the symbol $\leftarrow$ is omitted when writing facts in ASP. If $head(r) = \emptyset$, then r is called a constraint. Constraints act as filters on the possible answer sets. Indeed, answer set programs will follow a generate-and-test methodology, in which a set of rules is used to generate candidate solutions and constraints are then used to filter these candidates. The keyword *not* represents *negation-as-failure* in ASP, where *not a* intuitively holds whenever we cannot derive that a holds. An answer set program Π is a set of ASP rules of the form r. A set of atoms X is closed under Π if for any rule $r \in \Pi$, $head(r) \in X$ whenever $body^+(r) \subseteq X$. The smallest set of atoms closed under Π is denoted by $Cn(\Pi)$. The reduct Π^X of Π relative to X is defined by

$$\Pi^X = \{head(r) \leftarrow body^+(r) \mid r \in \Pi \text{ and } body^-(r) \cap X = \emptyset\}.$$

A set X of atoms is called an *answer set* (i.e. stable model) of Π if $Cn(\Pi^X) = X$. For example, let Π be the answer set program formed by the rule $c \leftarrow not\ b$ and the fact a. This program has one answer set $\{a, c\}$.

In practice, it is often easier to encode ASP programs using first-order rules like $R(X_1, X_2, X_3) \leftarrow Q(X_1, X_2), not\ S(X_3)$. Such rules should be seen as a compact representation of a set of ASP rules, called the groundings of the first-order rule, which are obtained by considering all possible instantiations of the variables by constants appearing in the program. There are ASP grounders (e.g. *gringo*) that combine a set of constant (ground) facts and ungrounded rules to give us an equivalent ground program. These programs are then solved using an ASP solver (e.g. *clasp*) to give us answer sets that correspond to solutions of our problem.

3 Case Study: Biological Networks

Biological networks are an established application of ASP [9,11]. Such networks offer a good opportunity for assessing how expert knowledge can help with repairing inconsistencies, as several rules of thumb that could be derived about such networks have been described in the biological literature. In this section, we briefly recall what Gene Regulatory Networks are, and present a setup where inconsistencies arise. We also provide a summary of the relevant properties found in the literature that we can formulate into rules of thumb.

A Gene Regulatory Network (GRN) is a network that represents the interactions between a group of cell genes. The nodes of the network are the genes, whereas the edges of the network encode the interactions between the genes. There are two types of possible interactions between a pair of genes: a gene either *activates* another gene, or *inhibits* another gene. This means that if gene A activates gene B, and A is active at time step t, then B becomes active at time step t+1. Likewise, if gene A inhibits gene B, and A is active at time step t, then B becomes inactive at time step t+1. In the case where a gene is activated and inhibited simultaneously, different activation rules may be applied to determine its subsequent state [23]. Different kinds of experimental observations can be used to automatically construct GRNs [8,20].

Our setup consists of the following. We have an automatically generated GRN describing a cell cycle. Since it has been automatically constructed, it is likely to be imperfect, in the sense that we may obtain observations that are inconsistent with the behaviour predicted by the network, which would mean that it needs to be revised. We also have a time-series table that lists the state of the corresponding genes in this GRN at consecutive time steps. At every time step, a given gene can either be *active* or *inactive*. The table we have at our disposal corresponds to data that has been experimentally observed, but was not available during the GRN generation process. Our problem then comes down to checking whether the GRN is inconsistent with the data from the time-series table, i.e. whether it fails to correctly predict how the states of the genes evolve, and in that case, repair it.

Several methods have been developed that use ASP to repair inconsistencies found in GRNs [10,11,22]. These methods often consist of finding some kind of minimal repair. However, expert knowledge about GRNs can be used to derive rules of thumb that help in finding more plausible repairs, which are not necessarily minimal. In [16], it is stated that every gene network converges to a final stable state (Property 1). This allows us to create an extra check to find whether the GRN we are trying to repair converges to a stable state that indeed corresponds to the final time step in the table. In [13], Kauffman found that a genetic network will behave chaotically unless there is a restriction on the number of regulatory inputs and outputs per node (Property 2). This can be encoded as a rule of thumb where the number of input and output edges of every node should be limited. Another rule of thumb can be derived from the fact that various biological properties in a gene network depend on the number of non-zero interactions between the nodes of this network, as is discussed in [14] (Property 3). This allows us to derive that similar gene networks would more likely have a similar number of total interactions. Also in [14], it is observed that nodes tend to be positively regulated by nodes that are active at earlier states of a cell cycle and negatively regulated by nodes that are active later in the process (Property 4). In [5], it is stated that the diameter (i.e. the length of the shortest path between the two nodes that are furthest apart in the network) of GRN graphs tend to be very small (Property 5). In [15], the idea of dominant motifs (i.e. sub-graphs) is discussed, where these motifs tend to occur frequently in multiple kinds of GRNs. This allows us to formulate a rule of thumb that similar networks are likely to share the same dominant motifs (Property 6). Finally, [27] states that the size of the basin of attractors (i.e. the stable states to which most initial states of the network converge to) in a GRN is a vital quantity in terms of understanding network behaviour and may relate to other network properties such as stability (Property 7). This allows us to check whether the state of the repaired network with the largest basin size corresponds indeed to the final stable state in the time-series table.

4   Repairing ASP Programs Using Rules of Thumb

In this section, we show how inconsistent ASP programs can be repaired. In Sect. 4.1, we encode the facts of our program, which correspond in our case to a GRN and a table with observed data, and show the rule that checks our program's consistency. We then recall in Sect. 4.2 how a minimal repair of an inconsistent ASP program can be found, using the meta-programming technique proposed in [10]. Finally, in Sect. 4.3 we improve the baseline method from Sect. 4.2 by taking into account rules of thumb. We illustrate the main idea by focusing on the biological properties of GRNs that we discussed in the previous section.

4.1 Encoding GRNs in ASP

In this section, we recall how a GRN and corresponding time-series table can be encoded in ASP, as presented in [9,11,22,23]. This includes the observed time-series table data, as well as the GRN graph that might be inconsistent with the table. For every gene i, we introduce the fact $gene(i)$. For every edge from gene i to gene j, we introduce the fact $activates(i,j)$ if i activates j, or $inhibits(i,j)$ if i inhibits j. As for the time-series table, we include facts of the form $active(i,t)$ and $inactive(i,t)$ which indicate that gene i is active at time t and that gene i is inactive at time t respectively. We also represent every time step with the fact $time(i)$ with $0 \leq i \leq t_{\text{final}}$, with t_{final} representing the final time step that the gene regulatory network converge to.

Then, to check for consistency between the graph and the table, three things need to be done. First, we write activation and inhibition rules for the graph that determine whether a gene is activated or inhibited (or neither) at each time step. We use the following activation rule: if a gene is positively regulated by at least one other gene, and it is not negatively regulated by any other gene, then it is activated. A similar rule is used to determine when a gene is inhibited. This is shown in (2). Second, we determine the state for every gene at every time step based on its state given by the table, and on the activation and inhibition rules from the graph. This is shown in (3). Third, we check if the states of the genes generated by the activation and inhibition rules of the graph correspond with the states of the genes in the time-series table, shown in (4),

$$receivesAct(Y,T) \leftarrow activates(X,Y),\ active(X,T).$$
$$receivesInh(Y,T) \leftarrow inhibits(X,Y),\ active(X,T).$$
$$activated(Y,T) \leftarrow receivesAct(Y,T-1),\ not\ receivesInh(Y,T-1). \quad (2)$$
$$inhibited(Y,T) \leftarrow receivesInh(Y,T-1),\ not\ receivesAct(Y,T-1).$$
$$\leftarrow activated(Y,T),\ inhibited(Y,T).$$

$$inactive(Y,T) \leftarrow active(Y,T-1),\ inhibited(Y,T).$$
$$active(Y,T) \leftarrow active(Y,T-1),\ not\ inhibited(Y,T).$$
$$active(Y,T) \leftarrow inactive(Y,T-1),\ activated(Y,T). \quad (3)$$
$$inactive(Y,T) \leftarrow inactive(Y,T-1),\ not\ activated(Y,T).$$

$$\leftarrow active(Y,T),\ inactive(Y,T). \quad (4)$$

resulting in an inconsistent program if and only if the data from the graph and the time-series table do not correspond to one another.

4.2   Minimal Repair

The repair operations consist of either adding or removing an edge between two genes. Thus, we generate four possible repair choices for every pair of nodes: add a new activation edge, add a new inhibition edge, remove an existing edge or do nothing. This introduces the facts $addActEdge(U, V)$, $addInhEdge(U, V)$ and $removeEdge(U, V, S)$ respectively. Then, to take the generated repair into account, we create the facts $activates(i, j)$ and $inhibits(i, j)$ using the following

$$
\begin{aligned}
activates(U, V) &\leftarrow edge(U, V, 1), \; not \; removeEdge(U, V, 1). \\
activates(U, V) &\leftarrow addActEdge(U, V). \\
inhibits(U, V) &\leftarrow edge(U, V, -1), \; not \; removeEdge(U, V, -1). \\
inhibits(U, V) &\leftarrow addInhEdge(U, V).
\end{aligned}
\tag{5}
$$

The ASP program constructed so far has one answer set for each possible repair of the original GRN that will make it consistent with the table. To consider minimal repairs only, we first define the cost of a repair, using the following rules

$$
\begin{aligned}
addEdge(U, V, 1) &\leftarrow addActEdge(U, V). \\
addEdge(U, V, -1) &\leftarrow addInhEdge(U, V). \\
costAdding(X) &\leftarrow X = \#count\{addEdge(U, V, S)\}. \\
costRemoving(Y) &\leftarrow Y = \#count\{removeEdge(U, V, S)\}. \\
repairCost(Z) &\leftarrow costAdding(X), \; costRemoving(Y), \; Z = X + Y. \\
&\#minimize[repairCost(Z) = Z].
\end{aligned}
\tag{6}
$$

These rules contain *aggregates* and *conditions* supported by the ASP solver *clasp*. Aggregates behave like built-in functions in the ASP solver. For example, in (6), the aggregate *#count* intuitively counts the number of instances of the literals *addEdge* and *removeEdge*, and stores the results in variables X and Y respectively. Whereas the aggregate *#minimize* adds an optimization value that minimizes the number held by the variable Z in the literal *repairCost(Z)*. This intuitively minimizes the cost of the repair. In addition, the 5th rule in (6) contains the condition $Z = X + Y$. This condition can be added to the body of a rule, and has to be satisfied for the rule to be satisfied.

4.3   Using Rules of Thumb for Identifying Plausible Repairs

While the idea of finding a minimal repair, as explained for GRNs in Sect. 4.2, is defensible in cases where we have no further information, it is far from optimal. First, there is no reason why the correct repair has to be minimal; as we will see in Sect. 5, in the case of GRNs the correct repair is actually rarely minimal. Second, there can be exponentially many minimal repairs, and without further knowledge we would have to select one arbitrarily. We assume that in most real-world applications, however, we have access to some kind of background knowledge that could help us identify plausible repairs. Such background knowledge usually comes in the form of rules of thumb. In this section, we will consider the seven principles about GRNs that we found in the literature (see Sect. 3). We will briefly explain how each of these principles can be encoded as some kind of soft constraint in ASP. These soft constraints will introduce penalty weights if they are not satisfied by correct repairs. These individual penalties are then added up and included in the repair cost that already contains the cost of adding and removing an edge. The repair that makes the graph consistent with the time-series table and has the lowest overall cost will be selected as the best repair. Note that we cannot realistically expect any training data to be available to learn weights (e.g. reflecting the importance of each principle), making our approach quite different from e.g. approaches for repairing using soft constraints in Markov logic [21,25]. Therefore, we instead set the weights in a uniform manner, i.e. they are chosen such that each principle roughly has the same impact on the choice of repair. In other words, our approach is only based on a direct encoding of available expert knowledge. The effectiveness of such a strategy will be experimentally analyzed in Sect. 5.

Property 1: Last Time Step as Fixed State. We create a new time step $(t_{\text{final}+1})$ in the table with the state of the genes identical to their states at (t_{final}). Then, we perform a similar consistency check as we did in rules (2)–(4). If the repair is still correct, i.e. if the graph is still consistent with the table after the addition of the time step $(t_{\text{final}+1})$, then the time step (t_{final}) is indeed a fixed state, and no cost is added to the repair. Otherwise, we increase the cost of the repair by a constant value equal to the total number of initial edges in the network. Since we are repairing by adding and removing edges, we choose the maximum penalty for every property to be equal to the total number of initial edges in the network.

Property 2: Degree of a Gene. We need to find the degree of a gene, given by $k = k_{\text{in}} + k_{\text{out}}$ with k_{in} being the number of incoming edges and k_{out} the number of outgoing edges of the gene. We then need to make sure that these degrees fall within a certain range. We explain how we obtain this range in Sect. 5. We then call $kBadGenes$ the number of genes that have a k degree that falls outside the range limits. The penalty from this property is then multiplied by the ratio of initial edges per gene for every "bad gene" found. This makes sure that the maximum penalty is equal to the total number of initial edges in the network (if all the genes of the network are "bad genes", the maximum penalty is $\text{penalty}_{\text{max}} = (\text{genes}_{\text{total}}) \times (\text{edges}_{\text{initial}}/\text{genes}_{\text{total}}) = \text{edges}_{\text{initial}})$. This property is added using the following rules:

$$edgeAfterRepair(U,V) \leftarrow activates(U,V).$$
$$edgeAfterRepair(U,V) \leftarrow inhibits(U,V).$$
$$kOut(C,X) \leftarrow X = \#count\{edgeAfterRepair(C,D)\}, \ gene(C).$$
$$kIn(C,X) \leftarrow X = \#count\{edgeAfterRepair(D,C)\}, \ gene(C).$$

$$kDegree(C, Z) \leftarrow kIn(C, X), \ kOut(C, Y), \ Z = X + Y. \qquad (7)$$
$$kBadGene(C) \leftarrow kDegree(C, Z), \ Z < k_{\min}.$$
$$kBadGene(C) \leftarrow kDegree(C, Z), \ Z > k_{\max}.$$
$$kBadGenes(X) \leftarrow X = \#count\{kBadGene(C)\}.$$

Property 3: Total Number of Edges. To encode this property, we count the total number of interactions between the genes and check whether this number falls within a certain range limit (see Sect. 5). If the number is outside the range we set, a penalty equal to the total number of initial edges is added to the repair cost. Otherwise, the penalty is zero.

Property 4: Likely Interactions Based on Gene State. For this property, we divide the genes into likely activators and likely inhibitors based on whether they are active during the first half or the second half of the cycle respectively. The same gene can be both a likely activator and a likely inhibitor. We then check the outgoing edges of every gene, and increase the cost of the repair every time a likely activator (that is not also a likely inhibitor) inhibits another gene, or a likely inhibitor (that is not also a likely activator) activates another gene. The penalty is increased by 1 for every "bad" edge found, with the maximum penalty being the total number of initial edges in the network. We encode this property using the following rules:

$$likelyAct(C) \leftarrow active(C, T), \ T <= t_{\text{half}}.$$
$$likelyInh(C) \leftarrow active(C, T), \ T > t_{\text{half}}.$$
$$badEdge(C, D) \leftarrow likelyAct(C), \ inhibits(C, D), \ not \ likelyInh(C), \ C \ != D.$$
$$badEdge(C, D) \leftarrow likelyInh(C), \ activates(C, D), \ not \ likelyAct(C), \ C \ != D.$$
$$badEdges(X) \leftarrow X = \#count\{badEdge(C, D)\}. \qquad (8)$$

Property 5: Network Diameter. To encode this property, we first need to make sure that every gene of the network is reachable, using the following rules:

$$link(X, Y) \leftarrow edgeAfterRepair(X, Y), \ X \ != Y.$$
$$link(Y, X) \leftarrow edgeAfterRepair(X, Y), \ Y \ != X.$$
$$reachable(X) \leftarrow link(1, X). \qquad (9)$$
$$reachable(Y) \leftarrow reachable(X), \ link(X, Y).$$
$$\leftarrow gene(X), \ not \ reachable(X).$$

Then, we find the shortest distance between every pair of genes by finding all the possible paths between them, and minimizing the number of path links. The greatest value of these shortest distances is the diameter of the network.

$$dist(X, Y, 1) \leftarrow link(X, Y), \ X \ ! = Y.$$
$$dist(X, Y, 2) \leftarrow link(X, A), \ link(A, Y), \ X \ ! = Y.$$
$$dist(X, Y, 3) \leftarrow link(X, A), \ link(A, B), \ link(B, Y), \ X \ ! = Y.$$
$$\ldots \tag{10}$$
$$smallestDist(X, Y, D) \leftarrow D = \#min[dist(X, Y, C) = C], \ dist(X, Y, Z).$$
$$diameter(D) \leftarrow D = \#max[smallestDist(X, Y, C) = C].$$

The penalty cost that is added depends on the diameter that was found. Again, if the diameter falls within a certain range limit (see Sect. 5), no penalty is added to the repair cost. Otherwise, the cost is increased by a penalty equal to the total number of initial edges in the network.

Property 6: Dominant Motifs. A motif is a small pattern with usually 3 or 4 nodes that is found repeatedly in a network graph. It does not matter which genes these nodes correspond to, or the type of the edges between the nodes. For this property, we use an external program described in [26] to find the dominant motifs of popular GRNs in the literature. The GRN that we are repairing is not used during this step. We then encode these motifs in our program and try to maximize the number of their instances in the repaired network. For every instance of dominant motif that we find in the repaired network, we decrease the penalty of this property by 1, starting with the maximum penalty equal to the total number of initial edges in the network (the minimum penalty is zero).

Property 7: Size of Basin of Attractors. To use this property, we need to find the final state of every possible initial state of a network. To do this, we use a standalone program described in [3]. We then need to make sure that the most popular final state of the network given by the output of this program corresponds indeed to its state at the final time step (t_{final}) given by the time-series table. To apply this property, we adapt the answer sets of our program in the following way. For each repaired network (i.e. for each answer set), we add a penalty equal to the total number of initial edges in the network if its most popular final state does not correspond to the state at the final time step (t_{final}) given by the table. Otherwise, we do not add any penalty.

5   Experimental Results

To test our approach, we use the following 5 GRNs: *Budding Yeast, Fission Yeast, C. Elegans, Arabidopsis* and *Mammalian Cell Cycle*. We corrupt each of these GRNs by adding and removing edges, and then try to repair them. Every time we corrupt a network, we remove R randomly chosen edges, and subsequently add N randomly chosen edges (choosing between activation and inhibition edges with equal probability). We set N and R as percentages of the initial number of edges for each network that we are corrupting. For our experiments, we consider 7 corruption setups by varying the percentages N and R in the following way:

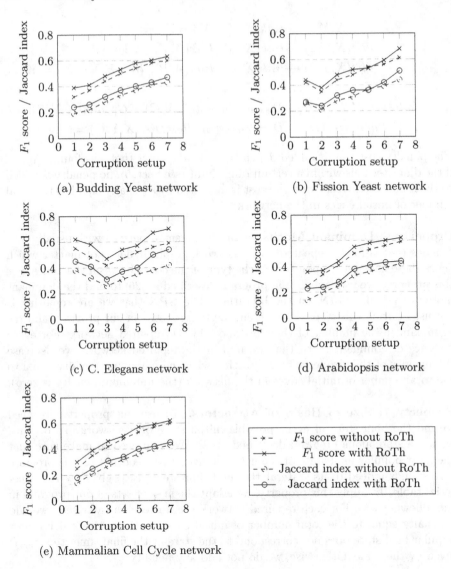

(a) Budding Yeast network

(b) Fission Yeast network

(c) C. Elegans network

(d) Arabidopsis network

(e) Mammalian Cell Cycle network

Fig. 3. Average F_1 score and Jaccard index for 7 corruption setups of 5 GRNs, with and without the addition of Rules of Thumb (RoTh).

$N = 20\,\% / R = 80\,\%$, $N = 30\,\% / R = 70\,\%$, $N = 40\,\% / R = 60\,\%$, $N = 50\,\% / R = 50\,\%$, $N = 60\,\% / R = 40\,\%$, $N = 70\,\% / R = 30\,\%$ and $N = 80\,\% / R = 20\,\%$.

Every time we select a network to corrupt and repair, we learn the relevant parameters of the rules of thumb from the other four, uncorrupted networks. For Property 2, we learn the degrees k_{min} and k_{max} from the other four networks by setting k_{min} as the smallest degree value of the other four networks and k_{max} as the largest degree value. The range [diameter$_{min}$, diameter$_{max}$] in Property 5 is learned similarly, where $diameter_{min}$ is the smallest diameter value of the other four networks, and $diameter_{max}$ is the largest diameter value. For Property 3, the range of the total number of edges is calculated as follows. We learn from the other four networks the ratio of number of edges per node, and we keep the minimum ($ratio_{min}$) and maximum ($ratio_{max}$) values that we find. Then, we determine what the expected number of edges should be for the test network by multiplying these two ratios with the number of nodes in the test network.

To evaluate our results, we use the F_1 score and Jaccard index which we calculate as follows. Let A be the set of edges of the repaired network and B the set of edges of the original network. We write $|A|$ and $|B|$ for the number of edges of the repaired and original network respectively. The F_1 score is given by $F_1 = 2 \times (precision \times recall)/(precision + recall)$, with $precision = |A \cap B| / |B|$ and $recall = |A \cap B| / |A|$. The Jaccard index is given by $J(A, B) = |A \cap B| / |A \cup B|$. We run every experiment (i.e. every corruption setup on every network) 10 times and report the average F_1 score and Jaccard Index of the best repair that was found. In the case where multiple repairs with the same minimum cost were found, we select the first repair that we get from the solver as best repair. We have used the grounder *gringo* and the solver *clasp* to run our experiments.

The results of our experiments are shown in Fig. 3. Each graph corresponds to a different GRN. The dashed lines represent the average F_1 score and Jaccard index of the best repair without the addition of rules of thumb (i.e. best minimal repair), and the solid lines represent the same values after the addition of rules of thumb. We notice a consistent improvement in both metrics with the addition of rules of thumb. Instead of only minimizing the number of applied repair operations, the addition of rules of thumb also focuses on preserving the biggest number of properties that were found in similar GRNs. This allows the repairing process to avoid many mistakes. For example, while a minimal repair can have a node detached from the rest of the graph, a more plausible repair that keeps all the nodes of a GRN connected can be achieved by simply following a rule that describes this property.

6   Conclusion and Discussion

In this paper, we have explored how expert knowledge, in the form of rules of thumb, can be used to find better ways of repairing an inconsistent ASP program. As a case study, we have focused our experiments on a biological setup where a GRN and a time-series table are in conflict with each other. Our experiments have shown that our method of repairing by using rules of thumb leads to a better performance in terms of F_1 score and Jaccard measure. This leads to more plausible repairs than when simply selecting the minimal one, with only very limited access to training data.

The idea of adding soft constraints to prefer a model over another is not new. Many applications of Markov logic strongly utilize this concept [21, 25]. However, combining the idea of rules of thumb with the ease by which ASP can model systems provides a framework that elegantly takes advantage of both soft and hard constraints. Compared to Markov logic, ASP offers us more flexibility in the term of what we optimize, e.g. we are not restricted to minimizing the sum of penalties, although that is how we used the rules of thumb in this paper. In the future, it would be interesting to do an experimental comparison between our ASP approach and its Markov logic counterpart.

References

1. Arenas, M., Bertossi, L., Chomicki, J.: Answer sets for consistent query answering in inconsistent databases. Theory Pract. Logic Program. **3**, 393–424 (2003)
2. Arieli, O., Denecker, M., Van Nuffelen, B., Bruynooghe, M.: Database repair by signed formulae. In: Seipel, D., Turull-Torres, J.M. (eds.) FoIKS 2004. LNCS, vol. 2942, pp. 14–30. Springer, Heidelberg (2004)
3. Berntenis, N., Ebeling, M.: Detection of attractors of large boolean networks via exhaustive enumeration of appropriate subspaces of the state space. BMC Bioinform. **14**(1), 361 (2013)
4. Chen, T., Filkov, V., Skiena, S.S.: Identifying gene regulatory networks from experimental data. Parallel Comput. **27**(1), 141–162 (2001)
5. Cohen, R., Havlin, S.: Scale-free networks are ultrasmall. Phys. Rev. Lett. **90**(5), 058701 (2003)
6. De Jong, H.: Modeling and simulation of genetic regulatory systems: a literature review. J. Comput. Biol. **9**(1), 67–103 (2002)
7. Delgrande, J.P., Schaub, T.: A consistency-based approach for belief change. Artif. Intell. **151**(1), 1–41 (2003)
8. Ellis, J.J., Kobe, B.: Predicting protein kinase specificity: predikin update and performance in the DREAM4 challenge. PLoS One **6**(7), e21169 (2011)
9. Fayruzov, T., Janssen, J., Vermeir, D., Cornelis, C.: Modelling gene and protein regulatory networks with answer set programming. Int. J. Data Mining Bioinform. **5**(2), 209–229 (2011)
10. Gebser, M., Guziolowski, C., Ivanchev, M., Schaub, T., Siegel, A., Thiele, S., Veber, P.: Repair and prediction (under inconsistency) in large biological networks with answer set programming. In: KR (2010)
11. Gebser, M., König, A., Schaub, T., Thiele, S., Veber, P.: The BioASP library: ASP solutions for systems biology. In: ICTAI (1), pp. 383–389 (2010)
12. Gelfond, M., Lifschitz, V.: The stable model semantics for logic programming. In: ICLP/SLP, vol. 88, pp. 1070–1080 (1988)
13. Kauffman, S.A.: The Origins of Order: Self-organization and Selection in Evolution. Oxford University Press, Oxford (1993)
14. Lau, K.Y., Ganguli, S., Tang, C.: Function constrains network architecture and dynamics: a case study on the yeast cell cycle boolean network. Phys. Rev. E **75**(5), 051907 (2007)
15. Lee, T.I., Rinaldi, N.J., Robert, F., Odom, D.T., Bar-Joseph, Z., Gerber, G.K., Hannett, N.M., Harbison, C.T., Thompson, C.M., Simon, I., et al.: Transcriptional regulatory networks in saccharomyces cerevisiae. Science **298**(5594), 799–804 (2002)
16. Li, F., Long, T., Lu, Y., Ouyang, Q., Tang, C.: The yeast cell-cycle network is robustly designed. Proc. Natl. Acad. Sci. U.S.A. **101**(14), 4781–4786 (2004)
17. Lifschitz, V.: Action languages, answer sets, and planning. In: Apt, K.R., et al. (eds.) The Logic Programming Paradigm, pp. 357–373. Springer, Heidelberg (1999)

18. Lifschitz, V.: Answer set programming and plan generation. Artif. Intell. **138**(1), 39–54 (2002)
19. Lifschitz, V.: What is answer set programming? In: AAAI, vol. 8, pp. 1594–1597 (2008)
20. Menéndez, P., Kourmpetis, Y.A., ter Braak, C.J., van Eeuwijk, F.A.: Gene regulatory networks from multifactorial perturbations using graphical lasso: application to the DREAM4 challenge. PloS One **5**(12), e14147 (2010)
21. Merhej, E., Schockaert, S., De Cock, M., Blondeel, M., Alfarone, D., Davis, J.: Repairing inconsistent taxonomies using map inference and rules of thumb. In: Proceedings of the 5th International Workshop on Web-scale Knowledge Representation Retrieval & Reasoning, pp. 31–36. ACM (2014)
22. Mobilia, N., Rocca, A., Chorlton, S., Fanchon, E., Trilling, L.: Logical modeling and analysis of regulatory genetic networks in a non monotonic framework. In: Ortuño, F., Rojas, I. (eds.) IWBBIO 2015, Part I. LNCS, vol. 9043, pp. 599–612. Springer, Heidelberg (2015)
23. Mushthofa, M., Torres, G., Van de Peer, Y., Marchal, K., De Cock, M.: ASP-G: an ASP-based method for finding attractors in genetic regulatory networks. Bioinformatics **30**(21), 3086–3092 (2014)
24. Shmulevich, I., Dougherty, E.R., Zhang, W.: From boolean to probabilistic boolean networks as models of genetic regulatory networks. Proc. IEEE **90**(11), 1778–1792 (2002)
25. Singla, P., Kautz, H., Luo, J., Gallagher, A.: Discovery of social relationships in consumer photo collections using markov logic. In: IEEE Computer Society Conference on Computer Vision and Pattern Recognition Workshops, CVPRW 2008, pp. 1–7. IEEE (2008)
26. Wernicke, S., Rasche, F.: Fanmod: a tool for fast network motif detection. Bioinformatics **22**(9), 1152–1153 (2006)
27. Yang, L., Meng, Y., Bao, C., Liu, W., Ma, C., Li, A., Xuan, Z., Shan, G., Jia, Y.: Robustness and backbone motif of a cancer network regulated by miR-17-92 cluster during the G1/S transition. PloS One **8**(3), e57009 (2013)

Applications

Applications

Fuzzy XPath for the Automatic Search of Fuzzy Formulae Models

Jesús M. Almendros-Jiménez[1], Miquel Bofill[2], Alejandro Luna-Tedesqui[3],
Ginés Moreno[3(✉)], Carlos Vázquez[3], and Mateu Villaret[2]

[1] Department of Languages and Computation, University of Almería,
Almería, Spain
jalmen@ual.es

[2] Department of Computer Science, Applied Mathematics and Statistics,
University of Girona, Girona, Spain
{Miquel.Bofill,Mateu.Villaret}@udg.edu

[3] Department of Computing Systems, University of Castilla-La Mancha,
Albacete, Spain
{Alejandro.Luna,Gines.Moreno,Carlos.Vazquez}@uclm.es

Abstract. In this paper we deal with propositional fuzzy formulae containing several propositional symbols linked with connectives defined in a lattice of truth degrees more complex than *Bool*. Instead of focusing on satisfiability (i.e., proving the existence of at least one model) as usually done in a SAT/SMT setting, our interest moves to the problem of finding the whole set of models (with a finite domain) for a given fuzzy formula. We reuse a previous method based on fuzzy logic programming where the formula is conceived as a goal whose derivation tree, provided by our FLOPER tool, contains on its leaves all the models of the original formula, together with other interpretations. Next, we use the ability of the FuzzyXPath tool (developed in our research group with FLOPER) for exploring these derivation trees once exported in XML format, in order to discover whether the formula is a tautology, satisfiable, or a contradiction, thus reinforcing the bi-lateral synergies between FuzzyXPath and FLOPER.

Keywords: Fuzzy logic programming · Automatic theorem proving · Fuzzy XPath

1 Introduction

Research on SAT (Boolean Satisfiability) and SMT (Satisfiability Modulo Theories) [9] represents a successful and large tradition in the development of highly efficient automatic theorem solvers for classic logic. More recently there also

This work has been partially supported by the EU (FEDER), and the Spanish MINECO Ministry (*Ministerio de Economía y Competitividad*) under grants TIN2013-44742-C4-4-R, TIN2012-33042 and TIN2013-45732-C4-2-P.

C. Beierle and A. Dekhtyar (Eds.): SUM 2015, LNAI 9310, pp. 385–398, 2015.
DOI: 10.1007/978-3-319-23540-0_26

exist attempts for covering fuzzy logics, as occurs with the approaches presented in [6,21]. Moreover, if automatic theorem solving supposes too a starting point for the foundations of logic programming as well as one of its important application fields [7,12,16,20], in [11] we showed some preliminary guidelines about how fuzzy logic programming [8,13,14,17,19] can face the automatic proving of fuzzy theorems by making use of the FLOPER environment developed in our research group [18] (visit http://dectau.uclm.es/floper/). The main goal of the present paper is to reinforce this last method of [11] by means of the FuzzyX-Path tool developed too with FLOPER as described in [2,4,5] (the application is freely available from http://dectau.uclm.es/fuzzyXPath).

Let us start our discussion with an easy motivating example. Assume that we have a very simple digital chip with just a single input port and just one output port, such that it reverts on Out the signal received from In. The behaviour of such chip can be represented by the following propositional formula $F : (\neg In \wedge Out) \vee (In \wedge \neg Out)$. Depending on how we interpret each propositional symbol, we obtain the following final set of interpretations for the whole formula:

$$I1 : \{In = 0, Out = 0\} \Rightarrow F = 0 \qquad I2 : \{In = 0, Out = 1\} \Rightarrow F = 1$$
$$I3 : \{In = 1, Out = 0\} \Rightarrow F = 1 \qquad I4 : \{In = 1, Out = 1\} \Rightarrow F = 0$$

A SAT solver easily proves that F is satisfiable since, in fact, it has two models (i.e., interpretations of the propositional variables In and Out that assign 1 to the whole formula) represented by $I2$ and $I3$. An alternative way for explicitly obtaining such interpretations consists of using the fuzzy logic environment FLOPER developed in our research group. As we will explain in the rest of the paper, when FLOPER executes the following goal representing formula F "(@not(i(In)) & i(Out)) | (i(In) & @not(i(Out)))" with respect to a fuzzy logic program composed by just two rules: "i(1) with 1" and "i(0) with 0", it generates an execution tree where models $I2$ and $I3$ appear as leaves (see [11]). Each branch in the tree starts by interpreting variables In and Out and continues with the evaluation of operators (connectives) appearing in F.

Note that whereas formula F describes the behaviour of our chip in an "implicit way", the whole set of models $I2$ and $I3$ "explicitly" describes how the chip successfully works (any other interpretation not being a model, represents an abnormal behaviour of the chip), hence the importance of finding the whole set of models for a given formula.

Assume now that we plan to model an "analogic" version of the chip, where both the input and output signals might vary in an infinite range of values between 0 and 1, such that Out will simply represent the "complement" of In. The new behaviour of the chip can be expressed again by the same previous formula, but taking into account now that connectives involved in F could be defined in a fuzzy way as follows (see also Fig. 1 afterwards):

$$\neg x \quad = 1 - x \qquad \text{Product logic's negation}$$
$$x \wedge y = min(x, y) \qquad \text{Gödel logic's conjunction}$$
$$x \vee y = min(x + y, 1) \text{ Lukasiewicz logic's disjunction}$$

Here we could use an SMT solver to prove that F is satisfiable, as done in [6,11], but the goal of this paper is to use techniques based on fuzzy logic programming for discovering models.

On the other hand, the eXtensible Markup Language (XML) is widely used in many areas of computer software to represent machine readable data. XML provides a very simple language to represent the structure of data, using tags to label pieces of textual content, and a tree structure to describe the hierarchical content. XML emerged as a solution to data exchange between applications where tags permit to locate the content. XML documents are mainly used in databases. The XPath language [10] was designed as a query language for XML in which the path of the tree is used to describe the query. XPath expressions can be adorned with boolean conditions on nodes and leaves to restrict the number of answers of the query. XPath is the basis of a more powerful query language (called XQuery) designed to join multiple XML documents and to give format to the answer. In [2,4,5] we have presented an XPath interpreter (together with a debugger, as documented in [1,3]) extended with fuzzy commands which somehow rely on the implementation based on fuzzy logic programming by using FLOPER.

In [5] we illustrated the mutual benefits between the FLOPER programming environment and the FUZZYXPATH interpreter. Initially FLOPER was conceived as a tool for implementing flexible software applications – as it is the case of FUZZYXPATH – coded with the fuzzy logic language MALP and offering options for compiling fuzzy rules to standard Prolog clauses, running goals and drawing execution trees. Such trees, once modeled in XML format inside the proper FLOPER tool, can be then analyzed by the FUZZYXPATH interpreter – by means of simple XPath queries augmented with fuzzy commands – in order to discover details (such as fuzzy computed answers, possible infinite branches and so on) of the computational behaviour of MALP programs after being executed into FLOPER. The main goal of this paper is to use FUZZYXPATH for to automate the process of directly extracting the set of models contained on the proof trees associated to fuzzy formulae explained before, once such trees have been exported by FLOPER in XML format.

2 Fuzzy Logic Programming and FLOPER

In what follows we describe a very simple *subset of the Multi-Adjoint Logic Programming language*, MALP in brief, (see [17] for a complete formulation of this framework), which in essence consists of a first-order language, $\mathcal{L}$, containing variables, constants, function symbols, predicate symbols, and several (arbitrary) connectives to increase language expressiveness: implication connectives (denoted by $\leftarrow_1, \leftarrow_2, \dots$); conjunctive connectives ($\wedge_1, \wedge_2, \dots$), disjunctive connectives ($\vee_1, \vee_2, \dots$), and hybrid operators (usually denoted by $@_1, @_2, \dots$), all of them are grouped under the name of "aggregators". Although these connectives are usually binary operators, our framework also admits aggregators of any arity denoted as $@(x_1, \dots, x_n)$. By definition, the truth function for an n-ary aggregation operator $[\![@]\!] : L^n \to L$ is required to be monotonous.

$$\&_{\text{P}}(x,y) \triangleq x * y \qquad\qquad |_{\text{P}}(x,y) \triangleq x + y - x * y \qquad \leftarrow_{\text{P}}(x,y) \triangleq \min(1, x/y)$$

$$\&_{\text{G}}(x,y) \triangleq \min(x,y) \qquad |_{\text{G}}(x,y) \triangleq \max\{x,y\} \qquad \leftarrow_{\text{G}}(x,y) \triangleq \begin{cases} 1 & \text{if } y \leq x \\ x & \text{otherwise} \end{cases}$$

$$\&_{\text{L}}(x,y) \triangleq \max(0, x + y - 1) \quad |_{\text{L}}(x,y) \triangleq \min\{x+y,1\} \quad \leftarrow_{\text{L}}(x,y) \triangleq \min\{x - y + 1, 1\}$$

Fig. 1. Conjunctors, disjunctors and implications from *Product, Gödel* and *Łukasiewicz* logics.

Additionally, our language $\mathcal{L}$ contains the values of a lattice $(L, \leq)$ and a set of connectives interpreted over such lattice. In general, L may be the carrier of any complete bounded lattice where a L-expression is a well-formed expression composed by values of L, as well as variable symbols, connectives and *primitive operators* (i.e., arithmetic symbols such as $*$, $+$, min, etc.). In what follows, we assume that the truth function of any connective @ in L is given by its corresponding *connective definition*, that is, an equation of the form $@(x_1, \ldots, x_n) \triangleq E$, where E is a L-expression not containing variable symbols apart from $x_1, \ldots, x_n$. For instance, some fuzzy connective definitions in the lattice $([0,1], \leq)$ are presented in Fig. 1 (from now on, this lattice will be called $\mathcal{V}$ along this paper), where labels L, G and P mean respectively *Łukasiewicz logic*, *Gödel logic* and *product logic* (with different capabilities for modeling *pessimistic, optimistic* and *realistic scenarios*, respectively).

This subset of MALP is intended to cope with fuzzy propositional formulae like $P \wedge Q \rightarrow P \vee Q$, where propositions P and Q are interpreted as values of the lattice. To this end, a *program* is defined as a set of rules (also called "facts") of the form "H *with* v", where H is an atomic formula or atom (usually called *head*), and v is its associated *truth degree* (i.e., a value of L). More precisely, in our application, heads have always the form "$i(v)$" and each program rule looks like "$i(v)$ *with* v". It is noteworthy to point out that even when we use the same names for constants (building data terms) and truth degrees, the Herbrand Universe of each program and the carrier set of its associated lattice should never be confused, since they are in fact disjoint sets.

A *goal* is a formula built from atomic formulas $B_1, \ldots, B_n$ ($n \geq 0$), truth values of L, conjunctions, disjunctions and aggregations, submitted as a query to the system. In this subset of MALP, the atomic formulas of a *goal* have always the form "$i(P)$", being P a variable symbol. In this way, when running a simple goal like "$i(P)$" (as done in Fig. 2), we could obtain several answers meaning something like "when $P = v$, then the resulting truth degree is v", representing all possible interpretations in L for proposition P in the original formula.

The procedural semantics of this subset of the MALP language consists of an operational phase (based on admissible steps that exploits the atoms in the goal), followed by an interpretive phase (that performs arithmetic operations to interpret the resulting formula on the lattice). In the following, $\mathcal{C}[A]$ denotes a formula where A is a sub-expression which occurs in the – possibly empty – context $\mathcal{C}[]$. Moreover, $\mathcal{C}[A/A']$ means the replacement of A by A' in context $\mathcal{C}[]$.

Definition 1 (Admissible Step). *Let $\mathcal{Q}$ be a goal and let σ be a substitution. The pair $\langle \mathcal{Q}; \sigma \rangle$ is a state. Given a program $\mathcal{P}$, an admissible computation is formalized as a state transition system, whose transition relation $\rightarrow_{AS}$ is defined as the least one satisfying $\langle \mathcal{Q}[A]; \sigma \rangle \quad \rightarrow_{AS} \quad \langle (\mathcal{Q}[A/v])\theta; \sigma\theta \rangle$, where A is the selected atom in $\mathcal{Q}$, $\theta = mgu(\{H = A\})^1$ and "H with v" in $\mathcal{P}$. An admissible derivation is a sequence $\langle \mathcal{Q}; id \rangle \rightarrow_{AS} \cdots \rightarrow_{AS} \langle \mathcal{Q}'; \theta \rangle$.*

If we exploit all atoms of a given goal, by applying admissible steps as much as needed during the operational phase, then it becomes a formula with no atoms (a L-expression) which can be then interpreted w.r.t. lattice L as follows.

Definition 2 (Fuzzy Computed Answer). *Let $\mathcal{P}$ be a program, $\mathcal{Q}$ a goal and σ a substitution. Assume that $[\![@]\!]$ is the truth function of connective @ in the lattice $(L, \leq)$ associated to $\mathcal{P}$, such that, for values $r_1, \ldots, r_n, r_{n+1} \in L$, we have that $[\![@]\!](r_1, \ldots, r_n) = r_{n+1}$. Then, we formalize the notion of interpretive computation as a state transition system, whose transition relation $\rightarrow_{IS}$ is defined as the least one satisfying: $\langle \mathcal{Q}[@(r_1, \ldots, r_n)]; \sigma \rangle \quad \rightarrow_{IS} \quad \langle \mathcal{Q}[@(r_1, \ldots, r_n)/r_{n+1}]; \sigma \rangle$. An interpretive derivation is a sequence $\langle \mathcal{Q}; \sigma \rangle \rightarrow_{IS} \cdots \rightarrow_{IS} \langle \mathcal{Q}'; \sigma \rangle$. When $\mathcal{Q}' = r \in L$, the state $\langle r; \sigma \rangle$ is called a fuzzy computed answer (f.c.a.) for that derivation.*

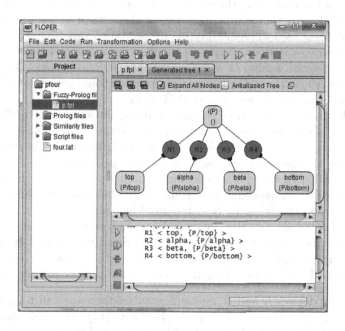

Fig. 2. A work-session with FLOPER solving goal i(P).

The parser of our FLOPER tool [18] has been implemented by using the Prolog language. Once the application is loaded inside a Prolog interpreter, it

---

[1] Here $mgu(E)$ denotes the *most general unifier* of an equation set E [15].

shows a menu which includes options for loading/compiling, parsing, listing and saving MALP programs, as well as for executing/debugging fuzzy goals. Moreover, FLOPER has been recently equipped with new options, called "lat" and "show", for allowing the possibility of respectively changing and displaying the lattice associated to a given program.

A very easy way to model truth-degree lattices for being included into the FLOPER tool is based on the following guidelines. All relevant components of each lattice are encapsulated inside a Prolog file which must necessarily contain the definitions of a minimal set of predicates defining the set of valid elements (member/1 predicate), the top and bottom elements (top/1 and bot/1 predicates), the full or partial ordering established among them (leq/2 predicate), as well as the repertoire of fuzzy connectives which can be used for their subsequent manipulation. If we have, for instance, some fuzzy connectives of the form $\&_{label_1}$ (conjunction), $|_{label_2}$ (disjunction) or $@_{label_3}$ (aggregation) with arities n_1, n_2 and n_3 respectively, we must provide clauses defining the *connective predicates* "and_$label_1$/(n_1+1)", "or_$label_2$/(n_2+1)" and "agr_$label_3$/(n_3+1)", where the extra argument of each predicate is intended to contain the result achieved after the evaluation of the proper connective. Finally, for the purposes of the current work, we also require for each lattice a Prolog fact of the form members(L) being the L a list containing the set of truth degrees belonging to the modeled lattice (or at least a representative subset of them when working with infinite lattices) for being used when interpreting propositional variables of fuzzy formulae. For instance, a lattice defining the simplest notion of binary lattice should implement predicate member/1 with facts member(0) and member(1) (including also members([0,1])) and the Boolean conjunction could be defined by the pair of facts and_bool(0,_,0) and and_bool(1,X,X).

Consider now the following partially ordered lattice $\mathcal{F}$ in the diagram of Fig. 3, which is equipped with conjunction, disjunction and implication connectives based on the *Gödel* logic described in Fig. 1, but with the particularity that now, in the general case, the *Gödel*'s conjunction must be expressed as $\&_G(x, y) \triangleq inf(x, y)$, where it is important to note that we must replace the use of "*min*" by "*inf*" in the connective definition (and similarly for the disjunction connective, where "*max*" must be substituted by "*sup*").

To this end, observe in the Prolog code accompanying the graphic in Fig. 3 that we have introduced clauses defining the primitive operators "pri_inf/3" and "pri_sup/3" which are intended to return the *infimum* and *supremum* of two elements. Related with this fact, we must point out the following aspects:

- Since truth degrees α and β are incomparable, then any call to goals of the form "?- leq(alpha,beta)." or "?- leq(beta,alpha)." will always fail.
- The goal "?- pri_inf(alpha,beta,X).", instead of failing, successfully produces the desired result "X=bottom".
- Note anyway that the implementation of the "pri_inf/3" predicate is mandatory for coding the general definition of "and_godel/3" (a similar reasoning follows for "pri_sup/3" and "or_godel/3").

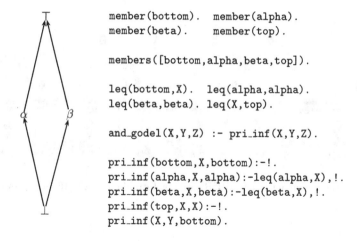

```
member(bottom).   member(alpha).
member(beta).     member(top).

members([bottom,alpha,beta,top]).

leq(bottom,X).   leq(alpha,alpha).
leq(beta,beta). leq(X,top).

and_godel(X,Y,Z)  :- pri_inf(X,Y,Z).

pri_inf(bottom,X,bottom):-!.
pri_inf(alpha,X,alpha):-leq(alpha,X),!.
pri_inf(beta,X,beta):-leq(beta,X),!.
pri_inf(top,X,X):-!.
pri_inf(X,Y,bottom).
```

Fig. 3. Lattice of truth degrees $\mathcal{F}$ modeled in Prolog.

3   Looking for Models with FuzzyXPath

The subset of the MALP language detailed in Sect. 2 suffices for developing
a simple fuzzy theorem prover, where it is important to remark that our tool
can cope with different lattices (not only the real interval [0,1]) containing a
finite number of elements -marked in "members"- maintaining full or partial
ordering among them. Hence, we can use FLOPER for enumerating the whole
set of interpretations and models of fuzzy formulae. To this end, only a concrete
lattice L is required in order to automatically build a program composed by a
set of facts of the form "$i(\alpha)$ *with* α", for each $\alpha \in L$. For instance, the MALP
program associated to lattice $\mathcal{F}$ in Fig. 3 looks like:

```
i(top)       with    top.
i(alpha)     with    alpha.
i(beta)      with    beta.
i(bottom)    with    bottom.
```

Once the lattice and the residual program have been loaded into FLOPER, the
formula to be evaluated is introduced as a goal to the system following some
conventions:

- If P is a propositional variable in the original formula, then it is denoted as
  "i(P)" in the goal F.
- If & is a conjunction of a certain logic "label" in the original formula, then it
  is denoted as "&label" in goal F.
- For disjunctions, negations and implications, use respectively the patterns
  "|label", "@no_label" and "@im_label" in F.
- For other aggregators use "@label" in F.

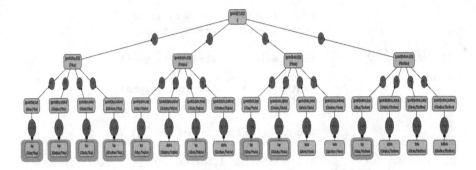

Fig. 4. A work-session with FLOPER solving formula $P \vee Q$ (16 interpretations, 9 models).

In what follows we discuss some examples related with the lattice shown in Fig. 3 and its residual MALP program just seen before. Firstly, if we execute goal "i(P)" into FLOPER, we obtain the four interpretations for P shown in Fig. 2. On the other hand, consider now the propositional formula $P \vee Q$, which is translated into the MALP goal "(i(P) | i(Q))" and after being executed with FLOPER, the tool returns a tree as seen in Fig. 4 whose 16 leaves represent the whole set of interpretations, where 9 of them -inside blue clouds- are models (see part of the corresponding XML file produced by FLOPER in Fig. 5). Here, each state contains its corresponding goal and substitution components and they are drawn inside yellow circles. Admissible steps, coloured in blue, are labelled with the program rule they exploit. Finally, those blue circles annotated with word "*is*", correspond to interpretive steps. Sometimes we include blue circles labelled with "*result*" which represents a chained sequence of interpretive steps.

Let us recall now that XPath was designed as a query language for XML text in which the path of the underlying tree of any XML document is used to describe the query (subsequent nodes on XPath expressions are separated by one slash '/' or a double slash '//', being this last case useful for overriding several nodes). Moreover, XPath expressions can be adorned with Boolean conditions (between square brackets '[]') on nodes and leaves to restrict the number of answers of the query. In our fuzzy version of XPath, a FUZZYXPATH expression defines, w.r.t. an XML document, a sequence of subtrees of the XML document where each subtree has an associated *retrieval status value, rsv*. XPath conditions, which are defined as fuzzy operators applied to XPath expressions, compute a new rsv from the rsv's of the involved XPath expressions, which at the same time, provides a rsv to the node. We consider three fuzzy versions for each one of the classical conjunction and disjunction operators describing *pessimistic, realistic* and *optimistic* scenarios, see Fig. 1. In XPath expressions the fuzzy versions of the connectives make harder to hold conditions, and therefore can be used to debilitate/force conditions. Furthermore, assuming two given *rsv*'s r_1 and r_2, the *avg* operator is obviously defined with a fuzzy taste as $(r_1 + r_2)/2$, whereas its *priority-based* variant, i.e. $avg\{p_1, p_2\}$, is defined as $(p_1 * r_1 + p_2 * r_2)/p_1 + p_2$.

```
<node>
  <rule>R0</rule>
  <goal>or_godel(i(P),i(Q))</goal>
  <substitution>{}</sub>
  <children>
    <node>
      <rule>R1</rule>
      <goal>or_godel(bottom,i(Q))</goal>
      <sub>{P/bottom}</sub>
      <children>
        <node>
          <rule>R1</rule>
          <goal>or_godel(bottom,bottom)</goal>
          <sub>{Q/bottom,P/bottom}
          </sub>
          <children>
            <node>
              <rule>result</rule>
              <goal>bottom</goal>
              <sub>{Q/bottom,P/bottom}
              </sub>
              <children>
              </children>
            </node>
          </children>
        </node>
      </children>
      </node>
      . . .
```

Fig. 5. Part of the XML file representing the execution tree shown in Fig. 4.

With our FUZZYXPATH tool we have executed "//node[goal='top']/sub"
against the XML file shown in Fig. 5, which was generated by FLOPER when
producing the proof tree drawn in Fig. 4, thus returning as output the new
XML document listed in Fig. 6. As illustrated in Fig. 5, note that the XML files
representing proof trees exported by FLOPER, are always rooted with the **node**
label, whose children are based on four kinds of 'tags' (this structure is nested
as much as needed):

- **rule**, which indicates the program rule exploited to reach the current node
  (the virtual rule R0 is pointed out only in the initial node),
- **goal**, which contains the MALP expression under evaluation, that is, the formula that the system is trying to prove on its current initial/intermediate/final
  step. Note that, when in our example such value is **top**, then we have found a
  model, where the values assigned to the propositional symbols of the formula
  are collected in the following tag...
- **sub**, acronym of "substitution", which accumulates the variable bindings performed along a fuzzy logic derivation (i.e., chain of computational steps along
  a branch of the execution tree) and whose meaning in our target setting,

```
<result>
  <sub rsv=1>{Q/top,P/top}</sub>
  <sub rsv=1>{Q/alpha,P/top}</sub>
  <sub rsv=1>{Q/beta,P/top}</sub>
  <sub rsv=1>{Q/bottom,P/top}</sub>
  <sub rsv=1>{Q/top,P/alpha}</sub>
  <sub rsv=1>{Q/beta,P/alpha}</sub>
  <sub rsv=1>{Q/top,P/beta}</sub>
  <sub rsv=1>{Q/alpha,P/beta}</sub>
  <sub rsv=1>{Q/top,P/bottom}</sub>
</result>
```

Fig. 6. XML file obtained after evaluating an XPath query.

reveals the way of interpreting the propositions contained on a formula for obtaining its models. See for instance Fig. 6, where the nine solutions labeled with this tag and reported in the output XML document, indicate each one the truth values for the propositional variables that satisfy the formula with the maximum truth degree. And finally,

- `children`, which contains the set of underlying nodes of the tree in a nested way.

Consider now the more involved formula $P \wedge Q \rightarrow P \vee Q$ which becomes into the MALP goal "`(i(P) & i(Q)) @impl (i(P) | i(Q))`". When interpreted by FLOPER, the system returns a list of answers having all them the maximum truth degree "*top*", which proves that this formula is a tautology, as wanted. Assume now a more general version with the following shape $F_n = P_1 \wedge \ldots \wedge P_n \rightarrow P_1 \vee \ldots \vee P_n$. With respect to the efficiency of the method presented here, we have studied the behaviour of formula F_n in the table of Fig. 7. In the horizontal axis we represent the number n of different propositional variables appearing in the formula, whereas the vertical axis refers to the number of seconds needed to obtain the whole set of interpretations (all them are models in this case) for the formula. The benchmarks have been performed using a computer with processor Intel Core Duo, with 2 GB RAM and Windows Vista. Both the red and blue lines refers to the method just commented along this paper, but whereas the red line indicates that the derivation tree has been produced by performing admissible and interpretive steps according Definitions 1 and 2, respectively, the blue line refers to the execution of the Prolog code obtained after compiling with FLOPER the MALP program and goal associated to our intended formula.

The results achieved in Fig. 7 show that our method has a nice behaviour in both cases, even for formulae with a big number of propositional variables. Of course, the method does not try to compete with SAT techniques (which are always faster and can deal with more complex formulae containing many more propositional variables), but it is important to remark again that in our case, we face the problem of finding the whole set of models for a given formula, instead of only focusing on satisfiability.

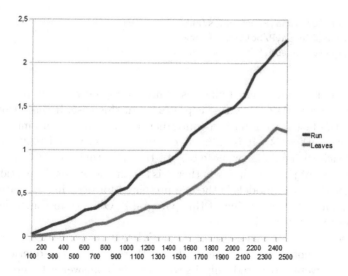

Fig. 7. Efficiency of the method.

We address now formula F_n because it illustrates one key point of this paper. Note that there are $|L|^n$ interpretations for that formula, where $|L|$ is the cardinality of the carrier set of lattice L that models truth degrees. For our example lattice of Fig. 3, with four elements, we have 4^n interpretations. Consider, for example, that we are interested in proving that a certain formula, say F_5, is a tautology. In [11] we would have to search at least one interpretation that is not model of F_5 to prove that it is not a tautology, but since there exist $4^5 = 1024$ interpretations, this task is not suitable to be made by hand. To overcome this problem we use FUZZYXPATH to automatically search in the XML file generated by FLOPER. The manual task, then, is reduced to designing the FUZZYXPATH query. In this case, since we are interested in proving that F_5 is a tautology, our FUZZYXPATH query should be //node[rule='result' & goal<>'top']/sub, that is, the system searches nodes whose rule tag contain the text "result" (i.e., we are looking for leaves in the tree) and whose tag goal is not "top" (in order to exclude models). If the output of this query is an empty list of nodes, as it actually is, the formula F_5 is proven to be a tautology, as desired.

FUZZYXPATH can also be used for determining the satisfiability of a formula. Consider again formula $P \lor Q$ whose set of interpretations are shown in Fig. 4. The query //node[rule='result' & goal<>'top']/sub seen above, shows that this formula is not a tautology, since its further evaluation returns the non-empty set:

```
<result>
  <sub rsv=1>{Q/alpha,P/alpha}</sub>
  <sub rsv=1>{Q/bottom,P/alpha}</sub>
  <sub rsv=1>{Q/beta,P/beta}</sub>
  <sub rsv=1>{Q/bottom,P/beta}</sub>
```

```
<sub rsv=1>{Q/alpha,P/bottom}</sub>
<sub rsv=1>{Q/beta,P/bottom}</sub>
<sub rsv=1>{Q/bottom,P/bottom}</sub>
</result>
```

Consider now the new query (which is almost antagonist to the previous one) //node[rule='result' & goal='top']/sub. In this case, if the output is the empty set, the tested formula is a contradiction (i.e., there is no interpretation satisfying it). Otherwise, it is satisfiable. Furthermore, with FUZZYXPATH we can come back to the main purpose of [11], that is listing the set of models of a formula instead of just deciding whether it is satisfiable or not. In particular, the query to list the set of models is the one presented for deciding the satisfiability of the formula at the beginning of this paragraph. Observe in Fig. 6 the output of this query w.r.t. formula $P \vee Q$.

Until now we have made use of FUZZYXPATH to decide immediately the satisfiability or not of a certain formula. With respect to the queries we have presented, we were interested only in whether their answer set were empty or not. Now we present a query which, by making use of the fuzzy capabilities of FUZZYXPATH, returns the list of interpretations together with extra information (into the rsv attribute) about the extent in which they satisfy the formula or not. Consider again formula $P \vee Q$, part of whose derivation tree is represented in the form of the XML file provided by FLOPER in Fig. 5. This formula is satisfiable but not a tautology, that is, some of its interpretations satisfy it but other ones do not.

Let us focus now on query //node[rule='result'&(goal='top' avg{3,1}, goal<>'top')]/sub for such formula. Here, we ask for those states which are leaves of the tree (condition rule='result') and which are either models (condition goal='top') or not (condition goal<>'top'), with the particularity that if the leaf is a model, it fulfils the query at a 75 % and, if it is not, with a 25 %. The result is the set of interpretations with a *rsv* value (the degree in which they fulfil the query) between 0.75 and 0.25, as shown in the following table:

```
<result>
  <sub rsv=0.75>{Q/top,P/top}</sub>
  <sub rsv=0.75>{Q/alpha,P/top}</sub>
  <sub rsv=0.75>{Q/beta,P/top}</sub>
  <sub rsv=0.75>{Q/bottom,P/top}</sub>
  <sub rsv=0.75>{Q/top,P/alpha}</sub>
  <sub rsv=0.75>{Q/beta,P/alpha}</sub>
  <sub rsv=0.75>{Q/top,P/beta}</sub>
  <sub rsv=0.75>{Q/alpha,P/beta}</sub>
  <sub rsv=0.75>{Q/top,P/bottom}</sub>
  <sub rsv=0.25>{Q/alpha,P/alpha}</sub>
  <sub rsv=0.25>{Q/bottom,P/alpha}</sub>
  <sub rsv=0.25>{Q/beta,P/beta}</sub>
  <sub rsv=0.25>{Q/bottom,P/beta}</sub>
  <sub rsv=0.25>{Q/alpha,P/bottom}</sub>
  <sub rsv=0.25>{Q/beta,P/bottom}</sub>
```

```
<sub rsv=0.25>{Q/bottom,P/bottom}</sub>
</result>
```

This set of answers briefly show the set of interpretations of the formula. For formulas like F_5, whose XML file of 5.5 MB would be impossible to check by hand, this method offers a quick look of the answers, even when they are very numerous.

4 Conclusions and Future Work

In this paper we have recasted from our previous works [5,11], two applications developed with our fuzzy logic programming environment FLOPER in order to feedback and reinforce themselves. In the first paper we proposed a technique for evaluating propositional fuzzy formulae in an alternative way than fuzzy SAT/SMT methods, while in the second work we used the FuzzyXPath interpreter for analyzing derivation trees exported by FLOPER in XML format in order to help the analysis of fuzzy logic computations. In the current paper we have applied this last capability of FuzzyXPath focusing exclusively on derivation trees associated to fuzzy formulae developed according the methodology proposed in [11]. As a result, we have presented an automatic technique useful for determining important features of such formulae (tautology, contradiction, etc.) by making use of XPath queries with a fuzzy taste. As future work, we are nowadays introducing fuzzy thresholding techniques in our application for improving the efficiency of the tool.

References

1. Almendros-Jiménez, J.M., Luna, A., Moreno, G.: A XPath debugger based on fuzzy chance degrees. In: Herrero, P., Panetto, H., Meersman, R., Dillon, T. (eds.) OTM-WS 2012. LNCS, vol. 7567, pp. 669–672. Springer, Heidelberg (2012)
2. Almendros-Jiménez, J.M., Luna, A., Moreno, G.: Fuzzy logic programming for implementing a flexible XPath-based query language. Electron. Notes Theor. Comput. Sci. **282**, 3–18 (2012)
3. Almendros-Jiménez, J.M., Luna Tedesqui, A., Moreno, G.: Annotating "fuzzy chance degrees" when debugging XPath queries. In: Rojas, I., Joya, G., Cabestany, J. (eds.) IWANN 2013, Part II. LNCS, vol. 7903, pp. 300–311. Springer, Heidelberg (2013)
4. Almendros-Jiménez, J.M., Luna, A., Moreno, G.: Fuzzy XPath through fuzzy logic programming. New Gener. Comput. **33**(2), 173–209 (2015)
5. Almendros-Jiménez, J.M., Luna, A., Moreno, G., Vázquez, C.: Analyzing fuzzy logic computations with fuzzy XPath. In: Proceedings of XIII Spanish Conference on Programming and Languages, PROLE 2013, pp. 136–150 (extended version to appear in ECEASST). Complutense University of Madrid (2013)
6. Ansótegui, C., Bofill, M., Manyà, f., Villaret, M.: Building automated theorem provers for infinitely-valued logics with satisfiability modulo theory solvers. In: Proceedings of the 42nd IEEE International Symposium on Multiple-Valued Logic, ISMVL 2012, pp. 25–30 (2012)

7. Apt, K.R.: Introduction to logic programming. In: van Leeuwen, J. (ed.) Handbook of Theoretical Computer Science. Volume B: Formal Models and Semantics, pp. 493–574. Elsevier, MIT Press, Amsterdam, Cambridge (1990)

8. Arcelli, F., Formato, F.: A similarity-based resolution rule. Int. J. Intell. Syst. **17**(9), 853–872 (2002)

9. Barrett, C.W., Sebastiani, R., Seshia, S.A., Tinelli, C.: Satisfiability modulo theories. In: Biere, A., Heule, M., van Mareen, H., Walsh, T. (eds.) Handbook of Satisfiability. Frontiers in Artificial Intelligence and Applications, vol. 185, pp. 825–885. IOS Press, Amsterdam (2009)

10. Berglund, A., Boag, S., Chamberlin, D., Fernandez, M.F., Kay, M., Robie, J., Siméon, J.: XML path language (XPath) 2.0. In: W3C (2007)

11. Bofill, M., Moreno, G., Vázquez, C., Villaret, M.: Automatic proving of fuzzy formulae with fuzzy logic programming and SMT. In: Proceedings of XIII Spanish Conference on Programming and Languages, PROLE 2013, pp. 151–165 (extended version to appear in ECEASST). Complutense University of Madrid (2013)

12. Bratko, I.: Prolog Programming for Artificial Intelligence. Addison Wesley, Harlow (2000)

13. Gerla, G.: Fuzzy control as a fuzzy deduction system. Fuzzy Sets Syst. **121**(3), 409–425 (2001)

14. Kifer, M., Subrahmanian, V.S.: Theory of generalized annotated logic programming and its applications. J. Logic Program. **12**, 335–367 (1992)

15. Lassez, J.L., Maher, M.J., Marriott, K.: Unification revisited. In: Minker, J. (ed.) Foundations of Deductive Databases and Logic Programming, pp. 587–625. Morgan Kaufmann, Los Altos (1988)

16. Lloyd, J.W.: Foundations of Logic Programming. Springer, Berlin (1987)

17. Medina, J., Ojeda-Aciego, M., Vojtáš, P.: Similarity-based unification: a multi-adjoint approach. Fuzzy Sets Syst. (Elsevier) **146**, 43–62 (2004)

18. Moreno, G., Vázquez, C.: Fuzzy logic programming in action with FLOPER. J. Softw. Eng. Appl. **7**, 273–298 (2014)

19. Muñoz-Hernández, S., Pablos-Ceruelo, V., Strass, H.: Rfuzzy: syntax, semantics and implementation details of a simple and expressive fuzzy tool over prolog. Inf. Sci. **181**(10), 1951–1970 (2011)

20. Stickel, M.E.: A prolog technology theorem prover: implementation by an extended prolog compiler. J. Autom. Reasoning **4**(4), 353–380 (1988)

21. Vidal, A., Bou, F., Godo, L.: An SMT-based solver for continuous t-norm based logics. In: Hüllermeier, E., Link, S., Fober, T., Seeger, B. (eds.) SUM 2012. LNCS, vol. 7520, pp. 633–640. Springer, Heidelberg (2012)

ERBlox: Combining Matching Dependencies with Machine Learning for Entity Resolution

Zeinab Bahmani[1], Leopoldo Bertossi[1(✉)], and Nikolaos Vasiloglou[2]

[1] Carleton University, School of Computer Science, Ottawa, Canada
zeinabbahmani@cmail.carleton.ca, bertossi@scs.carleton.ca
[2] LogicBlox Inc., Atlanta, GA 30309, USA

Abstract. Entity resolution (ER), an important and common data cleaning problem, is about detecting data duplicate representations for the same external entities, and merging them into single representations. Relatively recently, declarative rules called *matching dependencies* (MDs) have been proposed for specifying similarity conditions under which attribute values in database records are merged. In this work we show the process and the benefits of integrating three components of ER: (a) Classifiers for duplicate/non-duplicate record pairs built using machine learning (ML) techniques, (b) MDs for supporting both the blocking phase of ML and the merge itself; and (c) The use of the declarative language *LogiQL* -an extended form of Datalog supported by the *LogicBlox* platform- for data processing, and the specification and enforcement of MDs.

Keywords: Entity resolution · Matching dependencies · Support-vector machines · Classification · Datalog

1 Introduction

Entity resolution (ER) is a common and difficult problem in data cleaning that has to do with handling unintended multiple representations in a database of the same external objects. Multiple representations lead to uncertainty in data and the problem of managing it. Cleaning the database reduces uncertainty. In more precise terms, ER is about the identification and fusion of database records (think of rows or tuples in tables) that represent the same real-world entity [8,15]. As a consequence, ER usually goes through two main consecutive phases: (a) detecting duplicates, and (b) merging them into single representations.

For duplicate detection, one must first analyze multiple pairs of records, comparing the two records in them, and discriminating between: *pairs of duplicate records* and *pairs of non-duplicate records*. This classification problem is approached with machine learning (ML) methods, to learn from previously known or already made classifications (a training set for supervised learning), building a *classification model* (a classifier) for deciding about other record pairs [10,15].

In principle, in ER every two records (forming a pair) have to be compared, and then classified. Most of the work on applying ML to ER work at the record

© Springer International Publishing Switzerland 2015
C. Beierle and A. Dekhtyar (Eds.): SUM 2015, LNAI 9310, pp. 399–414, 2015.
DOI: 10.1007/978-3-319-23540-0_27

level [10,11,21], and only some of the attributes, or their features, i.e. numerical values associated to them, may be involved in duplicate detection. The choice of relevant sets of attributes and features is application dependent.

ER may be a task of quadratic complexity since it requires comparing every two records. To reduce the large number two-record comparisons, *blocking techniques* are used [2,19,23]. Commonly, a single record attribute, or a combination of attributes, the so-called *blocking key*, is used to split the database records into blocks. Next, under the assumption that any two records in different blocks are unlikely to be duplicates, only every two records in a same block are compared for duplicate detection.

Although blocking will discard many record pairs that are obvious non-duplicates, some true duplicate pairs might be missed (by putting them in different blocks), due to errors or typographical variations in attribute values. More interestingly, similarity between blocking keys alone may fail to capture the relationships that naturally hold in the data and could be used for blocking. Thus, entity blocking based only on blocking key similarities may cause low recall. This is a major drawback of traditional blocking techniques.

In this work we consider different and coexisting entities. For each of them, there is a collection of records. Records for different entities may be related via attributes in common or referential constraints. Blocking can be performed on each of the participating entities, and the way records for an entity are placed in blocks may influence the way the records for another entity are assigned to blocks. This is called "collective blocking". Semantic information, in addition to that provided by blocking keys for single entities, can be used to state relationships between different entities and their corresponding similarity criteria. So, blocking decision making forms a collective and intertwined process involving several entities. In the end, the records for each individual entity will be placed in blocks associated to that entity.

Example 1. Consider two entities, Author and Paper. For each of them, there is a set of records (for all practical purposes, think of database tuples in a single table). For Author we have records of the form $\mathbf{a} = \langle name, \ldots, affiliation, \ldots, paper\ title, \ldots \rangle$, with {*name, affiliation*} the blocking key; and for Paper, records of the form $\mathbf{p} = \langle title, \ldots, author\ name, \ldots \rangle$, with *title* the blocking key. We want to group Author and Paper records at the same time, in an entwined process. We block together two Author entities on the basis of the similarities of authors' names and affiliations.

Assume that Author entities $\mathbf{a}_1, \mathbf{a}_2$ have similar names, but their affiliations are not. So, the two records would not be put in the same block. However, $\mathbf{a}_1, \mathbf{a}_2$ are authors of papers (in Paper records) $\mathbf{p}_1, \mathbf{p}_2$, resp., which have been put in the same block (of papers) on the basis of similarities of paper titles. In this case, additional *semantic knowledge* might specify that if two papers are in the same block, then corresponding Author records that have similar author names should be put in the same block too. Then, $\mathbf{a}_1$ and $\mathbf{a}_2$ would end up in the same block.

In this example, we are blocking Author and Paper entities, separately, but collectively and in interaction. ∎

Collective blocking is based on blocking keys and *the enforcement* of semantic information about the *relational closeness* of entities Author and Paper, which is captured by a set of *matching dependencies* (MDs). So, we propose "MD-based collective blocking" (more on MDs right below).

After records are divided in blocks, the proper duplicate detection process starts, and is carried out by comparing every two records in a block, and classifying the pair as "duplicates" or "non-duplicates" using the trained ML model at hand. In the end, records in duplicate pairs are considered to represent the same external entity, and have to be *merged* into a single representation, i.e. into a single record. This second phase is also application dependent. MDs were originally proposed to support this task.

Matching dependencies are declarative logical rules that tell us under what conditions of similarity between attribute values, any two records must have certain attribute values merged, i.e. made identical [16,17]. For example, the MD

$$Dept_B[dept] \approx Dept_B[dept] \; \rightarrow \; Dept_B[city] \doteq Dept_B[city] \tag{1}$$

tells us that for any two records for entity (or relation or table) $Dept_B$ that have similar values for attribute *dept* attribute, their values for attribute *city* should be matched, i.e. made the same.

MDs as introduced in [17] do not specify how to merge values. In [6,7], MDs were extended with *matching functions* (MFs). For a data domain, an MF specifies how to assign a value in common to two values. We adopt MDs with MFs in this work. In the end, the enforcement of MDs with MFs should produce a duplicate-free instance (cf. Section 2 for more details).

MDs have to be specified in a declarative manner, and at some point enforced, by producing changes on the data. For this purpose, we use the *LogicBlox* platform, a data management system developed by the LogicBlox[1] company, that is centered around its declarative language, *LogiQL*. *LogiQL* supports relational data management and, among several other features [1], an extended form of Datalog with stratified negation [9]. This language is expressive enough for the kind of MDs considered in this work.[2]

In this paper, we describe our *ERBlox* system. It is built on top of the *LogicBlox* platform, and implements entity resolution (ER) applying to *LogiQL*, ML techniques, and the specification and enforcement of MDs. More specifically, *ERBlox* has three main components: (a) MD-based collective blocking, (b) ML-based duplicate detection, and (c) MD-based merging. The sets of MDs are fixed and different for the first and last components. In both cases, the set of MDs are *interaction-free* [7], which results, for each entity, in the unique set of blocks, and eventually into a single, duplicate-free instance [7]. We use *LogicQL* to declaratively implement the two MD-based components of *ERBlox*.

The blocking phase uses MDs to specify the blocking strategy. They express conditions in terms of blocking key similarities and also relational closeness (the

---

[1] www.logicblox.com.

[2] For arbitrary sets of MDs, we need higher expressive power [7], such as that provided by answer set programming [3].

semantic knowledge) to assign two records to a same block (by making the block identifiers identical). Then, under MD-based collective blocking different records of possibly several related entities are simultaneously assigned to blocks through the enforcement of MDs (cf. Sect. 5 for details).

On the ML side, the problem is about detecting pairs of duplicate records. The ML algorithm is trained using record-pairs known to be duplicates or non-duplicates. We independently used three established classification algorithms: *support vector machines* (SVMs) [24], *k-nearest neighbor* (K-NN) [14], and *non-parametric Bayes classifier* (NBC) [4]. We used the Ismion[3] implementations of them due to the in-house expertise at LogicBlox. Since the emphasis of this work is on the use of *LogiQL* and MDs, we will refer only to our use of SVMs.

We experimented with our *ERBlox* system using as dataset a snapshot of Microsoft Academic Search (MAS)[4] (as of January 2013) including 250 K authors and 2.5 M papers. It contains a training set. The experimental results show that our system improves ER accuracy over traditional blocking techniques [18], which we will call *standard* blocking, where just blocking-key similarities are used. Actually, MD-based collective blocking leads to higher precision and recall on the given datasets.

This paper is structured as follows. Section 2 introduces background on matching dependencies and their semantics, and SVMs. A general overview of the *ERBlox* system is presented in Sect. 3. The specific components of *ERBlox* are discussed in Sects. 4, 5, and 6. Experimental results are shown in Sect. 7. Section 8 presents conclusions.

2 Preliminaries

2.1 Matching Dependencies

We consider an application-dependent relational schema $\mathcal{R}$, with a data domain U. For an attribute A, Dom_A is its finite domain. We assume predicates do not share attributes, but different attributes may share a domain. An instance D for $\mathcal{R}$ is a finite set of ground atoms of the form $R(c_1, \ldots, c_n)$, with $R \in \mathcal{R}$, $c_i \in U$.

We assume that each entity is represented by a relational predicate, and its tuples or rows in its extension correspond to records for the entity. As in [7], we assume records have unique, fixed, global identifiers, *rids*, which are positive integers. This allows us to trace changes of attribute values in records. Record ids are placed in an extra attribute for $R \in \mathcal{R}$ that acts as a key. Then, records take the form $R(r, \bar{r})$, with r the rid, and $\bar{r} = (c_1, \ldots, c_n)$. Sometimes we leave rids implicit, and sometimes we use them to denote whole records: if r is a record identifier in instance D, $\bar{r}$ denotes the record in D identified by r. Similarly, if $\mathcal{A}$ is a sublist of the attributes of predicate R, then $r[\mathcal{A}]$ denotes the restriction of $\bar{r}$ to $\mathcal{A}$.

---

[3] http://www.ismion.com.

[4] http://academic.research.microsoft.com. For comparison, we also tested our system with data from DBLP and Cora.

MDs are formulas of the form: $R_1[\bar{X}_1] \approx R_2[\bar{X}_2] \rightarrow R_1[\bar{Y}_1] \doteq R_2[\bar{Y}_2]$ [16,17]. Here, $R_1, R_2 \in \mathcal{R}$ (and may be the same); and $\bar{X}_1, \bar{X}_2$ are lists of attribute names of the same length that are *pairwise comparable*, that is, X_1^i and X_2^i, and also $\bar{Y}_1, \bar{Y}_2$, share the same domain.[5] The MD says that, for every pair of tuples (one in relation R_1, the other in relation R_2) where the LHS is true, the attribute values in them on the RHS have to be made identical. Symbol $\approx$ denotes generic, reflexive, symmetric, and application/domain dependent similarity relations on shared attribute domains.

A *dynamic, chase-based semantics* for MDs with matching functions (MFs) was introduced in [7]. Given an initial instance D, the set Σ of MDs is iteratively enforced until they cannot be applied any further, at which point a *resolved instance* has been produced. In order to *enforce* (the RHSs of) MDs, there are binary *matching functions* (MFs) $m_A : Dom_A \times Dom_A \rightarrow Dom_A$; and $m_A(a, a')$ is used to replace two values $a, a' \in Dom_A$ that have to be made identical. MFs are idempotent, commutative, and associative, and then induce a partial-order structure $\langle Dom_A, \preceq_A \rangle$, with: $a \preceq_A a' :\Leftrightarrow m_A(a, a') = a'$ [5,6]. It always holds: $a, a' \preceq_A m_A(a, a')$. In this work, MFs are treated as built-in relations.

There may be several resolved instances for D and Σ. However, when (a) MFs are similarity-preserving (i.e., $a \approx a'$ implies $a \approx m_A(a', a'')$); or (b) Σ is interaction-free (i.e., each attribute may appear in either the RHS or LHS of MDs in Σ), there is a unique resolved instance that is computable in polynomial time in $|D|$ [7].

2.2 Support Vector Machines

The SVMs technique [24] is a form of kernel-based learning. SVMs can be used for classifying vectors in an inner-product vector space $\mathcal{V}$ over $\mathbb{R}$. Vectors are classified in two classes, with a label in $\{0, 1\}$. The algorithm learns from a training set, say $\{(\mathbf{e}_1, f(\mathbf{e}_1)), (\mathbf{e}_2, f(\mathbf{e}_2)), (\mathbf{e}_3, f(\mathbf{e}_3)), \dots, (\mathbf{e}_n, f(\mathbf{e}_n))\}$. Here, $\mathbf{e}_i \in \mathcal{V}$, and for the *feature* (function) f: $f(\mathbf{e}_i) \in \{0, 1\}$.

SVMs find an optimal hyperplane, $\mathcal{H}$, in $\mathcal{V}$ that separates the two classes where the training vectors are classified. Hyperplane $\mathcal{H}$ has an equation of the form $\mathbf{w} \bullet \mathbf{x} + b$, where $\bullet$ denotes the inner product, $\mathbf{x}$ is a vector variable, $\mathbf{w}$ is a weight vector of real values, and b is a real number. Now, a new vector $\mathbf{e}$ in $\mathcal{V}$ can be classified as positive or negative depending on the side of $\mathcal{H}$ it lies. This is determined by computing $h(\mathbf{e}) := sign(\mathbf{w} \bullet \mathbf{e} + b)$. If $h(\mathbf{e}) > 0$, $\mathbf{e}$ belongs to class 1; otherwise, to class 0.

It is possible to compute real numbers $\alpha_1, \dots, \alpha_n$, such that the classifier h can be computed through: $h(\mathbf{e}) = sign(\sum_i \alpha_i \cdot f(\mathbf{e}_i) \cdot \mathbf{e}_i \bullet \mathbf{e} + b)$ (cf. Fig. 3).

3 Overview of *ERBlox*

A high-level description of the components of *ERBlox* is given in Fig. 1. It shows the workflow supported by *ERBlox* when doing ER. *ERBlox*'s three main compo-

---

[5] A more precise notation for the MD would be: $\forall x_1^1 \cdots \forall y_2^m (\bigwedge_j R_1[x_1^j] \approx_j R_2[x_2^j] \longrightarrow \bigwedge_k R_1[y_1^k] \doteq R_2[y_2^k])$.

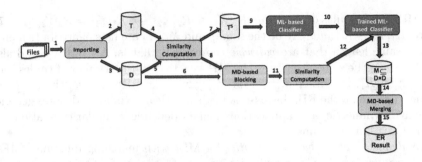

Fig. 1. Overview of *ERBlox*

nents are: (1) MD-based collective blocking (path **1, 3, 5, {6, 8}**), (2) ML-based record duplicate detection (the whole initial workflow up to task **13**, inclusive), and (3) MD-based merging (path **14, 15**). In the figure, all the boxes in light grey are supported by *LogiQL*. As just done, in the rest of this section, numbers in boldface refer to the edges in this figure.

The initial input data is stored in structured text files. (We assume these data are already standardized and free of misspellings, etc., but duplicates may be present.) Our general *LogiQL* program that supports the whole workflow contains some rules for importing data from the files into the extensions of relational predicates (think of tables, this is edge **1**). This results in a relational database instance T containing the training data (edge **2**), and the instance D on which ER will be performed (edge **3**).

The next main task is blocking, which requires similarity computation of pairs of records in D (edge **5**). For record pairs $\langle r_1, r_2 \rangle$ in T, similarities have to be computed as well (edge **4**). Similarity computation is based on similarity functions, $Sf_i : Dom_{A_i} \times Dom_{A_i} \rightarrow [0, 1]$, each of which assigns a numerical value, called similarity weight, to the comparisons of values for a record attribute A_i (from a

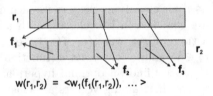

$w(r_1, r_2) = <w_1(f_1(r_1, r_2)), \ ... >$

Fig. 2. Feature-based similarity

pre-chosen subset of attributes) (cf. Fig. 2). A weight vector $w(r_1, r_2) = \langle \cdots, Sf_i(r_1[A_i], r_2[A_i]), \cdots \rangle$ is formed by similarity weights (edge **7**). For more details on similarity computation see Sect. 4.

Since some pairs in T are considered to be duplicates and others non-duplicates, the result of this process leads to a "similarity-enhanced" database T^s of tuples of the form $\langle r_1, r_2, w(r_1, r_2), L \rangle$, with label $L \in \{0, 1\}$ indicating if the two records are duplicates ($L = 1$) or not ($L = 0$). The labels are consistent with the corresponding weight vectors. The classifier is trained using T^s, leading to a classification model (edges **9, 10**).

For records in D, similarity measures are needed for blocking, to decide if two records r_1, r_2 go to the same block. Initially, every record has its rid assigned as block (number). To assign two records to the same block, we use matching dependencies that specify and enforce (through their RHSs) that their blocks have to be identical. This happens when certain similarities between pairs of attribute values appearing in the LHSs of the MDs hold. For this reason, similarity computation is also needed before blocking (workflow **5, 6, 8**). This similarity computation process is similar to the one for T. However, in the case of D, this does not lead directly to the same kind of weight vector computation. Instead, the computation of similarity measures is only for the similarity predicates appearing in the LHSs of the blocking-MDs. (So, as the evaluation of the LHS in (1) requires the computation of similarities for *dept*-string values.)

Notice that these blocking-MDs may capture semantic knowledge, so they could involve in their LHSs similarities of attribute values in records for different kinds of entities. For example, in relation to Example 1, there could be similarity comparisons involving attributes for entities Author and Paper, e.g.

$$Author(x_1, y_1, bl_1) \wedge Paper(y_1, z_1, bl_3) \wedge Author(x_2, y_2, bl_2) \wedge$$
$$Paper(y_2, z_2, bl_4) \wedge x_1 \approx_1 x_2 \wedge z_1 \approx_2 z_2 \;\; \rightarrow \;\; bl_1 \doteq bl_2, \tag{2}$$

expressing that when the similarities on the LHS hold, the blocks bl_1, bl_2 have to be made identical.[6] The similarity comparison atoms on the LHS are considered to be true when the similarity values are above predefined thresholds (edges **5, 8**).[7]

This is the *MD-based collective blocking* stage that results in database D enhanced with information about the blocks to which the records are assigned. Pairs of records with the same block form *candidate duplicate record pairs*, and any two records with different blocks are simply not tested as possible duplicates (of each other).

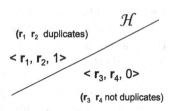

Fig. 3. Classification hyperplane

After the records have been assigned to blocks, pairs of records $\langle r_1, r_2 \rangle$ in the same block are considered for the duplicate test. As this point we proceed as we did for T: the similarity vectors $w(r_1, r_2)$ have to be computed (edges **11, 12**).[8] Next, tuples $\langle r_1, r_2, w(r_1, r_2) \rangle$ are used as input for the trained classification algorithm (edge **12**).

---

[6] These MDs are more general than those introduced in Sect. 2.1: they may contain regular database atoms, which are used to give context to the similarity atoms in the same antecedent.

[7] At this point, since all we want is to do blocking, and not yet decisions about duplicates, we could, in comparison with what is done with pairs in T, compute less similarity measures and even with low thresholds.

[8] Similarity computations are kept in appropriate program predicates. So similarity values computed before blocking can be reused at this stage, or whenever needed.

The result of the trained ML-based classifier, in this case obtained through SVMs as a separation hyperplane $\mathcal{H}$, is a set M of record pairs $\langle r_1, r_2, 1 \rangle$ that come from the same block and are considered to be duplicates (edge **13**).[9] The records in these pairs will be merged on the basis of an *ad hoc* set of MDs (edge **15**), different from those used in edges **6, 8**.

Informally, the merge-MDs are of the form: $r_1 \approx r_2 \rightarrow r_1 \doteq r_2$, where the antecedent is true when $\langle r_1, r_2, 1 \rangle$ is an output of the classifier. The RHS is a shorthand for: $r_1[A_1] \doteq r_2[A_1] \wedge \cdots \wedge r_1[A_m] \doteq r_2[A_m]$, where m is the total number of record attributes. Merge at the attribute level uses the matching functions m_{A_i}.

We point out that MD-based merging takes care of transitive cases provided by the classifier, e.g. if it returns $\langle r_1, r_2, 1 \rangle$, $\langle r_2, r_3, 1 \rangle$, but not $\langle r_1, r_3, 1 \rangle$, we still merge r_1, r_3 (even when $r_1 \approx r_3$ does not hold). Actually, we do this by merging all the records r_1, r_2, r_3 into the same record. Our system is capable of recognizing this situation and solving it as expected. This relies on the way we store and manage -via our *LogiQL* program- the positive cases obtained from the classifier (details can be found in Sect. 6). In essence, this makes our set of merging-MDs *interaction-free*, and leads to a unique resolved instance [7].

The following sections provide more details on *ERBlox* and our approach to ER.

4   Initial Data and Similarity Computation

We describe now some aspects of the MAS dataset, highlighting the input for- and output of each component of the *ERBlox* system. The data is represented and provided as follows. The Author relation contains authors names and their affiliations. The Paper relation contains paper titles, years, conference IDs, journal IDs, and keywords. The PaperAuthor relation contains papers IDs, authors IDs, authors names, and their affiliations. The Journal and Conference relations contain short names, full names, and home pages of journals and conferences, respectively. By using *ERBlox* on this dataset, we determine which papers in MAS data are written by a given author. This is clear case of ER since there are many authors who publish under several variations of their names. Also the same paper may appear under slightly different titles, etc.[10]

From the MAS dataset, which contains the data in structured files, extensions for intentional, relational predicates are computed by *LogiQL*-rules of the general program, e.g.

$$\_file\_in(x1, x2, x3) \rightarrow string(x1), string(x2), string(x3). \tag{3}$$

$$lang : physical : filePath['\_file\_in] = \text{"}author.csv\text{"}. \tag{4}$$

$$+author(id1, x2, x3) \leftarrow \_file\_in(x1, x2, x3), string : int64 : convert[x1] = id1. \tag{5}$$

---

[9] The classifier also returns pairs or records that come from the same block, but are not considered to be duplicate. The set thereof in not interesting, at least as a workflow component.

[10] For our experiments, we independently used two other datasets: DBLP and Cora Citation.

Here, (3) is a predicate schema declaration (metadata uses "→"), in this case of the "_file_in" predicate with three string-valued attributes,[11] which is used to store the contents extracted from the source file, whose path is specified by (4). Derivation rules, such as (5), use the usual "←". In this case, it defines the author predicate, and the "+" in the rule head inserts the data into the predicate extension. The first attribute is made an identifier [1]. Figure 4 illustrates a small part of the dataset obtained by importing data into the relational predicates. (There may be missing attributes values.)

Author	AID	Name	Affiliation	Bl#
	659	Jean-Pierre Olivier de	Ecole des Hautes	659
	2546	Olivier de Sardan	Recherche Scientifique	2546
	612	Matthias Roeckl	German Aerospace Center	612
	4994	Matthias Roeckl	Institute of Communications	4994

Paper	PID	Title	Year	CID	JID	Keyword	Bl#
	123	Illness entities in West Africa	1998	179		West Africa, Illness	123
	205	Illness entities in Africa	1998	179		Africa, Illness	205
	769	DLR Simulation Environment m3	2007	146		Simulation m3	769
	195	DLR Simulation Environment	2007	146		Simulation	195

PaperAuthor	PID	AID	Name	Affiliation
	123	659	Jean-Pierre Olivier de	Ecole des Hautes
	205	2546	Olivier de Sardan	Recherche Scientifique
	769	612	Matthias Roeckl	German Aerospace Center
	195	4994	Matthias Roeckl	Institute of Communications

Fig. 4. Relation extensions from MAS using LogiQL rules

As described above, in *ERBlox*, similarity computation generates similarity weights, which are used to: (a) compute the weight vectors for the training data T and the data in D under classification; and (b) do the blocking, where similarity weights are compared with predefined thresholds for the similarity conditions in the LHSs of blocking-MDs.[12]

We used three well-known similarity functions [13], depending on the attribute domains. "TF-IDF cosine similarity" [22] used for computing similarities for text-valued attributes, whose values are string vectors. It assigns low weights to frequent strings and high weights to rare strings. It was used for attribute values that contain frequent strings, such as affiliation. For attributes with short string values, such as author name, we applied "Jaro-Winkler similarity" [25]. Finally, for numerical attributes, such as publication year, we used "Levenshtein distance" [20], which computes similarity of two numbers on the basis of the minimum number of operations required to transform one into the other.

---

[11] In *LogiQL*, each predicate has to be declared, unless it can be inferred from the rest of the program.

[12] As described at the end of Sect. 3, these similarity computations are *not* used with the MDs that support the final merging process (cf. Sect. 6).

Similarity computation for *ERBlox* is supported by *LogiQL*-rules that define similarity functions. In particular, similarity computations are kept in extensions of program predicates. For example, if the similarity weight of values a_1, a_2 for attribute *Title* is above the threshold, a tuple $TitleSim(a_1, a_2)$ is created by the program.

5   MD-Based Collective Blocking and Duplicate Detection

Since every record has an identifier, *rid*, initially each record uses its *rid* as its block number, in an extra attribute $Bl\#$. In this way, we create the *initial blocking instance* from the initial instance D, also denoted with D. Now, blocking strategies are captured by means of (blocking) MDs of the form:

$$R_i(\bar{X}_1, Bl_1) \wedge R_i(\bar{X}_2, Bl_2) \wedge \psi(\bar{X}_3) \;\; \rightarrow \;\; Bl_1 \doteq Bl_2. \tag{6}$$

Here Bl_1, Bl_2 are variables for block numbers, and R_i is a database (record) predicate. The lists of variables $\bar{X}_1, \bar{X}_2$ stand for all the attributes in R_i, but $Bl\#$. Formula ψ is a conjunction of relational atoms and comparison atoms via similarity predicates; but it does not contain similarity comparisons of blocking numbers, such as $Bl_3 \approx Bl_4$.[13] The variables in the list $\bar{X}_3$ appear in R_i or in another database predicate or in a similarity atom. It holds that $(\bar{X}_1 \cup \bar{X}_2) \cap \bar{X}_3 \neq \emptyset$. For an example, see (2), where R_i is Author.

In order to enforce these MDs on two records, we use a binary matching function $m_{Bl\#}$, to make two block numbers identical: $m_{Bl\#}(i,j) := i$ if $j \leq i$. More generally, for the application-dependent set, Σ^{Bl}, of blocking-MDs we adopt the chase-based semantics for entity resolution [7]. Since this set of MDs is interaction-free, its enforcement results in a single instance D^{Bl}, where now records may share block numbers, in which case they belong to the same block. Every record is assigned to a single block.

Example 2. These are some of the blocking-MDs used for the MAS dataset:

$$Paper(pid_1, x_1, y_1, z_1, w_1, v_1, bl_1) \wedge Paper(pid_2, x_2, y_2, z_2, w_2, v_2, bl_2) \wedge \tag{7}$$
$$x_1 \approx_{Title} x_2 \;\wedge\; y_1 = y_2 \;\wedge\; z_1 = z_2 \;\rightarrow\; bl_1 \doteq bl_2.$$

$$Author(aid_1, x_1, y_1, bl_1) \wedge Author(aid_2, x_2, y_2, bl_2) \wedge \tag{8}$$
$$x_1 \approx_{Name} x_2 \;\wedge\; y_1 \approx_{Aff} y_2 \;\rightarrow\; bl_1 \doteq bl_2.$$

$$Paper(pid_1, x_1, y_1, z_1, w_1, v_1, bl_1) \;\wedge\; Paper(pid_2, x_2, y_2, z_2, w_2, v_2, bl_2) \wedge \tag{9}$$
$$PaperAuthor(pid_1, aid_1, x_1', y_1') \;\wedge\; PaperAuthor(pid_2, aid_2, x_2', y_2') \wedge$$
$$Author(aid_1, x_1', y_1', bl_3) \wedge Author(aid_2, x_2', y_2', bl_3) \wedge x_1 \approx_{Title} x_2 \;\rightarrow\; bl_1 \doteq bl_2.$$

$$Author(aid_1, x_1, y_1, bl_1) \wedge Author(aid_2, x_2, y_2, bl_2) \;\wedge\; x_1 \approx_{Name} x_2 \wedge \tag{10}$$
$$PaperAuthor(pid_1, aid_1, x_1, y_1) \;\wedge\; PaperAuthor(pid_2, aid_2, x_2, y_2) \wedge$$
$$Paper(pid_1, x_1', y_1', z_1', w_1', v_1', bl_3) \wedge Paper(pid_2, x_2', y_2', z_2', w_2', v_2', bl_3) \;\rightarrow\; bl_1 \doteq bl_2.$$

---

[13] Actually, this natural condition makes the set of blocking-MDs interaction-free, i.e. for every two blocking-MDs m_1, m_2, the set of attributes on the RHS of m_1 and the set of attributes on the LHS of m_2 on which there are similarity predicates, are disjoint [7].

Informally, (7) tells us that, for every two Paper entities $\mathbf{p}_1, \mathbf{p}_2$ for which the values for attribute *Title* are similar and with same publication year, conference ID, the values for attribute *Bl#* must be made the same. By (8), whenever there are similar values for name and affiliation in Author, the corresponding authors should be in the same block. Furthermore, (9) and (10) collectively block Paper and Author entities. For instance, (9) states that if two authors are in the same block, their papers $\mathbf{p}_1$, $\mathbf{p}_2$ having similar titles must be in the same block. Notice that if papers $\mathbf{p}_1$ and $\mathbf{p}_2$ have similar titles, but they do not have same publication year or conference ID, we cannot block them together using (7) alone.   ∎

We now show how these MDs are represented in *LogiQL*, and how we use *LogiQL* programs for declarative specification of MD-based collective blocking.[14] In *LogiQL*, an MD takes the form:

$$R_i[\bar{X}_1] = Bl_2, \quad R_i[\bar{X}_2] = Bl_2 \quad \longleftarrow \quad R_i[\bar{X}_1] = Bl_1, \; R_i[\bar{X}_2] = Bl_2, \; \psi(\bar{X}_3), \; Bl_1 < Bl_2,$$
$$(11)$$

subject to the same conditions as in (6). An atom $R_i[\bar{X}]=Bl$ states that predicate R_i is functional on $\bar{X}$ [1]. It means each record in R_i can have only one block number $Bl\#$.

Given an initial instance D, a *LogiQL* program $\mathcal{P}^B(D)$ that specifies MD-based collective blocking contains the following (kind of) rules:

1. For every atom $R(rid, \bar{x}, bl) \in D$, the fact $R[rid, \bar{x}] = bl$. (Initially, $bl := rid$.)
2. For every attribute A of R_i, facts of the form $A\text{-}Sim(a_1, a_2)$, with $a_1, a_2 \in Dom_A$, the finite attribute domain. They are obtained by similarity computation.
3. The blocking-MDs as in (11).
4. Rules to represent the consecutive versions of entities during MD-enforcement:

$$R\text{-}OldVersion(r_1, \bar{x}_1, bl_1) \quad \longleftarrow \quad R[r_1, \bar{x}_1] = bl_1, \; R[r_1, \bar{x}_1] = bl_2, \; bl_1 < bl_2.$$

For each *rid*, r, there could be several atoms of the form $R[r, \bar{x}] = bl$, corresponding to the evolution of the record identified by r due to MD-enforcement. The rule specifies that versions of records with lower block numbers are old.
5. Rules that collect the latest versions of records. They are used to form blocks:

$$R\text{-}MDBlock[r_1, \bar{x}_1] = bl_1 \quad \longleftarrow \quad R[r_1, \bar{x}_1] = bl_1, \; !\, R\text{-}OldVersion(r_1, \bar{x}_1, bl_1).$$

In *LogiQL*, "!", as in the body above, is used for negation [1]. The rule collects R-records that are not old versions.

---

[14] Notice that since we have interaction-free sets of blocking-MDs, stratified Datalog programs are expressive enough to express and enforce them [3]. *LogiQL* supports stratified Datalog.

Programs $\mathcal{P}^B(D)$ as above are stratified (there is no recursion involving negation). Then, as expected in relation to the blocking-MDs, they have a single
model, which can be used to read the final block number for each record.

Example 3. (ex. 2 cont.) Considering only MDs (7) and (9), the portion of
$\mathcal{P}^B(D)$ for blocking Paper entities has the following rules:

2. Facts such as: $TitleSim(Illness\ entities\ in\ West\ Africa, Illness\ entities\ in\ Africa).$
$$TitleSim(DLR\ Simulation\ Environment\ m3, DLR\ Simulation\ Environment).$$

3. $Paper[pid_1, x_1, y_1, z_1, w_1, v_1] = bl_2, Paper[pid_2, x_2, y_2, z_2, w_2, v_2] = bl_2 \leftarrow$
$$Paper[pid_1, x_1, y_1, z_1, w_1, v_1] = bl_1, Paper[pid_2, x_2, y_2, z_2, w_2, v_2] = bl_2,$$
$$TitleSim(x_1, x_2), y_1 = y_2, z_1 = z_2, bl_1 < bl_2.$$
$Paper[pid_1, x_1, y_1, z_1, w_1, v_1] = bl_2, Paper[pid_2, x_2, y_2, z_2, w_2, v_2] = bl_2 \leftarrow$
$$Paper[pid_1, x_1, y_1, z_1, w_1, v_1] = bl_1, Paper[pid_2, x_2, y_2, z_2, w_2, v_2] = bl_2, TitleSim(x_1, x_2),$$
$$PaperAuthor(pid_1, aid_1, x_1', y_1'), PaperAuthor(pid_2, aid_2, x_2', y_2'),$$
$$Author[aid_1, x_1', y_1'] = bl_3, Author[aid_2, x_2', y_2'] = bl_3, bl_1 < bl_2.$$

4. $PaperOldVersion(pid_1, x_1, y_1, z_1, w_1, v_1, bl_1) \leftarrow Paper[pid_1, x_1, y_1, z_1, w_1, v_1] = bl_1,$
$$Paper[pid_1, x_1, y_1, z_1, w_1, v_1] = bl_2, bl_1 < bl_2.$$

5. $PaperMDBlock[pid, \bar{x}_1] = bl_1 \leftarrow Paper[pid_1, x_1, y_1, z_1, w_1, v_1] = bl_1,$
$$PaperOldVersion(pid_1, x_1, y_1, z_1, w_1, v_1, bl_1).$$

Restricting the model of the program to the relevant attributes of predicate
PaperMDBlock returns: $\{\{123, 205\}, \{195, 769\}\}$, i.e. the papers with *pids* 123
and 205 are blocked together; similarly for those with *pids* 195 and 769.   ∎

As described above, the input to the trained classifier is a set of tuples of the
form $\langle r_1, r_2, w(r_1, r_2)\rangle$, with $w(r_1, r_2)$ the computed weight vector for records
(with ids) r_1, r_2 in a same block.[15]

Example 4. (ex. 3 cont.) Consider the blocks for entity Paper. If the "journal
ID" values are null in both records, but not the "conference ID" values, "journal
ID" is not considered for a feature. Similarly, when the conference ID values are
null. However, the values for "journal ID" and "conference ID" are replaced by
"journal full name" and "conference full name" values, found in Conference and
Journal records, resp. In this case then, attributes *Title, Year, ConfFullName* or
JourFullName and *Keyword* are used for corresponding feature for weight vector
computation.

Considering the previous Paper records, the input to the classifier con-
sists of: $\langle 123, 205, w(123, 205)\rangle$, with $w(123, 205) = [0.8, 1.0, 1.0, 0.7]$, and
$\langle 195, 769, w(195, 769)\rangle$, with $w(195, 769) = [0.93, 1.0, 1.0, 0.5]$ (actually the contents of the two square brackets only).   ∎

Several ML techniques are accessible from *LogicBlox* platform through the
BloxMLPack library, that provides a generic Datalog interface. Then, *ERBlox*
can call an ML-based record duplicate detection component through the general
LogiQL program. In this way, the SVMs package is invoked by *ERBlox*.

---

[15] The features considered in a weight vector computation depend on whether they
have a strong discrimination power, i.e. do not contain missing values.

The output is a set of tuples of the form $\langle r_1, r_2, 1 \rangle$ or $\langle r_1, r_2, 0 \rangle$, where r_1, r_2 are ids for records of entity (table) R. In the former case, a tuple R-*Duplicate*(r_1, r_2) is created (as defined by the *LogicQL* program). In the previous example, the SVMs method return $\langle [0.8, 1.0, 1.0, 0.7], 1 \rangle$ and $\langle [0.93, 1.0, 1.0, 0.5], 1 \rangle$, then *PaperDuplicate*$(123, 205)$ and *PaperDuplicate*$(195, 769)$ are created.

6   MD-Based Merging

When *EntityDuplicate*(r_1, r_2) is created, the corresponding full records $\bar{r}_1, \bar{r}_2$ have to be merged via record-level merge-MDs of the form $R[r_1] \approx R[r_2] \longrightarrow R[\bar{r}_1] \doteq R[\bar{r}_2]$, where $R[r_1] \approx R[r_2]$ is true when R-*Duplicate*(r_1, r_2) has been created according to the output of the SVMs classifier. The RHS means that the two records are merged into a new full record $\bar{r}$, with $\bar{r}[A_i] := m_{A_i}(\bar{r}_1[A_i], \bar{r}_2[A_i])$ [7].

Example 5. (ex. 4 cont.) We merge duplicate Paper entities enforcing the MD:
$Paper\ [pid_1] \approx Paper[pid_2] \longrightarrow Paper[Title, Year, CID, Keyword] \doteq Paper[Title, Year, CID, Keyword]$. ∎

The portion, $\mathcal{P}^M$, of the general *LogiQL* program that represents MD-based merging contains rules as in **1.**–**4.** below:

1. The atoms of the form R-*Duplicate* mentioned above, and those representing the matching functions (MFs) m_A.
2. For an MD $R[r_1] \approx R[r_2] \longrightarrow R[\bar{r}_1] \doteq R[\bar{r}_2]$, the rule:

$$R[r_1, \bar{x}_3] = bl,\ R[r_2, \bar{x}_3] = bl \longleftarrow R\text{-}Duplicate(r_1, r_2),\ R[r_1, \bar{x}_1] = bl,$$
$$R[r_2, \bar{x}_2] = bl,\ m(\bar{x}_1, \bar{x}_2) = \bar{x}_3,$$

which creates two records (one of them can be purged afterwards) with different ids but all the other attribute values the same, and computed componentwise according to the MFs for m. Here, $\bar{x}_1, \bar{x}_2, \bar{x}_3$ stand each for all attributes of relation R, except for the id and the block number (represented by bl). (Block numbers play no role in merging.)
3. As for program $\mathcal{P}^B(D)$ given in Sect. 5, rules specify the old versions of a record:

$$R\text{-}OldVersion(r_1, \bar{x}_1) \leftarrow R[r_1, \bar{x}_1] = bl,\ R[r_1, \bar{x}_2] = bl,\ \bar{x}_1 \prec \bar{x}_2.$$

Here, $\bar{x}_1$ stands for all attributes other than the id and the block number; and on the RHS $\bar{x}_1 \prec \bar{x}_2$ means componentwise comparison of values according to the partial orders defined by the MFs.
4. Finally, rules to collect the latest version of each record, building the final resolved instance: $R\text{-}ER(r_1, \bar{x}_1) \leftarrow R[r_1, \bar{x}_1] = bl,\ !\, R\text{-}OldVersion(r_1, \bar{x}_1)$.

Notice that the derived tables R-*Duplicate* that appear in the LHSs of the MDs (or in the bodies of the corresponding rules) are all computed before (and kept fixed during) the enforcement of the merge-MDs. In particular, a duplicate relationship between any two records is not lost. This has the effect of making the set of merging-MDs interaction-free, which results in a unique resolved instance.

7 Experimental Evaluation

We now show that our approach to ER can improve accuracy in comparison
with standard blocking. In addition to the MAS, we used datasets from DBLP
and Cora Citation.

In order to emphasize the importance of semantic knowledge in blocking, we consider standard blocking and
two different sets of MDs, (1) and
(2), for MD-based collective blocking.
Under (1), we define blocking-MDs for
all the blocking keys used for standard blocking, but under (2) we have
MDs for only some of the used blocking keys. In both cases, in addition to
properly collective blocking MDs.

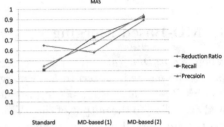

Fig. 5. The experiments (MAS)

We use three measures for the comparisons of blocking techniques. One is
reduction ratio, which is the ratio (minus 1) of the number of candidate record-
pairs over the initial number of records. The higher this value, the less candidate
record-pairs are being generated, but the quality of the generated candidate
record pairs is not taken into account. We also use recall and precision measures.
The former is the number of true duplicate candidate record-pairs divided by the
number of true duplicate pairs, and precision is the number of true candidate
duplicate record-pairs divided by the total number of candidate pairs [12].

Figures 5, 6 and 7 show the comparative performance of *ERBlox*. They show
that standard blocking has higher reduction ratio than MD-based collective
blocking version (1). This means that less candidate record-pairs are being generated by standard blocking. However, the precision and recall of MD-based
blocking version (1) are higher than standard blocking, meaning that MD-based
blocking version (1) can lead to improved ER results at the cost of larger blocks,
and thus more candidate record pairs that need to be compared.

In blocking, this is a common
tradeoff that needs to be considered.
On the one hand, having a large
number of smaller blocks will result
in fewer candidate record-pairs that
will be generated, probably increasing
the number of true duplicate record-
pairs that are missed. On the other
hand, blocking techniques that result
in larger blocks generate a higher number of candidate record-pairs that will
likely cover more true duplicate pairs,
at the cost of having to compare more

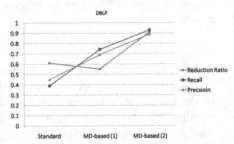

Fig. 6. The experiments (DBLP)

candidate pairs [12]. The experiments are all done before MD-based merging.

Interestingly, MD-based blocking version (2) has higher reduction ratio, recall, and precision than standard blocking. This emphasizes the importance of MDs supporting collective blocking, and shows that blocking based on string similarity alone fails to capture the relationships that naturally hold in the data.

As expected, the experiments show that different sets of MDs for MD-based collective blocking have different impact on reduction ratio, so as standard blocking depends on the choice of blocking keys. However, the quality of MD-based collective blocking, in its two versions, dominates standard blocking for the three datasets.

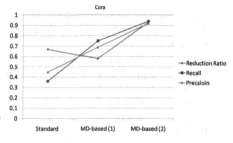

Fig. 7. The experiments (Cora)

8 Conclusions

We have shown that matching dependencies, a new class of data quality/cleaning semantic constraints in databases, can be profitably integrated with traditional ML-methods, in our case for entity resolution. They play a role not only in the intended goal of merging duplicate representations, but also in the record blocking process that precedes the learning task. At that stage they allow to declaratively capture semantic information that can be used to enrich the blocking activity. MDs declaration and enforcement, data processing in general, and machine learning can all be integrated using the *LogiQL* language.

Acknowledgments. Part of this research was funded by an NSERC Discovery grant and the NSERC Strategic Network on Business Intelligence (BIN). Z. Bahmani and L. Bertossi are very much grateful for the support from LogicBlox during their internship and sabbatical visit.

References

1. Aref, M., ten Cate, B., Green, T.J., Kimelfeld, B., Olteanu, D., Pasalic, E., Veldhuizen, T.L., Washburn, G.: Design and Implementation of the LogicBlox System. In: Proceeding SIGMOD 2015, pp. 125–141 (2015)
2. Baxter, R., Christen, P., Churches, T.: A comparison of fast blocking methods for record linkage. In: Proceeding ACM SIGKDD Workshop on Data Cleaning, Record Linkage, and Object Identification , pp. 234–256 (2003)
3. Bahmani, Z., Bertossi, L., Kolahi, S., Lakshmanan, L.: Declarative entity resolution via matching dependencies and answer set programs. In: Proceeding KR 2012, pp. 380–390 (2012)
4. Baudat, G., Anouar, F.: Generalized discriminant analysis using a kernel approach. Neural Comput. **12**(3), 2385–2404 (2000)

5. Benjelloun, O., Garcia-Molina, H., Menestrina, D., Su, Q., EuijongWhang, S., Widom, J.: Swoosh: a generic approach to entity resolution. VLDB J. **18**(1), 255–276 (2009)
6. Bertossi, L., Kolahi, S., Lakshmanan, L.: Data cleaning and query answering with matching dependencies and matching functions. In: Proceeding ICDT 2011. ACM Press (2011)
7. Bertossi, L., Kolahi, S., Lakshmanan, L.: Data cleaning and query answering with matching dependencies and matching functions. Thoer. Comp. Syst. **52**(3), 441–482 (2013)
8. Bleiholder, J., Naumann, F.: Data fusion. ACM Comput. Surv. **41**(1), 1–41 (2008)
9. Ceri, S., Gottlob, G., Tanca, L.: Logic Programming and Databases. Springer, Heidelberg (1989)
10. Christen, P., Goiser, K.: Quality and complexity measures for data linkage and deduplication. In: Guillet, F., Hamilton, H. (eds.) Quality Measures in Data Mining. SCI, pp. 127–151. Springer, Heidelberg (2007)
11. Christen, P.: Automatic record linkage using seeded nearest neighbour and support vector machine classification. In: Proceeding SIGKDD 2008, pp. 151–159 (2008)
12. Christen, P.: A survey of indexing techniques for scalable record linkage and deduplication. IEEE Trans. Knowl. Data Eng. **19**(1), 1–16 (2011)
13. Cohen, W., Ravikumar, P., Fienberg, S.: A comparison of string metrics for matching names and records. In: Proceeding Workshop on Data Cleaning and Object Consolidation 2003, pp. 123–134 (2003)
14. Cover, T.M., Hart, P.E.: Nearest neighbor pattern classification. IEEE Trans. Inf. Theor. **13**(1), 21–27 (1967)
15. Elmagarmid, A., Ipeirotis, P., Verykios, V.: Duplicate record detection: a survey. IEEE Trans. Knowl. Data Eng. **19**(1), 1–16 (2007)
16. Fan, W.: Dependencies revisited for improving data quality. In: Proceeding PODS (2008)
17. Fan, W., Jia, X., Li, J., Ma, S.: Reasoning about Record Matching Rules. PVLDB **2**(1), 407–418 (2009)
18. Fellegi, I.P., Sunter, A.B.: A theory for record linkage. J. Am. Stat. Soc. **64**(1), 328–339 (1969)
19. Herzog, T.N., Scheuren, F.J., Winkler, W.E.: Data Quality and Record Linkage Techniques. Springer, New York (2007)
20. Navarro, G.: A guided tour to approximate string matching. ACM Comput. Surv. **33**(1), 31–88 (2001)
21. Rastogi, V., Dalvi, N.N., Garofalakis, M.N.: Large-scale collective entity matching. PVLDB **4**(4), 208–218 (2011)
22. Salton, G., Buckley, C.: Term-weighting approaches in automatic text retrieval. Inf. Process. Manage. **24**(5), 513–523 (1988)
23. Euijong Whang, S., Menestrina, D., Koutrika, G., Theobald, M., Garcia-Molina, H.: Entity resolution with iterative blocking. In: Proceeding SIGMOD 2009, pp. 219–232 (2009)
24. Vapnik, V.N.: Statistical Learning Theory. Wiley (1998)
25. Winkler, W.E.: The State of record linkage and currentresearch problems. Technical Report, U.S. Census Bureau (1999)

Matching Uncertain Identities
Against Sparse Knowledge

Steven Horn[1]([⊠]), Anthony Isenor[2], Moira MacNeil[1], and Adrienne Turnbull[1]

[1] Centre for Operational Research and Analysis,
Defence Research and Development Canada, Ottawa, Canada
{steven.horn,moira.macneil,adrienne.turnbull}@forces.gc.ca
[2] Atlantic Research Centre,
Defence Research and Development Canada, Halifax, Canada
anthony.isenor@drdc-rddc.gc.ca

Abstract. This paper presents a method for fast matching of data attributes contained in a high-volume data stream against an incomplete database of known attribute values. The method is applied to vessel observational data and databases of known vessel characteristics, with emphasis on vessel identity attributes. Due to the large quantity of streaming observations, it is desirable to compute the best matching identity to a sufficient confidence level rather than include all possible identity information in the matching result. The question of which observed attributes to use in the calculation is addressed using information theory and the combination of the information conveyed by each attribute is addressed using evidence theory. An algorithm is developed which matches observations to known identities with a configurable level of desired confidence, represented as a χ^2 value for statistical significance.

Keywords: Entropy · Transferrable belief model · Generalized Bayes theorem · Database · Intelligence · Information · Data errors

1 Introduction

Data quality is a continual issue when dealing with an automated processing system. Introducing the requirement for real-time processing magnifies the problem by creating an environment where pausing and reflecting on the quality issue impacts the time criticality of the system.

Here we consider data quality as related to both errors (i.e., incorrect values) and inconsistencies (e.g., CA as compared to CAN for Canada; syntax issues; differing vocabularies issues). Inconsistencies are often related to vocabularies issues that sometime require semantic level matching [1]. Such data quality issues influence the system's ability to process the incoming data stream. Some simplistic views of how to deal with data inconsistency have been reported. For example, [2] examined inconsistencies in vessel information reported from a selection of open websites. This investigation illustrated the complexities associated with vocabulary matching when aggregating multiple information sources.

© Springer International Publishing Switzerland 2015
C. Beierle and A. Dekhtyar (Eds.): SUM 2015, LNAI 9310, pp. 415–420, 2015.
DOI: 10.1007/978-3-319-23540-0_28

Data inconsistencies are also influenced by the data volume. As volume increases, the ability of a user to identify and correct the data drops dramatically. Of course each data stream will be unique and in some sense these differences are related to the data attributes in the stream. Also, as fewer restrictions are placed on the data attribute, the automatic identification of quality issues becomes more problematic. An example of this case is a name text field. The observed and reported name of a person, object or thing, has such variation that few restrictions can be placed on the attribute content. A system's ability to learn and correct is influenced by the data attributes and the content permitted in those attributes.

2   Problem Formulation

Commercially provided databases of ship identities are available, containing identifying information based on regulation and registration information. This knowledge base is represented as a database table, herein referred to as the Reference Table (RT). An incoming data stream of vessel observations is considered the target data and each incoming target record consists of numerous target attributes. In order to verify the identity of the target, one must associate these target attributes with the values in the RT entries. Figure 1 illustrates this problem.

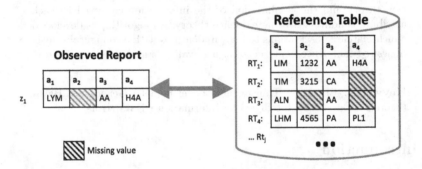

Fig. 1. Illustration of the sparse attribute association problem. a_i are the attribute types and RT_j are the reference table identities. Note that there are missing attribute values in both the report and the reference table rows.

Let A be the set of possible attributes, and a_i be the elements of set A. For the results presented here, the set of attribute element labels in the data are: vessel name, Maritime Mobile Service Identifier (MMSI), International Radio Call-sign (IRCS), and International Maritime Organization (IMO) number. To formalize the description of observations, a target observation at time k, denoted by z_k consists of a set of attributes: $z_k = \{a_i\}\ a_i \in A$. The reference table contains the set of known targets, each with a set of attributes $RT_j = \{a_i\}\ a_i \in A$.

An additional challenge arises since the accuracy of the values of a_i in z_k is not ideal due to possible errors either from observation or transmission faults. However, it is assumed that the rate of error is known for a specific source or sensor. For example, an analysis of 561,771 distinct real-world observations resulted in IMO values that were incorrect in 2.37 % of the cases, and MMSI values that were incorrect 6.64 % of the cases. These values provide a priori indications on the accuracy of information contained in each attribute. The problem described here is similar to the information retrieval problem with missing data [3] with the additional complexities of an uncertain query and query speed requirement.

3   Cost of Attribute Search

The entire RT must be searched for each attribute, since one must assume that there are errors in the attribute values and therefore the first result may not be the only, nor correct result. Furthermore, this means that one cannot necessarily use a previous attribute value match to narrow the RT search space. In order to reduce the number of attribute comparisons required to declare a match, the proposed approach is to achieve a significant level of confidence as early as possible in matching against the RT, even if not all observed attributes are used in the association.

Each attribute a_i in z_k carries with it some measure of information when considered against the RT. In this case, the quantification of information gain as the change in information entropy by the attribute is used as the attribute selection criteria. A similar approach has been used by others [4] to quantify database vulnerabilities to deriving hidden fields from a subset of known fields. In this case, the hidden field is the identifier RT_j. Information gain is also used for decision trees in machine learning [5]. The information gain (IG) of an attribute can be calculated as the difference between the entire RT (first sum in Eq. 1), and the conditional entropy from the attribute (second sum in Eq. 1).

$$IG(RT, a_i) = -\sum_v p_t \cdot log_e p_t - \sum_{v \in a_i} \frac{|\{t \in RT | t_i = v\}|}{|RT|} \cdot H(\{t \in RT | t_i = v\}) \quad (1)$$

where p_t is the probability of a value occurring in the RT, and can be calculated in the frequentist approach from the RT as $p_t = \frac{1}{|RT|}$. The set entropy is defined as $H(X) = -\sum_x p_x \cdot log_e p_x$. Alternatively, the Laplace correction [6] can be used to estimate p_t by assuming p_t as a posterior Bayesian estimate, which better accounts for the possibility that the RT is not complete. The information gain $IG(RT, a_i)$ for each attribute is an indication of the value of that attribute for discriminating the value of RT_j. When an observation z_k is evaluated, those attributes with higher information gain should be searched first as they provide the strongest evidence (for or against) the association. The following section will discuss how each attribute comparison is combined using evidence theory.

4   Uncertainty Model

The potential outcomes for each observed attribute are expressed in Table 1 as four possible cases, with a joint likelihood $p_1 \cdot p_2 \cdot p_3$ as a function of the target birth rate λ_b, observation rate λ_o, and error rate λ_e. The probability p_1 relates to the likelihood that the observed target is or is not described in the RT, p_2 relates to the likelihood that the RT has or has not a non-null value for the attribute, and p_3 relates to the likelihood that the observed attribute is correct or incorrect. Note that the values for λ_e and λ_o are conditional on which source and sensor is providing the observation.

Table 1. Joint likelihood calculation for attribute match cases.

Case	Condition(s)	p_1	p_2	p_3
I:	$z \notin RT$	$\frac{\lambda_b}{\lambda_o}$		
II:	$z \in RT$, $a_i \notin RT_j$	$1 - \frac{\lambda_b}{\lambda_o}$	$1 - \frac{|RT_j|}{|RT|}$	
III:	$z \in RT$, $a_i \in RT_j$, a_i error	$1 - \frac{\lambda_b}{\lambda_o}$	$\frac{|RT_j|}{|RT|}$	λ_e
IV:	$z \in RT$, $a_i \in RT_j$, a_i correct	$1 - \frac{\lambda_b}{\lambda_o}$	$\frac{|RT_j|}{|RT|}$	$1 - \lambda_e$

The generalized Bayesian Theorem (GBT) in the transferrable belief model (TBM) is a standard method for the combination of evidence [7]. The pmf $f(RT_j|z)$ is used in the basic belief assignment (bba), as explained later in the text. The bba is a function on the set of hypotheses (Ω) which consists of elements of the power set for the possible combinations for values of RT_j on the set space 2^Ω. The bba function, generates the basic belief mass, which has the property that the sum of the values equals 1 [12], i.e. $\sum_{A \in \Omega} m(A) = 1$.

In our work, the open-world assumption is adopted, which relaxes this sum such that the hypothesis $m(\emptyset)$ can also have a non-zero mass which represents the evidence (or lack of evidence) that the match is not contained in the RT. This assumption accepts the fact that it is possible the RT is incomplete and that no solution may be possible. The lower bound on the combined evidence is represented by the belief function defined as $bel(A) = \sum_{\emptyset \neq B \subseteq A} m(B), \forall A \subseteq \Omega$, and the upper bound by the plausibility function defined as $pl(A) = \sum_{B:A \cap B \neq \emptyset} m(B)$.

In GBT, the mass function is created assuming that $pl(RT_j|p) = P(RT_j|p)$ and therefore, the mass bba is assigned such that $pl(RT_j) = p(RT_j)$ for the singleton hypotheses [7]. In our application of the GBT, the probability mass function (pmf) $f(RT_j|z)$ used in the bba is calculated using the joint likelihoods $(p_1 \cdot p_2 \cdot p_3)$ in Table 1. The null set receives the combined likelihoods from cases I and II, and possible matches receive the combined likelihoods from cases III and IV, which are then normalized. Other techniques exist for generating the bba, such as the use of Akaike information criterion [11], or expert training sets [5].

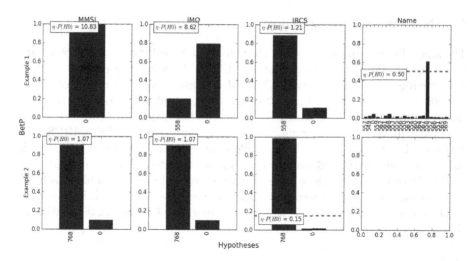

Fig. 2. Two examples of searching the RT. Attributes are searched in order from left to right. The horizontal dashed red line and text box indicate the cut-off probability to declare a match with $\alpha = 0.001$.

Each set of evidence generated by an attribute of the observed report z_k, the set of matches, will not necessarily intersect. This is a typical case where each source of evidence is over a non-exhaustive frame of discernment. There are many approaches to deal with this situation [8]. The approach adopted here is to use the disjunctive rule of combination to combine the evidence [9]. This supports the combination of evidence from multiple attributes which may provide evidence for non-intersecting hypotheses [10].

To declare a confident match, a statistical hypothesis test is set up to determine if there is enough evidence to make a decision. If there is not enough evidence, more attributes must be included. Since the mass functions are represented on the set space 2^{Ω}, the pignistic transform [7] is used to collapse the evidence onto the singular hypothesis set for decision making.

$$BetP(A) = \sum_{X \subseteq \Omega} \frac{|A \cap X|}{|X|} \frac{m(X)}{1 - m(\emptyset)} \qquad (2)$$

The pignistic probability for each potential matching RT_j at each stage is calculated. To indicate confidence in the result, a likelihood ratio test is used. Here, we take H_0 as RT_{H0} where $BetP(RT_{H0}) = sup(BetP(RT_j))$ and H_1 as the next highest $RT_{H1} \neq RT_{H0}$. The test statistic is defined as $\Lambda(t) = \mathcal{L}(H_0|z)/\mathcal{L}(H_1|z)$ using $BetP$ as the likelihood, and a threshold η is chosen to reject the null hypothesis according to the desired confidence in the match. The higher the desired confidence, the more information required to achieve the level of confidence and the more computational burden to select and combine the evidence.

The χ^2 distribution percent point function (quantile) is used with significance level α to reject the null hypothesis. H_0 is rejected if $\Lambda(t) < \eta = Q(\alpha, \chi^2(1))$. If H_0 is rejected, then another attribute must be included in the estimate. If there are no more attributes to check, then the observation is declared as not matched.

5   Results and Conclusion

The algorithm was implemented in Python 2.7 and used the open source library for Dempster-Shafer theory calculations [13]. Real unclassified production data from the Royal Canadian Navy (RCN), and a commercial RT consisting of almost 600,000 records was used. Figure 2 shows two examples of observations being associated to the RT. Note that vessel identities have been obfuscated. The first example achieves the desired confidence after considering the IMO, IRCS, and Name of the vessel. The second example is missing an IMO value in z_k but is able to achieve the desired confidence using the reported MMSI and IRCS.

The implementation was evaluated against real data and was able to achieve a processing rate of up to tens of thousands of records per second. Future work involves an in-depth analysis of the Receiver Operating Characteristics and inclusion of fuzzy comparisons such as soundex, edit distance, and Metaphones.

References

1. Graybeal, J., Isenor, A.W., Reuda, C.: Semantic mediation of vocabularies for ocean observing systems. Comput. Geosci. **40**, 120–131 (2012)
2. ST-Hilaire, M.-O., Isenor, A.W.: Determining the consistency of information between multiple subsystems used in maritime domain awareness. NATO Science for Peace and Security Series-E: Human and Societal Dynamics, NATO Advanced Science Institutes Series (2011)
3. Jousselme, A.-L., Maupin, P.: Comparison of uncertainty representations for missing data in information retrieval. In: Proceedings of the 16th International Conference on Information Fusion, pp. 1902–1909 (2013)
4. Unger, E.A., Harn, L.: Entropy as a measure of database information. In: Proceedings of the Sixth Annual Computer Security Applications Conference, pp. 80–77 (1990)
5. Elouedi, Z., Mellouli, K., Smets, P.: Belief decision trees: theoretical foundations. Int. J. Approx. Reason. **28**(2–3), 91–124 (2001)
6. Wolpert, D., Wolf, D.: Estimating functions of probability distributions from a finite set of samples; part i: Bayes estimators and the Shannon entropy. Santa Fe Institute 1993–07-046 (1993)
7. Smets, P.: Data fusion in the transferable belief model. In: Proceedings of the 3rd International Conference on Information Fusion, pp. 21–33 (2000)
8. Janez, F., Appriou, A.: Theory of evidence and non-exhaustive frames of discernment Plausibilities correction methods. Int. J. Approx. Reason. **18**, 1–19 (1998)
9. Smets, P.: Belief functions: the disjunctive rule of combination and the generalized Bayes theorem. Int. J. Approx. Reason. **9**, 1–35 (1993)
10. Denoeux, T.: Conjunctive and disjunctive combination of belief functions induced by nondistinct bodies of evidence. J. Artif. Intell. **172**(2–3), 234–264 (2008)
11. Lefevre, E., Vannoorenberghe, P., Colot, O.: Using information criteria in Dempster-Shafer's basic belief assignment. In: Proceedings of the 1999 IEEE International Fuzzy Systems Conference, pp. 173–178 (1999)
12. Shafer, G.: A Mathematical Theory of Evidence. Princeton University Press, Princeton (1976)
13. Reineking, T.: A Python library for performing calculations in the Dempster-Shafer theory of evidence. https://github.com/reineking/pyds

Author Index

Printed in the United States
By Bookmasters